LOCAL AREA NETWORKS

Patrick Regan

Institute of Technology

Upper Saddle River, New Jersey
Columbus, Ohio

Library of Congress Cataloging-in-Publication Data
Regan, Patrick E.
  Local area networks / by Patrick Regan.
    p. cm.
  Includes index.
  ISBN 0-13-046577-1
  1. Local area networks (Computer networks) I. Title.

TK5105.7R43 2004
004.6'8--dc21

Editor in Chief: Stephen Helba
Assistant Vice President and Publisher: Charles E. Stewart, Jr.
Assistant Editor: Mayda Bosco
Production Editor: Alexandrina Benedicto Wolf
Production Coordination: Carlisle Publishers Services
Design Coordinator: Diane Ernsberger
Cover Designer: Thomas Borah
Cover art: Digital Vision
Production Manager: Matt Ottenweller
Marketing Manager: Ben Leonard

This book was set in 10/12 Times by Carlisle Communications, Ltd. It was printed and bound by R.R. Donnelley & Sons Company. The cover was printed by Phoenix Color Corp.

Pearson Education LTD.
Pearson Education Singapore, Pte. Ltd.
Pearson Education Canada, Ltd.
Pearson Education-Japan, *Toyoko*

Pearson Education Australia PTY. Limited
Pearson Education North Asia Ltd.
Pearson Educación de Mexico, S.A. de C.V.
Pearson Education Malaysia, Pte. Ltd.

10 9 8 7 6 5 4 3 2 1
ISBN 0-13-046577-1

Dedicated to Natali, the woman whom I love with all of my heart

Brief Contents

Contents

CHAPTER **3**

Media and Topologies   74

PART II   PROTOCOLS   223

CHAPTER **TCP/IP   224**

CHAPTER **11**   **Networking with Novell NetWare   436**

CHAPTER **12** **Networking with Linux   464**

PART IV   MAINTAINING A NETWORK   555

CHAPTER **13   Network Security   556**

CHAPTER **14** **Disaster Prevention and Recovery 596**

Preface

Most people will agree that the world of computers and networking is a fast-paced, changing environment that requires constant learning. Therefore, a person who learns how to fix computers must become familiar with new and old technology to become an effective technician.

This book is intended as an introduction to local area networks, starting with the OSI model and going on through the popular network protocols, technology, and services. The book does not focus on one specific network operating system, but instead covers the three most popular network operating systems: Microsoft Windows Servers, Novell NetWare, and Linux. This gives you an introduction to all three and lets you compare and contrast them. To reinforce key concepts, there are questions at the end of the chapters and several lab exercises throughout the book. Also, for those interested in pursuing COMPTIA's Network+ exam, the book covers everything that is on the exam, and it prepares you to excel within the network environment.

To use this book, you should be A+ certified or have equivalent knowledge and have a general understanding of Microsoft Windows, including file management and the control panel.

Each chapter begins with basic concepts and ends with questions. In addition, many of the chapters also have hands-on exercises to reinforce basic skills and troubleshooting practices. In addition, the book lists important websites that will allow you to go beyond the book and to find more information when you encounter problems.

As you go through the book, you will notice that a lot of information is repeated. This gives instructors the flexibility to teach chapters out of order if they desire. In addition, I have found that the best way to teach (and learn) this material is to repeat the information several times and combine it with hands-on exercises.

PART I

BUILDING A NETWORK

CHAPTER 1

Introduction to Networks

Topics Covered in this Chapter

Introduction

A network is a group of computers connected together using cables or some method of wireless technology. In today's business world, a network is an important tool to many businesses. It allows a user to access large databases, access centralized located files, or access expensive equipment. This chapter introduces networks, network operating systems, and servers. In addition, it describes a general troubleshooting strategy that can be used to troubleshoot networking problems.

Objectives

- Differentiate the LANs, MANs, and WANs.
- List and describe the components that make up a network.
- Define NOS and explain how it relates to the LAN.
- List and explain the different server roles.
- Differentiate client/server and peer-to-peer networks.
- List the responsibilities of the administrator.
- Given a network problem scenario, select an appropriate course of action based on a general troubleshooting strategy.
- Describe the importance and use of documentation.

1.1   NETWORK BEGINNINGS

Early computers, before the PC, were computers called **mainframes,** large centralized computers used to store and organize data. To access a mainframe, you would use a **dumb terminal,** which consists of a monitor to display the data and a keyboard to input the data. Different from a PC, the dumb terminal does not process the data; instead all of the processing is done by the mainframe computer (centralized computing).

Eventually as computers became smaller and less expensive, the **personal computer (PC)** was introduced. Unlike the mainframe computer, personal computers are meant to be used by one person and contain their own processing capabilities. As PCs became more popular, a need grew for people to move data from one computer to another. This was initially done with "sneaker-net," where a person would hand-carry the data from computer to computer on a floppy disk. Unfortunately, this method was relatively expensive and time-consuming when the disk had to be transported over long distance.

As the need to share data between different computers grew, **distributed computing** was developed. Distributed computing consists of taking several computers and connecting them together with cable. Different from what happens in centralized computing, the processing in a distributed computing system is done by the individual PCs. In fact, the distributing computing system is typically more powerful than the centralized computing system because the sum of all the processing power of the individual PCs is more than that of a single mainframe.

1.2   DEFINITION OF NETWORK

A **network** is two or more computers connected together to share resources such as files or a printer. For a network to function, it requires a network service to share or access a common medium or pathway to connect the computers. To bring it all together, protocols give the entire system common communication rules. (See figure 1.1.)

1.2.1   Network Services

The most common two services provided by a network are file sharing and print sharing. **File sharing** allows you to access files on another computer without using a floppy disk or other forms of removable media. To ensure that the files are secure, most networks can limit the access to a directory or file and can limit the kind of access (permissions or rights) that a person or a group of people have. For example, if you have full access to your home directory (personal directory on the network to store files), you can list, read, execute, create, change, and delete files in your home directory.

Depending on the contents of a directory or file, you could specify who has access to the directory or file, and you can specify what permissions or rights those people have over the directory or file. For example, you could specify that one group of people will not be able to see or execute the files, while giving a second group of people the ability to see or execute the files but not make changes to the files and not delete them. Lastly, you could give rights to a third group so that they can see, execute, and change the files.

Figure 1.1   Computers Networked Together

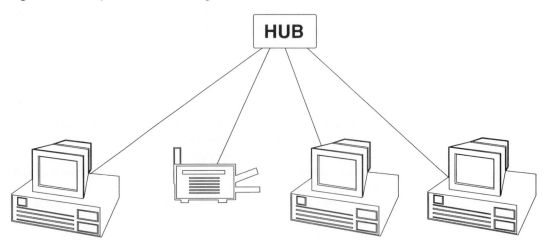

Print sharing allows several people to send documents to a centrally located printer in the office. Therefore, not everyone requires his or her own personal laser printer. Much like they can with files, networks can limit who has access to the printer. For example, if you have two laser printers (a standard laser printer and an expensive high-resolution color laser printer), you can assign everyone access to the standard laser printer while only assigning a handful of people access to the expensive printer.

Internet services provide important tools to business. Email and the World Wide Web access are two popular services. **Electronic mail,** or **email,** is a powerful, sophisticated tool that allows you to send text messages and file attachments (documents, pictures, sound, and movies) to anyone with an email address. In 1999, it was estimated that there were over 83.2 million email users sending 394 billion messages per year in the United States. In 2000, it was estimated that 96.6 million users sent 536.3 billion messages. This makes a striking comparison to the 202 billion pieces of mail delivered by the U.S. Postal Service for the same year. (See figure 1.2.)

Much like the mail from the post office, email is delivered to a mailbox (delivery location or holding area for your electronic messages). An Internet mail address will include the user name, followed by the @ symbol, followed by the name of the mail server. When you connect to the network, you can then access your email messages. Other features that are useful for the sender may include return receipt, so that you know that the email was read or delivered, and the ability to send the message to several people at the same time. Features helpful for the recipient might include the capability to reply to the email messages by clicking on the reply button or option and forwarding the message to someone else or to several people at the same time.

Since the Internet is essentially a huge network, it is possible to make your network part of the Internet or to provide a common connection to the Internet for many users. You can create your own web page on the Internet to provide products and services to the public, or you can perform research on the Internet.

Figure 1.2   Email with Microsoft Outlook

Figure 1.3   A Network Card with an Unshielded Twisted-pair (UTP) Cable Attached to an RJ-45 Connector and an Unused BNC Connector

1.2.2   Network Media

After you have something to share such as a file or a printer, you must then have a pathway to access the network resource. Computers connect to the network by using a special expansion card (or interface built into the motherboard) called a **network interface card (NIC).** The network card will then communicate by sending signals through a cable (twisted pair, coaxial, or fiber optics) or by using wireless technology (infrared or radio waves). The role of the network card is to prepare and send data to another computer, receive data from another computer, and control the flow of data between the computer and the cabling system. (See figure 1.3.)

1.2.3   Protocols

Protocols (TCP/IP, IPX, and NetBEUI) are the rules or standards that allow computers to connect to one another and that enable computers and peripheral devices to exchange information with as little error as possible. Common protocol suites (also referred to as stack) are TCP/IP and IPX. A suite is a set of protocols that work together.

1.3   NETWORK CHARACTERISTICS

All networks can be characterized by the following:

1. Client/server network or peer-to-peer network
2. LAN, MAN, or WAN

1.3.1   Servers, Clients, and Peers

A computer on the network can provide services or request services. A **server** is a service provider that allows access to network resources, and a **client** is a computer that requests services. A network that is made up of servers and clients is known as a **client/server network,** which is typically used on a medium or large network. A server-based network is the best network for sharing resources and data, while providing centralized network security for those resources and data. In addition, since the data files can be centrally located on a server, a server-based network allows for centralized backup of those files. Windows NT Server, Windows 2000 Server, Windows Server 2003, Linux, and Novell NetWare networks are primarily client/server networks.

A **peer-to-peer network,** sometimes referred to as a **workgroup,** has no dedicated servers. Instead, all computers are equal. Therefore, they provide services and request services. Since a person's resources are kept on his or her own machine, a user manages his or her own shared resources. Windows 9X and Linux can be used to form a peer-to-peer network.

When talking about server and server applications, you may hear the terms *front end* and *back end.* In client/server applications, the client part of the program is often called the **front end,** and the server part is called the **back end.** The front end is the interface that is provided to a user or another program, and it is the part that the user or program will see and interact with. The back end is services provided by the server. All that the user and program see and all that they need to worry about is that they perform a request using the front end and the front end provides the response. The user does not need to be concerned with what happens in the back end to get that result. For example, when you access a website, you type in a URL in a browser and the web page is displayed. You do not have to worry that the computer has to first contact a DNS server to get an IP address for that website, and you do not have to worry about setting up a connection with that web server to download the HTML instructions to display the web page.

1.3.2   Server's Roles

Since different companies have different needs for their networks, there are several different roles that a server can assume. They are:

■ **File server**—A server that manages user access to files stored on a server. When a file is accessed on a file server, the file is downloaded to the client's RAM. For example, if you are working on a report using a word processor, the word processor files will be executed from your client computer and the report will be stored on the server. As the report is accessed from the server, it would be downloaded or copied to the RAM of the client computer.

NOTE: All the processing done on the report is done by the client's microprocessors.

Other advantages of a file server allow easy access to the data files from any computer and allow for easy backup of data files on the server.

- **Print server**—A server that manages user access to printer resources connected to the network, allowing one printer to be used by many people.
- **Application server**—A server that is similar to a file and print server except that the application server also does some of the processing.
- **Mail server**—A server that manages electronic messages (email) between users.
- **Fax server**—A server that manages fax messages sent into and out of the network through a fax modem.
- **Remote access server**—A server that hosts modems for inbound requests to connect to the network. Remote access servers provide remote users that are working at home or on the road with a connection to the network.
- **Telephony server**—A server that functions as an intelligent answering machine for the network. It can also perform call center and call-routing functions.
- **Web server**—A server that runs WWW and FTP services for access by users of the intranet or the Internet.
- **Proxy server**—A server that performs a function on behalf of other computers. It is typically used to provide local intranet clients with access to the Internet while keeping the local intranet free from intruders.
- **Directory services servers**—A server used to locate information about the network such as domains (logical divisions of the network) and other servers.

NOTE: A single server could have several roles.

1.3.3   LANs, MANs, and WANs

Today, networks are broken into three main categories: a **local area network (LAN),** a **metropolitan area network (MAN),** or a **wide area network (WAN).** A LAN has computers that are connected within a geographically close network, such as a room, a building, or a group of adjacent buildings. A MAN is a network designed for a town or city, usually using high-speed connections such as fiber optics. A WAN is a network that uses long-range telecommunication links to connect the network computers over long distances; it often consists of two or more smaller LANs. Typically, the LANs are connected through public networks, such as the public telephone system.

The WAN can be either an enterprise WAN or a global WAN. An **enterprise WAN** is a WAN that is owned by one company or organization. A **global WAN** is not owned by any one company and could cross national boundaries. The best-known example of a global WAN is the Internet, which connects millions of computers. As of February 2002, there were over 544.2 million users on the Internet, and the number is growing rapidly. (See figures 1.4 and 1.5.)

Other Network Terms

Internetwork—A network that is internal to a company and is private. It is often a network consisting of several LANs that are linked together. The smaller LANs are known as **subnetworks** or **subnets.**

Intranet—A network based on TCP/IP, the same protocol that the Internet uses. Unlike the Internet, the intranet belongs to a single organization, accessible only by

Figure 1.4   WAN

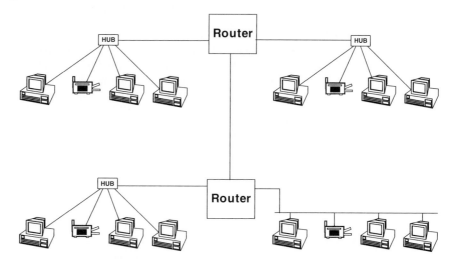

Figure 1.5   The Internet

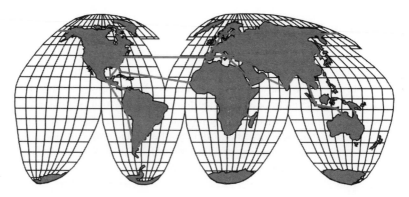

the organization's members. An intranet's websites look and act just like any other websites, but they are isolated by a firewall to stop illegal access.

NOTE: An intranet could have access to the Internet, but does not require it.

Enterprise—Any large organization that utilizes computers, usually consisting of multiple LANs.

Small Office/Home Office (SOHO)—A small network used primarily in home offices that might be part of a larger corporation but yet remains apart from it. SOHO networks are usually peer-to-peer networks.

Value-Added Network (VAN)—A network with special services such as electronic data interchange (EDI) or financial services such as credit card authorization or ATM transactions.

1.4   NETWORK SOFTWARE

A **network operating system (NOS)** is an operating system that includes special functions for connecting computers and devices into a LAN, to manage the resources and services of the network and to provide network security for multiple users. The most common client/server network operating systems are Windows NT Server, Windows 2000 Server, Windows Server 2003, Novell NetWare, UNIX, and Linux. (See table 1.1.)

NOTE: Some operating systems such as Windows 9X, Windows NT Workstation, Windows 2000 Professional, and Windows XP can provide network resources such as file and printer access although they are not servers.

The workstations attached to the network communicate through the use of **client software** called **shells, redirectors,** or **requesters.** While the network protocol enables data transmission across the LAN, the client software resides on top of the network protocol so that it can communicate with other network computers. In Windows 9X and Windows NT Workstation, the client software is installed and configured using the network dialog box accessed by clicking on the network applet in the control panel or by accessing the shortcut menu of the Network Neighborhood and selecting the Properties option.

1.4.1   Windows NT Family

Microsoft's early attempt at a network operating system began with a combined effort between Microsoft and IBM as OS/2. After some disagreement, Microsoft abandoned OS/2 to develop Windows NT. **Windows NT** (NT stands for "new technology") is an advanced, high-performance network operating system, which is robust in features and services, security, performance, and upgradability. While Windows NT provides a good file and print server, it also makes an excellent application server.

The early versions of Windows NT (such as Windows NT 3.51) used the Windows 3.XX Program Manager interface. Newer versions of Windows NT use the popular Windows 95 or the Internet Explorer Active Desktop interface.

Table 1.1   Worldwide Server OS Shipments

NOS	Unit Shipments, 2000	Unit Shipments, 2001
Linux	9.4K (2.2%)	285K (6.4%)
NetWare	615K (14.2%)	508K (11.5%)
UNIX	672K (15.5%)	530K (12.0%)
Windows	2842K (65.7%)	3035K (68.6%)
Other	103K (2.4%)	67K (1.5%)

The next version of Windows NT is **Windows 2000** (NT 5.0). There are four versions of Windows 2000: Windows 2000 Professional (workstation version), Windows 2000 Server, Windows 2000 Advanced Server, and Windows 2000 DataCenter Server. Windows 2000 servers provide the following:

- **File and print sharing**—All Microsoft operating systems since Windows 95 support file and print sharing through File and Print Sharing for Microsoft Networks and Client for Microsoft Networks services.
- **Internet information server (IIS)**—Provides web services (HTTP and FTP).
- **SQL server**-Provides database services.
- **Microsoft Exchange**—Provides email services.
- **NTFS**-Offers a robust, flexible file system that provides file security.
- **Active Directory (AD)**—Similar to NDS in NetWare, AD uses the "tree" concept for managing resources on a network. Its benefits will especially be realized in Enterprise applications where network administration and management will be almost painless. Everything is treated as an object that can be moved or edited across servers and domains.
- **Internet connection sharing (ICS)**—Microsoft Windows since Windows 98SE will allow a single dial-up connection to be shared across the network. This has great implications for SOHOs and home users with multiple machines since they will no longer need to purchase an application to provide this feature for them.
- **Kerberos security**—A security protocol that is used for distributed security within a domain tree/forest. This allows for transitive trusts and a single logon to provide access to all domain resources.
- **Clustering**—Windows 2000 Advanced Server and Windows 2000 Datacenter Server support clustering. Clustering enables two or more servers to work together to keep server-based applications available.

.NET is a Microsoft operating system platform that incorporates distributed applications that bring users into the next generation of the Internet by conquering the deficiencies of the first generation and giving users a more enriched experience in using the web for both personal and business applications. It proposes to integrate web applications with the desktop PC seamlessly by "hosting" desktop applications (primarily) on the Net. So instead of having an application like Microsoft Word installed on an end user's computer, it would be, in theory, "housed" on the web. Microsoft's .NET strategy revolves around utilizing the Internet as a pseudo-operating system, by storing data "on the Net" instead of on a PC. Developers can thus create programs that theoretically "live" on the Net, and end users can accordingly utilize these applications, with their own personal data, from anywhere.

There are four main principles of .NET from the perspective of the user:

- It erases the boundaries between applications and the Internet. Instead of interacting with an application or a single website, .NET will connect the user to an array of computers and services that will exchange and combine objects and data.
- Software will be rented as a hosted service over the Internet instead of purchased on a store shelf. Essentially, the Internet will be housing all your applications and data.

- Users will have access to their information on the Internet from any device, anytime, anywhere.
- There will be new ways to interact with application data, such as speech and handwriting recognition.

.NET allows the user to move from one site to another and not worry about multiple user names and passwords. It remembers your preferences and delivers the appropriate data at the appropriate time to any device you choose, including PCs and hand-held devices such as cellular phones. In addition, it provides a set of services that let you manage your personal information and control access to it.

.NET uses the XML language, which is designed for web pages, to enable the definition, transmission, validation, and interpretation of data between applications and between organizations. Using the XML in a modular approach, .NET attempts to remove barriers to data sharing and software integration. Therefore, businesses can benefit because the modular approach allows them to expand their online applications by adding more modules as needed and to have data move from one module or application to another as needed. Therefore, a lot of applications can easily interface with each other and share information.

With the introduction of the .NET, Microsoft released a new family of .NET servers: Windows .NET Web Server, Windows .NET Standard Server, Windows .NET Enterprise Server, and Windows .NET Datacenter Server.

NOTE: Windows XP is the client version of the Windows .NET server has.

But Windows Server 2003 is still based on much of the same code as that used in Windows 2000, especially for the deployment and administration of Active Directory. In addition, IIS version 6.0 is said to be up to 30 percent faster than IIS version 5.0. In addition, all .NET servers include the .NET Framework, a run-time environment for XML-based web services.

1.4.2   UNIX

UNIX, a multiuser, multitasking operating system, is the grandfather of network operating systems; it was developed at Bell Labs in the early 1970s. Although UNIX is a mature, powerful, reliable operating system, it has been traditionally known for its cryptic commands and its general lack of user-friendliness. It is designed to handle high-usage loads while having support for common Internet services such as web server, FTP server, terminal emulation (telnet), and database access. In addition, it can use the Network File System (NFS), which allows various network clients running different operating systems to access shared files stored on a UNIX machine.

Different from the other network operating systems, UNIX is produced by many manufacturers. The two main dialects of UNIX are AT&T's System V and Berkeley University's BSD4.x. Popular manufacturer versions include Digital Equipment Corporation UNIX, Hewlett-Packard HP-UX, SCO OpenServer, and Sun Microsystems Solaris. Another popular NOS, known as Banyan Vines, is based on a highly modified UNIX System V core. The "official" trademarked UNIX is now owned by the The Open Group, an industry standards organization, which certifies and brands UNIX implementations.

POSIX is a set of IEEE and ISO standards that define an interface between programs and UNIX. By following the POSIX standard, developers have some assurance that their software can be easily ported or translated to a POSIX-compliant operating system. The POSIX standard is now maintained by an arm of the IEEE called the Portable Applications Standards Committee (PASC).

1.4.3   Linux

Linux (pronounced LIH-nuhks with a short "i") is a UNIX-like operating system that was designed to provide personal computer users a free or very low-cost operating system comparable to traditional and usually more expensive UNIX systems. Linux comes in versions for all the major microprocessor platforms including the Intel, PowerPC, Sparc, and Alpha platforms. Because Linux conforms to the POSIX standard user and programming interfaces, developers can write programs that can be ported to other platforms running Linux (or UNIX, which also conforms to the POSIX standard).

The Linux kernel acts as a mediator for your programs and your hardware. Like the UNIX kernel, the Linux kernel is designed to do one thing well. It handles low-level tasks like managing memory, files, programs that are running, networking, and various hardware devices. For example, it arranges for the memory management of all of the running programs (processes), and makes sure they all get a fair share of the processor clock cycles. In addition, it provides a nice, fairly portable interface for programs to talk to your hardware. Therefore, you can often think of the kernel as a cop directing traffic. Unlike Windows, it does not include a windowing system or Graphical User Interface (GUI). Instead, Linux users can choose among a number of X servers and Window Managers.

By the time Linux version 1.0 was released, several Linux distributors had begun packaging the Linux kernel with basic support programs, the GNU utilities and compilers, the X Window system, and other useful programs. In 1995, the Linux 1.2 kernel was released, which supported kernel modules, the PCI bus, kernel-level firewalls, and non-TCP/IP networking protocols. By this time, Linux had become as stable as any commercial version of UNIX on the Intel x86 platforms.

Since Linux is often used as a server, it has a complete implementation of TCP/IP networking software, including drivers for the popular Ethernet cards and the ability to use serial line protocols such as SLIP and PPP to provide access to a TCP/IP network via a modem. With Linux, TCP/IP, and a connection to the network, you can communicate with users and machines across the Internet via electronic mail, USENET news, file transfers, and more.

In fact, Linux owes its versatility to the wide availability of software that runs on it. Understanding what software to use to solve what problem is key to maximizing the utility of Linux. Some of these software packages include:

- ■ **Web server**—Apache (http://www.apache.org) is the most popular web server.
- ■ **Web proxy**—To better control web usage and to allow for caching of frequently accessed pages, Squid (http://www.squid-cache.org) is used.
- ■ **File sharing**—Linux can be made to look like an NT server with respect to file and print sharing. Samba (http://www.samba.org) is the software that does this.

- **Email**—Linux excels at handling email. Sendmail (http://www.sendmail.org) is the most widely used mail transfer agent (MTA). Qmail (http://www.qmail.org) and PostFix (http://www.postfix.org) are alternatives.
- **DNS**—The Domain Name System provides mappings between names and IP addresses, along with distributing network information (i.e., mail servers). BIND (http://www.isc.org/products/BIND/) is the most widely used name server.

1.4.4   Novell NetWare

Several years ago, **Novell NetWare** was the standard LAN-based NOS. Since it was one of the primary players that brought networking to the PC arena, it helped replace the dumb terminals with a PC, allowing a much easier way to share data files, applications, and printers. Novell NetWare is known as a strong file and print server. Unlike the other primary network operating systems, Novell NetWare runs as a dedicated stand-alone server.

Novell NetWare dominance was lost because of the popularity of the Windows GUI versus NetWare's command/menu interface. In addition, you could not perform common administration tasks, such as creating users and giving a user access to a network resource, at the server, only at a client computer. Lastly, Novell initially missed key opportunities during the development of the Internet.

Early versions of NetWare (NetWare 3.11, 3.12, and 3.2) use the NetWare Bindery for security. The Bindery is a flat-based database that resides on a single server and contains profiles of the network users for that server.

Newer versions of NetWare use **Novell Directory Services (NDS),** a global, distributed, replicated database that keeps track of users and resources and provides controlled access to network resources. It is global because it spans an enterprise or multiple-server network. It is a distributed database because the database is kept close to the users rather than in a single central location, and it is replicated to several servers for fault tolerance. Since the database exists for multiple servers, a user has to be only created once, and the user only has to log in to access all of the servers. Within the NDS design, the network objects such as users, printers, and storage units are grouped into containers (used to organize the network objects) much like files are divided into subdirectories or folders on your hard drive.

The latest version of NetWare, NetWare 5 and 6, comes with support for both Novell's own IPX network protocol and the Internet's TCP/IP. NetWare has integrated its own Novell Directory Services with the TCP/IP's Domain Name System, Dynamic Host Configuration Protocol (DHCP), and application-level support for a web server. In addition, NetWare 5 and 6 supports Java applications and includes an enhanced file system, enhanced printing services, and advanced security.

1.4.5   OS/2

OS/2 was one of the first 32-bit GUI operating systems originally developed by IBM and Microsoft before Microsoft's Windows NT, but it is now owned and controlled solely by IBM. It uses the same domain concept used by Windows NT. Domain controllers are used to store the network user and group database, which is replicated from a PDC to one or more BDCs. OS/2 uses NetBEUI as its native protocol, but supports TCP/IP as well. It is compatible with DOS, all Windows versions, Macintosh, and OS/2 clients.

1.4.6   Apple Macintosh Computers

The Apple Macintosh interface is considered to be the easiest to use of all graphical user interfaces. Different from the previous operating systems, with very few exceptions, the MacOS will not run on any hardware platform except the Macintosh.

The MacOS can perform many functions on a network. In addition to being a client, a Macintosh can be a file and print server using AppleShare (Apple's proprietary networking software) as well as an Internet server using various Apple and third-party software. The advantage of having a Macintosh as a server is that it is extremely easy to administer while offering reliable security.

1.5   INTRODUCTION TO TROUBLESHOOTING

When comparing PC problems with network problems, network problems can have a much more devastating effect to those who use the network. This is because users in corporations tend to store many of their documents on the network, and they require access to many of their network services such as file and print sharing and email. In addition, a network problem can affect many users, not just one.

The problems you're bound to come across will likely span a wide range. You might have to troubleshoot a server problem, a network problem, a client problem, or a user problem. The only way to get good at troubleshooting computer and network problems is to practice, practice, practice.

When encountering a problem, you should follow certain guidelines. First, be sure your mind is clear and rested. You must be able to concentrate on the problem. If not, you may overlook something that was obvious such as the power cord not being plugged in.

Next, don't panic, don't get frustrated, and do allow enough time to do the job right. If you panic, you may do something that will make the situation worse. If you start to get frustrated, take a break. You will be amazed how 5 or 10 minutes away from the problem will clear your mind and will allow you to look at the problem a little differently when you come back. Lastly, make sure you have enough time to properly analyze the problem, fix the problem, and test the system after the repair. Again, if you rush a job, you may make the problem worse, or you may overlook something simple.

1.5.1   Computer Problem

Keep in mind that everything that applies to a PC problem also applies to a server problem or a client problem. The problem could be hardware failure, a compatibility problem,

Table 1.2   Computer Problem Classifications

Problem	Description
Hardware failure	This occurs when one or more components fail inside the computer.
Hardware compatibility	Although this may appear to be a hardware failure, the real reason it occurs is because a component is not compatible with another component.
Improper hardware configuration	This error also will often appear to be a hardware failure. Instead, though, the reason for the failure is that the hardware has not been installed or configured properly. This happens often when the user does not read the manual or does not have the knowledge to make use of the manual.
Improper software configuration	This error too may appear to be hardware failure, but it happens when the software (operating system or application software) is not installed or configured properly. This occurs often when the user does not read the manual or does not have the knowledge to make use of the manual.
Software failure	Again, this may appear as a hardware failure. The cause is a glitch in the software. This can range from corrupted data to a flaw in the programming.
Software compatibility	This may also appear to be a hardware failure. The software may not be compatible with the hardware or other software.
Environment	As well, this problem may appear as a hardware failure—but, again, it's not. The location of the computer and its environment (temperature, airflow, electromagnetic interference, magnetic fields) may affect the reliability of the PC and has a direct impact on the PC's life.
User error	This is a very common situation where the user hits the wrong keys or is not familiar with the computer and/or software. It could be something as simple as the user hitting the zero (0) key rather than the letter O.

improper configuration, a software glitch, an environment factor, or a user error. (See table 1.2.) For example, there might be a faulty component such as a hard drive, floppy drive, power supply, cable, or modem. In addition, problems can be caused by a virus, software that isn't compatible with a screen saver, BIOS setup program settings, power management features, control panel settings, software drivers, hardware settings, power fluctuations, or electromagnetic interference.

1.5.2   Network Problem

A network problem would include problems with the medium that connects a server and/or client to the network and anything else the medium connects to. For example, most computers are connected to a network through a cable that is connected to a hub. Then the smaller networks can be connected together to form a larger network by using routers. Therefore, in the event of a problem, the cables, the hub, or the router could be faulty—a cable may not be connected properly, or a hub or router may not be configured properly.

In addition, for a computer and a server to operate properly, the computer's protocol has to be configured properly. Also, you must be given proper permissions to use a network resource. Or the network may require services provided by other servers—perhaps a DNS server provides name resolution for your network to function properly. For example, when you type in http://www.acme.com, a DNS server will look up the IP address for www.acme.com so that it knows how to contact that web server. Also, network communications can be interrupted by external environmental factors such as electromagnetic interference.

1.5.3   General Troubleshooting Strategy

COMPTIA, the organization responsible for the A+, Network+, and Server+ exams, uses a general troubleshooting strategy that consists of the following eight steps:

1. Establish symptoms.
2. Identify the affected area.
3. Establish what has changed.
4. Select the most probable cause.
5. Implement a solution.
6. Test the result.
7. Recognize the potential effects of the solution.
8. Document the solution.

Before trying to fix the problem, you need to gather information. First, you need to clarify exactly what is happening. This should include checking to see if the problem is localized to either one machine or a small group of machines, or if the problem is of a more far-reaching nature throughout the network. Second, make sure that you can duplicate the problem and also make sure that the user is not part of the problem. In addition, determine if the problem is always repeatable or is an intermittent problem. If it is an intermittent problem, does the problem follow a certain pattern (for example, does it occur when the computer is on for a while), or does it occur completely randomly?

The more difficult problems are the intermittent ones. Since intermittent errors do not happen on a regular basis, whatever is causing the problem may work fine when you test it, but as soon as the customer takes the equipment home, it fails again. When dealing with intermittent problems, you make a change that might fix the system. You must then thoroughly test the system over a period of time to see if the problem actually goes away.

After the problem has been repeated and verified, you need to test the system further to see the extent of the problem. For example, if your system doesn't boot properly, can you boot from a floppy drive and access the hard drive? If you can't print from a program, can you print using another program? If the mouse doesn't work in one environment such as Windows, does it work using DOS? Is the problem exhibited on one computer or on several computers on the network? If several computers have the same problem, what is common among those computers that is not common to the computers that do work?

You can gather additional information by trying to use software utilities to test your system and by using a digital multimeter (DMM) and cable-testing equipment. Some of the utilities include software to test the computer components, check for viruses, look for formatting errors on a disk, or check software and protocol configuration.

When gathering information about the problem, you should also find out what has changed recently on the computer and the network. Many times servicing or changes can cause other problems.

After you have gathered as much information as you can, you are now ready to make the repair or fix. Sometimes, you will know exactly what to change or replace. Other times, you will have the problem narrowed down to several causes, which will require you to try several probable solutions. If you suspect a faulty component, you can replace the component with a known good component. If the system works with the new component, the problem was the item you just removed.

NOTE: When a new item is taken off the shelf, it does not mean that the item is always good.

In addition, you can try "reverse swapping," which is trying a suspected component in a second working system.

Other solutions include reconfiguring the software or hardware; reloading the operating system, application software, or drivers; or making changes to the BIOS Setup program. Whatever course of action you choose, you should only make one change at a time. If the problem still exists after you make the change, you will then make another change, repeating this process until the problem no longer exists. When determining which item to check or swap, you should first try to check items that are likely to cause the problem and are the easiest and quickest to check. In addition, you should always check the obvious first. For example, if the hub is not working at all and there are no lights, make sure the AC adapter is plugged in.

After you fix the problem, you should always thoroughly test the computer and/or network. This will ensure that the problem did go away and that you did not cause another problem when fixing the first problem. It is probably more important with networking than with a computer to understand what other problems your solution may cause so that you can select the best solution.

Lastly, you should keep a log of changes made to a system and list any problems and their solutions that you encounter. The log can be particularly useful in two ways. It will let you know if the system has gone through any changes recently, especially if you work for a department that has several PC support people. In addition, the log will help you look for trends so that you can make plans to minimize the problem in the future and to have the resources available when the problem occurs again.

1.5.4   **Where to Get Help**

What is particularly funny is when you get a new error message or research a network problem and you are told to contact your system administrator to get the problem straightened out. Unfortunately, it turns out that you are the system administrator and you don't have a clue what to do. Fortunately, there are several places where you can get help. They include looking at product documentation (including resource kits), Help and readme files, product support, and Internet websites.

When troubleshooting computers, you should also note that there is lots of help available on your PC, your server, and many of your networking components such as routers. Most of these produce log files which you can analyze, which may give you a new insight into a problem. So before giving up, always be sure to utilize whatever resources are available.

SUMMARY

1. Early computers, before the PC, were computers called mainframes, large centralized computers used to store and organize data. To access a mainframe, you would use a dumb terminal, which consists of a monitor to display the data and a keyboard to input the data.

2. Different from a PC, the dumb terminal does not process the data; instead all of the processing is done by the mainframe computer (centralized computing).

3. Distributed computing consists of taking several computers and connecting them together with cable. Different from what happens in centralized computing, the processing in a distributed computing system is done by the individual PCs.

4. A network is two or more computers connected together to share resources such as files or a printer. For a network to function, it requires a network service to share or access a common medium or pathway to connect the computers. To bring it all together, protocols give the entire system common communication rules.

5. Computers connect to the network by using a special expansion card (or interface built into the motherboard) called a network interface card (NIC). The network card will then communicate by sending signals through a cable (twisted pair, coaxial, or fiber optics) or by using wireless technology (infrared or radio waves).

6. The role of the network card is to prepare and send data to another computer, receive data from another computer, and control the flow of data between the computer and the cabling system.

7. Protocols (TCP/IP, IPX, and NetBEUI) are the rules or standards that allow computers to connect to one another and that enable computers and peripheral devices to exchange information with as little error as possible.

8. Common protocol suites (also referred to as stack) are TCP/IP and IPX. A suite is a set of protocols that work together.

9. A computer on the network can provide services or request services.

10. A server is a service provider that provides access to network resources, and a client is a computer that requests services.

11. A network that is made up of servers and clients is known as a client/server network, which is typically used on a medium or large network. A server-based network is the best network for sharing resources and data, while providing centralized network security for those resources and data.

12. A peer-to-peer network, sometimes referred to as a workgroup, has no dedicated servers. Instead, all computers are equal. Therefore, they provide services and request services.

13. A LAN has computers that are connected within a geographically close network, such as a room, a building, or a group of adjacent buildings.

14. A MAN is a network designed for a town or city, usually using high-speed connections such as fiber optics.

15. A WAN is a network that uses long-range telecommunication links to connect the network computers over long distances; it often consists of two or more smaller LANs. Typically, the LANs are connected through public networks, such as the public telephone system.

16. A network operating system (NOS) is an operating system that includes special functions for connecting computers and devices into a LAN, to manage the resources and services of the network and to provide network security for multiple users.

17. The workstations attached to the network communicate through the use of client software called shells, redirectors, or requesters.

18. When comparing PC problems with network problems, network problems can have a much more devastating effect to those who use the network.

19. The only way to get good at troubleshooting computer and network problems is to practice, practice, practice.

20. COMPTIA, the organization responsible for the A+, Network+, and Server+ exams, uses a general troubleshooting strategy that consists of the following eight steps:
    1. Establish symptoms.
    2. Identify the affected area.
    3. Establish what has changed.
    4. Select the most probable cause.
    5. Implement a solution.
    6. Test the result.
    7. Recognize the potential effects of the solution.
    8. Document the solution.

QUESTIONS

1. A network needs all of the following components except:
   a. a protocol
   b. services
   c. media
   d. a dedicated server

2. The _____ is used to connect to a network's media.
   a. hard drive controller
   b. motherboard
   c. SCSI card
   d. network card

3. What are the rules or standards that allow the computers to communicate with each other?
   a. Media
   b. Client software
   c. Services
   d. Protocols

4. Which type of network is made for a single building or campus?
   a. LAN
   b. Global WAN
   c. MAN
   d. Enterprise WAN

5. A _____ is a network that uses long-range telecommunication links to connect the network computers over long distances.
   a. LAN   b. WAN
   c. MAN   d. NOS

6. An operating system that includes special functions for connecting computers and devices into a local area network (LAN), to manage the resources and services of the network and to provide network security for multiple users, is a _____.
   a. client   b. NOS
   c. server   d. WAN

7. You are the IT administrator for a small company. You currently have 15 users, and you have plans to hire another 10 users. You have files that need to be located by all 15 users, and you must maintain different security requirements for the users. What type of network would you implement?
   a. A peer-to-peer network
   b. A server-based network
   c. A workgroup network
   d. A client/server network, with each computer acting as both a client and a server

8. A peer-to-peer network can sometimes be a better solution than a server-based network. Which of the following statements best describes a peer-to-peer network?
   a. It requires a centralized and dedicated server.
   b. It provides extensive user and resource security.
   c. It gives each user the ability to manage his or her shared resources.
   d. It requires at least one administrator to provide centralized resource administration.

9. Which of the following are characteristics of a peer-to-peer network? (Choose all that apply.)
   a. It has centralized security and administration.
   b. A computer can be both a client and a server.
   c. A limited number of computers are involved.
   d. It does not require a hub.

10. In a client/server environment, tasks are divided between the client computer and the server. Which role does the server play in a client/server environment?
   a. The server satisfies requests from the client computer for data and processing resources.
   b. The server stores data for the client, but all data processing occurs on the client computer.
   c. The server stores data and performs all the data processing so that the client computer functions primarily as an intelligent display device.
   d. The server satisfies requests from the client for remote processing resources, but data is stored on the client computer.

11. A client/server approach uses what type of security model?
   a. Centralized   b. Decentralized
   c. Server   d. Distributed

12. Which of the following are characteristics of a true client/server environment? (Choose all that apply.)
   a. It does require a hub.
   b. A computer can be both a client and a server.
   c. It has centralized security and administration.
   d. It has a centralized backup.

13. Which of the following statements is true of peer-to-peer networks?
   a. They are also called workgroups.
   b. They are best suited for sharing many resources and data.
   c. They can support thousands of users in different geographical areas.
   d. They provide greater security and more control than server-based networks.

14. Most networks operate in the client/server environment. Which of the following is an example of client/server computing?
   a. A terminal accessing information in a mainframe database
   b. A workstation application accessing information in a remote database
   c. A workstation application accessing information in a database on the local hard drive
   d. A workstation application processing information obtained from a local database

15. Which of the following is *not* a NOS?
   a. Windows NT
   b. Windows 2000
   c. Windows 98
   d. UNIX
   e. Novell NetWare
   f. Linux

16. For an operating system to connect to a network resource, the PC must _____.
   a. load the appropriate server driver
   b. share a drive or directory
   c. be an administrator or server operator
   d. load the appropriate client software

17. Patrick works for a large Internet bookseller. Patrick's manager has directed him to install a computer on the company's local area network (LAN) that will provide customers with the company's ecommerce web pages. Which type of network node should Patrick install?
    a. A bridge   b. A workstation
    c. A client   d. A server

18. Which network node provides file and printer services?
    a. A workstation   b. A router
    c. A server   d. A hub

19. What term can be used to describe multiple networks that are connected?
    a. LAN   b. Transnetwork
    c. WAN   d. Internetwork

20. Which of the following are indicative of a client/server environment as opposed to a peer-to-peer environment? (Select two answers.)
    a. Backups are centralized.
    b. Client machines often take the role of a server.
    c. Security and administration are managed at the server level.
    d. The requirement for a router.

21. Which of the following is NOT a function of a NOS?
    a. Management of printer and file resources.
    b. Data backup to tape.
    c. Server performance monitoring.
    d. Management of a user directory.
    e. All of the choices listed are functions of a NOS.

22. What is the function of a print server?
    a. To advertise print jobs on the LAN
    b. To provide print sharing on the network
    c. To give a printer the ability to render high-quality graphics on the network
    d. To give a printer the ability to be locally connected to a computer

23. You have five network printers that each need to function as an independent node, allowing multiple workstations access to each. What device on the network provides this functionality?
    a. Print server
    b. Printer hub
    c. Print router
    d. Printer data link

24. In network terminology, what is the back end?
    a. The workstation component in a client-server configuration
    b. The network core
    c. The backbone
    d. The server component in a client/server configuration

25. True or false? The eight troubleshooting steps in order are:
    a. Verify the symptoms. See if you can repeat them.
    b. Identify the area(s) of trouble (entire network or only one workstation, etc.).
    c. Determine if anything has been changed (network settings, ports, cabling, software, etc.).
    d. Deduce the most likely cause.
    e. Engage a solution.
    f. Test the results of the solution.
    g. See if the solution has any side effects.
    h. Write down the solution.

26. What should be your first step when troubleshooting a network problem?
    a. Select the most probable cause.
    b. Establish what has changed.
    c. Establish symptoms.
    d. Implement the result.

27. What three items are included in the basic troubleshooting model?
    a. Apply all of the latest service packs.
    b. Test the results.
    c. Develop a plan to correct the problem.
    d. Establish symptoms, including duplicating the problem.
    e. Replace all questionable hardware.

28. Maintaining documentation as you progress through the troubleshooting process _____. (Select two answers.)
    a. is not as useful as doing all the documentation after the problem has been resolved
    b. helps organize the identification and isolation process
    c. helps in identifying the root problem and not just the symptoms
    d. is usually not cost-justifiable
    e. adds too much time to the project

29. What is the first thing you need to determine when a failure on the network occurs?
   a. Whether it is server- or workstation-related
   b. Whether it is hardware or software
   c. Whether it is a cable problem or a computer problem
   d. Whether it is an isolated problem or networkwide

30. You have intermittent problems on the network. Which of the following would NOT generally be one of the things you'd check for first?
   a. Exposed network cable
   b. EMI interference
   c. Crashed server
   d. Virus on the server

CHAPTER 2

Introduction to the OSI Model

Topics Covered in this Chapter

Introduction

In the early days of networking, networking software was created in haphazard fashion. When networks grew in popularity, the need to standardize the by-products of network software and hardware became evident. Standardization allows vendors to create hardware and software systems that can communicate with one another, even if the underlying architecture is dissimilar. For example, you can run TCP/IP protocol to access the Internet on an IBM-compatible PC running Windows or an Apple Macintosh running OS X.

Objectives

- Describe the OSI model and how it relates to networks.
- Specify the main features of 802.2 LLC and MAC.
- Identify the purpose, features, and functions of hubs, switches, bridges, routers, gateways, and network interface cards.
- Given an example, identify a MAC address.
- Identify the seven layers of the OSI model and their functions.
- Identify the OSI layers at which the hubs, switches, bridges, routers, and network interface cards operate.
- Explain the need for standardization in networks.

- List the differences between the de jure and de facto standards.
- Define *virtual circuit*.
- Define *session* and explain how it relates to networks.
- List and define the types of dialogs.
- List and define the Project 802 standards.
- Define *signaling, modulation,* and *encoding*.
- Define *bandwidth* and compare *baseband* and *broadband*.
- List and describe the different connection services.

2.1 THE NEED FOR STANDARDIZATION

To overcome compatibility issues, hardware and software often follow **standards** (dictated specifications or most popular). Standards exist for operating systems, data formats, communication protocols, and electrical interfaces. If a product does not follow a widely used standard, the product will usually not be widely accepted in the computer market and will often cause problems with your PC. As far as the user is concerned, standards help you determine what hardware and software to purchase, and they allow you to customize a network made of components from different manufacturers.

As new technology is introduced, manufacturers are usually rushing to get their product out so that their product has a better chance of becoming the standard. Often, competing computer manufacturers introduce similar technology at the same time. Until one product is designated as the standard, other companies and customers are sometimes forced to take sides. Since it is sometimes difficult to determine what will emerge as the true standard and since the technology sometimes needs time to mature, it is best to wait a little to see what happens.

There are two main types of standards. The first standard is called the de jure standard. The **de jure standard,** or the "by law" standard, is a standard that has been dictated by an appointed committee such as the International Standards Organization (ISO). Some of the more common standard committees are shown in table 2.1.

The other type of standard is the de facto standard. The **de facto standard,** or the "from the fact" standard, is a standard that has been accepted by the industry just because it was the most common. De facto standards are not recognized by a standard committee. For example, the de facto standards for microprocessors are those produced by Intel, and the de facto standards for sound cards are those produced by Creative Labs.

Some systems or standards have an open architecture. The term **open architecture** indicates that the specification of the system or standard is public. This category includes approved standards as well as privately designed architecture whose specifications are made public by the designers. The advantage of an open architecture is that anyone can design products based on the architecture and design add-on products for it. Of course, this also allows other manufacturers to duplicate the product.

The opposite of an open architecture is a proprietary system. A **proprietary system** is privately owned and controlled by a company, and the specifications that would allow other companies to duplicate the product are not divulged to the public. Proprietary ar-

Table 2.1   Common Standard Committees

American National Standards Institute (ANSI) http://www.ansi.org	ANSI is primarily concerned with software. ANSI has defined standards for a number of programming languages, including C language and the SCSI interface.
Electronics Industry Alliance (EIA) http://www.eia.org	EIA is a trade organization composed of representatives from electronics manufacturing firms across the United States. EIA is divided into several subgroups: the Telecommunications Industry Association (TIA); the Consumer Electronics Manufacturers Association (CEA); the Joint Electron Device Engineering Council (JEDEC); the Solid State Technology Association, the Government Division; and the Electronic Information Group (EIG).
International Telecommunications Union (ITU) http://www.itu.int/	ITU defines international standards, particularly communications protocols. The group was formerly called the Comité Consultatif Internationale Télégraphique et Téléphonique (CCITT).
Institute of Electrical and Electronic Engineers (IEEE) http://www.ieee.org	IEEE sets standards for most types of electrical interfaces, including RS-232C (serial communication interface) and network communications.
International Standards Organization (ISO) http://www.iso.ch/	ISO is an international standard-setting group for communications and information exchange.

chitectures often do not allow mixing and matching products from different manufacturers and may cause hardware and software compatibility problems.

2.2   OSI REFERENCE MODEL

With the goal of standardizing the network world, the International Organization for Standardization (ISO) began development of the **Open Systems Interconnection (OSI) reference model.** The OSI reference model was completed and released in 1984.

Today, the OSI reference model is the world's prominent networking architecture model. It is a popular tool for learning about networks. The OSI protocols, on the other hand, have had a long "growing-up period." While OSI implementations are not unheard of, the OSI protocols have not yet attained the popularity of many proprietary and de facto standards.

The OSI reference model adopts a layered approach, where a communication subsystem is broken down into seven layers, each one of which performs a well-defined function. The OSI reference model defines the functionality that needs to be provided at each layer but does not specify actual services and protocols to be used at each one of the layers. From this reference model, actual protocol architecture can be developed. (See figure 2.1 and table 2.2.)

To facilitate this, ISO has defined internationally standardized protocols for each one of the seven layers. The seven layers are divided into three separate groups: application-oriented (upper) layers, an intermediate layer, and network-oriented (lower) layers.

NOTE: For an easy way to remember the order of the OSI reference model, you should use one of the following mnemonics:

Please	Do	Not	Take	Sales	Peoples'	Advice
Physical	Data Link	Network	Transport	Session	Presentation	Application

or

All	People	Seem	To	Need	Data	Processing
Application	Presentation	Session	Transport	Network	Data Link	Physical

When a computer needs to communicate with another computer, it will start with a network service, which is running in the application layer. The actual data that needs to be sent is generated by the software and is sent to the presentation layer. The presentation layer then adds its own control information, called a header, which contains the presentation layer's requests and/or information. Next, the packet is sent to the session layer, where another header is added. It keeps going down the OSI model until it reaches the physical layer, which means that the data is sent on the network media by the network interface card (NIC). (See figure 2.2.) The concept of placing data behind headers (and before trailers) for each layer is called **encapsulation.**

When the data packet gets to the destination computer, the network interface card sends the data packet to the data link layer. The data link layer then strips the first header off. As the packet goes up the model, each header is stripped away until the data reaches the application layer. At this time, only the original data is left, which is processed by the network service.

What is so great about this system is that it allows you to communicate with different computer systems. For example, a Windows NT server can send information to a UNIX server or an Apple Macintosh client.

Figure 2.1   The OSI Reference Model

OSI Reference Model

Application
Concerned with the support of end-user application processes

- Supports the local operating system via a redirector/shell
- Provides access for different file systems
- Provides common APIs for file, print, and message services
- Defines common data syntax and semantics
- Converts to format required by computer via data encoding and conversion functions

Presentation
Provides for the representation of the data

Session
Performs administrative tasks and security

- Establishes sessions between services
- Handles logical naming services
- Provides checkpoints for resynchronization

Transport
Ensures end-to-end error-free delivery

- Breaks up blocks of data on send or reassembles on receive
- Has end-to-end flow control and error recovery
- Provides a distinct connection for each session

Network
Responsible for addressing and routing between subnetworks

- Forms internetwork by providing routing functions
- Defines end-to-end addressing (logical − Net ID + Host ID)
- Provides connectionless datagram services

Data Link
Responsible for the transfer of data over the channel

- Sends frames; turns received bits into frames
- Defines the station address (physical); provides link management
- Provides error detection across the physical segment

Physical
Handles physical signaling, including connectors, timing, voltages, and other matters

- Provides access to the media.
- Defines voltages and data rates for sending binary data.
- Defines physical connectors.

Let's follow the steps of communication between two computers from initial contact to data delivery. A computer wants to request a file. The process would start with a network application on the client computer. Before the request can be made, the client has to first determine the address of the computer. Therefore, it sends a request out to resolve a computer name to a network address.

If the client already knows the server's network address, it will make its request in the form of a data packet and send it to a presentation-layer protocol. The presentation protocol would encrypt and compress the packet and send it to a session-layer protocol, which would establish a connection with the server. During this time the packet would include the type of dialog, such as half-duplex connection (discussed later in this chapter), and determine how long the computer can transmit to the server. The session protocol will then send the packet to the transport layer, which will divide the packet into smaller packets so that they can be sent over the physical network. The

Table 2.2   Common Technologies as They Relate to the OSI Model

OSI Model	TCP/IP	Novell NetWare	Microsoft Windows
Application	FTP, SMTP, telnet	NDS	SMB
Presentation	ACSII, MPEG, GIF, JPEG	NCP	NetBIOS
Session		SAP	NetBEUI
Transport	TCP, UDP	SPX	NetBEUI
Network	IP	IPX	NetBEUI
Data link	Ethernet, 802.3, 802.5, FDDI, Frame Relay, ISDN	Ethernet, 802.3, 802.5, FDDI, Frame Relay, ISDN	Ethernet, 802.3, 802.5, FDDI, Frame Relay, ISDN
Physical	10BaseT, 100BaseT, UTP 4/16 unshielded twisted pair, SONET	10BaseT, 100BaseT, UTP 4/16 unshielded twisted pair, SONET	10BaseT, 100BaseT, UTP 4/16 unshielded twisted pair, SONET

Figure 2.2   Layer Interaction of the OSI Reference Model

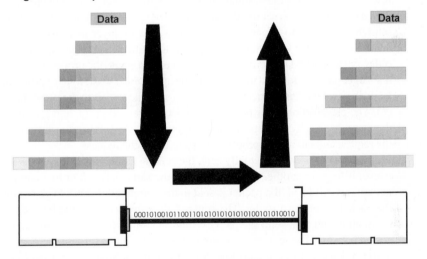

transport-layer protocol will send the packets to the network layer, where the source network and destination network addresses are added. The network-layer protocol will then send the packets to a data link layer to add the source and destination address and the port number that identifies the service requested and to prepare the packets to be sent over the media. As you can see in figure 2.2, the packet is bigger than when it started. The packet is then converted to electrical or electromagnetic signals that are sent on the network media.

Figure 2.3   Encapsulation that Actually Occurs on Today's Networks

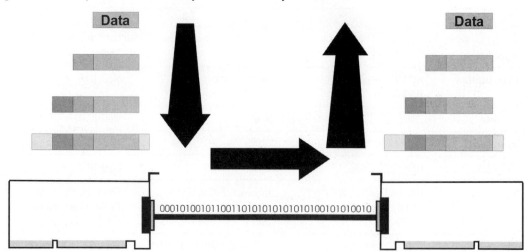

When the packets get to the destination server, the network card sees the signals and interprets the signals as 0s and 1s. It then takes the bits and groups them together into frames. It will then determine the destination address of the packet to see if the packet was meant for the server. After the server has determined that it was, it will remove the data link header and send the packet up to the network layer. The network layer will remove the network layer and send the packet to the transport layer. If the data packets have reached the server out of order, the transport-layer protocol will put them back into the proper order, merge them into one larger packet, and send the packet to the appropriate session-layer protocol. The session-layer protocol will authenticate the user and send the packet to the presentation-layer protocol. The presentation-layer protocol will decompress, decrypt, and reformat the packet so that it can be read by the application-layer protocol. The application-layer protocol will read the request and take the appropriate steps to fulfill the request.

Before the packet is actually sent, the sending computer or host must determine if the packet is to go to another computer within the same network or to a computer in another network. If it is local, the packet is sent to the computer. If it has to go to another network, the packet is sent to a router. The router will then determine the best way to get it to its destination and then send the packet from one router to another until it gets to the destination network. If one network route is congested, it can reroute the packet another way; and if it detects errors, it can slow down the transmission of the packet in the hopes that the link will become more reliable.

In reality, encapsulation does not occur for all seven layers. Layers 5 through 7 use headers during initialization, but in most flows, there is no specific layer 5, 6, or 7 header. This is because there is no new information to exchange for every flow of data. Therefore, layers 5, 6, and 7 can be grouped together, and the layers actually used can be simplified into five layers: the application (equivalent to the application, presentation, and session layers), transport, Internet (equivalent to the network layer), network interface (equivalent to the data link layer), and physical layers. (See figure 2.3.)

Figure 2.4   Information Blocks Associated with the Simplified OSI Model

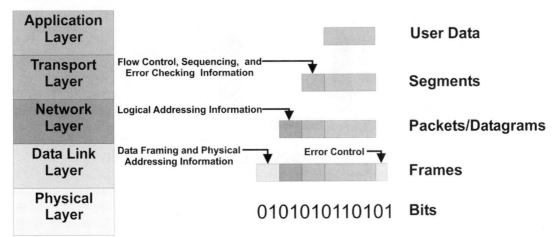

When encapsulation with TCP/IP is used, the order in which information blocks are created is data, segments, packets, frames, and bits. The user application creates the data and a variety of parameters and passes the segment to the network layer. The network layer places the destination network address in a header, puts the data behind it, and transmits a packet, or datagram, to the data link layer. The data link layer creates the data link header, which includes the destination MAC address. The datagram is converted to a frame and is passed to the physical layer. The physical layer transmits the bits. (See figure 2.4.)

Many protocols will have the destination computer send back an acknowledgment stating that the packet arrived intact. If the source computer does not receive an acknowledgment after a certain amount of time, it will resend the packet.

Example:

Let's say you are a user that is running a word processor, and you decide to access a file that is located on a remote computer's shared directory. A shared directory is a directory on a computer that provides the directory to clients over the network. You click on the Open button so you can view the files of the remote computer. The word processor initiates the entire process by generating a network request. The request is sent through a client/redirector, which forwards the request to a network protocol such as TCP/IP. The packet is then forwarded to the NIC driver and sent out through the network card, where it is sent through the network to the remote computer. Of course, as the request is being sent down from the word processor to the network card, the request becomes bigger as the packet is encapsulated. (See figure 2.5.)

When the packet is received at the remote computer, it is processed by the driver on the remote computer. It is then forwarded to the network protocol, which is then forwarded to the server service (file and print sharing). The server service uses the local file system services to access the file. As the packet goes from the network card to the local file system services, the packet is stripped back to the original request. The file is then sent back to the requesting computer.

Figure 2.5   A Network Request Going Through the Simplified OSI Model

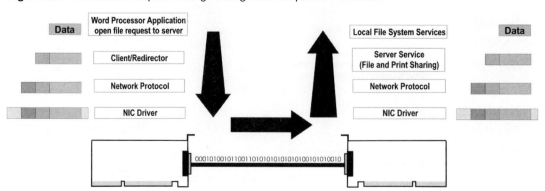

2.3   BITS AND BYTES

The most commonly used numbering system is the **decimal number system.** In a decimal number system, each position contains 10 different possible digits. Since there are 10 different possible digits, the decimal number system consists of numbers with base 10. These digits are 0, 1, 2, 3, 4, 5, 6, 7, 8, and 9. In order to count values larger than 9, each position away from the decimal point in a decimal number increases in value by a multiple of 10. (See table 2.3.)

Example:

Decimal number: 234

2	3	4
2×10^2	3×10^1	4×10^0
200	30	4

Therefore, the value is $200 + 30 + 4 = 234$.

Table 2.3   Decimal Number System

7th Place	6th Place	5th Place	4th Place	3rd Place	2nd Place	1st Place
10^6	10^5	10^4	10^3	10^2	10^1	10^0
1,000,000	100,000	10,000	1000	100	10	1

Table 2.4   One-Digit Binary Number

Wire 1	Binary Equivalent	Decimal Equivalent
Off	0	0
On	1	1

Table 2.5   Two-Digit Binary Number

Wire 1	Wire 2	Binary Equivalent	Decimal Equivalent
Off	Off	0 0	0
Off	On	0 1	1
On	Off	1 0	2
On	On	1 1	3

2.3.1   Binary Number System

The **binary number system** is another way to count. The binary system is less complicated than the decimal system because it has only two digits: zero (0) and 1. A computer represents a binary value with an electronic switch known as a transistor. If the switch is on, it allows current to flow through a wire or metal trace to represent a binary value of 1. If the switch is off, it does not allow current to flow through a wire, representing a value of zero (0). (See table 2.4.) The on switch is also referred to as a high signal, whereas the off switch is referred to as a low signal.

If you use two wires to represent data, the first switch can be on or off and the second switch can be on or off, giving you a total of four combinations or four binary values. (See table 2.5.) If you use four wires to represent data, you can represent 16 different binary values. (See table 2.6.) Since each switch represents two values, each switch used doubles the number of binary values. Therefore, the number of binary values can be expressed with the following equation:

$$\text{Number of binary numbers} = 2^{\text{number of binary digits}}$$

Thus one wire allows $2^1 = 2$ binary numbers, 0 and 1. Two wires allow $2^2 = 4$ binary numbers, 0, 1, 2, and 3. And four wires allow $2^4 = 16$ binary numbers.

Question:

How many values does 8 bits (binary digits) represent?

Answer:

Since a byte has 8 binary digits, a byte can represent $2^8 = 256$ characters.

Much like decimal numbers, binary digits have placeholders that represent certain values, as shown in table 2.7.

Table 2.6   Four-Digit Binary Number

Wire 1	Wire 2	Wire 3	Wire 4	Binary Equivalent	Decimal Equivalent
Off	Off	Off	Off	0 0 0 0	0
Off	Off	Off	On	0 0 0 1	1
Off	Off	On	Off	0 0 1 0	2
Off	Off	On	On	0 0 1 1	3
Off	On	Off	Off	0 1 0 0	4
Off	On	Off	On	0 1 0 1	5
Off	On	On	Off	0 1 1 0	6
Off	On	On	On	0 1 1 1	7
On	Off	Off	Off	1 0 0 0	8
On	Off	Off	On	1 0 0 1	9
On	Off	On	Off	1 0 1 0	10
On	Off	On	On	1 0 1 1	11
On	On	Off	Off	1 1 0 0	12
On	On	Off	On	1 1 0 1	13
On	On	On	Off	1 1 1 0	14
On	On	On	On	1 1 1 1	15

Table 2.7   Binary Number System

8th Place	7th Place	6th Place	5th Place	4th Place	3rd Place	2nd Place	1st Place
2^7	2^6	2^5	2^4	2^3	2^2	2^1	2^0
128	64	32	16	8	4	2	1

Example:

Convert the binary number 11101010 to a decimal number.

1	1	1	0	1	0	1	0
1×2^7	1×2^6	1×2^5	0×2^4	1×2^3	0×2^2	1×2^1	0×2^0
128	64	32	0	8	0	2	0

Therefore, the binary number 11101010 is equal to the decimal number $128 + 64 + 32 + 8 + 2 = 234$.

Example:

Convert the decimal number 234 to a binary number.

Referring to table 2.7, you can see that the largest power of 2 that will fit into 234 is 2^7 (128). This leaves the value of $234 - 128 = 106$. The next largest power of 2 that will fit into 106 is 2^6 (64). This leaves a value of $106 - 64 = 42$. The next largest power of 2 that will fit into 42 is 2^5 (32), which gives us $42 - 32 = 10$. The next largest power of 2 that will fit into 10 is 2^3 (8), which gives us $10 - 8 = 2$. The next largest power of 2 that will fit into 2 is 2^1 (2), which gives us $2 - 2 = 0$.

$$
\begin{array}{rl}
234 & \\
-128 & 2^7 \\
\hline
106 & \\
-64 & 2^6 \\
\hline
42 & \\
-32 & 2^5 \\
\hline
10 & \\
-8 & 2^3 \\
\hline
2 & \\
-2 & 2^1 \\
\hline
0 &
\end{array}
$$

Therefore, the binary equivalent is 11101010.

1	1	1	0	1	0	1	0
2^7	2^6	2^5	2^4	2^3	2^2	2^1	2^0

In computers, one of these switches is known as a **bit,** or **binary digit.** When several bits are combined together, they can signify a letter, a digit, a punctuation mark, a special graphical character, or a computer instruction. Eight bits make up a **byte.**

Since bytes are such a small unit, larger units of measure are used: kilobytes (KB), megabytes (MB), gigabytes (GB), and terabytes (TB). The prefix *kilo* indicates a thousand, *mega* indicates a million, *giga* indicates a billion, and *tera* indicates a trillion. With computers, the measurement is not exactly equal to the number the prefix represents. A kilobyte is actually 1024 bytes, not 1000. This is because 2^{10} is equal to 1024. And similar to the kilobyte, a megabyte is 1024 kilobytes, a gigabyte is 1024 megabytes, and a terabyte is 1024 gigabytes.

1 kilobyte = 1024 bytes
1 megabyte = 1024 kilobytes = 1,048,576 bytes

Table 2.8 Hexadecimal Digit

Decimal	Binary	Hexadecimal
0	0000	0
1	0001	1
2	0010	2
3	0011	3
4	0100	4
5	0101	5
6	0110	6
7	0111	7
8	1000	8
9	1001	9
10	1010	A
11	1011	B
12	1100	C
13	1101	D
14	1110	E
15	1111	F

1 gigabyte = 1024 megabytes = 1,048,576 kilobytes = 1,073,741,824 bytes
1 terabyte = 1024 gigabytes = 1,048,576 megabytes = 1,073,741,824 kilobytes
  = 1,099,511,627,776 bytes

2.3.2   Hexadecimal Number System

The **hexadecimal number system** has 16 digits. One hexadecimal digit is equivalent to a four-digit binary number (4 bits or a nibble), and two hexadecimal digits are used to represent a byte (8 bits). Therefore, it is very easy to translate between hexadecimal and binary, and so the hexadecimal system is used as a "shorthand" way of displaying binary numbers. (See table 2.8.) To designate a number as a hexadecimal number, the number will often end with the letter H. In order to count values larger than 15, each position away from the decimal point in a decimal number increases in value by a multiple of 16. (See table 2.9.)

Question:

What is the hexadecimal number of the binary number of 1001 1010?

Table 2.9 Hexadecimal Number System

7th Place	6th Place	5th Place	4th Place	3rd Place	2nd Place	1st Place
16^6	16^5	16^4	16^3	16^2	16^1	16^0
16777216	1048576	65536	4096	256	16	1

Answer:

Since 1001 is equivalent to 9 and 1010 is equivalent to A, the hexadecimal equivalent is 9AH.

To convert a hexadecimal number to a decimal number, you could first convert the hexadecimal number to binary and then convert the binary number to decimal. Another way to convert is to multiply the decimal value of each hexadecimal digit by its weight or value place and then take the sum of these products.

Example 1:

Convert EAH to a decimal number.
To do this, you would multiply the A by the 1s and the E by 16 and then add them up.

E	A
$E \times 16^1$	$A \times 16^0$
14×16^1	10×16^0
14×16	10×1
224	10

Therefore, the hexadecimal number EA is equal to the decimal number $224 + 10 = 234$.

Example 2:

Convert the decimal number 234 to a hexadecimal number. Referring to table 2.9, you can see that the largest power of 16 that will fit into 234 is 16^1 (16). The number of times that 16 goes into 234 is 14 (E) times, leaving a 10 (A).

$234/16 = \mathbf{14}.625$
$234 - (\mathbf{14} \times 16) = 10$
$14 = EH$

16^1	16^0
14×16^1	10×16^0
E	A

Figure 2.6   A Byte Representing an ASCII Character

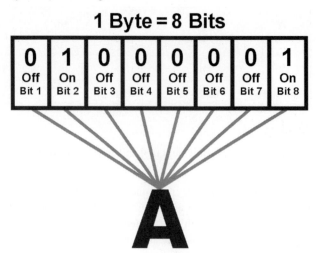

Today, TCP/IP addresses (IPv4) are based on 32 bits divided as 4 groups of 8 bits. The groupings of 8 bits is sometimes referred to as an **octet.** Of course, since a hexadecimal digit consists of 4 bits, it takes 2 hexadecimal digits to express an octet.

2.3.3   ASCII Character Set

In order to communicate, you need numbers, letters, punctuation, and other symbols. An alphanumeric code represents these characters and various instructions necessary for conveying information. One commonly used alphanumeric code is the **ASCII (American Standard Code for Information Interchange)** character set. Since ASCII is based on 8 bits, there is a total of 256 different combinations of 0s and 1s, allowing 256 ($2^8 = 256$) different characters. A partial list of the ASCII characters is shown in table 2.10. As an example, if a byte had the binary code 01000001, the byte would represent the capital letter A (see figure 2.6) and 01100001 would represent lowercase a.

2.4   PHYSICAL LAYER

The network-oriented (lower) layers are concerned with the protocols associated with the physical part of the network which allow two or more computers to communicate. The **physical layer** is responsible for the actual transmission of the bits sent across a physical medium. It allows signals, such as electrical signals, optical signals, or radio signals, to be exchanged among communicating machines. Therefore, it defines the electrical, physical, and procedural characteristics required to establish, maintain and deactivate physical links. This includes how the bits (0s and 1s) of data are represented and transmitted. It is not concerned with how many bits make up each unit of data, nor is it concerned with the meaning of the data being transmitted. In the physical layer, the sender simply transmits a signal and the receiver detects it. Lastly, the physical layer is also responsible for the

Table 2.10 A Partial List of the ASCII Character Set

DEC	BIN	HEX	ASCII	DEC	BIN	HEX	ASCII
32	00100000	20	space	80	01010000	50	P
33	00100001	21	!	81	01010001	51	Q
34	00100010	22	"	82	01010010	52	R
35	00100011	23	#	83	01010011	53	S
36	00100100	24	$	84	01010100	54	T
37	00100101	25	%	85	01010101	55	U
38	00100110	26	&	86	01010110	56	V
39	00100111	27	'	87	01010111	57	W
40	00101000	28	(	88	01011000	58	X
41	00101001	29	)	89	01011001	59	Y
42	00101010	2A	*	90	01011010	5A	Z
43	00101011	2B	+	91	01011011	5B	[
44	00101100	2C	'	92	01011100	5C	\
45	00101101	2D	-	93	01011101	5D	]
46	00101110	2E	.	94	01011110	5E	^
47	00101111	2F	/	95	01011111	5F	_
48	00110000	30	0	96	01100000	60	`
49	00110001	31	1	97	01100001	61	a
50	00110010	32	2	98	01100010	62	b
51	00110011	33	3	99	01100011	63	c
52	00110100	34	4	100	01100100	64	d
53	00110101	35	5	101	01100101	65	e
54	00110110	36	6	102	01100110	66	f
55	00110111	37	7	103	01100111	67	g
56	00111000	38	8	104	01101000	68	h
57	00111001	39	9	105	01101001	69	i
58	00111010	3A	:	106	01101010	6A	j
59	00111011	3B	;	107	01101011	6B	k
60	00111100	3C	<	108	01101100	6C	l
61	00111101	3D	=	109	01101101	6D	m
62	00111110	3E	>	110	01101110	6E	n
63	00111111	3F	?	111	01101111	6F	o
64	01000000	40	@	112	01110000	70	p
65	01000001	41	A	113	01110001	71	q
66	01000010	42	B	114	01110010	72	r
67	01000011	43	C	115	00111011	73	s
68	01000100	44	D	116	01110100	74	t
69	01000101	45	E	117	01110101	75	u
70	01000110	46	F	118	01110110	76	v
71	01000111	47	G	119	01110111	77	w
72	01001000	48	H	120	01111000	78	x
73	01001001	49	I	121	01111001	79	y
74	01001010	4A	J	122	01111010	7A	z
75	01001011	4B	K	123	01111011	7B	{
76	01001100	4C	L	124	01111100	7C		
77	01001101	4D	M	125	01111101	7D	}
78	01001110	4E	N	126	01111110	7E	~
79	01001111	4F	O	127	01111111	7F	Delete

physical topology or actual network layout. The physical layer includes the network cabling, hubs, repeaters, and network interface card (NIC).

NOTE: Most network problems do occur at the physical layer, whether it is a failed NIC, cable, transceiver, router or switch port, EMI or RFI, or just a loose connection.

2.4.1   Signaling

Signaling is the method for using electrical, light energy, or radio waves to communicate. The process of changing a signal to represent data is often called **modulation** or **encoding.** The two forms of signaling are digital signaling and analog signaling.

Digital signals (the language of computers) are based on a binary signal system produced by pulses of light or electric voltages. The site of the pulse is either on/high or off/low to represent 1s and 0s. Binary digits (bits) can be combined to represent different values.

A digital signal can be measured in one of two ways. The first method, the **current state** method, periodically measures the digital signal for the specific state. The second method is the **transition state.** It represents data by how the signal transitions from high to low or low to high. A transition indicates a binary 1, and the absence of a transition represents a binary 0. (See figure 2.7.)

An **analog signal** is the opposite of a digital signal. Instead of having a finite number of states, it has an infinite number of values, which are constantly changing. Analog signals are typically sinusoidal waveforms, which are characterized by their amplitude and frequency. The **amplitude** represents the peak voltage of the sine wave. The **frequency** indicates the number of times that a single wave will repeat over any period. It is measured in hertz (Hz), or cycles per second. Another term used when talking about modulation is its *time reference* or *phase.* A **phase** is measured in degrees. Phase can be measured in 0°, 90°, 180°, and 270° or in 0°, 45°, 90°, 135°, 180°, 225°, 270°, 315°, and 360°. (See figure 2.8.)

By varying the amplitude, frequency, and phase, data is sent over the telephone lines. The earliest form of encoding data over telephone lines is frequency shift keying (FSK), which is very similar to frequency modulation used with FM radios. FSK sends a logical 1 at one particular frequency (usually 1750 Hz), and a logical 0 is sent at another frequency (often 1080 Hz). FSK was usually used with 300-baud modems.

Phase shift keying (PSK) is the varying of the phase angle (0°, 90°, 180°, and 270°) to represent data. Since the phase has four different values, it represents 2 bits of data. Therefore, a 1200-baud modem using PSK can transmit data at 2400 bits per second (bps). (See table 2.11.)

Much like digital signals, analog signals can be measured as current state or state transition. For example, they can measure the amplitude at set intervals, or they can measure the transition from one amplitude to another.

Noise is interference or static that destroys the integrity of signals. Noise can come from a variety of sources, including radio waves, nearby electric wires, lightning and other power fluctuations, and bad connections. One example is when you are talking to someone over a phone and you hear static on the line; another example is when you can hear another conversation on your phone line.

Figure 2.7 Measuring Data Bits by Using Current State and State Transition

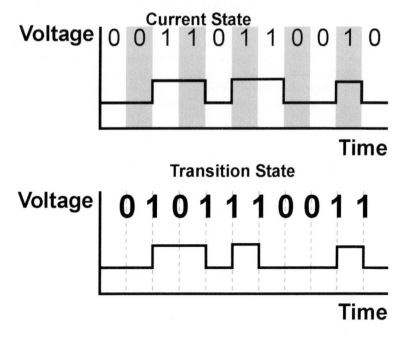

Figure 2.8 A Sine Wave Showing the Amplitude, Frequency (Cycles per Second), and Phase

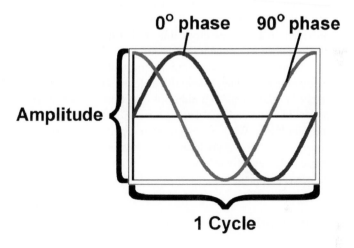

Table 2.11 Data Bits Interpreted from Phase Signals

Phase	Data Bits
0°	0 0
90°	0 1
180°	1 0
270°	1 1

41

Analog signals are not optimal for data transmission since they are more prone to errors than data signals. This is because as analog signals are amplified so that they can travel longer distances, the noise is also amplified. Since digital signals are discrete values (such as a bit being either a 0 or a 1), the signal is regenerated or re-created with a repeater. Therefore, every time the signal goes through a repeater, the signal starts out fresh with no noise.

2.4.2   Synchronous and Asynchronous Connections

As mentioned earlier in this chapter, data bits are encoded on a network medium, and the receiving network card interprets the signal by taking measurements of the signal. Therefore, the receiving network card must use a clock or a timing method to determine when to measure and decode the signal and when to decode the data bits. The two methods used to do this are synchronous and asynchronous.

Synchronous devices use a timing or clock signal to coordinate communications between the two devices. If the send and receiving devices were both supplied by exactly the same clock signal, then transmission could take place forever with the assurance that the signal sampling at the receiver was always in perfect synchronization with the transmitter.

In synchronous communications, data is not sent in individual bytes, but as frames of large data blocks. For example, frame sizes vary from a few bytes to many bytes. Ethernet uses 1500-byte packets. The clock is embedded in the data stream encoding, or it is provided on a separate clock line so that the sender and receiver are always in synchronization during a frame transmission.

Asynchronous signals are intermittent signals—they can occur at any time and at irregular intervals. They do not use a clock or timing signal. The data frame that is sent consists of a start signal, a number of data bits, and a stop signal. The start signal is sent to notify the other end that data is coming, and the stop signal is sent to indicate the end of the data frame. While the asynchronous signals are simpler electronic devices than synchronous devices, asynchronous signals are not as efficient as synchronous signals because of the extra overhead of the start and stop signal. An example of an asynchronous device is a modem.

2.4.3   Bandwidth

Bandwidth refers to the amount of data that can be carried on a given transmission medium. Larger bandwidth means greater data transmission capabilities. Bandwidth is typically expressed in bits per second. Of course, since a bit is such a small unit, you will usually see bandwidth measured in kilobits per second, megabits per second, and gigabits per second. Kilobits per second is often abbreviated as Kbits/second or Kb/s, as opposed to kilobytes/second, Kbytes/second, or KB/s.

Bandwidths use schemes based upon the availability and utilization of channels. A channel is a part of the medium's total bandwidth. It can be created by using the entire bandwidth for one channel or by splitting up multiple frequencies to accommodate multiple channels. For example, if a medium can support 10 megabits per second (Mbps), two channels can be created at 5 Mbps each.

Baseband systems use the transmission medium's entire capacity for a single channel. Baseband networks can use either analog or digital signals, but digital is much more common.

A **broadband system** uses the transmission medium's capacity to provide multiple channels by using **frequency-division multiplexing (FDM).** Each channel uses a carrier signal, which runs at a different frequency than the other carrier signals used by the other channels. The data is embedded within the carrier channel. As data is sent onto the transmission channel, one multiplexer (mux) sends the several data signals at different frequencies. When the various signals reach the end of the transmission medium, another multiplexer separates the frequencies so that the data can be read.

Although a baseband system can only support one signal at a time, it can be made into a broadband system by using time-division multiplexing. Time-division multiplexing (TDM) divides a single channel into short time slots, allowing multiple devices to be assigned time slots.

The TDM method works well in many cases, but it does not always account for the varying data transmission needs of different devices or users. Regardless of the needs of the devices that have to transmit data, the same duration of time to transmit or receive data is allocated. A derivative of the TDM is the **statistical time-division multiplexing (STDM),** which analyzes the amount of data that each device needs to transmit and determines on-the-fly how much time each device should be allocated for data transmission on the cable or line. As a result, the STDM uses the bandwidth more efficiently than the TDM.

2.4.4   Point-to-Point and Multipoint Connections

All physical topologies are variations of two fundamental methods of connecting devices, point-to-point and multipoint connections. (See figure 2.9.) **Point-to-point topology** connects two nodes directly together. Examples include two computers connected together using modems, a PC communicating with a printer using a parallel cable, or WAN links connected together to two routers using a dedicated T1 line. In a point-to-point link, the

Figure 2.9   A Point-to-point and a Multipoint Connection

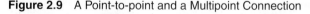

two devices monopolize the communication medium between the two nodes. Because the medium is not shared, nothing is needed to identify the two nodes. Whatever data is sent from one device is sent to the other device.

A multipoint connection links three or more devices together through a single communication medium. Because multipoint connections share a common channel, each device needs a way to identify itself and the device to which it wants to send information. The method used to identify senders and receivers is called **addressing.** SCSI devices connected on a single ribbon cable are identified with their SCSI ID numbers. Network cards connected on a network are identified with their media access control (MAC) address.

2.5   DATA LINK LAYER

The **data link layer** is responsible for providing error-free data transmission and establishes local connections between two computers or hosts. It divides data it receives from the network layer into distinct frames that can then be transmitted by the physical layer, and it packages raw bits from the physical layer into blocks of data called frames. A **frame** is a structured package for moving data that includes not only the raw data, or "payload," but also the sender's and receiver's network addresses and error-checking and control information. The control information provides necessary synchronization and flow control. The error-checking information includes a checksum (CRC) or some other form of error-checking information. Since a package contains the data and destination address, each packet travels the network independent from other packets.

The data link layer is divided into two sublayers, the logical link control sublayer (IEEE 802.2) and the media access control sublayer. The **media access control (MAC) sublayer** is the lower sublayer that communicates directly with the network adapter card. It defines the network logical topology, which is the actual pathway (ring or bus) of the data signals being sent. In addition, it allows multiple devices to use the same media, and it determines how the network card gets access or control of the network media so that two devices don't trample over each other. Also, it maintains the physical device address, known as the MAC address, which is used to identify each network connection. Lastly, it builds frames from bits that are received from the physical layer. Some examples of the media access control sublayer protocols include CSMA/CA, CSMA/CD, token passing, and demand priority.

The **logical link control (LLC)** sublayer manages the data link between two computers within the same subnet. A **subnet** is a simple network or smaller network that is used to form a larger network. In addition, if we were operating a multiprotocol LAN, each network layer protocol would have its own **service access point (SAP),** which is used by the LLC to identify which protocol it is. For example, TCP/IP, IPX/SPX, and NetBIOS would all have different SAPs so that they can be identified by a lower protocol. In addition, when a packet is received by the network card, the LLC checks the destination MAC address in the frame. If it matches the host, the host system accepts the frame. If it does not match, the host system ignores the frame. Lastly, if a destination LLC receives a bad

Figure 2.10   Three Networks Connected Together with a Router. Each Computer is Identified by Its 8-Hexadecimal-digit MAC Address. Notice the MAC Address 22-33-A3-34-43-43 is Used Repeatedly but on Different Networks.

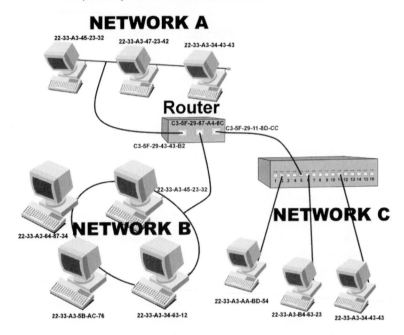

frame, it discards it and does not send an acknowledgment that it received the frame. Since the sending LLC does not receive the acknowledgment, it resends the frame.

2.5.1   Physical Device Addressing

Since many network devices share the same transmission channel, the data link layer must have some way to identify itself from the other devices. The physical device address or **media access control (MAC) address** is a unique hardware address (unique on the LAN/subnet) burned onto a ROM chip assigned by the hardware vendors or selected with jumpers or DIP switches. A MAC address is a lot like a street address that identifies a house or building within a city. And much like street addresses within a city, you cannot have two network cards or nodes with the same MAC address on the same network. You can have the same MAC address on two separate networks. For Ethernet and Token Ring cards, the MAC address is embedded onto the ROM chip on the card. (See figure 2.10.)

NOTE: If two computers or network interface are separated by only one router, the router could be confused since it sees the same MAC address on two networks.

The MAC address is 48 bits, or 6 bytes, in length and is usually represented in format. The first 24 bits of a MAC address are referred to as the Organizationally Unique Identifier, or OUI. OUIs are sold and assigned to network hardware vendors by the IEEE.

The last 24 bits are assigned by the individual vendor. A MAC address is ordinarily expressed as six pairs of hexadecimal characters, separated by colons. An example of a MAC address is 00:53:AD:B2:13:BA.

2.5.2   Media Access Methods

The set of rules defining how a computer puts data onto the network cable and takes data from the cable is called the **access method,** sometimes referred to as **arbitration.** While multiple computers share the same cable system, only one device can access a cable at a time. If two devices try to use the same cable at the same time, both data packets that are sent by the card become corrupted. The typical cable access methods are:

- Contention
- Token passing
- Polling
- Demand priority

Contention is when two or more devices contend (compete) for network access. Any device can transmit whenever it needs to send information. To avoid data collisions (two devices sending data at the same time), specific contention protocols were developed, requiring the device to listen to the cable before transmitting data.

The most common form of contention is the **carrier sense multiple access (CSMA)** network. Even though each station listens for network traffic before it attempts to transmit, it remains possible for two transmissions to overlap on the network or to cause a collision. As a result of collisions, access to a CSMA network is somewhat unpredictable, and CSMA networks can be referred to as random or statistical access networks. To avoid collisions, CSMA will use one of two specialized methods of collision management: collision detection (CD) and collision avoidance (CA).

The collision detection approach listens to the network traffic as the card is transmitting. By analyzing network traffic, it detects collisions and initiates retransmissions. **Carrier sense multiple access with collision detection (CSMA/CD)** is the access method utilized in Ethernet and IEEE 802.3. Collision avoidance uses time slices to make network access smarter and avoid collisions. **Carrier sense multiple access with collision avoidance (CSMA/CA)** is the access mechanism used in Apple's LocalTalk network. When a collision occurs, it is the logical link control sublayer of the data link layer that will retransmit a frame if it does not receive an acknowledgment from the destination LLC.

While contention is a very simple access method that has low administrative overhead requirements, high traffic levels cause more collisions, which cause a lot of retransmitting, which causes even slower network performance.

Token passing uses a special authorizing packet of information to inform devices that they can transmit data. These packets, called tokens, are passed around the network in an orderly fashion from one device to the next. Devices can transmit only if they have control of the token, which distributes the access control among all the devices. Token Ring uses a ring topology—each station passes the token to the next station in the ring. ARCnet uses a token passing bus as it passes the token to the next higher hardware address (MAC address), regardless of its physical location on the network.

Token passing is deterministic because you can calculate the maximum time before a workstation can grab the token and begin to transmit. In addition, you can assign priorities to certain network devices that will use the network more frequently. If a workstation has a priority equal to or higher than the priority value in the token, it can take possession of the token.

Polling has a single device (sometimes referred to as a channel-access administrator) such as a mainframe front-end processor designated as the primary device. The primary device polls or asks each of the secondary devices, known as slaves, if they have information to be transmitted. Only when it is polled does the secondary computer have access to the communication channel. To make sure that a slave doesn't hog all of the bandwidth, each system has rules pertaining to how long each secondary computer can transmit data.

The newest access method is called demand priority. In **demand priority,** a device makes a request to the hub, and the hub grants permission. High-priority packets are serviced before any normal-priority packets. To effectively guarantee bandwidth to time-sensitive applications like voice, video, and multimedia applications, the normal-priority packets are promoted to a high priority after 200–300 ms.

2.6   THE 802 PROJECT MODEL

In the late 1970s, when LANs first began to emerge as a valuable potential business tool, the IEEE realized that there was a need for certain LAN standards, specifically for the physical and data link layers of the OSI reference model. To accomplish this, the IEEE launched Project 802, which was named for the year and month it began (1980, February—the second month of the year). These standards have several areas of responsibility, including the network card, the wide area network components, and the media components. (See table 2.12 and figure 2.11.)

Of these, probably the two most popular discussed 802 standards are 802.3 [Carrier Sense Multiple Access with Collision Detection (CSMA/CD) LAN (Ethernet)] and 802.5 (Token Ring). When these two standards were originally created, the 802.3 standard had better performance on smaller networks, and the 802.5 standard had better performance on larger networks. While both networks will, of course, slow down as more computers are added because of the increased traffic, the 802.5 network runs more efficiently than the 802.3, as the number of collisions increase with the 802.3 network. Fortunately, the 802.3 network performance has been enhanced with increased bandwidth and with switches.

NOTE: Recently the 802.3u working group updated 802.3 to include Ethernet 100BaseT implementation, and 802.3z describes the gigabit Ethernet.

Another more recently popular 802 standard is the 802.11, which covers wireless networks. (See table 2.13.)

Table 2.12   Project 802 Standards

Standard	Category
802.1	Overview and architecture of internetworking including bridging and virtual LAN (VLAN)
802.2	Logical link control (LLC), which is the data link layer sublayer
802.3	Carrier sense multiple access with collision detection (CSMA/CD) LAN (Ethernet)
802.4	Token Bus LAN
802.5	Token Ring LAN
802.6	Metropolitan area network (MAN), which includes the Distributed Queue Dual Bus (DQDB)
802.7	Broadband local area networks
802.8	Fiber-Optic Technical Advisory Group covering fiber-optic LANs and MANs, which include FDDI and 10BaseFL
802.9	Integrated voice/data networks including ISDN and DSL lines
802.10	Network security, which includes virtual private network (VPN)
802.11	Wireless networks
802.12	Demand priority access LAN, 100BaseVG-AnyLAN
802.14	Coaxial and fiber cables such as those found on cable television to support two-way communications

Figure 2.11   The IEEE Standards as They Relate to the OSI Model

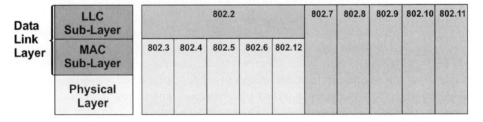

2.7   NETWORK LAYER

The **network layer** is concerned with the addressing and routing processes necessary to move data (known as packets or datagrams) from one network (or subnet) to another. This includes establishing, maintaining, and terminating connections between networks, making routing decisions, and relaying data from one network to another. The OSI model classifies a **host** as the computer or device that connects to the network and is the source or

Table 2.13   Popular 802 Standards

Technology	Speed(s)	Access Method	Topologies	Media
Ethernet (IEEE 802.3)	10, 100, or 1000 Mbps	CSMA/CD	Logical bus	Coax or UTP
Token Ring (IEEE 802.5)	4 or 16 Mbps	Token passing	Physical star, logical ring	STP or UTP
FDDI (IEEE 802.8)	100 or 200 Mbps	Token passing	Physical star, logical ring	Fiber optic or UTP (implemented as CDDI)
Wireless (IEEE 802.11)	1, 22, or 54	CSMA/CA	Cellular	Wireless

final destination of data. Since routers (intermediate systems) perform routing and relaying functions that link the individual networks, routers are network-layer devices.

2.7.1   Network Addressing

For networks to operate on an internetwork and to identify themselves to each other, each network must be assigned a network ID. Therefore, at the network layer, networks are identified with a unique **network address.** This means that every computer on an individual LAN must use the same network address. Of course, if the LAN is connected to other LANs, its address must be different from all the other LANs. Physical devices (MAC addresses) and logical network addresses are used jointly to move data between devices on an internetwork.

When planning out a network, assign logical addresses to each of your hosts. For example, when you use the TCP/IP, the administrator assigns IP addresses to each computer. The IP address contains the network address for the subnet that the host is on and a host address to represent the host on the subnet.

When a packet is sent to a remote host several networks away, the packet is sent to the first router. The router looks at the destination by looking at the logical address. The router then determines which way to send the packet, strips off and rebuilds the data link layer source and destination addresses information, and sends the packet to the next router. When the packet gets to the next router, the router reads the logical address to determine the destination. It then determines which way to send the packet, strips off and rebuilds the data link layer source and destination address information, and sends it to the next router. It will keep doing this until it gets to the destination network and then the final router sends the packet to the destination computer.

Remember, the data link layer provides for the transmission of frames within the same LAN, and the network layer provides for the much more complex task of transmitting packets between two network computers or devices in the network, regardless of how many data links or routers exist between the two. Examples of network-layer protocols that connect the networks together include IP and IPX, and examples of protocols that determine which route to take include RIP and OSPF.

A network computer or other network device can perform several roles simultaneously. The term *entity* identifies the hardware and software that fulfill each individual role. Every entity must have its service address so that it can send and receive data. This address is usually referred to as a **port** or **socket,** which is used to identify a specific upper-layer software process or protocol. Multiple service addresses can be assigned to any computer on which several network applications are running.

For example, when you try to read a web page, the packet from the Internet is routed to your computer using a TCP/IP network. Since today's operating systems are multitasking environments, they can be running or accessing several different network services. The port, which is included in the packet, would be used to identify the type of packet so that the system would know which software on the host is needed for the packet.

Since a network computer can handle multiple conversations with other computers at the same time, the system must be able to keep track of the different conversations. The two methods of distinguishing conversations are the use of connection identifiers and the use of transaction identifiers.

A term you will hear a lot when dealing with Windows and networking is *Windows socket,* or *WinSock.* A Windows socket is a Windows implementation of the UC Berkeley Sockets Application Programming Interface (API), which is used in TCP/IPs to connect to the appropriate TCP/IP service such as HTTP, FTP, or telnet. The WINSOCK.DLL file is the Microsoft Windows interface for the TCP/IP.

2.7.2   Data-Switching Techniques

Since large internetworks can have multiple paths linking source and destination devices, information is switched as it travels through the various communication channels. The data-switching techniques can be divided into circuit switching and packet switching. **Circuit switching** is a technique that connects the sender and the receiver by a single path for the duration of a conversation. Once a connection is established, a dedicated path exists between both ends; it is always consuming network capacity, even when there is no active transmission taking place (such as when a caller is put on hold). Once the connection has been made, the destination device acknowledges that it is ready to carry on a transfer. When the conversation is complete, the connection is terminated. (See figure 2.12.) Therefore, circuit-switching networks are sometimes called connection-oriented networks. Examples of circuit switching include phone systems and data that needs to be transmitted in live video and sound.

In **packet switching,** messages are broken into small parts called packets. Each packet is tagged with source, destination, and intermediary node addresses as appropriate. Packets can have a defined maximum length and can be stored in RAM instead of hard disk. Packets can take a variety of possible paths through the network in an attempt to keep the network connections filled at all time. However, because the message is broken into multiple parts, sending packets via different paths adds to the possibility that packet order could get scrambled. Therefore, a sequencing number is added to each packet. Packets are sent over the most appropriate path. Each device chooses the best path at that time for every packet. Thus if one path is too busy, the sending device can send the packet through another path. Since some packets may be delayed, which will cause the packets to arrive out

Figure 2.12   Circuit Switching

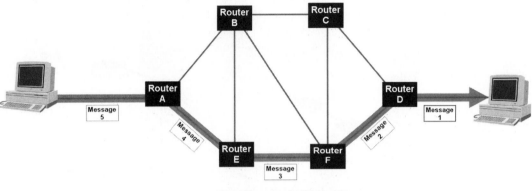

Dedicated Circuit
During Conversation

Figure 2.13   Packet Switching

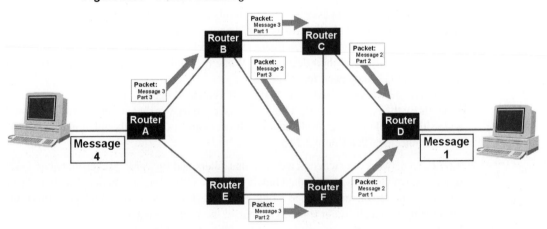

of order, the device will reorder them by sequence number to reconstruct the original message. (See figure 2.13.) The Internet is based on a packet-switching protocol. Other examples include Asynchronous Transfer Mode (ATM), Frame Relay, Switched Multimegabit Data Service (SMDS), and X.25. Message switching is typically used to support services such as email, web pages, calendaring, and workflow information.

Connections between two hosts in a packet-switching network are often described as a virtual circuit. A **virtual circuit** is a logical circuit created to ensure reliable communications between two network devices. To do this, the circuit provides a bidirectional communications path from one device to another and is uniquely identified by some type of identifier. A number of virtual circuits can be multiplexed into a single physical circuit for transmission across the network. This capability often can reduce the equipment and network complexity required for multiple-device connections.

Figure 2.14   Virtual Circuit as a Cloud

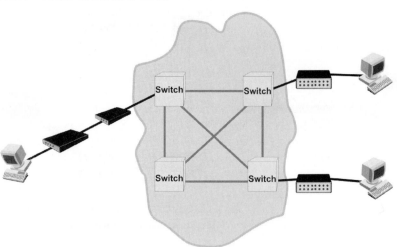

A virtual circuit can pass through any number of intermediate devices or switches located within the virtual circuit. In this case, the two hosts can communicate as though they have a dedicated connection, even though the packets might actually travel very different routes before arriving at their destination. While the path may vary, the computers do not know this, nor do they need to.

Virtual networks are sometimes depicted as a cloud because the user does not worry about the path taken through the cloud. The user is only worried about entering the cloud and exiting the cloud. The circuit acts as a direct connection even though it may not be directly connected. (See figure 2.14.)

NOTE: In networks, a switch is a device that filters and forwards packets between LAN segments, which operate at the data link layer of the OSI reference model. A switch (layer 2 switch) can also be referred to as a routing switch (layer 3 switch), which performs routing operations much like a router. Of course, the routing switch performs many of the functions found in the network layer of the OSI model. Since the routing switch has its routing capabilities implemented with hardware and not software, it is faster. Routing switches are not as powerful as full-fledged routers.

Virtual circuits can be either **permanent virtual circuits (PVCs)** or **temporary virtual circuits (SVCs).**

NOTE: SVC is short for switched virtual circuits.

A PVC is a permanently established virtual circuit that consists of one mode: data transfer. PVCs are used in situations in which data transfer between devices is constant. PVCs decrease the bandwidth use associated with the establishment and termination of virtual circuits, but they increase costs due to constant virtual circuit availability. Of course, PVCs are more efficient for connections between hosts that communicate frequently. PVCs play a central role in Frame Relay and X.25 networks.

SVCs are virtual circuits that are dynamically established on demand and terminated when transmission is complete. Communication over an SVC consists of three phases: circuit establishment, data transfer, and circuit termination. The establishment phase involves creating the virtual circuit between the source and destination devices. Data transfer involves transmitting data between the devices over the virtual circuit, and the circuit termination phase involves tearing down the virtual circuit between the source and destination devices. SVCs are used in situations in which data transmission between devices in sporadic, largely because SVCs increase bandwidth use due to the circuit establishment and termination phases, but they decrease the cost associated with constant virtual circuit availability.

When leasing a PVC or SVC end-to-end circuit, you will have to pay monthly whether you send data or not. If you choose not to lease the line but sign up with a network provider, the provider then bills you only for the amount of data packets you send and the distance the packets travel. Of course, the cost will also be based on the technology used.

NOTE: There is often a minimum charge per call.

2.7.3   Routing Protocols

To determine the best route, the routes use complex routing algorithms that take into account a variety of factors, including the speed of the transmission media, the number of network segments, and the network segment that carries the least amount of traffic. Routers then share status and routing information with other routers so that they can provide better traffic management and bypass slow connections. In addition, routers provide additional functionality, such as the ability to filter messages and forward them to different places based on various criteria. Most routers are multiprotocol routers because they can route data packets using many different protocols.

A metric is a standard of measurement, such as a hop count, that is used by routing algorithms to determine the optimal path to a destination. A hop is the trip a data packet takes from one router to another router or from a router to another intermediate point to another intermediate point in the network. On a large network, the number of hops a packet takes toward its destination is called the hop count. When a computer communicates with another computer and the computer has to go through four routers, the hop count would be 4. With no other factors taken in account, a metric of 4 would be assigned. If a router had a choice between a route with 4 metrics and a route with 6 metrics, it would choose the route with 4 metrics. Of course, if you want the router to choose the route with 6 metrics, you can overwrite the metric for the route with 4 hops in the routing table to a higher value.

To keep track of the various routes in a network, the routers will create and maintain routing tables. The routers communicate with one another to maintain their routing tables through a routing update message. The routing update message can consist of all or a portion of a routing table. By analyzing routing updates from all other routers, a router can build a detailed picture of network topology.

The various routing protocols use different metrics, including path length, hop counts, routing delay, bandwidth, load, reliability, and cost. Path length is the most common routing metric. Some routing protocols allow network administrators to assign arbitrary costs to each network link. In this case, path length is the sum of the costs associated with each link traversed. Other routing protocols define hop counts.

Routing algorithms can be differentiated based on several key characteristics. First, the capability of the routing algorithm to optimally choose the best route is very important. One routing algorithm may use the number of hops and the length of delays, but may weigh the length of delays more heavily in the calculation. To maintain consistency and predictability, routing protocols use strict metric calculations.

Routing algorithms are designed to be as simple as possible. In other words, the routing algorithm must offer efficiency with a minimum of software and utilization overhead. But while being efficient, the routing algorithm must be robust so that it can quickly change routes when a route goes down because of hardware failure or when a route has a high load (high amount of traffic). Of course, when this is happening, the routing algorithm must be stable.

Routing algorithms must converge rapidly. Convergence is the process of agreement by all routers of which routes are the optimal routes. When a route goes down, routers distribute the new routes by sending routing update messages. The time it takes for all routers to get new routes in which they all agree on should be quick. If not, routing loops or network outages can occur. A routing loop is created when a packet is sent back and forth between several routers without ever getting to its final destination.

2.8   TRANSPORT LAYER

The **transport layer** can be described as the middle layer that connects the lower and upper layers together. In addition, it is responsible for reliable transparent transfer of data (known as segments) between two end points. Since it provides end-to-end recovery of lost and corrupted packets and flow control, it deals with end-to-end error handling, division of messages into smaller packets, numbers of the messages, and repackaging of messages. In addition, if packets arrive out of order at the destination, the transport layer is responsible for reordering the packets back to the original order. Examples of transport-layer protocols include SPX, TCP, UDP, and NetBEUI.

Computers and network devices often use long strings of numbers to identify themselves. The transport layer is responsible for name resolution, where it can take a more meaningful name and translate it to a computer or network address. Examples of name resolution protocols include the Domain Name System (DNS) and the Windows Internet Naming Service (WINS).

2.8.1   Connection Services

Networks are divided into connection-oriented and connectionless-oriented networks. In a **connection-oriented network,** you must establish a connection using an exchange of messages or must have a preestablished pathway between a source point and a destination point before you can transmit packets. Establishing a connection before transmitting packets is similar to making a telephone call. You must dial a number, the destination telephone must ring, and someone must pick up the telephone receiver before you can begin speaking. An example of a connection-oriented protocol is TCP.

Connection-oriented services provide flow, error, and packet sequence control through acknowledgments. An **acknowledgment (ACK)** is a special message that is sent back when a data packet makes it to its destination. You can compare this to a letter sent at the post office with a return receipt that is sent back to the sender to indicate that the letter got to its destination.

TCP is connection-oriented because a set of three messages must be completed before data is exchanged. Likewise, SPX is connection-oriented. Frame Relay, when using permanent virtual circuits (PVCs), does not require that any messages be sent ahead of time, but it does require predefinition in the Frame Relay switches, establishing a connection between two Frame Relay attached devices. ATM PVCs are connection-oriented for similar reasons.

The disadvantage of using a connection-oriented network is that it takes time to establish a connection before transmitting packets. The advantage of using a connection is that the connection can reserve bandwidth for specific connections. As a result, connection-oriented networks can guarantee a certain **quality of service (QoS).** By using quality of service to guarantee bandwidth, connection-oriented networks can provide sufficient bandwidth for audio and video without the jitters or pauses and with the transfer of important data within a timely manner. Lastly, the connection-oriented network can better manage network traffic and prevent congestion by refusing traffic that it cannot handle.

Connectionless protocols do not require an exchange of messages with the destination host before data transfer begins, nor do they make a dedicated connection, or virtual circuit, with a destination host. Instead, connectionless protocols rely upon upper-level, not lower-level, protocols for safe delivery and error handling. Since data is segmented and delivered in the order the data is received from the upper OSI layers, connectionless protocols do not sequence data segments. Therefore connectionless protocols are best suited for situations where a high data transfer rate and low data integrity are needed. An example of a connectionless protocol is UDP.

The exchange of messages before data transfer begins is called a call setup or a three-way handshake:

1. The first "connection agreement" segment is a request for synchronization.
2. The second and third segments acknowledge the request and establish connection parameters (the rules) between hosts.
3. The final segment is also an acknowledgment. It notifies the destination host that the connection agreement has been accepted and that the actual connection has been established.

The data is then transferred, and when the transfer is finished, a call termination takes place to tear down the virtual circuit.

The data link layer, network layer, and transport layer offer some form of connection service, which defines how two devices form a connection. The connection services determine the level of error detection and recovery and flow control that is used in communicating data between two network devices. The data link layer is concerned with connection services from one device to another device within the LAN, whereas the network layer is concerned with connection services from one device to another device on different LANs. The transport layer is concerned with an end-to-end connection service (a type of overall quality assurance).

NOTE: To speed communication, most network-layer protocols are connectionless-oriented and do not bother with the sending or receiving of acknowledgments. Instead, they let the transport layer handle the acknowledgments and provide the error control.

2.8.2   Error Control

Error control refers to the notification of lost or damaged data frames. This includes the following:

- The destination does not receive the data packet.
- The checksum does not match. A checksum value is mathematically generated before the data packet is sent. It is attached to the data packet and sent with the packet. When the data packet gets to its destination, the same mathematical calculation is performed. If that value matches the value that was sent, the data is assumed intact.
- The packet size does not configure to a minimum or maximum size requirement for the frame type used.
- There is noise, interference, and distortion, which scramble the data.
- The capacity of a channel or network device is exceeded, causing a buffer overflow.

Before continuing on, you need to differentiate between error detection and error recovery. Any header or trailer with a frame check sequence (FCS) or similar field can be used to detect bit errors in a protocol data unit. Error detection uses the FCS to detect the error, which results in discarding the PDU. Error recovery implies that the protocol reacts to the lost data and somehow causes the data to be retransmitted.

Regardless of which protocol specification performs the error recovery, all work in basically the same way. The transmitted data is labeled or numbered. After receipt, the receiver signals back to the sender that the data was received, using the same label or number to identify the data. If the data was not received, either the receiver will tell the sender to resend the data, or if no acknowledgment is received, the sender will assume that the data was not received and will resend the data.

2.8.3   Flow Control

Flow control is the process of controlling the rate at which a computer sends data. Depending on the particular protocol, both the sender and the receiver of the data (as well as any intermediate routers, bridges, and switches) might participate in the process of controlling the flow from sender to receiver.

Flow control is needed because data is discarded when congestion occurs. A sender of data might be sending the data faster than the receiver can receive the data, and so the receiver discards the data. Also, the sender might be sending the data faster than the intermediate switching devices (switches and routers) can forward the data, also causing discards. Packets can be lost due to transmission errors as well. This happens in every network, sometimes temporarily and sometimes regularly, depending on the network and the traffic patterns. The receiving computer can have insufficient buffer space to receive the next incoming frame, or possibly the processor is too busy to process the incoming frame. In addition, intermediate routers might need to discard the packets based on a temporary lack of buffers (holding space) or processing.

Flow control attempts to reduce unnecessary discarding of data. Flow control protocols do not prevent the loss of data due to congestion; these protocols simply reduce the amount of lost data, which in turn reduces the amount of retransmitted traffic, which, it is hoped, reduces overall congestion. However, with flow control, the sender is artificially slowed or throttled so that it sends data less quickly than it could without flow control.

There are three methods of implementing flow control:

- Buffering
- Congestion avoidance
- Windowing

Buffering simply means that the computer reserves enough buffer space in memory so that a burst of incoming data can be held until processed. No attempt is made to actually slow the transmission rate of the sender of data. In fact, buffering is such a common method of dealing with changes in the rate of arrival of data that most of us would probably just assume that it is happening.

Congestion avoidance is the second method of flow control. The computer receiving the data notices that its buffers are filling. This causes either a separate protocol data unit or field in a header to be sent toward the sender, signaling the sender to stop transmitting.

A preferred method might be to get the sender to simply slow down instead of stopping altogether. This method would still be considered congestion avoidance, but instead of telling the sender to stop, the signal would mean to slow down.

The third category of flow control methods is windowing (also called sliding windows). A window is the maximum amount of data the sender can send without getting an acknowledgment. When you have configured a window size of 1, the sending machine waits for acknowledgment for each data segment it transmits before transmitting another. If you have configured a window size of 3, the sending machine is allowed to transmit three data segments before an acknowledgment is received. This cuts down on the amount of traffic because the number of acknowledgments is reduced. If your network is stable, you will use a higher number for the window. If your network is not very stable, it's best to use a low number to focus more on reliability instead of performance.

2.8.4   Multiplexing and Demultiplexing

Transport-level communication protocols can use a technique called multiplexing and demultiplexing. This technique allows data from different applications to share a single data stream. When multiplexing occurs, the only way to tell one application from another is by identifying each unit of data with a transition mechanism such as a port number and a sequence number. Sequence numbers assure proper reassembly, and port numbers assure appropriate application destination. Demultiplexing occurs when the destination computer receives the data stream and separates and rejoins the application's segments.

2.9   SESSION LAYER

The application-oriented layers are concerned with the protocols that provide user network applications and services. The **session layer** allows remote users to establish, manage, and terminate a connection (session). A **session** is a reliable dialog between two

Figure 2.15   Simplex, Half-Duplex, and Full-Duplex Communications

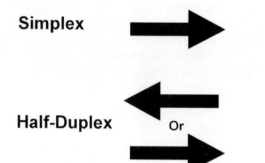

computers. In addition, it enables two users to organize and manage their data exchange and to implement dialog control between the source and destination network devices, including the type of dialog (simplex, half duplex, full duplex) and the length of time a computer transmits. (Simplex, half duplex, and full duplex are explained in the paragraph that follows.) When establishing these sessions, the session layer protocol creates checkpoints in the data streams. If a transmission fails, only the data after the last checkpoint has to be retransmitted. Lastly, session layer protocol implements security, such as the validation of user names (commonly known as authentication) and passwords. Examples of session-layer protocols include SQL, NFS, RPC, NetBIOS, FTP, NCP, and telnet.

Data can flow in one of three ways: simplex, half duplex, and full duplex. Simplex dialog allows communications on the transmission channel to occur in only one direction. Essentially, one device is allowed to transmit, and all the other devices receive. This is often compared to a PA system, where the speaker talks to an audience, but the audience does not talk back to the speaker.

Half-duplex dialog allows each device to both transmit and receive, but not at the same time. Therefore, only one device can transmit at a time. This is often compared to a CB radio or walkie-talkies. A full-duplex dialog allows every device to both transmit and receive simultaneously. For networks, the network media channels would consist of two physical channels, one for receiving and one for transmitting. (See figure 2.15.)

2.10   PRESENTATION LAYER

The **presentation layer** ensures that information sent by an application-layer protocol on one system will be readable by the application-layer protocol on the remote system. This layer, unlike its lower layers, is simply concerned with the syntax and the semantics of the information transmitted. Therefore, it acts as the translator between different data formats, protocols, and systems. It also provides encryption/decryption, compression/decompression of data, and network redirectors. Examples include NCP and SMB.

In addition, the presentation layer deals with character set translation. Not all computer systems use the same table to convert binary numbers into text. Most standard computer systems use the American Standard Code for Information Interchange (ASCII). Mainframe computers (and some IBM) network systems use the Extended Binary Coded Decimal Interchange Code (EBCDIC). ASCII and EBCDIC are totally different from each other. Protocols at the presentation layer can translate between the two.

Another form of translation performed by the presentation layer is byte-order translation. In big-endian architectures (which are used by PCs with Intel processors), when representing a multibyte value, the leftmost bytes are the most significant. In little-endian architecture (used by systems based on Motorola processors), the rightmost bytes are significant. For example, the big-endian architecture would represent a 16-bit binary number as 11111111 01010101, while little-endian would represent the same 16-bit binary number as 01010101 11111111. To allow both of these types of systems to connect to the same network, the presentation layer will translate from one format to the other.

2.11   APPLICATION LAYER

The **application layer** represents the highest layer of the OSI reference model, and it initiates communication requests. It is responsible for interaction between the operating systems of the sending and receiving computers and provides an interface to the systems. It provides the user interface to a range of networkwide distributed services including file transfer, printer access, and electronic mail. Note that it does not include software applications such as word processors and spreadsheets. Since the application layer provides the user interface to the network services, when the transport layer notifies the upper-layer protocols when a nonrecoverable data error is detected, the application layer will then notify the user via error messages. Some examples of upper-layer protocols include the File Transfer Protocol (FTP), Simple Mail Transfer Protocol (SMTP), AppleTalk, and NFS.

A **directory service** is a network service that identifies all resources on a network and makes those resources accessible to users and applications. Resources can include email addresses, computers, and peripheral devices (such as printers). Ideally, the directory service should make the physical network, including topology and protocols, transparent to the users on the network. Users should be able to access any resources without knowing where or how they are physically connected.

There are a number of directory services that are widely used. The most common is the **X.500**, which uses a hierarchical approach in which objects are organized similar to the way files and folders are organized on a hard drive. The root container is at the top of the structure, and the children are organized under it. X.500 is part of the OSI model; however, it does not translate well into a TCP/IP environment. Therefore, many of the protocols that are based on the X.500 do not fully comply with it.

The **Lightweight Directory Access Protocol (LDAP)** is a set of protocols for accessing information directories. LDAP is based on the standards contained within the X.500 standard, but it is significantly simpler. And unlike X.500, LDAP supports TCP/IP, which is necessary for any type of Internet access. Because it's a simpler version of the X.500, LDAP is sometimes called X.500-lite.

2.12   CONNECTIVITY DEVICES

To connect networks together, several devices can be used. These include network cards and cables, repeaters, hubs and MAUs, bridges and switching hubs, routers, brouters, and gateways. (See table 2.14.)

2.12.1   Network Cards and Cables

The network cards that allow computers to connect to the network, as well as the cables that they connect, work at the physical OSI layer. **Network cards,** or **network interface cards (NICs),** are found as expansion cards or are integrated into the motherboard. To communicate on LANs, network cards use **transceivers** (devices that both transmit and receive analog or digital signals). To get a better idea of the function of the network card,

Table 2.14   Connectivity Devices
and Their Associated OSI Layer

OSI Layer	Device
Application	Gateways
Presentation	Gateways
Session	Gateways
Transport	Gateways
Network	Routers Brouters Layer 3 switches
Data link	Bridges Layer 2 switches
Physical	NICs Transceivers Repeaters Hubs MAUs Cables

Figure 2.16   A Repeater

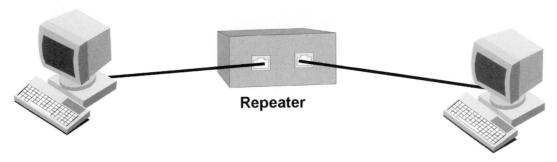

Repeater

you must see how the data is sent from one computer to another. The data is generated by the microprocessor and sent to the network card using the computer's parallel data bus. At this point, the data consists of alphanumeric characters represented as a series of on (1) and off (0) signals.

Within the network card, the data is processed and converted so that it can be transmitted on the network media. If it is being sent on a cable, the data is turned into a serial bit stream, amplified, and sent on to the cable. The electrical signals that travel over the network cable must follow specific rules on how the packet gains sole access to the cable and what specific signal is sent to represent the data onto the cable.

The network card on the receiving computer sees the packet and reads part of the packet to determine if the packet was meant for this computer. The network card then converts the data into digital signals and then sends the data to the microprocessor to be processed.

Cables, which also work at the physical OSI model, are used to carry electrical or light signals between computers and networks. Therefore, since most computers connect to the network using cables and cables connect most networks, cables are very common on networks.

2.12.2   Repeaters

A **repeater,** which works at the physical OSI layer, is a network device used to regenerate or replicate a signal or to move packets from one physical medium to another.

NOTE: A repeater cannot connect different network topologies or access methods.

A repeater can be used to regenerate analog or digital signals distorted by transmission loss, extending the length of a cable connection. Analog repeaters usually can only amplify the signal (including distortion), whereas digital repeaters can reconstruct a signal to near its original quality. (See figure 2.16.)

2.12.3   Hubs and MAUs

A **hub** (also known as a concentrator), which works at the physical OSI layer, is a multi-ported connection point used to connect network devices via a cable segment. When a PC

Figure 2.17 A Hub

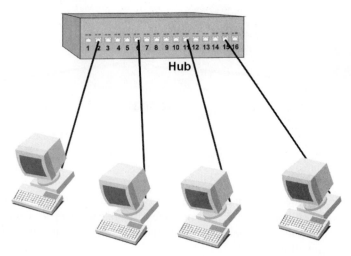

needs to communicate with another computer, it sends a data packet to the port, which the device is connected to. When the packet arrives at the port, it is forwarded or copied to the other ports so that all network devices can see the packet. In this way, all of the stations "see" every packet just as they do on a bus network. Of course, a standard hub is not very efficient on networks with heavy traffic since it causes a lot of collisions and retransmitted packets. (See figure 2.17.) See chapter 4 for more information on hubs.

The **multistation access unit (MAU** or **MSAU)** is a physical-layer device unique to Token Ring networks. The MAU resembles a hub. Yet while Token Ring networks use a physical star topology, they also use a logical ring topology. Whereas a hub defines a logical bus, the MAU defines a logical ring. Topologies, Token Ring, and MAUs are discussed in chapter 4.

2.12.4 Bridge and Switching Hub

A **bridge,** which works at the data link OSI layer, is a device that connects two LANs and makes them appear as one. It is also used to connect two segments of the same LAN. The two LANs being connected can be alike or dissimilar, such as an Ethernet LAN connected to a Token Ring LAN.

NOTE: A bridge can connect dissimilar network types (for example, Token Ring and Ethernet) as long as the bridge operates at the LLC sublayer of the data link layer. If the bridge operates only at the lower sublayer (the MAC sublayer), the bridge can connect only similar network types (Token Ring to Token Ring and Ethernet to Ethernet).

Different from a repeater or hub, a bridge analyzes the incoming data packet and will forward the packet if its destination is on the other side of the bridge. Many bridges today filter and forward packets with very little delay, making them good for networks with high traffic. (See figure 2.18.)

Figure 2.18   A Bridge

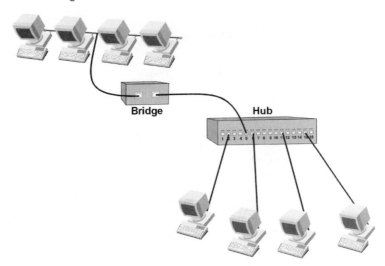

There are two kinds of bridges: basic and learning. A basic bridge is used to inter-connect LANs using one (or more) of the IEEE 802 standards. Packets received on one port may be retransmitted on another port. Unlike a repeater, a bridge will not start re-transmission until it has received the complete packet. As a consequence, stations on both sides of a bridge can transmit simultaneously without causing collisions.

A **switching hub** (sometimes referred to as a switch or a layer 2 switch) is a fast, mul-tiported bridge that builds a table of the MAC addresses of all the connected stations. It then reads the destination address of each packet and forwards the packet to the correct port. A major advantage of using a switching hub is that it allows one computer to open a connec-tion to another computer (or LAN segment). While those two computers communicate, other connections between the other computers (or LAN segments) can be opened at the same time. Therefore, several computers can communicate at the same time through the switching hub. As a result, the switches are used to increase performance of a network by segmenting large networks into several smaller, less congested LANs, while providing nec-essary interconnectivity between them. Switches increase network performance by pro-viding each port with dedicated bandwidth, without requiring users to change any existing equipment, such as NICs, hubs, wiring, or any routers or bridges that are currently in place.

Many switching hubs also support load balancing, so that ports are dynamically re-assigned to different LAN segments based on traffic patterns. In addition, some include fault tolerance, which can reroute traffic through other ports when a segment goes down.

2.12.5   Routers and Brouters

A **router,** which works at the network OSI layer, is a device that connects two or more LANs. In addition, it can break a large network into smaller, more manageable subnets. As multiple LANs are connected together, multiple routes are created to get from one LAN to another. Routers then share status and routing information with other routers so

Figure 2.19 A Router

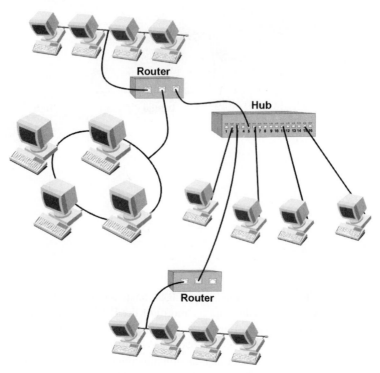

that they can provide better traffic management and bypass slow connections. In addition, routers provide additional functionality, such as the ability to filter messages and forward them to different places based on various criteria. Most routers are multiprotocol routers because they can route data packets using many different protocols. (See figure 2.19.)

NOTE: Routers cannot pass nonroutable protocols such as NetBEUI.

Broadcasting sends unaddressed packets to everyone on the network. Routers control broadcast storms by filtering these packets and stopping the broadcast from traveling to the other networks connected by the router.

A **brouter** (short for bridge router) is a device that functions as both a router and a bridge. A brouter understands how to route specific types of packets (routable protocols), such as TCP/IP packets. For other specified packets (nonroutable protocols), it acts as a bridge, which simply forwards the packets to the other networks.

When you send a packet from one computer to another computer, it first determines if the packet is sent locally to another computer on the same LAN or if the packet is sent to the router so that it can be routed to the destination LAN. If the packet is meant to go to a computer on another LAN, it is sent to the router (or gateway). The router will then determine what is the best route to take and forward the packet to that route. The packet will then go to the next router, and the entire process will repeat itself until the packet gets to the destination LAN. The destination router will then forward the packet to the destination computer.

A newer type of switch, the **layer 3 switch,** like a router, switches packets based on the logical addresses (layer 3) such as IP addresses. A layer 3 switch does many things that a traditional router does, including determining the forwarding path based on the layer 3 network-layer information, validating the integrity of the layer 3 header via checksum and verifying packet expiration and updates accordingly, and applying security controls if required. Since the layer 3 switch has been optimized for high-performance LAN support, it typically is much faster than a router. However, the layer 3 switch is not meant to service wide area connections.

2.12.6   Gateway

A **gateway** is hardware or software that links two different types of networks by repackaging and converting data from one network to another network or from one network operating system to another. An example of a gateway is a computer or device that connects a PC to a mainframe or minicomputer such as an AS400 midrange computer. One way this can be done is by adding an Ethernet card or Token Ring card to the AS400 computer, adding software to the client computers, and adding a special expansion card to the PC so that it can communicate with the AS400 directly. Another way is by using a gateway computer to act as a translator between the AS400 network and client networks.

NOTE: A gateway can be used at any layer of the OSI reference model, but it is usually identified with upper layers since they must communicate with an application, establish and manage sessions, translate encoded data, and interpret logical and physical addressing data.

Gateways can reside on servers, microcomputers, mainframes, or routers. They are typically more expensive than routers because of their vast capabilities, and they are almost always application-specific. Since they require more processing for their tasks, gateways transmit data more slowly than bridges or routers.

The three different types of gateways are:

- **Protocol gateway**—This functions like a router in that it connects networks that use different protocols. This is the most common type of gateway.
- **Address gateway**—This connects networks with different directory structures and file management techniques.
- **Format gateway**—This connects networks using different data format schemes. For example, a format gateway might connect two networks, one using ASCII and another using EBCDIC.

When working with today's networks, the most common gateways are:

- **Email gateway**—A gateway that translates messages from one type of system to another.
- **IBM host gateway**—A gateway that establishes and manages communications between a PC and an IBM mainframe computer.

- **Internet gateway**—A gateway that allows and manages access between LANs and the Internet. An Internet gateway can restrict the kind of access LAN users have to the Internet and vice versa.
- **LAN gateway**—A gateway (usually a router or a server) that allows segments of a LAN running different protocols or different network models to communicate with one another. The LAN gateway category might also include remote access servers that allow dial-up connectivity to a LAN.

SUMMARY

1. To overcome compatibility issues, hardware and software often follow standards (dictated specifications or most popular).

2. The de jure standard, or the "by law" standard, is a standard that has been dictated by an appointed committee.

3. The de facto standard, or the "from the fact" standard, is a standard that has been accepted by the industry just because it was the most common.

4. Some systems or standards have an open architecture. The term *open architecture* indicates that the specification of the system or standard is public.

5. A proprietary system is privately owned and controlled by a company, and the specifications that would allow other companies to duplicate the product are not divulged to the public.

6. The Open Systems Interconnection (OSI) reference model is the world's prominent networking architecture model.

7. The concept of placing data behind headers (and before trailers) for each layer is called encapsulation.

8. The physical layer is responsible for the actual transmission of the bits sent across a physical medium.

9. Signaling is the method for using electrical, light energy, or radio waves to communicate.

10. The process of changing a signal to represent data is often called modulation or encoding.

11. Digital signals (the language of computers) are based on a binary signal system produced by pulses of light or electric voltages. The site of the pulse is either on/high or off/low to represent 1s and 0s.

12. An analog signal is the opposite of a digital signal. Instead of having a finite number of states, it has an infinite number of values, which are constantly changing.

13. Synchronous devices use a timing or clock signal to coordinate communications between the two devices.

14. Asynchronous signals are intermittent signals—they can occur at any time and at irregular intervals. They do not use a clock or timing signal.

15. Bandwidth refers to the amount of data that can be carried on a given transmission medium.

16. Baseband systems use the transmission medium's entire capacity for a single channel.

17. A broadband system uses the transmission medium's capacity to provide multiple channels.

18. Point-to-point topology connects two nodes directly together.

19. A multipoint connection links three or more devices together through a single communication medium. Because multipoint connections share a common channel, each device needs a way to identify itself and the device to which it wants to send information. The method used to identify senders and receivers is called addressing.

20. The data link layer is responsible for providing error-free data transmission and establishes local connections between two computers or hosts.

21. The media access control (MAC) sublayer is the lower sublayer that communicates directly with the network adapter card.

22. The logical link control (LLC) sublayer manages the data link between two computers within the same subnet.

23. The physical device address or media access control (MAC) address is a unique hardware address (unique on the LAN/subnet).

24. The set of rules defining how a computer puts data onto the network cable and takes data from the cable is called the access method, sometimes referred to as arbitration.

25. Contention is when two or more devices contend (compete) for network access. Any device can transmit whenever it needs to send information. To avoid data collisions (two devices sending data at the same time), specific contention protocols were developed, requiring the device to listen to the cable before transmitting data.

26. The most common form of contention is called the carrier sense multiple access (CSMA) network.

27. Carrier sense multiple access with collision detection (CSMA/CD) is the access method utilized in Ethernet and IEEE 802.3.

28. Token passing uses a special authorizing packet of information to inform devices that they can transmit data.

29. Polling has a single device such as a mainframe front-end processor designated as the primary device.

30. The network layer is concerned with the addressing and routing processes necessary to move data (known as packets or datagrams) from one network (or subnet) to another.

31. Networks are identified with a unique network address.

32. A port or socket is used to identify a specific upper-layer software process or protocol.

33. Circuit switching is a technique that connects the sender and the receiver by a single path for the duration of a conversation.

34. In packet switching, messages are broken into small parts called packets.

35. A virtual circuit is a logical circuit created to ensure reliable communications between two network devices.

36. To determine the best route, the routes use complex routing algorithms that take into account a variety of factors, including the speed of the transmission media, the number of network segments, and the network segment that carries the least amount of traffic.

37. The transport layer can be described as the middle layer that connects the lower and upper layers together. In addition, it is responsible for reliable transparent transfer of data (known as segments) between two end points.

38. Flow control is the process of controlling the rate at which a computer sends data.

39. The session layer allows remote users to establish, manage, and terminate a connection (session).

40. A session is a reliable dialog between two computers.

41. The presentation layer ensures that information sent by an application-layer protocol on one system will be readable by the application-layer protocol on the remote system.

42. The application layer represents the highest layer of the OSI reference model.

43. A directory service is a network service that identifies all resources on a network and makes those resources accessible to users and applications.

44. A repeater, which works at the physical OSI layer, is a network device used to regenerate or replicate a signal or to move packets from one physical medium to another.

45. A hub, which works at the physical OSI layer, is a multiported connection point used to connect network devices via a cable segment.

46. A bridge, which works at the data link OSI layer, is a device that connects two LANs and makes them appear as one. It is also used to connect two segments of the same LAN.

47. A switching hub (sometimes referred to as a switch or a layer 2 switch) is a fast, multiported bridge that builds a table of the MAC addresses of all the connected stations. It then reads the destination address of each packet and forwards the packet to the correct port.

48. A router, which works at the network OSI layer, is a device that connects two or more LANs. In addition, it can break a large network into smaller, more manageable subnets.

49. A brouter (short for bridge router) is a device that functions as both a router and a bridge.

50. A gateway is hardware or software that links two different types of networks by repackaging and converting data from one network to another network or from one network operating system to another.

QUESTIONS

1. Which model is the most well-known model for networking architecture and technologies?
   a. OSI model
   b. Peer-to-peer model
   c. Client/server model
   d. IEEE model

2. Which describes the correct order of the OSI model layers from bottom to top?
   a. Physical, data link, network, transport, session, presentation, and application
   b. Data link, physical, network, transport, session, presentation, and application
   c. Physical, data link, network, transport, presentation, session, and application
   d. Application, presentation, session, transport, network, data link, and physical

3. Which is the fourth layer of the OSI model?
   a. Data link   b. Presentation
   c. Session   d. Transport

4. Which layer of the OSI model is between the session layer and the application layer?
   a. Presentation   b. Data link
   c. Network   d. Transport

5. What is the order in which information blocks are created when encapsulation with TCP/IP is used?
   a. Segments, packets or datagrams, frames, data, bits
   b. Data, segments, packets or datagrams, frames, bits
   c. Bits, frames, segments, packets or datagrams, data
   d. Packets or datagrams, frames, segments, bits, data

6. What happens to the data link layer source and destination addresses when packets are passed from router to router?
   a. They are stripped off and then re-created.
   b. They are stripped off and replaced with MAC (hardware) addresses.
   c. They are stripped off and replaced with NetBIOS names.
   d. They are reformatted according to the information stored in the routing table.

7. Which layer of the OSI model determines the route from source to destination computer?
   a. Transport   b. Network
   c. Session   d. Physical

8. Bridges are often called media access control bridges because they work at the media access control sublayer. In which OSI layer does the media access control sublayer reside?
   a. Transport   b. Network
   c. Physical   d. Data link

9. Which layer of the OSI model adds header information that identifies the upper-layer protocols sending the frame?
   a. Presentation layer
   b. MAC sublayer
   c. Network layer
   d. LLC sublayer

10. Which layer of the OSI model defines how a cable is attached to a network adapter card?
    a. Cable   b. Hardware
    c. Connection   d. Physical

11. Many applications use compression to reduce the number of bits to be transferred on the network. Which layer of the OSI model is responsible for data compression?
    a. Application   b. Network
    c. Session   d. Presentation

12. Which layer of the OSI model packages raw data bits into data frames?
    a. Physical
    b. Presentation
    c. Network
    d. Data link

13. Which layer of the OSI model is responsible for translating the data format?
    a. Application
    b. Presentation
    c. Network
    d. Data link

14. Which layer of the OSI model provides flow control and ensures messages are delivered error-free?
    a. Transport   b. Network
    c. Session   d. Physical

15. The IEEE 802 project divides the data link layer into two sublayers. Which sublayer of the data link layer communicates directly with the network adapter card?
    a. Logical link control
    b. Logical access control
    c. Media access control
    d. Data access control

16. The Project 802 model defines standards for which layers of the OSI model?
   a. The physical layer and the data link layer
   b. The network layer and the data link layer
   c. The transport layer and the network layer
   d. The application layer and the presentation layer

17. Which Project 802 model specification describes Ethernet?
   a. 802.2   b. 802.5
   c. 802.3   d. 802.10

18. Which Project 802 model specification adds header information that identifies the upper-layer protocols sending the frame and specifies destination processes for data?
   a. 802.2   b. 802.5
   c. 802.3   d. 802.10

19. What does the 802.5 specification specify?
   a. Ethernet   b. Fiber optics
   c. Token Ring   d. ARCnet

20. Which of the following allows for two devices to communicate at the same time?
   a. Simplex   b. Full duplex
   c. Half duplex   d. Complex

21. Which of the following allows for two devices to communicate to each other, but not at the same time?
   a. Simplex   b. Full duplex
   c. Half duplex   d. Complex

22. The address that identifies a network card to the network is known as the _____.
   a. MAC address
   b. port address
   c. MAC identity
   d. I/O address

23. Which of the following is a MAC address?
   a. 192.158.24.24
   b. 57204AB22
   c. 1423:3453
   d. 0A:55:BB:7A:FE:57

24. When you have different networks that are not physically separated, you must have a unique configuration for each network. What has to be unique besides the node address on the network to tell the networks apart?
   a. The network ID
   b. The IP address
   c. The computer name
   d. The workgroup or domain name

25. Connection-oriented communication and connectionless communication are the two ways that communication can be implemented on a network. Which of the following is often associated with connectionless communication?
   a. Fast but unreliable delivery
   b. Fiber-optic cable
   c. Error-free delivery
   d. Infrared technology

26. There are two ways to implement communication on networks: connection-oriented communication and connectionless communication. Which of the following is associated with connection-oriented communication?
   a. Fast but unreliable delivery
   b. Fiber-optic cable
   c. Assured delivery
   d. Infrared

27. What type of communication ensures reliable delivery from a sender to a receiver without any user intervention?
   a. Communication-oriented
   b. Connectionless
   c. Connection-oriented
   d. Physical

28. Which of the following best describes the function of a connectionless protocol? (Select the best choice.)
   a. A connectionless protocol requires an exchange of messages before data transfer begins.
   b. A connectionless protocol is a faster transfer method than a connection-oriented protocol.
   c. A connectionless protocol relies upon lower-level protocols for data delivery and error handling.
   d. A connectionless protocol creates a virtual circuit with the destination host.

29. Which of the following functions are performed by the media access control (MAC) sublayer of the data link layer? (Select two choices.)
   a. It adds a Source Service Access Point and a Destination Service Access Point to the header.
   b. It provides media access.
   c. It isolates upper-level protocols from details of the application layer.
   d. It builds frames from bits that are received from the physical layer.

30. Which of the following is a true statement regarding IP addresses and MAC addresses? (Select the best choice.)
   a. An IP address is 6 bytes long and is usually represented in decimal format.
   b. A MAC address is 48 bits long and is usually represented in hexadecimal format.
   c. An IP address is 48 bits long and is usually represented in hexadecimal format.
   d. A MAC address is 6 bytes long and is usually represented in decimal format.

31. Which of the following does TCP/IP implement at the transport layer to manage flow control? (Select all choices that are correct.)
   a. The use of buffering
   b. The use of "ready" and "not ready" indicators
   c. The use of packet switching
   d. The use of windowing

32. Which of the following describes frequency division multiplexing (FDM)?
   a. It is used in baseband data signaling and combines the signals in multiple cables into a single channel.
   b. It is used in baseband data signaling and divides the signal in a single cable into multiple channels.
   c. It is used in broadband data signaling and combines the signals in multiple cables into a single channel.
   d. It is used in broadband data signaling and divides the signal in a single cable into multiple channels.

33. Connectionless communication occurs when _____.
   a. acknowledgments are not sent for received frames
   b. packets are sent over wireless media
   c. packets are dropped due to memory buffer overruns
   d. there is a break in the network medium

34. Packet switching enables _____.
   a. each frame to be truncated in order to pass a router
   b. each frame to send an acknowledgment
   c. each frame to boost its signal when crossing a router
   d. each frame to follow a different path

35. Which OSI layer specifies how the electronic signals traveling the wire are converted into binary code?
   a. Data link   b. Network
   c. Presentation   d. Physical

36. What OSI layer is responsible for "windowing" or using sliding windows for data delivery?
   a. Session   b. Application
   c. Presentation   d. Transport

37. Network MAC (hardware) addresses are composed of how many octets?
   a. 6   b. 4
   c. 8   d. 2

38. What parts of the OSI model handle error flow control, detection, and correction? (Select two answers.)
   a. Physical   b. Network
   c. Logical link control   d. Data link

39. Data multiplexing is defined at what OSI layer?
   a. Network   b. Session
   c. Application   d. Transport

40. How many bits of a MAC address are used to identify the NIC's manufacturer?
   a. 32   b. 48
   c. 64   d. 24

41. The LLC inspects the MAC address of incoming frames for what purpose?
   a. To check the CRC header for error conditions
   b. To see if the frame's protocol matches that of the host
   c. To see if the frame needs to be routed to a different subnet
   d. To see if the frame belongs to that host

42. What OSI layer defines how communication between two applications on different computers is established?
   a. Application   b. Presentation
   c. Network   d. Session

43. What is the term for two computers acting as if there is a dedicated circuit between them even though there is not?
   a. Virtual circuit   b. Gateway
   c. Physical circuit   d. Dialog path

44. Which OSI layer's services are "connectionless"?
   a. Presentation   b. Session
   c. Application   d. Network

45. What is the purpose of a CRC?
   a. To ensure that frames are received intact
   b. To verify a NIC's functionality
   c. To check cable integrity
   d. None of the choices apply

46. When a network problem occurs, which OSI layer should you first suspect contains the problem?
   a. Application   b. Session
   c. Presentation   d. Physical

47. Which OSI layer specifies the mechanical requirements of a network medium?
   a. Session   b. Physical
   c. Application   d. Network

48. Why does the transport layer assign port numbers to data segments?
   a. So each segment can be reassembled in order.
   b. So each segment can be checked for errors.
   c. So each segment can be sent to its appropriate application.
   d. None of the choices apply.

49. The transport layer uses data segment sequence numbers for what purpose?
   a. To make sure that data is reassembled in the proper order
   b. To make sure that data passes the CRC check
   c. To make sure that data does not overrun memory buffers
   d. To verify destination addresses

50. Which OSI layer organizes datagrams into frames?
   a. Data link   b. Physical
   c. Network   d. Session

51. Which OSI layer provides for multiple applications to use the same data stream?
   a. Transport   b. Application
   c. Network   d. Session

52. Which OSI layer notifies the upper-layer protocols when a nonrecoverable data error is detected?
   a. Data link   b. Network
   c. Media access control d. Transport

53. At what OSI layer is data merely a stream of bits (ones and zeros)?
   a. Network   b. Data link
   c. Transport   d. Physical

54. What part of the OSI model is responsible for physical addressing?
   a. Media access control
   b. Network layer

   c. Logical link control
   d. Physical layer

55. Media access methods in a network—whether based on electrical, mechanical, optical, radio, or infrared transmission—are specified at which OSI layer?
   a. Media access control b. Network
   c. Physical   d. Data link

56. Which OSI layer has the function of providing for dialog control?
   a. Session   b. Network
   c. Presentation   d. Application

57. When a packet collision occurs, which of the following is responsible for frame retransmission?
   a. CRC   b. MAC
   c. ACK   d. LLC

58. Fragmentation and reassembly of data at a router are handled at which OSI layer?
   a. Network
   b. Media access control
   c. Logical link control
   d. Physical

59. If the LLC receives a bad frame, it discards it and does not send an ACK (acknowledgment). What is the result of this occurring?
   a. The sending LLC resends the frame.
   b. The receiving LLC sends a NACK to the MAC.
   c. The discarded frame is repackaged by the network layer.
   d. Data is lost.

60. Name recognition and security functions are defined at which OSI layer?
   a. Application   b. Transport
   c. Network   d. Session

61. In an internetwork, what layer of the OSI defines the path that data should take based on network conditions?
   a. Session   b. Physical
   c. Application   d. Network

62. Which layer initiates requests for services on a remote system such as FTP, DNS, or NFS?
   a. Presentation   b. Session
   c. Logical link control d. Application

63. What OSI layer is the window for application processes to access network services?
   a. Presentation   b. Session
   c. Network   d. Application

64. Standards for data encryption and decryption are defined at what OSI layer?
    - a. Transport
    - b. Session
    - c. Network
    - d. Presentation

65. What is the IEEE standard that defines Gigabit Ethernet?
    - a. 802.2
    - b. 804.5B
    - c. 801.9
    - d. 802.3Z

66. What is the OSI layer that controls which side transmits and for how long?
    - a. Session
    - b. Data link
    - c. Transport
    - d. Physical

67. The upper OSI layers have received communications about FTP or printing problems. What layer sent the communications?
    - a. Presentation
    - b. Physical
    - c. Network
    - d. Session

68. CSMA/CD is used in what IEEE specification?
    - a. 802.3
    - b. 802.2
    - c. 802.5
    - d. 802.11

69. Which OSI layer identifies and establishes whether an intended communication partner is available?
    - a. Presentation
    - b. Session
    - c. Data link
    - d. Application

70. The 802.5 standard has nodes using which of the following media access methods?
    - a. Polling
    - b. CSMA/CD
    - c. Contention
    - d. Token passing

71. A guaranteed capacity (no congestion) with fixed bandwidth and consistent network delay are characteristic of what network type?
    - a. Transport switching
    - b. Circuit switching
    - c. Data link routing
    - d. Packet switching

72. Broadband signals are of what type?
    - a. Serial only
    - b. Analog
    - c. Parallel only
    - d. Digital

73. Pat is setting up a multichannel voice/data network connection that will carry information through a single T-1 line. Which of the following transmission technologies will be necessary in order to implement this?
    - a. Broadband
    - b. Multiplex
    - c. Baseband
    - d. Single plex

74. Baseband signals are of what type?
    - a. Digital
    - b. Serial only
    - c. Analog
    - d. Parallel only

75. A data transmission scheme that requires signaling to coordinate timing is said to be _____.
    - a. asynchronous
    - b. analog
    - c. digital
    - d. synchronous

76. A channel-access administrator is used in what media access method?
    - a. Token passing
    - b. Contention
    - c. Digital sequencing
    - d. Polling

77. The specifications of low-level protocols such as FDDI, Ethernet, and Token Ring are the responsibility of what OSI layer?
    - a. Network
    - b. Physical
    - c. Transport
    - d. Data link

78. What is the IEEE specification for fast Ethernet?
    - a. 802.3u
    - b. 802.5
    - c. 802.2
    - d. 802.3

79. What is the IEEE specification for wireless LANs?
    - a. 802.2
    - b. 802.5
    - c. 802.3
    - d. 802.11

80. How many alphanumeric characters can be represented by a byte?
    - a. 2
    - b. 1
    - c. 4
    - d. 0

81. What is another name for a switch that implements layer 2 and layer 3 technology?
    - a. Bridge
    - b. Brouter
    - c. Repeater
    - d. Router

82. Which of the following connectivity devices typically work at the data link layer of the OSI model?
    - a. Routers
    - b. Repeaters
    - c. Bridges
    - d. Gateways

83. At what layer of the OSI model does a router function?
    - a. Data link
    - b. Network
    - c. Transport
    - d. Physical

84. Pat is the network administrator for a meatpacking plant in Sacramento, California. He has installed a Microsoft Exchange Server version 5.5 computer on the Windows NT 4.0 local area network at the plant. He tries to send an email message outside the LAN to test the Microsoft Exchange Server 5.5 computer. The email message never reaches its destination. Pat discovers that the IBM mainframe that uses the Simple Mail Transfer Protocol (SMTP) to transmit email messages over the Internet did not process his email message. Which device can Pat install to ensure that the meatpacking plant's Windows NT 4.0 LAN will

be able to communicate with the IBM mainframe email server?

   a. A router   b. A brouter

   c. A gateway   d. A hub

85. A hub (multiport repeater) operates at what OSI layer?

   a. Physical   b. Transport

   c. Network   d. Application

86. NICs operate at what layer of the OSI model?

   a. Data link   b. Network

   c. Transport   d. Physical

87. On one network segment, a router is allowing only IPX packets to pass. On another segment, a router is allowing only IP packets to pass. How can network nodes on both segments communicate with each other?

   a. Bridge   b. Router

   c. Brouter   d. Gateway

88. A Token Ring hub is also known as what?

   a. VAX   b. TPU

   c. BTU   d. MAU

89. What device will enable a mainframe computer to talk to a PC?

   a. Bridge

   b. Router

   c. Brouter

   d. Application gateway

90. Which of the following network devices doesn't do any packet examination as it passes the packets through?

   a. Router   b. Bridge

   c. Switch   d. Repeater

91. What will a brouter do with packets sent by non-routable protocols?

   a. It will drop them off the network.

   b. It will reroute them to a switch.

   c. It will ignore them.

   d. It will bridge them.

92. Your network suffers data loss due to attenuation. What device, at a minimum, must you install?

   a. Bridge   b. Router

   c. Brouter   d. Repeater

93. What number do you get when you convert the decimal value 6 to binary?

   a. 00000101   b. 00001111

   c. 00000111   d. 01010101

   e. 00000110

94. What number do you get when you convert the decimal value of 68 to binary?

   a. 10101010   b. 00001111

   c. 01000010   d. 11110000

   e. 01010101   f. 01000100

95. What number do you get when you convert the binary value of 10101010 to decimal?

   a. 170   b. 165

   c. 224   d. 128

96. What number do you get when you convert the hexadecimal number of FH to binary?

   a. 1000   b. 1110

   c. 1111   d. 1010

   e. 0001

CHAPTER 3

Media and Topologies

Topics Covered in this Chapter

Introduction

As mentioned in a previous chapter, one of the primary components of a network is the common medium used to connect the individual computers to a network by using a network card. To allow the computers to communicate with each other, you have many options in cable types, cable layouts, and protocols.

Objectives

- Recognize the following logical or physical network topologies; star, bus, mesh, ring and wireless, given a schematic diagram or description.
- Recognize the common media connectors (RJ-11, RJ-45, AUI, BNC, ST, and SC) and/or describe their uses.
- Choose the appropriate media type and connectors to add a client to an existing network.
- Identify the basic characteristics (speed, capacity, and media) for FDDI.
- Identify the main characteristics of VLANs.
- Given a wiring task, select the appropriate tool (wire crimper, media tester/certifier, punch-down tool, tone generator, optical tester, etc.).

- Given a network scenario, interpret visual indicators (e.g., link lights, collision lights, etc.) to determine the nature of the problem.
- Build and test a UTP cable.
- Build and test a coaxial (thinnet) cable.
- Explain the purpose of the plenum cable.
- Define the purpose of a MAC address.
- List and compare the different connectivity devices.
- List the different devices used for cable management.
- Identify a network troubleshooting scenario involving a wiring/infrastructure problem, and identify the cause of the problem (e.g., bad media, interference, or network hardware).

3.1 DATA AND THE NETWORK

To communicate on LANs, a **network card,** or **network interface card (NIC),** which is found as an expansion card or is integrated into the motherboard, uses transceivers (devices that both transmit and receive analog or digital signals). To get a better idea of the function of the network card, you must see how the data is sent from one computer to another. The data is generated by the microprocessor and sent to the network card using the computer's parallel data bus. At this point, the data consists of alphanumeric characters represented as a series of on (1) and off (0) signals.

Within the network card, the data is processed and converted so that it can be transmitted on the network media. If it is being sent on a cable, the data is turned into a serial bit stream, amplified, and sent on to the cable. When the network card has to send data, the network card will send specific electric signals to represent the data. In addition, the network card must follow specific rules on how it gains sole access to the cable.

The network card on the receiving computer sees the packet and reads part of the packet to determine if the packet was meant for this computer. The network card then converts the data into digital signals and sends the data to the microprocessor to be processed.

3.2   TOPOLOGIES

Topology describes the appearance or layout of the network. Depending on how you look at the network, there is the physical topology and the logical topology. The **physical topology** (part of the physical layer) describes how the network actually appears. The **logical topology** (part of the data link layer) describes how the data flows through the physical topology or the actual pathway of the data. While the physical topology is easy to recognize, the logical topology is not. The physical and logical topologies are not always the same.

3.2.1   Physical Topology

In essence, the physical topology describes how devices on a network are wired or connected together. The types of physical topologies used in networks are:

- Bus topology
- Ring topology
- Star topology
- Mesh topology
- Cellular topology

See table 3.1 for a brief look at some of their advantages and disadvantages.

A **bus topology** looks like a line. It consists of a single cable along which the data is sent. The two ends of the cable do not meet, and the two ends do not form a ring or loop. (See figure 3.1.) All **nodes** (devices connected to the computer including networked com-

Table 3.1   Physical Topologies

Topology	Installation	Expansion/ Reconfiguration	Troubleshooting	Media Failure
Bus	Relatively easy	Moderately difficult	Difficult	Break in the bus prevents transmission
Ring	Moderately easy	Moderately difficult	Easy	Break in the ring prevents transmission (partial for dual rings)
Star	Relatively easy, but may be time consuming	Easy	Easy	Break in one cable only affects one computer
Mesh	Difficult	Difficult	Easy	Very low
Cellular	Easy	Easy	Moderately easy	Very low

Figure 3.1   The Bus Topology

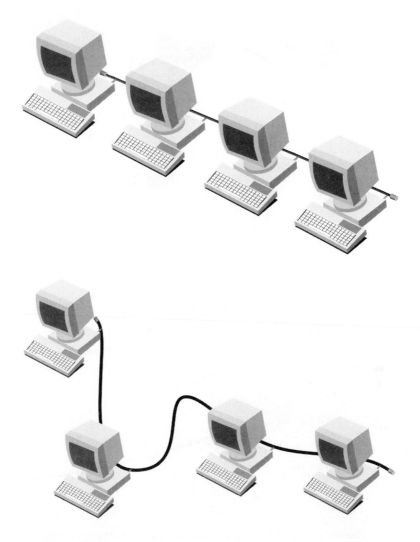

puters, routers, and network printers) listen to all the traffic on the network, but only accept the packets that are addressed to them. The single cable is sometimes referred to as a
segment, backbone cable, or **trunk.** Since all computers use the same backbone cable,
the bus topology is very easy to set up and install, and the cabling costs are minimized.
Unfortunately, traffic easily builds up on this topology, and it is not a recommended topology for large networks. Ethernet (10Base2 and 10Base5) is an example of a bus topology.

Typically with a bus topology network, the two ends of the cable must be terminated.
Otherwise, when signals get to the end of a cable segment, they are likely to bounce back,
colliding with new data packets. If there is a break anywhere or one system does not pass
the data along correctly, the entire network will go down. The reason is that a break divides the trunk into two pieces, each with an end that is not terminated. In addition, these

Figure 3.2   The Ring Topology

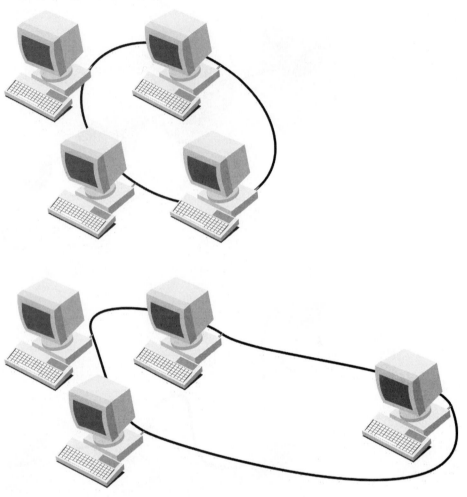

problems are difficult to troubleshoot since a break causes the entire network to go down with no indication of where the break is.

A **ring topology** has all devices connected to one another in a closed loop. (See figure 3.2a and b.) Each device is connected directly to two other devices. Typically in a ring, each node checks to see if the packet was addressed to it and acts as a repeater (duplicates the data signal, which helps keep the signal from degrading) for the other packets. This of course allows the network to span large distances. Though it might look inefficient, this topology sends data very quickly because each computer has equal access to communicate on the network.

There are two main disadvantages to a ring topology. The first is related to a break in the ring. Traditionally, a break in the ring will cause the entire network to go down and

Figure 3.3   The Star Topology

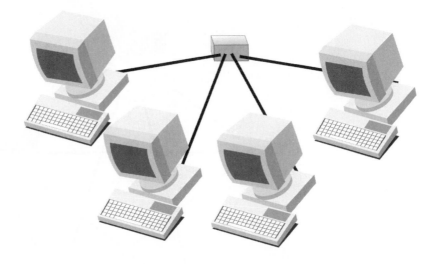

can be difficult to isolate. Today, some networks have overcome these pitfalls by allowing computers to still communicate with their connected partners, by using dual rings for fault tolerance, or by having a computer act as a beacon if it notices a break in the ring. The second disadvantage is cost. Since each node is a repeater, the networking device tends to be more expensive than the other topologies. IBM Token Ring and Fiber Distributed Data Interface (FDDI) are examples of a ring topology.

A **star topology** is the most popular topology in use. Each network device connects to a central point such as a hub or switch, which acts as a multipoint connector. Other names for a hub are a concentrator, multipoint repeater, and media access unit (MAU). (See figure 3.3.)

Star networks are relatively easy to install and manage, but they may take some time to install since each computer requires a cable that runs back to the central point. If a link fails (hub port or cable), the remaining workstations are not affected as they would be in the bus and ring topology. Unfortunately, if you are using a hub, bottlenecks can occur because all data must pass through the hub. Of course, since unmanaged switches are so inexpensive, there is no reason to purchase hubs. Examples of star networks include Ethernet 10BaseT and 100BaseTX.

Another topology is the **mesh topology.** In this type of topology, every computer is linked to every other computer. While this topology is not very common in LANs, it is common in WANs, where it connects remote sites over telecommunication links. This is of course the hardest to install and reconfigure since the number of cables increases geometrically with each computer that you add.

NOTE: Some networks will use a modified mesh topology, which has multiple links from one computer to another but doesn't necessarily have each computer linked to every other computer. (See figures 3.4a and 3.4b.)

Figure 3.4 Mesh Topology and Modified Mesh Topology

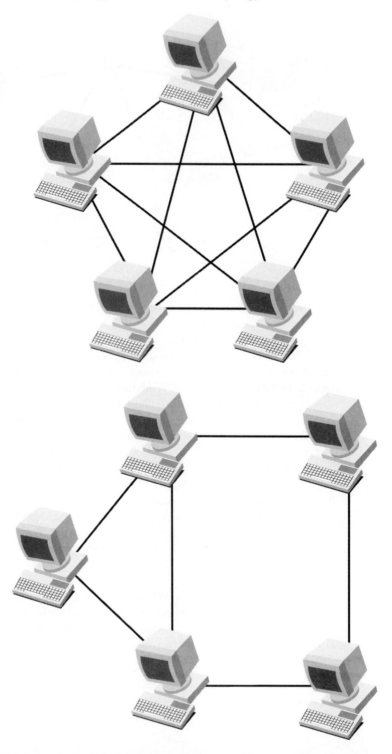

Figure 3.5   Cellular Topology

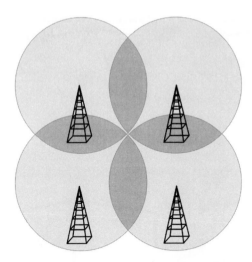

Many technologies that are wireless use a **cellular topology,** where an area is divided into cells. A broadcast device is located at the center and broadcasts in all directions to form an invisible circle (cell). All network devices located within the cell communicate with the network through the central station or hub, which is interconnected with the rest of the network infrastructure. If the cells are overlapped, devices may roam from cell to cell while maintaining connection to the network as the devices. (See figure 3.5.) The best-known example of cellular topology is cellular phones.

The **hybrid topology** scheme combines two of the traditional topologies, usually to create a larger topology. In addition, the hybrid topology allows you to use the strengths of the various topologies to maximize the effectiveness of the network. Examples of the hybrid topology would be the bus-star topology and the star-ring topology. (See figure 3.6a and b.)

3.2.2   Logical Topology

As mentioned earlier, the physical topology describes how the network actually appears, whereas the logical topology describes how the data flows through the physical topology or the actual pathway of the data. The two logical topologies are bus and ring.

Let's take Ethernet using a backbone cable. Here, you have a network that is physically a bus topology, and the pathway of the signals or the logical topology is also a bus topology. Now compare that with an Ethernet network using a star topology—the physical network is a star topology, and yet the pathway or logical topology is a bus. This reason is that each cable attached to the hub has two wires. One wire is used to carry data from the hub to the network device, and the other wire is used to carry the data from the network device to the hub. The hub then connects the pairs of wires with each other, creating a large bus pathway. Of course, the bus has two ends, which are both contained within the hub. (See figure 3.7a and b.)

Figure 3.6   The Hybrid Topology (Bus-star and Star-ring)

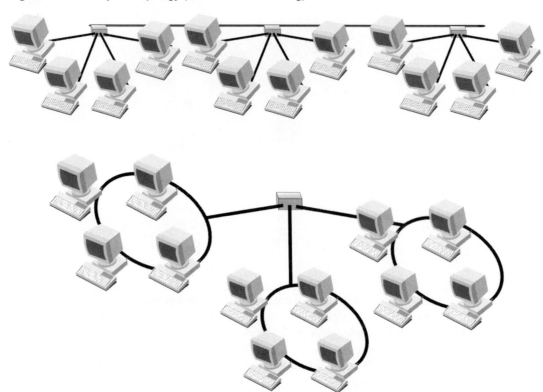

For another example, look at an IBM Token Ring network. One form of Token Ring is connecting the network devices using a star topology. Although this network has a physical star topology, the pathway or logical topology is a ring. The difference between the Ethernet hub and the Token Ring MAU is their hub configuration. Both have a hub that contains two cable ends, but the Token Ring MAU connects the two ends to form a ring. (See figure 3.8.)

3.2.3   Segment and Backbone

A segment could be a single cable such as a backbone cable, or it could be a cable that connects a hub and a computer. A logical segment contains all the computers on the same network and contains the same network address such as that of networks with a backbone cable or hub used in a logical bus topology.

A backbone can be used as a main cable segment such as that found in a bus topology network. This would include a long single cable, with patch cables used to attach the computers or smaller cables connected together with barrel and T-connectors. In addition, a backbone can refer to the main network connection through a building, campus, WAN, or the Internet.

Figure 3.7   Logical Bus Topology

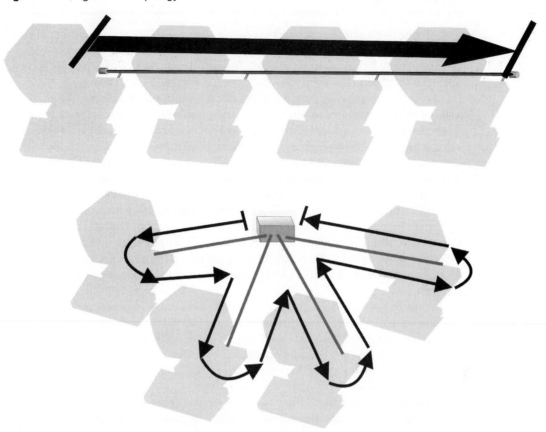

Figure 3.8   Logical Ring Topology

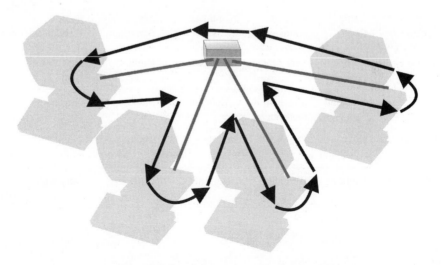

3.3 CABLING

The cabling system used in networks serves as the veins of the network that connect all the computers together and allow them to communicate with each other. The common types of cabling include unshielded twisted pair (UTP), shielded twisted pair (STP), coaxial, and fiber optics. (See table 3.2.)

Frequency (bandwidth) is the number of cycles that are completed per unit of time and is generally expressed in hertz (cycles per second). Data cabling is typically rated in kilohertz (KHz) or megahertz (MHz). The more cycles per second, the more interference the cable generates and the more susceptible to data loss the cable is.

The data rate (or information) capacity is defined as the number of bits per second that move through a transmission medium. When you choose a cable, you must choose one that will handle your current network traffic and will also allow for some growth.

With traditional cabling, there is a fundamental relationship between the number of cycles per second that a cable can support and the amount of data that can be pushed through the cable. For example, Category 5 and Category 5e cables are rated at 100 MHz. To implement a 100Base-TX network, you must use Category 5 cabling since the cable must support a frequency of 100 MHz. During each digital cycle, a single bit is pushed through.

While this is a good way to think of the relationship between information rate and cable bandwidth, the IEEE recently approved the 803.3ab standard (1000BaseT), which covers running Gigabit Ethernet over Category 5 cabling. Category 5 does not operate at 1 GHz. Instead, the cable uses four data pairs to transmit the data, and it uses sophisticated encoding (multiplexing) techniques to send more bits of data over the wire during each cycle.

In copper wiring, a signal loses energy as it travels because of electrical properties at work in the cable—there is opposition to the flow of current. This opposition to the flow

Table 3.2 Cable Types

Cable Type	Cable Cost	Ease of Installation	Installation Cost	EMI Sensitivity	Data Bandwidth	Comments
UTP	Lowest	Very simple	Lowest	Highest	Lowest to high	Used in more than 80% of LANs
STP	Medium	Simple to moderate	Moderate	Moderately low	Moderate	Usually found in older networks
Coaxial	Medium	Simple	Moderate	Moderate	High	Often used as a backbone cable
Fiber optic	Highest	Difficult	Highest	None	Very high	Uses light instead of electric signal

of current through a cable or circuit is expressed in impedance. Impedance is a combination of resistance, capacitance, and inductance and is expressed in ohms. A typical UTP cable is rated at between 100 and 120 ohms. Categories 3, 4, 5, and 5e cables are rated at 100 ohms.

Attenuation occurs as the strength of a signal falls off with distance over a transmission medium. This loss of signal strength is caused by several factors. One, for example, is that the signal converts to heat due to the resistance of the cable; another is that the energy is reflected as the signal encounters impedance changes throughout the cable. Low decibel values of attenuation are desirable since that means that less of the signal is lost on its way to the receiver.

Interference occurs when undesirable electromagnetic waves affect the desired signal. Interference can be caused by **electromagnetic interference (EMI)** produced by large electromagnets used in industrial machinery, motors, fluorescent lighting, and power lines. **Radio frequency interference (RFI)** is caused by transmission sources such as a radio station.

Another term used to describe instability in a signal wave is **jitter.** Jitter is caused by signal interference.

Besides looking at the capacity and electrical characteristics of the cable, you also need to look at two other factors when choosing a cable. First, you should look at cost. If you are installing a large network within a building, you will find that the cabling system can cost thousands of dollars.

NOTE: The cost of installing the cable (planning and hourly wages) is many times the cost of the cable itself. Therefore, you must make sure that you have the financial resources available to install such a system.

Second, you should look at the ease of installation of the cabling system since this affects your labor costs and can indirectly affect the reliability of the network. In addition, you should look at the ease of troubleshooting, including how troubleshooting is affected by media faults and if the cable system offers any fault tolerance.

3.3.1   Unshielded Twisted Pair

A **twisted pair** consists of two insulated copper wires twisted around each other. While each pair acts as a single communication link, twisted pairs are usually bundled together into a cable and wrapped in a protective sheath. (See figure 3.9.)

Figure 3.9   Twisted-pair Cable

Figure 3.10   UTP Cable

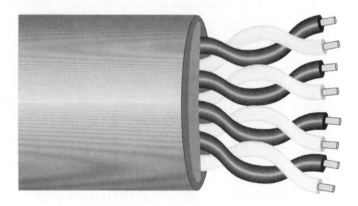

Question:

Why are the wires twisted around each other?

Answer:

The reason they are twisted is because copper wire does not constrain electromagnetic signals well. This means that if you have two copper wires next to each other, the signal will induct (law of induction) or transfer from one wire to the other. This phenomenon is called **crosstalk.**

The circuits that send data on the cable are differential amplifiers. Twisted wires ensure that the noise is the same on each wire. The common noise is then canceled at the differential amplifier. Since the data is inverted on the second wire, the data is not canceled. Therefore, by using twisted pair, the crosstalk is canceled.

Twisted pair can be **unshielded twisted pair (UTP)** or shielded twisted pair (STP). Unshielded twisted pair is the same type of cable that is used with telephones and is the most common cable used in networks. (See figure 3.10.) UTP cable consists of four pairs of wires in each cable. Each pair of wires is twisted about each other and used together to make a connection. Compared with other cable types (shielded twisted pair, coaxial cable, and fiber optics), UTP is inexpensive and is the easiest to install. The biggest disadvantages of UTP are its limited bandwidth of 100 meters and its susceptibility to interference and noise. Traditionally, UTP has had a limited network speed. However, more recently UTP has been used in networks running between 4 Mbps and 1 Gbps, and there are some companies such as Hewlett-Packard that are working on a 10-Gbps network standard.

In 1995, UTP cable was categorized by the Electronic Industries Association (EIA) based on the quality and number of twists per unit. The UTP categories are published in EIA-568-A. See table 3.3 for a summary of the categories.

Early networks that used UTP typically used Category 3, whereas today's newer high-speed networks typically use Category 5 or Enhanced Category 5 cabling. Category 3 has three to four twists per foot and could operate up to 16 MHz; while Category 5 uses three to four twists per inch, contains Teflon insulation, and can operate at 100 MHz. En-

Table 3.3   Unshielded Twisted Pair Cable Categories for Networks

Cable Type	Bandwidth, MHz	Function	Attenuation	Impedance	Network Usage
Category 3	16	Data	11.5	100 ohms	10BaseT (10 Mbps), Token Ring (4 Mbps), ARCnet, 100VG-ANYLAN (100 Mbps)
Category 4	20	Data	7.5	100 ohms	10BaseT (10 Mbps), Token Ring, ARCnet, 100VG-ANYLAN (100 Mbps)
Category 5	100	High-speed data	24.0	100 ohms	10BaseT (10 Mbps), Token Ring, Fast Ethernet (100 Mbps), Gigabit Ethernet (1000 Mbps), and ATM (155 Mbps)
Category 5E (Enhanced)	100	High-speed data	24.0	100 ohms	10BaseT (10 Mbps), Token Ring, Fast Ethernet (100 Mbps), Gigabit Ethernet (1000 Mbps), and ATM (155 Mbps)
Category 6	250	High-speed data	19.8	100 ohms	10BaseT (10 Mbps), Token Ring, Fast Ethernet (100 Mbps), Gigabit Ethernet (1000 Mbps), and ATM (155 Mbps)
Category 6E (Enhanced)	250	High-speed data	19.8	100 ohms	10BaseT (10 Mbps), Token Ring, Fast Ethernet (100 Mbps), Gigabit Ethernet (1000 Mbps), and ATM (155 Mbps)
Category 7 (not yet approved)	600	High-speed data			

hanced Category 5 is a higher-quality cable designed to reduce crosstalk even further and support applications that require additional bandwidth.

NOTE: Category 6 has been recently approved.

Category 6 and Category 7 have not been approved yet. Therefore, no vendor can promise complete compatibility for future networks and truthfully mean it.

As network applications increased network traffic, eventually there was a need for faster networks. One way to increase the performance of the network is to use more expensive, fast electronic devices. A cheaper way is to increase the amount of usable bandwidth by using all four pairs of the UTP cable instead of using two pairs.

Question:

If you are planning a network, what type of cable should you use?

Answer:

Standard Category 5 will handle standard network technologies for the next 5 years, including Fast Ethernet/100Base TX and Gigabit Ethernet 1000Base TX. If you are planning to support gigabit technology now or in the future, you should select Enhanced Category 5.

While your telephone uses a cable with two pairs (four wires) and an RJ-11 connector, a computer networks uses a cable with four pairs (eight wires) and an RJ-45 connector. (See figure 3.11.) In a simple network, one end of the cable attaches to the network card on the computer and the other end attaches to a hub (multiported connection).

With UTP wiring, you must be very careful about the quality of the RJ-45 connector crimping. Bad crimping can lead to intermittent connections or pulled-out wires. You also must make sure that the pairs stay twisted right down to the connector.

In a larger network, one end of the cable will connect the network card of the computer to a wall jack. The wall jack is connected to the back of a patch panel kept in a server room or wiring closet. A cable then is attached to the patch panel and connected to a hub. The cables that connect the computer to the wall jack and the cable that connects the patch panel and the hub are called patch cables.

UTP cable is commonly available in 22, 24, and 26 AWG (American wire gauge) using either solid or stranded cable. **Solid cable** is typically used for the cabling that exists throughout the building. This should include cables that lead from the wall jacks to the server room or wiring closet. **Stranded cable** is typically used as patch cables between patch panels and hubs and between the computers and wall jacks. Since the stranded wire isn't as firm as solid wire, it is a little easier to work with. Unfortunately, stranded cable has 20 percent more attenuation over solid wire.

3.3.2   Building UTP Cable

There are two types of Ethernet cables: a straight-through cable and a crossover cable. The **straight-through cable,** which can be used to connect a network card to a hub, has the same sequence of colored wires at both ends of the cable. A **crossover cable,** which can be used to connect one network card to another network card or a hub to a hub, reverses the transmit and receive wires. (See tables 3.4, 3.5, and 3.6.)

There are two wiring standards, 568A and 568B. The 568B wiring is used in almost 90 percent of installations. The pin numbers refer to a RJ-45 telephone type connector. Pin #1 is the one on the left when you hold the tab down with the wire facing you, as you would plug it in. Pin #2 is the next one on the right and lastly, pin #8 is the one all the way

Figure 3.11   A UTP Cable with an RJ-45 Connector and a UTP Cable
with an RJ-11 Connector

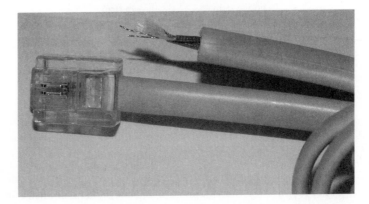

to the right. The 568A wiring is used when compatibility to telephone wiring systems
(USOC) is needed. It is not recommended, as parts are not as commonly available.

A good way of remembering how to wire a Crossover Ethernet cable is to wire one
end using the T-568A standard and the other end using the T-568B standard. Another way
of remembering the color coding is to simply switch the Green set of wires in place with
the Orange set of wires. Specifically, switch the solid Green (G) with the solid Orange,
and switch the green/white with the orange/white.

To create a Category 5 UTP patch cable, you will need the following:

- Category 5 stranded cable
- Two RJ-45 round wire connectors
- Small wire-cutter
- RJ-45 crimping tool

Table 3.4   Standard Color Scheme for UTP/RJ-45 Cables

Wire #	Pair #	IA/TIA 568A Color Standard	AT&T 258A or EIA/TIA 568B Color Standard	
Pin 1	**Pair 2**	White with green strip	White with orange strip	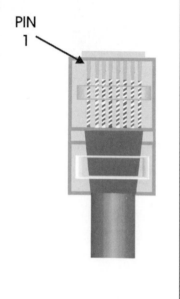
Pin 2	**Pair 2**	Green	Orange	
Pin 3	**Pair 3**	White with orange strip	White with green strip	
Pin 4	**Pair 1**	Blue	Blue	
Pin 5	**Pair 1**	White with blue strip	White with blue strip	
Pin 6	**Pair 3**	Orange	Green	
Pin 7	**Pair 4**	White with brown strip	White with brown strip	
Pin 8	**Pair 4**	Brown	Brown	

Table 3.5   Ethernet Crossover Cable Pinouts

End 1	End 2
Pin 1 (White with Green Strip)	Pin 1 (White with Orange Strip)
Pin 2 (Green)	Pin 2 (Orange)
Pin 3 (White with Orange Strip)	Pin 3 (White with Green Strip)
Pin 4 (Blue)	Pin 4 (Blue)
Pin 5 (White with Blue Strip)	Pin 5 (White with Blue Strip)
Pin 6 (Orange)	Pin 6 (Green)
Pin 7 (White with Brown Strip)	Pin 7 (White with Brown Strip)
Pin 8 (Brown)	Pin 8 (Brown)

Table 3.6   Pins Used in an RJ-45 Connector for the Various Technologies

Application	Pins 1/2	Pins 3/6	Pins 4/5	Pins 7/8
Analog voice	—	—	Tx/Rx	—
ISDN	Power	Tx	Rx	Power
10BaseT (802.3)	Tx	Rx	—	—
Token Ring (802.5)	—	Tx	Rx	—
100Base-TX (802.3u)	Tx	Rx	—	—
100Base-T4 (802.3u)	Tx	Rx	Bi*	Bi*
100Base-VG (802.12)	Bi*	Bi*	Bi*	Bi*
FDDI (TP-PMD)	Tx	Optional†	Optional†	Rx
ATM user device	Tx	Optional†	Optional†	Rx
ATM network equipment	Rx	Optional†	Optional†	Tx
1000Base-T (802.3ab)	Bi*	Bi*	Bi*	Bi*

*Bidirectional.
†May be required by some vendors.

Step 1   Measure and cut the needed length of the cable. Cut the cable about 3 inches longer than the final patch cable length.

Step 2   Take one end of the wire and remove the outer sheathing to reveal about ¾-inch twisted wires.

Step 3   Untwist all the exposed wires and straighten them out as much as you can. Rearrange the wires in a flat manner, as close together as possible in the proper color order. See table 3.4.

Step 4   Take all the wires and cut them to the same length of approximately ½ inch exposed. *NOTE:* Today, most connectors can act as a vampire tap since they can bite into the cable to get to the copper core. Therefore, when using these types of connectors, this step can be skipped.

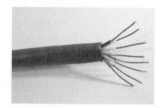

Step 5   Hold the RJ-45 connector and insert the wires from the bottom, making sure to keep the right color order. Each wire should slide into separate grooves toward the tip of the connector. Push the wires in as far as they will go. In addition, make sure that the outer sheathing is up into the opening as well and that the outer sheathing is under the crimping tab.

Step 6   Insert the RJ-45 connector into the crimping tool and crimp the connector.

Step 7   Repeat steps 2 through 6 for the other end of the cable.

Step 8   Using a cable tester or digital multimeter, test the cable.

3.3.3   Shielded Twisted Pair

Shielded twisted pair is similar to unshielded twisted pair except that the shielded twisted pair is usually surrounded by a braided shield that serves to reduce both EMI sensitivity and radio emissions. (See figure 3.12.) Shielded twisted-pair cable was required for all high-performance networks such as IBM Token Ring until a few years ago and is commonly used in IBM Token Ring networks (see figure 3.13) and Apple's LocalTalk network. STP is relatively expensive compared with UTP and is more difficult to work with.

The STP cables used by IBM networks are shown in table 3.7. Different from the other connectors, IBM connectors are hermaphroditic, which means they are not male or female. Therefore, if the cable does not connect properly, you only have to turn one around for it to connect.

Figure 3.12   Shielded Twisted-pair Cable

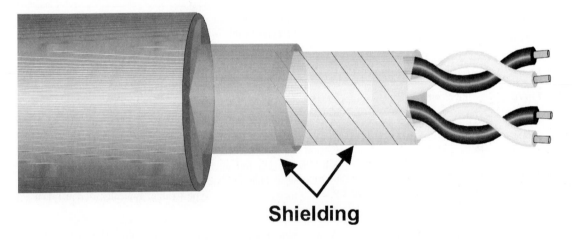

Shielding

Figure 3.13   IBM Shielded Twisted-pair Cable

Table 3.7   IBM Cabling System

IBM Type	Standard Label	Description
Type 1	Shielded twisted pair	Two pairs of 22-AWG shielded wires surrounded by an outer braided shield. Used for computers and MAU. The impedance of Type 1 cable is 150 ohms and is rated at 20 Mbps. Typically used in Token Ring applications.
Type 1A	Shielded twisted pair	Same as Type 1, except the cable is rated at 100 Mbps.
Type 2	Voice and data cable	A voice and data shielded cable with two twisted pairs of 22-AWG wires for data, an outer braided shield, and then four twisted pairs of 26 AWG for voice (unshielded). The impedance of Type 1 cable is 150 ohms. The shielded pairs are rated at 20 Mbps, and the unshielded pairs are rated at 4 Mbps.
Type 2A	Shielded twisted pair	Same as Type 2 except the cable is rated at 100 Mbps.
Type 3	Voice-grade cable	Consists of four solid, unshielded twisted-pair 22- or 24-AWG cables. The Type 3 cable is rated up to speeds of 4 Mbps and an impedance of 100 ohms.
Type 4	Not yet defined	
Type 5	Fiber-optic cable	Two 62.5/125-micron multimode optical fibers—the industry standard.
Type 6	Data patch cable	The Type 6 cable is thought of as a thinner and more flexible version of a Type 1 cable. Two 26-AWG twisted-pair stranded cables with a dual foil and braided shield. It is rated for speeds up to 20 Mbps and has an impedance of 150 ohms. Unfortunately, it has 1.33 times the loss of a Type 1 cable. Typically used in Token Ring applications.
Type 7	Not yet defined	A single-pair version of Type 6. It is a 150-ohm cable and is rated for speeds up to 20 Mbps.
Type 8	Carpet cable	Housed in a flat jacket for use under carpets. Two shielded twisted-pair 26-AWG cables. Limited to one-half the distance of Type 1 cable. Unfortunately, the Type 8 cable has twice the loss of a Type 1 cable.
Type 9	Plenum	Fire-safe. Two shielded twisted-pair cables with an impedance of 150 ohms and is rated up to 20 Mbps.

3.3.4   Coaxial Cable

Coaxial cable, sometimes referred to as coax, is a cable that has a center wire surrounded by insulation and then a grounded shield of braided wire (mesh shielding). (See figure 3.14.) The copper core carries the electromagnetic signal, and the braided metal shielding acts as both a shield against noise and a ground for the signal. The shield minimizes electrical and radio frequency interference and provides a connection to ground. Coaxial cable is the primary type of cable used by the cable television industry and is widely used for computer networks.

For computer networks, coaxial cables are usually used for the backbone cable for Ethernet networks. The network devices are attached by cutting the cable and using a T-connector or by applying a vampire tap (a mechanical device that uses conducting teeth to penetrate the insulation and attach directly to the wire conductor). To maintain the correct electrical properties of the wire, you must terminate both ends of the cable,

Figure 3.14   Coaxial Cable

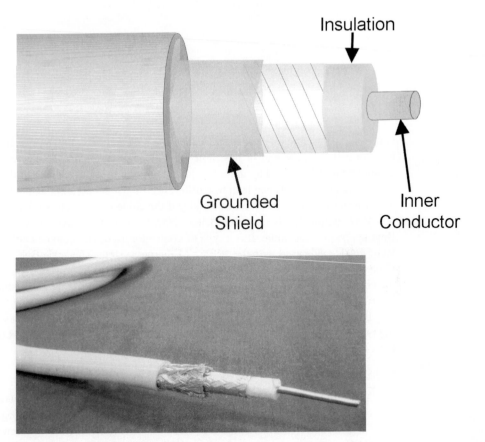

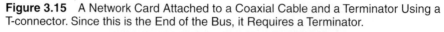

Figure 3.15 A Network Card Attached to a Coaxial Cable and a Terminator Using a T-connector. Since this is the End of the Bus, it Requires a Terminator.

and you must ground one end of the cable. The termination dampens signals that bounce back or reflect at the end of the cable. The ground completes the electric circuit. Not grounding at all can lead to an undesirable charge in the coax cable. On the other hand, grounding at both ends can lead to a difference in ground potential, causing an undesirable current on the coax cable, especially if the cable is between two buildings.

BNC connector is short for British Naval Connector, Bayonet Nut Connector, or Bayonet Neill Concelman. It is a type of connector used with coaxial cables, such as the RG-58 A/U cable used with the 10Base2 Ethernet system. The basic BNC connector is a male type mounted at each end of a cable. This connector has a center pin connected to the center cable conductor and a metal tube connected to the outer cable shield. A rotating ring outside the tube locks the cable to any female connector.

BNC T-connectors (used with the 10Base2 system) are female devices for connecting two cables to a network interface card. (See figure 3.15.) A **BNC barrel** connector allows two cables to be connected together.

Question:

You have two Ethernet hubs, which you need to link together as one network. Each hub contains a single BNC connector, which allows you to connect the hubs together. Therefore, you take a coaxial cable and connect it directly from the BNC connector to the other BNC connector. Unfortunately, the hubs do not function together. What is the problem?

Table 3.8   Coaxial Cables Commonly Used Today

RG Rating	Popular Name	Ethernet Implementation	Type of Cable	Center Wire Gauge	Resistance Value
RG-6/U	Thicknet	10Base5	Solid copper	18 AWG	75 ohms
RG-8/U	Thicknet	10Base5	Solid copper	10 AWG	50 ohms
RG-58/U	N/A	None	Solid copper	20 AWG	50 ohms
RG-58/AU	Thinnet	10Base2	Stranded copper	20 AWG	50 ohms
RG-59	Cable TV	None	Solid/stranded copper	20 AWG	75 ohms
RG-62	ARCnet	N/A	Solid copper	22 AWG	93 ohms

Answer:

Anytime you use a coaxial cable with Ethernet, you must always connect network devices using a T-connector. In addition, the two ends must be terminated with the proper terminating resistor.

There are several coaxial cables standards used in computer networking. Table 3.8 shows coaxial cables in common use today.

3.3.5   Building a Coaxial Cable

To create a coaxial 10Base2 Ethernet cable, you will need the following:

- RG-58 A/U coaxial cable
- Two RG-58/U BNC connectors
- Racheting coax crimping tool
- Two-step rotating stripper
- Cable cutter (straight cut)

Step 1   Cut the coaxial cable (straight cut) to the core wire using the cable cutter. Adjust the stripper to meet the desired cable diameter and stripping requirements. For best results, the stripper should be adjusted to expose ¼ inch of the conductor and ¼ inch of the insulation. Insert the coaxial cable into the coaxial wire stripper. Strip the ends of the coax cable using a rotary motion (three to five full turns

with the coax stripper). Always turn the stripper in the same direction. Do not cut completely through the jacket, to avoid nicking the shield. Flex the jacket to complete the separation. Pull the coax cable out of the stripper and inspect the cable for stripping quality. Ensure the center conductor is not nicked or scored and the braid is pushed away from the conductor.

Step 2   Seat the center pin of the BNC connector on the exposed conductor. Crimp the center pin to the end of the center conductor using the small-diameter pin crimp die on the ratchet crimping tool. Ensure that all strands of the center conductor are in the hole in the center pin before crimping.

Step 3   Slide the sleeve ferrule over the pin and exposed insulation. Place the BNC connector body on the cable end. Align the connector body on the cable end so that its shaft fits over the pin and between the braid and the insulation.

Step 4   Slip the connector body under the braided shield as far as it will go. Check for stray strands and push them out of the way. Ensure that the pin flange rests on the exposed insulation and that the top of the pin is flush with the top of the BNC body. Slide up the crimp ferrule sleeve to cover the exposed braided shield up to the BNC body shoulder.

Step 5   Ensure that the ratcheting crimping tool is fitted with the proper hex die. Place the crimping tool over the ferrule sleeve and squeeze it to evenly and completely crimp the ferrule to the BNC body.

Step 6   Inspect the neatness and tightness of the conductor. Pull and flex firmly on the BNC connector to make sure it is crimped tightly to the cable.

Step 7   Repeat steps 1 through 6 for the other end of the cable.

Step 8   Test the cable with a cable tester or digital multimeter.

3.3.6   Screened Twisted Pair

A newer type of cable that has been developed over the last few years is the screened twisted-pair (ScTP) cable. It is also sometimes referred to as a foil twisted-pair (FTP) cable because the foil surrounds all four conductors. It is a hybrid of the STP and UTP cable, which contains four pairs of 24-AWG, 100-ohm wire surrounded by a foil shield or wrapper and a drain wire for bonding purposes. This foil shield is not as large as the woven copper braided jacket used by some STPs such as IBM Types 1 and 1A. ScTP cable is basically STP cabling that does not shield the individual pairs; the shield may also be smaller than some varieties of STP cabling.

The foil shield is the reason ScTP is less susceptible to noise. In order to implement a completely effective ScTP system, however, the shield continuity must be maintained throughout the entire channel, including patch panels, wall plates, and patch cords. Like STP cabling, the entire system must be bonded to ground at each end of each cable run, or you will have created a massive antenna.

Standard eight-position modular jacks, commonly called RJ-45s, do not have the ability to do this. Instead, you must have special mating hardware, jacks, patch panels, and even tools to install an ScTP cabling system. ScTP is recommended for use in environments that have abnormally high ambient electromagnetic interference, such as hospitals, airports, or government/military communications centers.

Figure 3.16   Fiber-optics Cable and Common Connectors (ST, SC, and MT-RJ)

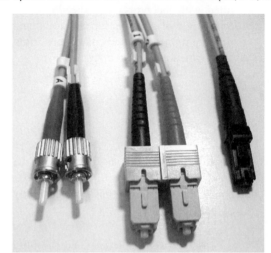

3.3.7   Fiber Optics

A **fiber-optic** cable consists of a bundle of glass or plastic threads, each of which is capable of carrying data signals in the form of modulated pulses of light. While glass can carry the light pulses (several kilometers) even farther than plastic, plastic is easier to work with. Since each thread can only carry a signal in one direction, a cable consists of two threads in separate jackets, one to transmit and one to receive. The fiber-optic cable uses cladding that surrounds the optical fiber core, which helps to reflect light back to the core and to ensure that little of the light signal is lost. Lastly, the cable contains Kevlar strands to provide strength. (See figure 3.16.)

The light signal used in a fiber-optic cable is generated by light-emitting diodes (LEDs) or by injection laser diodes (ILDs). ILDs are similar to LEDs but produce laser light. Since laser light is purer than normal light, it can increase both the data rates and transmission distances. Signals are received by photodiodes, solid-state devices that detect variations in light intensity.

Fiber-optic cables come in single-mode fiber and multimode fiber. **Multimode fiber (MMF)** is capable of transmitting multiple modes (independent light paths) at various wavelengths or phases. Unfortunately, the core's greater diameter makes light more likely to bounce off the sides of the core, resulting in dispersion and limiting the bandwidth to 2.5 Gbps and transmission distance between repeaters. MMF is usually the fiber-optic cable that is used with networking applications such as 10BaseFL, 10BaseF, FDDI, ATM, and other applications that require optical fiber cable for use as horizontal cable and backbone cable. Multimode cabling typically uses an orange sheath.

Single-mode fiber (SMF) can transmit light in only one mode, but the narrower diameter yields less dispersion, resulting in longer transmission distances. SMF is most commonly used by telephone companies and in data installations as backbone cable. Single-mode cabling typically uses a yellow sheath.

The types of fiber-optic cable are differentiated by mode, composition (glass or plastic), and core/cladding size. The size and purity of the core determines the amount of light that can be transmitted. Common types of fiber-optic cables include:

- 8.3-micron core/125-micron cladding single mode
- 62.5-micron core/125-micron cladding multimode
- 50-micron core/125-micron cladding multimode
- 100-micron core/140-micron cladding multimode

The typical multimode fiber-optic cable consists of two strands of fiber (duplex). The core is 62.5 microns in diameter, and the cladding is 125 microns in diameter. This is often simply referred to as 62.5/125 micron. The newest version of the TIA/EIA568 standard also recognizes the use of 50/125-micron multimode fiber-optic cable. A typical indoor fiber-optic cable consists of a core/cladding surrounded by a 250-micron acrylate coating. A 900-micron buffer surrounds the cladding. Then this buffer is surrounded by Kevlar strands.

Over the past 5 years, optical fibers have found their way into cable television networks—increasing reliability, providing a greater bandwidth, and reducing costs. In the local area network, fiber cabling has been deployed as the primary medium for campus and building backbones, offering high-speed connections between diverse LAN segments.

Fiber has the largest bandwidth (up to 10 GHz) of any media available. It can transmit signals over the longest distance (20 times farther than copper segments) at the lowest cost, with the fewest repeaters and the least amount of maintenance. In addition, since it has such a large bandwidth, it can support up to 1000 stations, and it can support the faster speeds that will be introduced during the next 15 to 20 years.

Fiber optics is extremely difficult to tap, making it very secure and highly reliable. Since fiber does not use electrical signals running on copper wire, interference does not affect fiber traffic, and as a result the number of retransmissions is reduced and the network efficiency is increased.

To transmit data, two fibers are required. One strand of fiber is used to send, and the other is used to receive. Fiber-optic cables can be divided into three categories based on the number of optical fibers:

- Simplex
- Duplex
- Multifiber

A simplex fiber-optic cable is a type of cable that has only one optical fiber inside the cable jacket. Since simplex cables only have one fiber inside them, there is usually a larger buffer and a thicker jacket to make the cable easier to handle. Duplex cables have two optical fibers inside a single jacket. The most popular use for duplex fiber-optic cable is as a fiber-optic LAN backbone cable. Duplex cables are perfect because all LAN connections need a transmission fiber and a reception fiber. Multifiber cable has anywhere from three to several hundred optical fibers in it, typically in a multiple of two.

The disadvantage of simplex connectors is that you have to keep careful track of polarity. In other words, you must always make sure that the plug on the send fiber is connected to the send jack and the receive plug is connected to the receive jack. When using duplex plugs and jacks, color coding and keying ensure that the plug will be inserted only one way in the jack and will always achieve correct polarity.

Fiber-optic cables use several connectors, but the two most popular and recognizable connectors are the straight tip (ST) and subscriber (SC) connectors. The ST fiber-optic connector, developed by AT&T, is probably the most widely used fiber-optic connector. It uses a BNC attachment mechanism similar to the thinnet connector mechanism.

The SC connector (sometimes known as the square connector) is typically a latched connector. This makes it impossible for the connector to be pulled out without releasing the connector's latch (usually by pressing some kind of button or release).

A new connector called the MT-RJ is similar to the RJ-45 connector. It offers a new small form factor two-fiber connector that is lower in cost and smaller than the duplex SC interface.

The main disadvantages of fiber optics are that the cables are expensive to install and that special skills and equipment are required to split or splice cables. In addition, fiber optics are more fragile than wire. Fortunately, in recent years, as fiber-optic products are being more mass-produced, the cost gap between the high grades of UTP have closed significantly and many premade products are available.

3.4   CABLE MANAGEMENT

With a physical star topology used in Ethernet networks, all the cables lead and connect to a hub or a set of hubs. While a small network will only use one or two hubs, a network consisting of a few hundred computers will use a few hundred cables. In these situations, it is important to have some form of cable management.

3.4.1   The EIA 568-A Wiring Standard

In a large network, you would establish a horizontal wiring and/or backbone wiring system. The **horizontal wiring system** has cables that extend from wall outlets throughout the building to the wiring closet or server room. **Backbone wiring,** which is often designed to handle a higher bandwidth, consists of cables used to interconnect wiring closets, server rooms, and entrance facilities (telephone systems and WAN links from the outside world). Backbone wiring is sometimes referred to as the **vertical cabling system.** Vertical connections between floors are known as **risers.** For example, the horizontal wiring system could be Category 5 UTP (or better), while the backbone wiring system could be Category 5 UTP (or better), coaxial, or fiber optics.

NOTE: Horizontal wiring and backbone wiring are usually done by a wiring contractor.

In 1991, TIA/EIA released their joint 568 Commercial Building Wiring Standard, also known as structured cabling, for uniform, enterprisewide, multivendor cabling systems. Structured cabling suggests how networking media can best be installed to maximize performance and minimize upkeep. The cabling infrastructure is described as follows:

Figure 3.17   EIA/TIA 568-A Wiring Summary

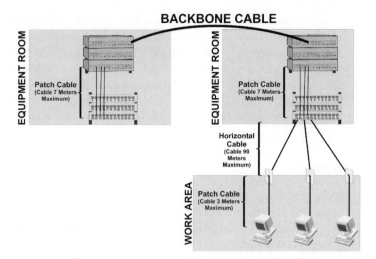

- **Work area**—The work area includes the station equipment, patch cable, and adapters (such as a media filter). The maximum recommended patch cable is 3 meters.
- **Telecommunications closet**—The telecommunications closet is considered to be the floor serving facilities for horizontal cable distribution and can be used for intermediate and main cross-connects. The closet must be designed with minimal cable stress from bends, cable ties, and tensions (as defined in the ANSI/EIA/TIA-569A standard).
- **Equipment room**—The equipment room is the area in a building where telecommunications equipment is located and the cabling system terminates.
- **Horizontal cabling**—The horizontal cabling extends from the work area receptacle to the horizontal cross-connect in the telecommunications closet. It includes the receptacle and optional transition connector (such as undercarpet cable connecting to a round cable). The maximum length of a horizontal cable should be 90 meters.
- **Backbone cabling**—The backbone cabling provides interconnections between telecommunications closets, equipment rooms, and entrance facilities and includes the backbone cables, intermediate and main cross-connects, terminations, and patch cords for backbone-to-backbone cross-connections. Cabling should be 30 meters or less for connection equipment to the backbone. If there are no intermediate cross-connects, a maximum of 90 meters is allowed. Hierarchical cross-connections cannot exceed two levels.
- **Entrance facility**—The entrance facility is defined as that area where the outside telecommunications service enters the building and interconnects with the building's telecommunications systems. In a campus or multibuilding environment, it may also contain the building's backbone cross-connections.

See figure 3.17.

Two outlets are required at the work area:

- A 100-ohm unshielded twisted pair (UTP).
- A 100-ohm shielded twisted pair (STP-A) or 62.5/125 micron multimode fiber.

The two outlets allow for both a data and a voice connection. Grounding needs to conform to applicable building codes.

The characteristics impedance is 100 ohms $\pm$ 15 percent from 1 MHz to the highest referenced frequency (16, 20, or 100 MHz) of a particular category. The recognized shielded twisted-pair cables are IBM Type 1A for backbone and horizontal distribution and IBM Type 6A for patch cables with 2-pair, 22-AWG solid with a characteristic impedance of 150 ohms $\pm$ 10 percent (3–300 MHz). The optical fiber medium for horizontal cabling is 62.5/125-micron multimode optical cable with a minimum of two fibers, and for backbone cabling it is 62.5/125-micron multimode and single-mode optical fiber.

3.4.2   Plenum

Most of the cables used contain halogens, chemicals that give off toxic fumes when they burn. It is these fumes that release acid gases that sear the eyes, nose, mouth, and throat; that cause a person to become disoriented and thus prevents the victim from escaping from a fire; and that cause severe respiratory damage and can kill.

In buildings, the **plenum** is the space above the ceiling and below the floor used to circulate air throughout the workplace. It can also be found in some walls. If there is a fire, it becomes evident that the plenum will also circulate the toxic fumes generated by the burning of cables.

While some international governments have already standardized on zero-halogen cabling, the U.S. National Electrical Code, which serves as the basis for local standards in most states, only forbids the use of halogen-sheathed cabling in the plenum spaces above ceilings and below floors in office buildings. Instead, a plenum cable is used.

A **plenum cable** is a special cable that gives off little or no toxic fumes when burned. Depending on the building's fire code, the building's wiring will include plenum cabling unless the wiring exists in a special sealed conduit that would prevent the toxic fumes from escaping into the open air. The major disadvantage of using a plenum cable is that it is significantly more expensive.

A **riser cable** is intended for use in the vertical shafts that run between floors. Many buildings have a series of equipment rooms that are placed vertically in a reinforced shaft for the purpose of enclosing power distribution equipment, HVAC units, telephone distribution, and other utility services used throughout the building. Cable placed in these shafts must not contribute to the spread of fire from floor to floor. Since the shafts are not plenums, the riser rating is less stringent than the plenum rating.

3.4.3   Wall Jacks

As the IT administrator, you may have to decide where to place the wall jacks that lead to the patch panel in the server room or wiring closet. If you plan out the location of the wall jacks while the building is being built, you will include jacks for any office computers (remember that some offices will be using two or more computers) that connect to the wall jacks directly and for any hubs (used to connect multiple computers using a single wall jack) that connect to the wall jacks directly. Also, you should add additional wall jacks for any printer

Figure 3.18   A 66 Punch-down Block

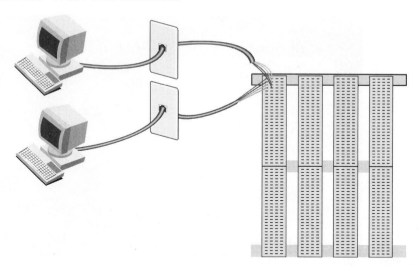

that needs to be connected to the network. Since the network cabling system is being built as the building is being built, it is best to plan for growth by adding extra jacks throughout the building.

3.4.4   Punch-Down Block

To help manage the hundreds of cables that may enter a wiring closet or server room, you would use a punch-down block or patch panel. A **punch-down block** is used to connect several cable runs to each other without going through a hub. Rather than using RJ-45 ports, the block attaches or is punched directly to the exposed wires by using a punch-down tool. Punch-down blocks are typically used for permanent wire connections. For example, if you have Category 5 cabling, you have four pairs, or eight wires. Each wire would then connect to a row of connectors on the punch-down block. You can then connect additional wire from other cables by using the same connectors in a row. (See figures 3.18 and 3.19.) To connect the wires to the punch-down block, you would use a **punch-down tool.**

The most common punch-down blocks are the 110 blocks (commonly referred to as insulation displacement connections, or IDCs) and 66 blocks. The 110 blocks are the termination of choice in Category 5 high-bandwidth environments, while the 66 blocks are mostly used with telephone systems.

3.4.5   Patch Panel

A **patch panels** fills the same purpose as a punch-down block. A patch panel is a panel with numerous RJ-45 ports. The wall jacks are connected on the back of the patch panel to the individual RJ-45 ports.

NOTE: The ports are labeled with the corresponding wall jack.

Figure 3.19   66 and 110 Punch-down Connectors

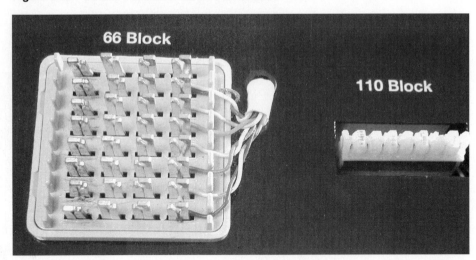

You can then use patch cables to connect the port in the front of the patch panel to a computer or a hub. As a result, you can connect multiple computers with a hub located in the wiring closet or server room. Compared with the punch-down block, the patch panel is easier to work with, allows for easier troubleshooting, makes temporary connections much easier to establish, and offers a much cleaner look. Therefore, patch panels are more common in LANs than punch-down blocks. (See figures 3.20 and 3.21.)

3.4.6   Cable Ties, Basket Systems, and Management Racks

Cable ties and basket systems are tools used to help manage the cables anywhere throughout the building. The **cable ties** are used to bundle cables traveling together and to pull the cables off the floor so they will not get trampled or run over by office furniture. While installing or laying cable, you should handle the cable with care. This includes not stepping on cables, not pinching tightly with the cable ties, and not making sharp bends or kinks in the cable.

Basket systems are baskets or trays used on the back of office furniture. They are used to hold and hide network cables without using cable ties. (See figure 3.22.)

3.4.7   Server Rooms

Within the server room and wiring closets, you will find that many cables will come together. While a single cable doesn't weigh much, many cables together add up quickly in weight, causing a lot of stress to the cables. For these situations, you should use some type of tray, conduit, trough, or ladder.

Figure 3.20   A Patch Panel (Front and Back)

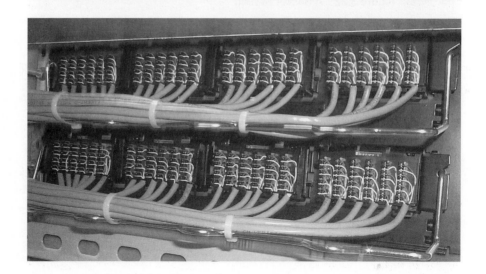

NOTE: Whether dealing with a small number of cables or a large number of cables, it is very important for you to always label both ends of the cables, especially those that are part of the backbone or horizontal wiring system. This will make your job easier in the future. For example, it will be invaluable in allowing you to quickly troubleshoot cable faults. In addition, it is very important that you document the cable system—that you have a blueprint showing where and how the cables are connected.

When you install cable, you must pay attention to the bend-radius limitations for the type of cable you are installing. **Bend radius** is the radius of the maximum arc into which you can loop a cable before you will impair data transmission. Generally, a cable

Figure 3.21   A Patch Panel Allows for Easy Connection of Computers

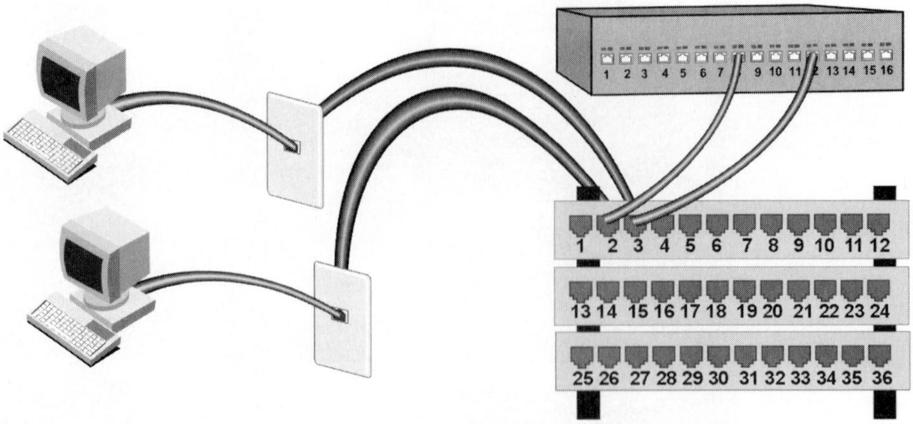

Figure 3.22   A Basket System

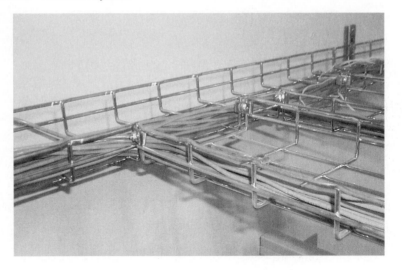

bend radius is less than four times the diameter of the cable. Therefore, be careful not to exceed it.

The server room is the work area of the IT department—it's where the servers and most of the communication devices reside. The room should be secure, with only a handful of people allowed to have access to it. In addition, the server should also be secure. Therefore, you should consider the following:

■ The computer should be locked when not in use.
■ You should always require user names and passwords.

- You should always log out when the server is not being used.
- You should restrict accounts so that they cannot log on directly (interactive) to the server.
- You should enable security monitoring such as auditing.

Besides securing the servers in the server room, you should also consider the following criteria. Since the computers and telecommunication systems can generate a lot of heat and a dry room is more susceptible to electrostatic discharge that can damage electronic components, you should consider separate environment controls for the server room. **Electrostatic discharge (ESD)** is electricity generated by friction such as when your arm slides on a tabletop or when you walk across a carpet. To reduce ESD, you should:

- consider an electrostatic discharge prevention program which would include using ESD wrist straps when opening computers and handling computer components, installing an anti-static flooring and/or by placing the servers on an anti-static mat.
- keep humidity low (less than 50%).

In addition, you should:

- for cleaner power, have the power outlets running directly to a breaker. You should also have the servers on their own circuit.
- install an uninterruptible power supply (UPS) to protect your system against power fluctuations including outages.
- keep your system cool (room temperature or less).
- always have a tape backup system or some other method to backup all important data files.

For the best results and longevity of the system, ensure that the server room is a clean environment. It is important that the server room have clean air. Most office environments are clean enough that the computer equipment only needs to be cleaned out annually or biannually as part of a regular preventive maintenance program.

However, if the server room is in an industrial environment or the air is dusty or full of pollutants, you should move the server to a better location. If that is not an option, at least maintain a regular cleaning schedule to keep it and the server room clear of air problems. One easy preventive measure is to use an air cleaner in the server room. You can also buy special cases and enclosures for server hardware designed for industrial environments to prevent damage from dirt and pollutants.

If you find that dust bunnies are continually causing the server to heat up and shut down, you should perform a weekly or at least monthly preventive maintenance schedule that includes cleaning the cooling system elements and the interior of the system case.

If you have several servers, you should also consider a **server rack** or **rackmount cabinet** to hold the servers. In addition, you can also purchase a switch box to connect a single keyboard, mouse, and monitor to several servers. This will allow for less equipment and for a more organized work environment.

With the growth of the Internet, there was a need to fit many servers into a small amount of space. Therefore, manufacturers developed server racks that follow the Network

Equipment Building Specification (NEBS), the same specification that is used in the telecommunications industry. The servers use the following form factors:

4u (7 inches high)
2u (3.5 inches high)
1u (1.75 inches high)

As you can imagine, when you stack these servers on top of one another, you can easily put a large number of servers in a small area. Of course, you must remember that when this many servers occupy a small area, they all require special electrical and cooling accommodations. The cabinet should have exhaust fans over each bay to pull the heat up and away from the servers and other devices, helping to maximize the cooling systems of this equipment. And of course, you must make sure that the ventilation fans are operating properly.

Rackmount cabinets can also be equipped with surge-suppressor power strips, but in most situations, you are much better off to install a UPS in the bottom of the rack and use it to power the devices mounted in the rack.

Lastly, when buying a server cabinet, you should buy a cabinet that has locking doors. These doors are a vital part of the server's security. In addition, you should avoid cabinets that use the same keys for the doors on every bay. And of course, for the lock to work, you must *keep* the cabinet locked.

A **server blade** is a single circuit board populated with components such as processors, memory, and network connections that are usually found on multiple boards. A server blade is designed to slide into an existing server (known as a **blade server**), much like an expansion slot is plugged into a PC. With server blades, many computers can be placed in a small area. As a result, server blades are more cost-efficient, are smaller, and consume less power than traditional box-based servers.

3.5 TESTING CABLE SYSTEMS

The best method for addressing a faulty cable installation is to avoid problems in the first place by purchasing high-quality components and installing them carefully. Of course, no matter how careful you are, problems do arise.

Troubleshooting your network's cable plant uses many of the same commonsense skills as any other troubleshooting task. You try to isolate the cause of the problem by asking questions like the following:

Has the cable ever worked properly?
When did the malfunctions start?
Do the malfunctions occur at specific times?
Is the cable brand new, or has it been used for a while?
What has changed since the cable functioned properly?

Once you have gathered all the information you can, the general troubleshooting process consists of steps such as the following:

1. Split the system into its logical elements.
2. Locate the element that is most likely the cause of the problem.

3. Test the element or install a substitute to verify it as the cause of the problem.

4. If the suspected element is not the cause, move on to the next likely element.

5. After locating the cause of the problem, repair or replace it.

Thus, you might begin troubleshooting by determining for sure that the cable is the source of the problem. You can do this by connecting different devices to both ends of the cable to see if the problem continues to occur. Once you verify that the cable is at fault, you can logically break it down into its component elements. For example, a typical cable run might consist of two patch cables, a wall plate, a punch-down block or patch panel, and the permanently installed cable itself.

In this type of installation, the easiest thing to do is test the patch cables, either by replacing them or by testing them with a cable scanner. Replacing components can be a good troubleshooting method, as long as you know that the replacements are good. The most accurate method is to test the individual components with some form of cable tester.

The types of problems that can occur with cables include:

- **Open**—A conductor with a break in it or wires that are unconnected, which prevents electricity to flow.
- **Short**—A circuit that has a zero or abnormally low-resistance path between two points, resulting in excessive current. In networking cables, a short is an unintentional connection made between two conductors (such as wires) or pins/contacts.
- **Crossed pair**—Two wires connected improperly, causing the two wires to be crossed.
- **Split pair**—Incorrect pinouts that cause data-carrying wires to be twisted together, resulting in additional crosstalk. Split pairs can be the result of mistakes during the installation. The solution is to reattach the connectors at both ends using either the T568-A or T568-B pinouts.
- **Excessive length**—Cables that are longer than the recommended maximum for the network protocol you plan to use.

  NOTE: Don't be overly concerned if the maximum length for a cable segment is 100 meters and you have a run that is 101 meters long. Most protocols have some leeway built into them that permits a little excess.

- **Excessive attenuation**—Excessive loss of signal typically caused by using an excessive length or by using a cable that is inferior or not designed for a particular network standard or speed. It can also be caused by faulty or substandard connectors, punch-down blocks, and patch panels.
- **Excessive NEXT**—Excessive near-end crosstalk typically caused by inferior cable, inferior components, improper patch cables, split pairs, loose twisting, or the sharing of cables for other signals such as voice communications.
- **Excessive noise**—Noise caused by AC power lines, light fixtures, electric motors, and other sources of EMI and RFI. Therefore, cabling should be routed away from these items.

Another problem that should be mentioned is a jabbering Ethernet network card. **Jabber** is when garbled bits of data are emitted within the frame sequence in a continuous transmission fashion. The packet length is usually more than 1,518 bytes. This can be identified by a protocol analyzer as a CRC error. As mentioned earlier, when nodes detect collisions they emit a normal JAM signal on the network segment to clear transmission. Sometimes certain nodes attempt to keep jamming the network due to excessive high collision rates; this also can be captured as high CRC or late collision error rate. The cause can be overloaded traffic levels. If the bandwidth-utilization levels are normal or low for the particular Ethernet segment, it is possible that the collision detection pair of a jamming node's NIC or transceiver cannot hear the network signal and may not know a collision has stopped. If this occurs, it continues to jam the network. If a certain node on an Ethernet segment emits a lot of jabber, the node's NIC and transceiver should be troubleshot through consecutive replacement and re-analysis. Note: If you use a protocol analyzer, you can capture the packets being transmitted and analyze them to determine which network interface is faulty.

As you will see, there are many tools and devices that can test cables. Some are simple and inexpensive, while others are elaborate and expensive. Some are easy to use as they supply easy-to-read pass/fail test results or simple numeric values, while others need to be analyzed. Of course, these tools can be used during installation to verify the installation and components used as well as in the future for troubleshooting problems.

3.5.1   Network Card Diagnostics

Today, most network cards come with diagnostic software that you can use to verify that the network card is functioning correctly. The tests that diagnostic software fall into two categories: hardware and software.

NOTE: If you can log into the network, generally speaking, the network card is functioning.

Hardware diagnostics examine the individual parts of the network cards and verify the function of each component. If there's a problem, the diagnostics will report it. Unfortunately, most diagnostics for UTP-based network cards cannot determine whether the network card is transmitting or receiving data successfully without a device known as a hardware loopback.

A hardware loopback is a special connector for Ethernet 10BaseT network cards. It functions similarly to a crossover cable, except that it connects the transmit pins directly to the receive pins (pin 1 connected to pin 3 and pin 2 connected to pin 6). It can then be used by the network cards software diagnostics to test transmission and reception capabilities. While the hardware loopback is no bigger than a single RJ-45 connection with a few small wires on the back, you just plug the loopback into the RJ-45 connector on the back of the network card, start the diagnostic software and start the diagnostic routine that requires the loopback. As a result, you will be able to tell if the network card can send and receive data.

Software diagnostics test the higher-level functions of the network card such as the network communication with out stations. These programs typically consist of a sender and receiver portion. Each portion is run on one of a pair of computers connected to the network. The sender sends a test packet out to the receiver, and when the receiver receives the packet, it immediately sends a response. While there may be similar tools available with protocol suites such as TCP/IP or IPX, these tests are protocol-independent.

3.5.2   Crossover Cable

As introduced earlier in the chapter, a crossover cable is a cable that can be used to connect two hubs. You can also use the crossover cable to test communications between two Ethernet stations without using a hub. By carrying a crossover cable in your tool bag, you can connect a portable computer directly to the server's network card using the cable. Assuming that you configured the network cards and logical addresses, you should be able to log in to the server.

3.5.3   Voltmeter and Ohmmeter

Since the computer is a sophisticated electronic device, a voltmeter and ohmmeter can be used to test certain aspects of the computer. A **voltmeter** can be used to see if a device is generating the correct voltage output or signal. An **ohmmeter** can be used to check wires and connectors and measure the resistance of an electronic device. A **digital multimeter (DMM)** combines several measuring devices including a voltmeter and an ohmmeter.

The ohmmeter can be used to test a wire. Since a wire or fuse is a conductor, essentially you should measure no resistance (0 ohms), showing there is no break in the wire or fuse. This is known as a **continuity check.**

3.5.4   Cable Testers

Before looking at cable testers, remember that most cables will only have two conductors, such as those found in a coaxial cable or a cable that contains several pairs of wires—for example, those found in unshielded twisted pair. Unfortunately, since the pinouts of the unshielded twisted pair vary depending on its implementation, it becomes more complicated to test the cables.

Of all the cable testers, the simplest type of tester performs continuity testing, which is designed to check a copper cable connection for basic installation problems, such as opens, shorts, and crossed pairs. A continuity tester consists of two separate units that you connect to each end of the cable to be tested. In many cases, the two units can snap together for storage and easy testing of patch cables. While these devices usually cannot detect more complicated twisted-pair wiring faults such as split pairs, they are sufficient for basic cable testing, especially for coaxial cables.

The more sophisticated form of cable tester is the wire map tester. A wire map tester is a device that transmits signals through each wire in a copper twisted-pair cable to determine if it is connected to the correct pin at the other end. Wire mapping is the most basic test for twisted-pair cables because the eight separate wire connections involved in each cable run are a common source of installation errors. Wire map testers detect transposed wires, opens, and shorts.

A wire map tester consists of a remote unit that you attach to the far end of a connection and the battery-operated, hand-held main unit that displays the results. Typically, the tester displays various codes to describe the type of faults it finds. In some cases, you can purchase a tester with multiple remote units that are numbered, so that one person can test several connections without constantly traveling back and forth from one end of the connection to the other to move the remote unit.

The one wiring fault that is not detectable by a dedicated wire map tester is split pairs, because even though the pinouts are incorrect, the cable is still wired straight through. To detect split pairs, you must use a device that tests the cable for the near-end crosstalk that split pairs cause.

3.5.5   Tone Generators

The simplest type of copper cable tester is also a two-piece unit and is called a tone generator and probe—also sometimes called a "fox and hound" wire tracer. This type of device consists of a unit you connect to a cable with a standard jack or an individual wire with alligator clips, which transmits a signal over the cable or wire. The other unit is a pen-like probe that emits an audible tone when touched to the other end of the cable or wire or even its insulating sheath.

This type of device is most often used to locate a specific connection in a punch-down block. For example, some installers prefer to run all the cables for a network to the central punch-down block without labeling them and then to use a tone generator to identify which block is connected to which wall plate and label the punch-down block accordingly. You can also use the device to identify a particular cable at any point between the two ends. Since the probe can detect the cable containing the tone signal through its sheath, you can locate one specific cable out of a bundle in a ceiling conduit or other type of raceway by connecting the tone generator to one end and touching the probe to each cable in the bundle until you hear the tone.

In addition, by testing the continuity of individual wires using alligator clips, you can use a tone generator and probe to locate opens, shorts, and miswires. An open wire will produce no tone at the other end, a short will produce a tone on two or more wires at the other end, and an improperly connected wire will produce a tone on the wrong pin at the other end.

This process is extremely time-consuming, however, and it's nearly as prone to errors as the cable installation itself. Either you have to continually travel from one end of the cable to the other to move the tone generator unit, or you have to use a partner to test each connection, keeping in close contact using phones, radios, or some other means of communication in order to avoid confusion. When you consider the time and effort involved, you will properly find that investing in a wire map tester is a more practical solution.

3.5.6   Time Domain Reflectometers

A **time domain reflectometer (TDR)** is the primary tool used to determine the length of a copper cable and to locate the impedance variations that are caused by opens, shorts, damaged cables, and interference with other systems. The TDR works much like radar, by transmitting a signal on a cable with the opposite end left open and measuring the amount of time it takes for the signal's reflection to return to the transmitter. When you have this elapsed-time measure, called the nominal velocity of propagation (NVP), and you know the speed at which electrons move through the cable, you can determine the length of the cable.

The NVP for a particular cable is usually provided by its manufacturer along with other specifications, measured as a percentage of light. Some manufacturers provide the NVP as a percentage, such as 75 percent, whereas others express it as a decimal value multiplied by the speed of light (c), such as 0.75c. Many cable testers compute the length internally, based on the results of the TDR test and an NVP value either that is preprogrammed or that you specify for the cable you are testing. Of course, since the NVP values for various cables can range from 60 to 90 percent, you can have a margin of error for the cable length results of up to 30 percent if you are using the wrong value.

There are two basic types of TDRs available, those that display their results as a waveform on an LCD or CRT screen and those that use a numeric readout to indicate the distance to a source of impedance. Of course, the numeric readout provides less detail but is easy to use and relatively inexpensive. Waveform TDRs are not often used for field testing these days, because they are much more expensive than the numeric type and require a great deal more expertise to use effectively.

You can use a TDR to test any kind of cable that uses metallic conductors, including the coaxial and twisted-pair cables used to construct LANs. A high-quality TDR can detect a large variety of cable faults; including open conductors; shorted conductors; loose connectors; sheath faults; water damage; crimped, cut, or smashed cables; and many other conditions. In addition, the TDR can measure the length of the cable and the distance to any of these faults.

3.5.7   Visual Fault Locators

The light that carries signals over fiber-optic cable is invisible to the naked eye, making it difficult, without a formal test, to ensure that installers have made the proper connections. A visual fault locator (sometimes called a cable tracer) is a quick and easy way to test the continuity of a fiber cable connection by sending visible light over a fiber-optic cable. A typical fault locator is essentially a flashlight that applies its LED or incandescent light source to one end of a cable, which is visible from the other end. This enables you to locate a specific cable out of a bundle and to ensure that a connection has been established.

More powerful units that use laser light sources can actually show points of high loss in the cable, such as breaks, kinks, and bad splices, as long as the cable sheath is not completely opaque. For example, the yellow- or orange-colored sheaths commonly used on a single-mode or multimode cable, respectively, usually admit enough of the light energy lost by major cable faults to make them detectable from the outside.

3.5.8   Fiber-Optic Power Meters

A fiber-optic power meter is a device that measures the intensity of the signal being transmitted over a fiber-optic cable. The meter is similar in principle to a multimeter that measures electric current, except that it works with light instead of electricity. The meter uses a solid-state detector to measure the signal intensity. There are different meters for different fiber-optic cables and applications, such as those used for short-wavelength systems and those for long-wavelength systems. More expensive units can measure both long- and short-wavelength signals.

In order to measure the strength of an optical signal, there must be a signal source at the other end of the cable. While you can use a fiber-optic power meter to measure the signal generated by your actual network equipment, accurately measuring the signal loss of a cable requires a consistent signal generated by a fiber-optic test source. Companion to the power meter in a fiber optic tool kit, the test source is also designed for use with a particular type of network. Sources typically use LEDs for multimode fiber or lasers for single-mode fiber to generate a signal at a specific wavelength, and you should choose a unit that simulates the type of signals used by your network equipment.

3.5.9   Optical Time Domain Reflectometers

An **optical time domain reflectometer (OTDR)** is the fiber-optic equivalent of the TDR that is used to test copper cables. The OTDR transmits a calibrated signal pulse over the cable to be tested and monitors the signal that returns to the unit. Instead of measuring signal reflections caused by electrical impedance as a TDR does, however, the OTDR measures the signal returned by backscatter, a phenomenon that affects all fiber-optic cables.

As with a TDR, the condition of the cable causes variances in the amount of backscatter returned to the OTDR, which is displayed on an LCD or CRT screen as a waveform. By interpreting the signal returned, it is possible to identify specific types of cable faults and other conditions. An OTDR can locate splices and connectors and measure their performance; identify stress problems caused by improper cable installation; and locate cable breaks, manufacturing faults, and other weaknesses. Knowing the speed of the pulse as it travels down the cable, the OTDR can also use the elapsed time between the pulse's transmission and reception to pinpoint the location of the specific condition of the cable.

You should not use OTDRs to measure a cable's signal loss or locate faults on LANs. To measure signal loss, you should use a power meter and light source, which are designed to simulate the actual conditions of the network. In addition, while it is possible to use a OTDR to compute a cable's length based on the backscatter, it is far less accurate in the short lengths found in LAN environment; it is used primarily on long distance connections, such as those used by telephone and cable television networks.

Another disadvantage of an OTDR is to be able to use an OTDR waveform requires a good deal of training and experience to interpret them. Lastly, full-featured OTDR units are quite expensive.

3.5.10   Fiber-Optic Inspection Microscopes

Fiber, used in fiber cabling is extremely thin, compared to any copper cable. Splicing and attaching connectors to fiber-optic cables require great precision. The best way to inspect cleaved fiber ends and polished connection ferrules is with a microscope.

Fiber-optic inspection microscopes are designed to hold cables and connectors in precisely the correct position for examination, enabling you to detect dirty, scratched, or cracked connectors and ensure that cables are cleaved in preparation for splicing. Good microscopes typically provide at least 100-power magnification, have a built-in light source (not a fiber-optic light source, but a source of illumination for the object under the scope), and are able to support various types of connectors using additional stages, which may or not be included.

3.5.11   Multifunction Cable Scanners

Multifunction cable scanners, sometimes called certification tools, are devices that are available for both copper and fiber-optic networks and perform a series of tests on a cable run, compare the results against either preprogrammed standards or parameters that you supply, and display the outcome as a series of pass or fail ratings. Most of these units perform the basic tests called for by the most commonly used standards, such as wire mapping, length, attenuation, and NEXT for copper cables, and optical power and signal loss for fiber-optic cables. Many of the copper cable scanners also go beyond the basics to perform a comprehensive battery of tests, including propagation delay, delay skew, various versions of crosstalk, and return loss.

The primary advantage of this type of device is that anyone can use it. You simply connect the unit to a cable, press a button, and read off the results after a few seconds. Many units can store the results of many individual tests in memory, download them to a PC, or output them directly to a printer.

3.6   WIRELESS LAN

The atmosphere can also be used as a medium to transport data over networks. For decades, radio and television stations have used the atmosphere to transport information via analog signals. The atmosphere is also capable of carrying digital signals. Networks that transmit signals through the atmosphere are known as wireless networks. Wireless networks typically use infrared or radio frequency signaling. (See table 3.9.) Wireless networks are suited to very specialized network environments that require mobility, long distances, or isolated locations.

NOTE: Radio and infrared are considered unbound media because they are not carried or bound with a physical cable.

A wireless LAN (WLAN) is a local area network without wires. WLANs have been around for more than a decade, but they are just beginning to gain momentum because of falling costs and improved standards. WLANs transfer data through the air using radio frequencies instead of cables. They can reach a radius of 500 feet indoors and 1000 feet outdoors, but antennas, transmitters, and other access devices can be used to widen that area. WLANs require a wired access point that plugs all the wireless devices into the wired network.

A wireless LAN system can provide LAN users with access to real-time information anywhere in their organization. Installing a wireless LAN system can be fast and easy and can eliminate the need to pull cable through walls and ceilings. In addition, wireless technology allows the network to go where wire cannot go. Thus, WLANs combined data connectivity with user mobility, and through simplified configuration, enable movable LANs.

Table 3.9   Wireless Technology

Media	Frequency Range	Cost	Ease of Installation	Capacity Range	Attenuation	Immunity from Interference and Signal Capture
Low-power single frequency	Entire RF; high GHz is most common	Moderate (depends on equipment)	Simple	<1 to 10 Mbps	High	Extremely low
High-power single frequency	Entire RF; high GHz is most common	Moderately expensive	Difficult	<1 to 10 Mbps	Low	Extremely low
Spread-spectrum radio	Entire RF; 902 to 928 MHz in United States; 2.4 GHz band is most common	Moderate (depends on equipment)	Simple to moderate	2 to 6 Mbps	High	Moderate
Terrestrial microwave	Low GHz; 4 to 6 or 21 to 23 is most common	Moderate to high (depends on equipment)	Difficult	<1 to 10 Mbps	Variable	Low
Satellite microwave	Low GHz; 11 to 14 GHz is most common	High	Extremely difficult	<1 to 10 Mbps	Variable	Low
Point-to-point infrared	100 GHz to 1000 THz	Low to moderate	Moderate to difficult	<1 to 16 Mbps	Variable	Moderate
Broadcast infrared	100 GHz to 1000 THz	Low	Simple	≤1 Mbp	High	Low

3.6.1   Radio Systems

Radio frequency (RF) resides between 10 KHz and 1 GHz on the electromagnetic spectrum. It includes shortwave radio, very high frequency (VHF), and ultrahigh frequency (UHF). Radio frequencies have been divided between regulated and unregulated bandwidths. Users of the regulated frequencies must get a license from the regulatory bodies that have jurisdiction over the desired operating area (the FCC in the United States and the CDC in Canada). While the licensing process can be difficult, licensed frequencies typically guarantee clear transmission within a specific area.

In unregulated bands, you must operate at regulated power levels (under 1 watt in the United States) to minimize interference with other signals. Of course, if a device broadcasts using less power, the effective area will be smaller. The Federal Communications Commission (FCC) allocated the following bands for unregulated broadcast: 902–928 MHz, 2400–2483.5 MHz, and 5752.5–5850 MHz. While you don't require a license from the FCC to use these frequencies, you must meet FCC regulations, which include power limits and the minimizing of interference (an antenna gain of 6 dBi and 1 watt of radiated power).

In radio network transmissions, a signal is transmitted in one or multiple directions, depending on the type of antenna used. The wave is very short in length, with a low transmission strength (unless the transmission operator has a special license for a high-wattage transmission), which means it is best suited to short-range line-of-sight transmissions. A line-of-sight transmission is one in which the signal goes from point to point, rather than bouncing off the atmosphere over great distances. Of course, a limitation of line-of-sight transmissions is that they are interrupted by land masses such as mountains.

A **band** is a contiguous group of frequencies that are used for a single purpose. Commercial radio stations often refer to the band of frequencies they are using as a single frequency. However, typical radio transmissions actually cover a range of frequencies and wavelengths. Because most tuning equipment is designed to address the entire bandwidth at the kilohertz or megahertz level, the distinction between one frequency and a band is often overlooked.

A **narrowband radio system** transmits and receives user information on a specific radio frequency. Narrowband radio keeps the radio signal frequency as narrow as possible—just wide enough to pass the information. Undesirable crosstalk between communications channels is avoided by carefully coordinating different users on different channel frequencies. In a radio system, privacy and noninterference are accomplished by the use of separate radio frequencies. The radio receiver filters out all radio signals except the ones on its designated frequency. Depending on the power and frequency of the radio signal, the range could be a room, an entire building, or long distances. A low-power (1–10 watts), single-frequency signal has a data capacity in the range of 1–10 Mbps.

Spread-spectrum signals are distributed over a wide range of frequencies and then collected onto their original frequency at the receiver. Different from narrowband signals, spread-spectrum signals uses wider bands, which transmit at a much lower spectral power density (measured in watts per hertz). Unless the receiver is not tuned to the right frequency or frequencies, a spread-spectrum signal resembles noise, making the signals harder to detect and harder to jam. As an additional bonus, spread-spectrum and narrowband signals can occupy the same band, with little or no interference. There are two types of spread spectrum: direct sequence and frequency hopping. Spread-spectrum frequency ranges are very high, in the 902–928-MHz range and higher and typically send data at a rate of 2–6 Mbps.

Direct-sequence spread spectrum (DSSS) generates a redundant bit pattern for each bit to be transmitted. This bit pattern is called a chip (or chipping code). The intended receiver knows which specific frequencies are valid and deciphers the signal by collecting valid signals and ignoring the spurious signals. The valid signals are then used to reassemble the data. Because multiple subsets can be used within any frequency range, direct-sequence signals can coexist with other signals. Although direct-sequence signals can be intercepted almost as easily as other RF signals, eavesdropping is ineffective because it is quite difficult just to determine which specific frequencies make up the bit pattern, not to mention how difficult it is to retrieve the bit pattern and interpret the signal. Because of modern error detection and correction methods, the longer the chip, the greater the probability that the original data can be recovered even if one or more bits in the chip are damaged during transmission.

Frequency hopping quickly switches between predetermined frequencies, many times each second. Both the transmitter and receiver must follow the same pattern and maintain complex timing intervals to be able to receive and interpret the data being sent. Similar to how hard it is to intercept data in the direct-sequence spread spectrum, intercepting the data being sent is extremely difficult unless the eavesdroppers know the signals to monitor and the timing pattern. In addition, dummy signals can be added to increase security and confuse eavesdroppers. The length of time that the transmitter remains on a given frequency is known as the dwell time.

3.6.2   Microwave Signals

Microwave communication is a form of electromagnetic energy that operates at a higher frequency (low-GHz frequency range) than radio wave communication. Since it provides higher bandwidths than those available using radio waves, it is currently one of the most popular long distance transmission technologies.

Microwave signals propagate in straight lines and are affected very little by the lower atmosphere. In addition, they are not refracted or reflected by ionized regions in the upper atmosphere. The attenuation of microwave systems is highly dependent on atmospheric conditions; for example, both rain and fog can reduce the maximum distance possible. Higher-frequency systems are usually affected most by such conditions. The systems are not particularly resistant to EMI, and protection for eavesdropping can only be achieved by encryption techniques.

There are two types of microwave systems:

■ Terrestrial systems are often used where cabling is difficult or the cost is prohibitively expensive. Relay towers are used to provide an unobstructed path over an extended distance. These line-of-sight systems use unidirectional parabolic dishes that must be aligned carefully.

■ Satellite systems provide far bigger areas of coverage than can be achieved using other technologies. The microwave dishes are aligned to geostationary satellites that can relay signals between sites either directly or via another satellite. The huge distances covered by the signal result in propagation delays of up to 5 seconds. The costs of launching and maintaining a satellite are enormous, and consequently customers usually lease the services from a provider.

The cost of these systems is relatively high, and technical expertise is required to install them, as accurate alignment is required. Often, this service is leased from a service provider, which reduces installation costs and provides access to the required expertise. However, this service can prove expensive in the long term.

3.6.3 Infrared Systems

Another form of wireless technology is an **infrared (IR) system,** which is based on infrared light (light that is just below the visible light in the electromagnetic spectrum). Similar to your TV or VCR remote controls, infrared links use light-emitting diodes (LEDs) or injection laser diodes (ILDs) to transmit signals and photodiodes to receive signals. An IR system transmits in light frequency ranges of 100 GHz–1000 THz. Unfortunately, IR only transmits up to 1 Mbps for omnidirectional communications and 16 Mbps for directional communications.

Since IR is essentially light, it cannot penetrate opaque objects. Infrared devices work using either directed or diffused technology. Directed IR uses line-of-sight or point-to-point technology. **Diffused** (also known as **reflective** or **indirect**) **IR** technology spreads the light over an area to create a cell, limited to individual rooms. Since infrared light can bounce off walls, ceilings, and any other objects in the path, indirect infrared is not confined to a specific pathway. Unfortunately, since it is not confined to a specific pathway, the transmission of data is not very secure. Lastly, since infrared signals are not capable of penetrating walls or other opaque objects and are diluted by strong light sources, infrared is most useful in small or open indoor environments.

3.6.4 Wireless LAN Configurations

WLAN configurations vary from simple, independent, peer-to-peer connections between a set of PCs, to more complex, intrabuilding infrastructure networks. There are also point-to-point and point-to-multipoint wireless solutions. A point-to-point solution is used to bridge between two local area networks and to provide an alternative to cable between two geographically distant locations (up to 30 miles). Point-to-multipoint solutions connect several separate locations to one single location or building. Both point-to-point and point-to-multipoint solutions can be based on the 802.11b standard or on more costly infrared-based solutions that can provide throughput rates up to 622 Mbps (OC-12 speed).

3.7   NETWORK CARDS

If you haven't figured it by now, **network interface cards** are media-dependent, media-access-method–dependent, and protocol-independent. This means that all the network cards on the network must use the same type of cable and must gain access to the cable the same way. In addition, it means that it does not matter what protocol you use. In fact, you can even use several protocols at the same time.

For example, suppose you are installing a network and you decide to start by installing a Novell NetWare 5.1 server. In this server, you choose an NE2000 network card (a popular ISA network card) connected to a twisted-pair cable running at 10 megabits running a protocol called IPX. If you install a second server, such as a Windows 2000 server, you must choose a network card that also connects to a twisted-pair

Figure 3.23   A Typical Network Card with a 10Base2/BNC and 10BaseT/RJ-45
Connection

cable that supports 10-megabits speed. Therefore, you could choose an Intel E100B
card (one of today's popular PCI network cards), which supports both 10 megabits- and
100-megabits speed. The Windows 2000 server is using TCP/IP. While the Novell
server can communicate with any computer on the network using IPX and the Windows 2000 server can communicate with any computer on the network using TCP/IP,
the two servers cannot communicate with each other even though they are using the
same cabling system.

3.7.1   Characteristics of a Network Card

When selecting a network card, you need to first look at the type of network you are connecting to. (See figure 3.23.) For example, is the network Ethernet, Token Ring, or ARC-
net, or are you connecting your computer using a WAN connection? Next, you would have
to look at the speed of the network. Are you going to be transmitting at 4 Mbps, 10 Mbps,
16 Mbps, 100 Mbps, or 1 Gbps? In addition, you must make sure that the network card
has the proper network connector. If you are using unshielded twisted pair (UTP), you
would most likely use an RJ-45 connector. If you are connecting to a 10Base2 coaxial cable, you will most likely use a BNC T-connector. Or if you are connecting to a 10Base5
coaxial cable, you would connect to an external transceiver using an **AUI (adapter unit
interface)** connector, a female 15-pin D-connector. The AUI connector is also known as
DIX (Digital-Intel-Xerox) connector.

Next, you should select a card based on the type of expansion bus (ISA, MCA, EISA,
VESA, PCI, PCI-X, PC-Card, or CardBus) available in your computer. Since the ISA slot
can transfer up to 8.33 megabytes per second and the PCI card can transfer 264 megabytes
per second, the PCI slot is needed for today's high-speed networks.

NOTE: If you are using a network card for a Windows NT machine, you should make sure
the card appears on the Windows NT Compatibility List (HCL).

After you have selected the basic characteristics of a network card, you can then choose a network card depending on the card's advanced features, such as how much processing the network card will do so that it can alleviate some of the work that the microprocessor has to do.

In addition, some cards offer various forms of fault tolerance, such as using two cards on the same network link. Therefore, if one network card or connection fails, the other card will still transmit and receive packets. Lastly, some of these cards offer port aggregation, such as asymmetric port aggregation and symmetric port aggregation. Asymmetric port aggregation is the capability to evenly distribute outbound LAN traffic among multiple NICs while inbound traffic passes through a single NIC. Symmetric port aggregation is similar to asymmetric port aggregation except that it combines the bandwidth of multiple NICs for both incoming and outgoing traffic to provide a single link of up to 800 Mbps.

3.7.2   MAC Address

A **media access control (MAC) address** is a hardware address identifying a node on the network, much like a street address identifies a house or building within a city. And much like a street address within a city, you cannot have two network cards or nodes with the same MAC address on the same network. You can, however, have the same MAC address on two separate networks. For Ethernet and Token Ring cards, the MAC address is embedded onto the ROM chip on the card.

The MAC address itself is a 6 byte (12-digit hexadecimal number). Each byte/two-digits is separated by colons. An example of a MAC address would be:

03:34:A4:B3:74:2D

NOTE: The broadcast address to all MAC addresses is FF:FF:FF:FF:FF:FF.

3.7.3   Installing a Network Card

To install a network card is not much different from installing other expansion cards. First, you must configure the card's I/O address, IRQ, DMA, and memory addresses and any options that are specific to the type of network card you selected. This would be done with jumpers, DIP switches, configuration software, or plug and play. After the card is physically installed and the network cable is attached, you must then load the appropriate driver (it will be included with the operating system, or it will be on a disk that comes with the network card, or it will have to be downloaded from the Internet) and client software. Depending on the type of card, the driver may need to have the hardware resources specified (I/O address, IRQ, DMA, and memory address). Of course, this needs to match the settings of the cards. Lastly, you should make sure that the LED for the expansion card is active.

In addition, you must **bind** the protocol stack to the network card and install any additional network software needed. If you are going to bind multiple protocols to a network card interface, for the best performance the most used protocol on the network should be listed first on the binding list. To reduce unnecessary packet traffic and to improve bandwidth for the protocols that are being used, you should unbind protocols for a NIC that are not being used.

NDIS (network device interface specification, developed by Microsoft) and **ODI** (open data link interface, developed by Apple and Novell) are sets of standards for protocols and network card drivers to communicate. Both allow multiple protocols to use a single network adapter card.

3.7.4   Diskless Workstations

A **diskless workstation** is a computer that does not have its own disk drive. Instead the computer stores files on a network file server. Diskless workstations can reduce the overall cost of a LAN because one large-capacity disk drive at the server is usually less expensive than several low-capacity drives at the workstations. In addition, workstations can provide an extra level of security because a user can't infect the network by inserting a virus-infected disk, nor can a user use a hackers/network snooper disk to bypass the network security, nor can a user save confidential/secure information to a floppy disk. The disadvantage of using a diskless workstation is that if the network is down, the diskless workstation does not function.

To enable a diskless workstation to connect to the network and to boot an operating system, you must have a network card that has a boot PROM chip that is inserted into a socket located on the network card. A PROM, short for programmable ROM, is a memory chip on which data can be written only once. Once a program has been written onto a PROM, it remains there forever. Unlike RAM, PROMs retain their contents when the computer is turned off. The type of PROM chip used today on the network cards is the Electrical Erasable PROM, or EEPROM.

The PROM chip includes enough instructions to attach the network and to read the necessary boot files from the network. For a TCP/IP network, the Bootstrap Protocol (BOOTP) enables a diskless workstation to discover its own IP address, the IP address of a BOOTP server on the network, and a file to be loaded into memory to boot the machine.

The Net PC is a special type of diskless computer; it was designed cooperatively by Microsoft and Intel. Different from most other diskless workstations, a Net PC has a small hard disk that is used for storing temporary cache to improve performance rather than for permanently storing data. Since the Net PC is designed to be an inexpensive computer and to discourage users from configuring the machines themselves, configuration and management of a Net PC is performed through a network server and a Microsoft's Zero Administration Windows (ZAW) system.

3.8   HUBS

As mentioned in chapter 2, a **hub,** which works at the physical OSI layer, is a multiported connection point used to connect network devices via a cable segment. When a PC needs to communicate with another computer, it sends a data packet to the port, which the device is connected to. When the packet arrives at the port, it is forwarded or copied to the other ports so that all network devices can see the packet. In this way, all the stations "see" every packet just as they do on a bus network. Of course, a standard hub is not very efficient on networks with heavy traffic since it causes a lot of collisions and retransmitted packets. (See figure 3.24.)

Figure 3.24   A Hub

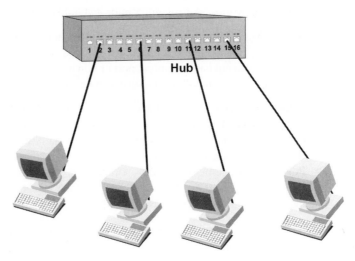

3.8.1   Types of Hubs

Hubs can be categorized as passive hubs or active hubs. A passive hub serves as a simple multiple connection point—it does not act as a repeater for the signal. An active hub, which always requires a power source, acts as a multiported repeater for the signal.

NOTE: When installing an active hub, if the hub has a fan (which provides cooling), make sure the fan is operating.

The most advanced hub is called the intelligent hub (also known as a manageable hub). An intelligent hub includes features that enable an administrator to monitor the traffic passing through the hub and to configure each port in the hub. For example, you can prevent certain computers from communicating with other computers, or you can stop certain types of packets from being forwarded. In addition, you can gather information on a variety of network parameters, such as how many packets pass through the hub and each of its ports, what types of packets they are, whether the packets contain errors, and how many collisions have occurred.

NOTE: All intelligent hubs are active hubs.

3.8.2   Parts of the Hub

The basic elements of a hub are:

- **Ports**—Receptacles where patch cables connect workstations or other devices to the hub. The number of ports (typically RJ-45 ports) on a hub generally ranges from 4 to 24, not including the uplink port.

■ **Uplink port**—The receptacle used to connect one hub to another hub. While an uplink port may look like any other port, it should only be used to interconnect hubs.

NOTE: Many hubs may designate one of the normal ports as the uplink port—you must use a push button to change between a normal port and an uplink port. In addition, if you do not have an uplink port or you need to connect additional hubs through the other ports, you will need to use a crossover cable.

■ **Backbone port**—The receptacle used to connect a hub to the network's backbone. For the 10BaseT networks, this type of connection is often made with short lengths of thinnet coaxial cabling. Today, hubs will usually use a fast RJ-45 connection or fiber optics for the backbone.

■ **Port for management console**—A receptacle used to connect a console (a desktop or portable PC) that enables you to view the hub's management information, such as the traffic load or number of collisions.

NOTE: Since not all hubs are intelligent hubs, not all hubs offer ports for the management consoles.

■ **Link LED**—The light on a port that indicates whether it is in use. If a connection is live, this light should be solid green. If you think you have everything connected properly and the connection should be live but the light is not on, you need to check connections and transmission speed settings for the hub and network device. In addition, you must make sure that the hub has power and that the network card has the correct driver loaded.

■ **Traffic (transmit or receive) LED**—The light on a port that indicates that traffic is passing through the port. Under normal data traffic situations, this light should blink. If your hub has traffic LEDs, they will normally be found adjacent to the link LEDs beside each data port.

■ **Power supply**—The device that provides power to the hub. Active hubs typically have a power-on light. If the power-on light is not lit, the hub has lost power.

■ **Ventilation fan**—Used to cool the device's internal circuitry. To function properly, hubs must cool their circuitry with a ventilation fan. When installing or moving hubs, you need to be careful not to block or cover the air-intake vents.

3.8.3   Installing a Hub

Hubs can be broken down into several types. Stand-alone hubs are hubs that serve a workgroup of computers that are isolated from the rest of the network. They may be connected to another hub by a coaxial, fiber-optic, or twisted-pair cable. Stand-alone hubs are best suited to small, independent departments, home offices, or test lab environments. Stand-alone hubs do not contain a standard number of ports, but usually contain 4, 8, 12, or 24 ports. A small, stand-alone hub that contains 4 ports may be called a hubby, a hublet, or a mini-hub.

Stackable hubs resemble stand-alone hubs, but they are physically designed to be linked with other hubs in a single telecommunications closet. Stackable hubs linked together logically represent one large hub to the network. Although many stackable hubs use Ethernet uplink ports, some use a proprietary high-speed cabling system to link the hubs together for better interhub performance.

Modular hubs provide a number of interface options within one chassis, making them more flexible than either stackable or stand-alone hubs. Similar to a PC, a modular hub contains a system board and slots into which you can insert different adapters that contain various types of ports.

To install a hub, you must connect the plug and hub in and then turn it on. Next, make sure the hub's power light goes on. Most hubs will perform self-tests when turned on. Blinking lights will indicate that these tests are in progress. When the tests are completed, attach the hub to the network by connecting a patch cable from it to the backbone or to a switch or router. Then connect patch cables from the computer or patch panel. Once the computer connects to the network through the newly installed hub, verify that the link and traffic lights act as they should. If you are installing a stackable hub or a rackmounted hub, you will need to use the screws and clamps that came with the hub to secure it to the rack or connect it to the other hub. If it is an intelligent hub, you will need to configure the hub using the software that is built into the hub's firmware. For example, you may need to assign an IP address to the hub. Of course, you should always check the documentation that comes with the hub before you start.

3.9   BRIDGES

A bridge, which is a layer 2 device, is used to break larger network segments into smaller network segments. It works much like a repeater, but because a bridge works solely with the layer 2 protocols and layer 2 MAC sublayer addresses, it operates at the data link layer.

Some bridges can also connect dissimilar network types (for example, Token Ring and Ethernet) as long as the bridge operates at the LLC sublayer of the data link layer. If the bridge operates only at the lower sublayer (the MAC sublayer), the bridge can connect only similar network types (Token Ring to Token Ring and Ethernet to Ethernet).

Several kinds of bridging have proved important as internetworking devices. Transparent bridging is found primarily in Ethernet environments, while source-route bridging occurs primarily in Token Ring environments. Translational bridging provides translation between the formats and transit principles of different media types (usually Ethernet and Token Ring). Finally, source-route transparent bridging combines the algorithms of transparent bridging and source-route bridging to enable communication in mixed Ethernet/ Token Ring environments.

There are two kinds of bridges, basic and learning (also called transparent).

3.9.1   Basic Bridges

A basic bridge is used to interconnect LANs using one (or more) of the IEEE 802 standards. Packets received on one port may be retransmitted on another port. Unlike a repeater, a bridge will not start retransmission until it has received the complete packet. As a consequence, stations on both sides of a bridge can transmit simultaneously without causing collisions. Bridges, like repeaters, do not modify the contents of a packet in any way, and they forward all broadcasts.

3.9.2   Transparent Bridge

A transparent (learning) bridge, the type used in Ethernet and documented in IEEE 802.1, is based on the concept of a spanning tree. Transparent bridging is called transparent because the end-point devices do not need to know that the bridge(s) exists. In other words, the computers attached to the LAN do not behave any differently in the presence or absence of transparent bridges.

The transparent bridge builds up its routing table by cataloging the network nodes that send out messages. A bridge examines the MAC address of a message's source or sending node. If this address is new to the bridge, it adds it to the routing table along with the network segment from which it originated. The bridge's routing table is stored in its RAM, and just like a PC's RAM, it is dynamic. Therefore, if the power goes off, the routing table goes away. When the power is restored, the bridge rebuilds the table. Because most network nodes send and receive packets continuously, it doesn't take long to completely rebuild the routing table.

The bridge operates with its Ethernet interfaces in promiscuous mode where it receives all frames sent on the network segments. The bridge "learns" by looking at the source and destination addresses of the frames sent on each network segment and builds a forwarding database (FDB) from the addresses as follows:

- which port the frame came from (source)
- which port to send the frame to (destination)

The bridge then applies three simple rules to an incoming frame:

- If its source and destination addresses are the same, ignore the frame.
- Find the destination address of the incoming frame in the FDB and send the frame to the port listed in the FDB for that destination address.
- If the destination address is not in the FDB (it is unknown), send the frame to all ports (flood) except the port it came in on.

NOTE: Broadcasts and multicast frames are forwarded by a bridge.

Several switching methods can be used for bridges and switches. They are:

- Store-and-forward
- Cut-through
- FragmentFree

A store-and-forward operation is typical in transparent bridging devices and is Cisco's primary LAN switching method. When in store-and-forward, the bridge or LAN switch copies the entire frame onto its onboard buffers and then computes the cyclic redundancy check (CRC). Because an entire frame is received before being forwarded, additional latency is introduced (as compared with a single LAN segment). The frame is discarded if it contains a CRC error, if it's too small (known as a runt—less than 64 bytes including the CRC), or if it's too large (known as a giant—more than 1518 bytes including the CRC).

If the frame doesn't contain any errors, the LAN switch looks up the destination hardware address in its forwarding or switching table to find the correct outgoing interface. If a given address has not been heard from in a specified period of time, then the address is deleted from the address table. Lastly, to create a loop-free environment with other bridges, use the Spanning Tree Protocol.

With cut-through processing, the first bits of the frame are sent out the outbound port before the last bit of the incoming frame is received, instead of waiting for the entire frame to be received. In other words, as soon as the switching port receives enough of the frame to see the destination MAC address, the frame is transmitted out the appropriate outgoing port to the destination device. This reduces the latency as compared with store-and-forward and FragmentFree processing. The unfortunate side effect is that because the frame check sequence (FCS) is in the Ethernet trailer, the forwarded frame may have bit errors that the switch would have noticed with store-and-forward logic. Of course, if the outbound port is busy, the switch will store the frame until the port is available. Very few switches use cut-through processing.

FragmentFree processing performs like cut-through, but the switch waits for 64 bytes to be received before forwarding the first bytes of the outgoing frame. According to Ethernet specifications, collisions should be detected during the first 64 bytes of the frame; frames in error due to collision will not be forwarded. Like cut-through processing, the FCS still cannot be checked.

3.9.3   Spanning Tree Algorithm

Complex topology can cause multiple loops to occur. For example, to make your network more fault-tolerant, you may have multiple bridge (or switch) connections from one segment to another. When a computer sends a message and the bridge does not know the destination computer, it will forward the messages to all the ports except the port the message came from. The only problem is when the message gets to the other segment and the bridge does not know where the computer is, it will forward the message to all the ports except the most recent port that the message came from—including the other connections to the other segment. So in reality, the message looped back to the other segment and will keep on looping. Since layer 2 has no mechanism to stop the loop, the Spanning Tree Protocol was created. (See figure 3.25.)

The **spanning tree algorithm (STA)** was developed by Digital Equipment Corporation and was revised by the IEEE 802 committee and published in the IEEE 802.1d specification.

NOTE: The Digital Equipment algorithm and the IEEE 802.1d algorithm are not compatible.

The STA designates a loop-free subset of the network's topology by placing those bridge ports that, if active, would create loops into a standby (blocking condition) mode. Blocking bridge ports can be activated in the event of the primary link failure, providing a new path through the internetwork. (See figure 3.26.)

3.9.4   Source-Route Bridging

Token Ring networks use source-route bridging (SRB). In this bridging method, the responsibility of determining the path to the destination node is placed on the sending node, not on the bridge. In an SRB environment, these steps are taken:

Figure 3.25   A Problem with Complex Topology is that a Nonstopping Loop Can Occur for Packets Destined for Unknown Computers

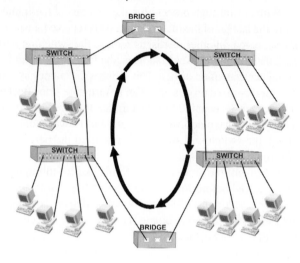

Figure 3.26   The Spanning Tree Algorithm Prevents Looping of Packets Destined to Unknown Computers

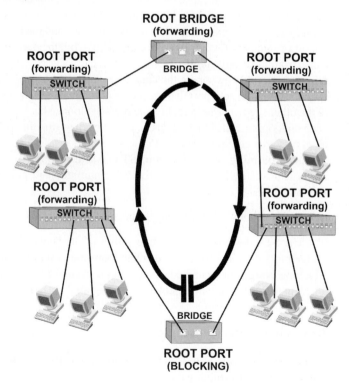

1. Token Ring devices send out a test frame to determine whether the destination node is on the local ring.
2. If no answer is forthcoming, which means the destination node is not on the local ring, the sending node sends out a broadcast message, which is called an explorer frame.
3. The bridge forwards the explorer frame across the network through the network's bridges. Each bridge adds its ring number and bridge number to the frame's routing information field (RIF), so it can retrace its route later.
4. The destination device, if it exists, receives and responds to the explorer frame. The sending node gets this response.
5. The sending node initiates communications between the two devices, with each intermediate bridge using the RIF value to determine the path between the two nodes.

Because SRB uses RIF information to determine its routes, no bridging table is created.

3.10   SWITCHES

In the most basic type of network found today, nodes are simply connected together using hubs. As a network grows, there are some potential problems with this configuration:

- **Scalability**—In a hub network, limited shared bandwidth makes it difficult to accommodate significant growth without sacrificing performance.
- **Latency**—This refers to the amount of time it takes a packet to get to its destination. Since each node in a hub-based network has to wait for an opportunity to transmit in order to avoid collisions, latency can increase significantly as you add more order to avoid collisions or as you add more nodes. Also, if someone is transmitting a large file across the network, all the other nodes will be waiting for an opportunity to send their own packets.
- **Network failure**—In a typical network, one device on a hub can cause problems for other devices attached to the hub due to wrong speed settings (100 Mbps on a 10-Mbps hub) or excessive broadcasts. Switches can be configured to limit broadcast levels.
- **Collisions**—A network with a large number of nodes on the same segment will often have a lot of collisions and therefore a large collision domain.

A **switching hub** (sometimes referred to as a switch or a layer 2 switch) is a fast, multiported bridge. Like a transparent bridge, a switch builds a table of the MAC addresses of all the connected stations. It then reads the destination address of each packet and forwards the packet to the correct port.

Switches can be used to segment a large network into several smaller, less congested segments (breaking up collision domains) while providing necessary interconnectivity between them. Switches increase network performance by providing each port with dedicated bandwidth, without requiring users to change any existing equipment, such as NICs, hubs, wiring, or any routers or bridges that are currently in place.

NOTE: Today, since an unmanaged Ethernet switch is so inexpensive, it does not make sense to purchase hubs.

A major advantage of using a switch is that it allows one computer (or LAN segment) to open a connection to another computer (or LAN segment). While those two computers communicate, other connections between the other computers (or LAN segments) can be opened at the same time. Therefore, several computers can communicate at the same time through the switching hub. In addition, if the Ethernet connections that connect to the switch are full duplex, a computer can send data to one computer and receive data from another computer at the same time.

NOTE: Remember that to transmit and receive simultaneously, full-duplex Ethernet requires a switch port, not a hub. Of course, this sets up a point-to-point connection, which also eliminates collisions.

LAN switches rely on packet switching. The switch establishes a connection between two segments just long enough to send the current packet. Incoming packets are saved to a temporary memory area known as a buffer, and the MAC address contained in the frame's header is read and then compared with a list of addresses maintained in the switch's lookup table. In an Ethernet-based LAN, an Ethernet frame contains a normal packet as the payload of the frame with a special header that includes the MAC address information for the source and destination of the packet.

While a switch works much like a bridge does, there are some important differences you should always keep in mind:

- Layer 2 switching is hardware-based, which means it uses the MAC addresses from the host's NIC cards to filter the network. Unlike bridges that use software to create and manage a filter table, switches use application-specific integrated circuits (ASICs) to build and maintain their filter tables. As a result, switches are faster than bridges.
- Bridges can only have one spanning tree instance per bridge, while switches can have many.
- Bridges can have up to 16 ports maximum. A switch can have hundreds.

Unfortunately, since the LAN switches review the MAC addresses of a frame, there is a small amount of latency added before the frame can be forwarded to the correct port. Therefore, one disadvantage of a switch would be the latency that is added.

3.11   VLANs

As networks have grown in size and complexity, many companies have turned to virtual local area networks (VLANs) to provide some way for structuring this growth logically. Basically, a VLAN is a collection of nodes grouped together in a single broadcast domain that is based on something other than physical location. A VLAN is a switched network that is logically segmented on an organizational basis, by functions, project teams, or applications, rather than on a physical or geographical basis. In a VLAN, member hosts can communicate as if they were attached to the same wire, when in fact they can be located on any number of physical LANs. Because VLANs form broadcast domains, members enjoy the connectivity, shared services, and security associated with physical LANs. Reconfiguration of the network can be done through software rather than by physically un-plugging and moving devices or wires.

In a traditional network with a hub connected to a router, you might often have your network divided by function. For example, let's say that the second floor of your building contains all the salespeople and their computers and the third floor contains all the marketing people and their computers. The computers on the second floor are connected to a hub (or hubs), which is connected to a router, and the computers on the third floor are connected to different hubs, which are also connected to the router. Suppose you run out of space on the second floor, but you need to add another salesperson. Unfortunately, if you connect the new salesperson's computer into the hubs used by the marketing people on the third floor, the new salesperson has to go through a slower router to get to the resources that have been assigned to the second floor staff, and the system will see all the broadcasts meant for the marketing people. Therefore, as you can see, this can be a security issue and a performance issue.

Some older applications have been rewritten to reduce their bandwidth needs. But today there is a new generation of multimedia applications that consume more bandwidth than ever before, consuming all it can find. These applications use broadcast and multicasts extensively. In addition, faulty equipment, inadequate segmentation, and poorly designed firewalls only serve to compound the problem that these broadcast-intensive applications create. All this has added a new dimension to network design. Making sure the network is properly segmented in order to isolate one segment's problems and keep those problems from propagating throughout the internetwork is imperative. The most effective way of doing this is through strategic switching and routing.

A broadcast domain is a set of NICs for which a broadcast frame sent by one NIC will be received by all other NICs in the broadcast domain. While a bridge and switch isolate collision domains on the same subnet, a VLAN (defined using a switch) will isolate broadcast domains.

NOTE: Since routers typically do not allow broadcasts, they also define broadcast domains. See figures 3.27 and 3.28.

Just as switches isolate collision domains for attached hosts and only forward appropriate traffic out a particular port, VLANs provide complete isolation between VLANs. A VLAN is a bridging domain, and all broadcast and multicast traffic is contained within it. Therefore, a VLAN can be thought of as a broadcast domain that exists within a defined set of switches. Since switches have become more cost-effective lately, many companies are replacing their flat hub networks with a pure switched network and VLANs environment. (See figure 3.29.)

VLANs also improve security by isolating groups. High-security users can be grouped into a VLAN, possibly on the same physical segment, and no users outside that VLAN can communicate with them. Administrators can now have control over each port and whatever resources that port could access. This prevents a user from plugging their workstation into any switch port and gaining access to network resources. To make it more flexible, servers can be members of multiple VLANs so that the users can get access to the resources they need.

Figure 3.27   A Broadcast Domain

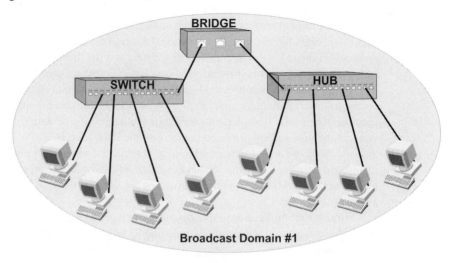

Figure 3.28   A Broadcast Domain is Defined by Routers

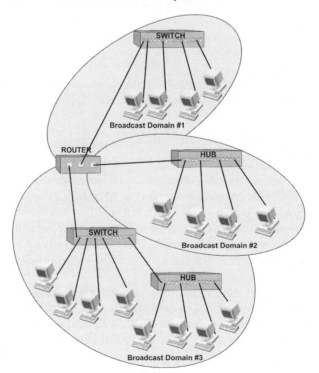

Figure 3.29   A Sample Network Using VLAN

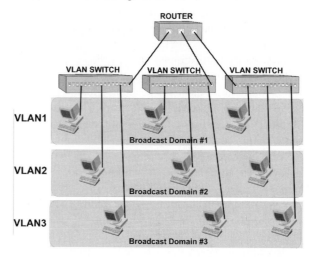

3.12   ROUTERS

A router, which works at the network ISO layer, is a device that connects two or more LANs. As multiple LANs are connected together, multiple routes are created to get from one LAN to another.

NOTE: Since a router needs to know which network to route to, each port must have a unique network address.

The primary role of a router is to transmit similar types of data packets from one local area network or wide area communications link (such as a T-1 or fiber link) to another. The second role of a router is to select the best path between the source and destination.

When you send a packet from one computer to another computer, it first determines if the packet is sent locally to another computer on the same LAN or if the packet is sent to a router so that it can be routed to the destination LAN. If the packet is meant to go to a computer on another LAN, it is sent to the router (or gateway). The router will determine what is the best route to take and forward the packet to that route. The packet will then go to the next router, and the entire process will repeat itself until the packet gets to the destination LAN. The destination router will then forward the packet to the destination computer.

To determine the best route, the routers use complex routing algorithms, which take into account a variety of factors including the number of the fastest set of transmission media, the number of network segments, and the network segment that carries the least amount of traffic. Routers then share status and routing information with other routers so that they can provide better traffic management and bypass slow connections. In addition, routers provide additional functionality, such as the ability to filter messages and forward them to different places based on various criteria. Most routers are multiprotocol routers because they can route data packets using many different protocols.

As noted in chapter 2, a metric is a standard of measurement, such as a hop count, that is used by routing algorithms to determine the optimal path to a destination. A hop is the trip a data packet takes from one router to another router or from a router to another intermediate point to another in the network. On a large network, the number of hops a packet takes toward its destination is called the hop count. When a computer communicates with another computer and the computer has to go through 4 routers, the hop count would be 4. With no other factors taken in account, a metric of 4 would be assigned. If a router had a choice between a route with 4 metrics and a route with 6 metrics, it would choose the route with 4 metrics. Of course, if you want the router to choose the route with 6 metrics, you can overwrite the metric for the route with 4 hops in the routing table to a higher value.

To keep track of the various routes in a network, the routers will create and maintain routing tables based on both network and node addresses (layer 3–network layer of the OSI model). The routers communicate with one another to maintain their routing tables through a routing update message. The routing update message can consist of all or a portion of a routing table. By analyzing routing updates from all other routers, a router can build a detailed picture of network topology.

In addition to performing these basic functions, routers may perform any of the following options:

- Filter out broadcast transmissions to alleviate network congestion.
- Prevent certain types of traffic from getting to a network, enabling customized segregation and security.
- Monitor network traffic and report statistics to a management information base (MIB).
- Diagnose internal or other connectivity problems and trigger alarms.

For all intents and purposes, routers are purpose-built computers dedicated to internetwork processing. Since they have the capabilities of handling the traffic for hundreds, and maybe even thousands, of users, when a router goes down or slows down, it can be devastating to a company.

A typical router has an internal processor, its own memory and power supply, input and output jacks for different types of network connectors, and usually a management console interface. High-powered, multiprotocol routers may have several slot bays to accommodate multiple-interface interfaces. A router with multiple slots that can hold different interface cards or other devices is called a modular router.

Just like switches, routers don't come with a monitor, keyboard, or mouse. Instead, you connect and manage the router using one of the following three methods:

- By using a terminal (usually a PC or workstation running in terminal mode) connection through a cable to the back of the router.
- By using a terminal that is in a different location from that of the router and is connected to it via a modem that calls a modem connected to the router with a cable.
- Via the network on which the router sits.

Since routers on large networks are often hidden away from the normal users, routers are often accessed through the network. Unfortunately, if the router is unreachable due to a network problem, or if there is a problem with the router itself, someone must go to the location of the router and use one of the terminals to reconfigure or troubleshoot the router.

Another routing hybrid, a **layer 3 switch** or routing switch, combines a router and a switch. It has been optimized for high-performance LAN support and is not meant to service wide area connections (although it could easily satisfy the requirements for high-performance MAN connectivity, such as SONET). Because it is designed to handle high-performance LAN traffic, a layer 3 switch can be placed anywhere within a network core or backbone, easily and cost-effectively replacing the traditional collapsed backbone router. The layer 3 switch communicates with the WAN router using industry-standard routing protocols like RIP and OSPF. Therefore, if you need a router to segment your network, you can use a layer 3 switch, in which it could use its high performance. If you need to connect through a WAN connection, you would use a router. Of course, for your campus or large building, you can use the layer 3 switches to connect your segments and then link the layer 3 switches to the router to connect to the WAN.

3.13   ROUTING PROTOCOLS

There are various types of routing algorithms, and each algorithm has a different impact on network and router resources. The routing algorithms use a variety of metrics that affect calculation of optimal routes.

Routing algorithms can be differentiated based on several key characteristics. First, the capability of the routing algorithm to optimally choose the best route is very important. One routing algorithm may use the number of hops and the length of delays, but it may weigh the length of delay more heavily in the calculation. To maintain consistency and predictability, routing protocols use strict metric calculations.

Routing algorithms are designed to be as simple as possible. In other words, the routing algorithm must offer efficiency with a minimum of software and utilization overhead. But while being efficient, the routing algorithm must be robust so that it can quickly change routes when a route goes down because of hardware failure or when a route has a high load (high amount of traffic). Of course, when this is happening, the routing algorithm must be stable.

Routing algorithms must converge rapidly. Convergence is the process of agreement by all routers of which routes are the optimal routes. When a route goes down, routers distribute the new routes by sending routing update messages. The time that it takes for all routers to get new routes on which they all agree should be quick. If not, routing loops or network outages can occur. A routing loop is created when a packet is sent back and forth between several routers without ever getting to its final destination.

3.13.1   Static versus Dynamic Routes

Static routing algorithms are hardly algorithms at all, but are table mappings established by the network administrator prior to the beginning of routing. These mappings do not change unless the network administrator alters them. Algorithms that use static routes are simple to design, and they work well in environments where network traffic is relatively predictable and where network design is relatively simple.

Because static routing systems cannot react to network changes, they generally are considered unsuitable for today's large, changing networks. Most of the dominant routing algorithms are dynamic routing algorithms, which adjust to changing network circumstances by analyzing incoming routing update messages. If the message indicates that a network change has occurred, the routing software recalculates routes and sends out new routing update messages. These messages flow through the network, stimulating routers to rerun their algorithms and change their routing tables accordingly.

Dynamic routing algorithms can be supplemented with static routes where appropriate. A router of last resort (a router to which all unroutable packets are sent), for example, can be designated to act as a repository for all unroutable packets, ensuring that all messages are at least handled in some way.

3.13.2   Distance-Vector versus Link-State Algorithm

Dynamic router algorithms are divided into categories, distance vector-based routing protocols and link-state routing protocols. Routers use distance-vector–based routing protocols to periodically advertise or broadcast the routes in their routing tables, but they only send the information to their neighboring routers. Routing information exchanged between typical distance-vector–based routers is unsynchronized and unacknowledged. Distance-vector–based routing protocols are simple and easy to understand and easy to configure. One disadvantage is that multiple routes to a given network can reflect multiple entries in the routing table, which leads to a large routing table. In addition, if you have a large routing table, network traffic increases as it periodically advertises the routing table to the other routers, even after the network has converged. Lastly with distance-vector protocols, the convergence of large internetworks can take several minutes. RIP for TCP/IP and RIP for IPX are examples of distance-vector routing protocols.

Link-state algorithms are also known as shortest-path-first algorithms. Instead of using broadcasts, link-state routers send updates directly (or by using multicast traffic) to all routers within the network. Each router, however, sends only the portion of the routing table that describes the state of its own links. In essence, link-state algorithms send small updates everywhere. Because they converge more quickly, link-state algorithms are somewhat less prone to routing loops than distance-vector algorithms. In addition, link-state algorithms do not exchange any routing information when the internetwork has converged. They have small routing tables since they store a single optimal route for each network ID. On the other hand, link-state algorithms require more CPU power and memory than distance-vector algorithms. Link-state algorithms, therefore, can be more expensive to implement and support and are considered harder to understand. OSPF for TCP/IP and NLSP for IPX are examples of link-state routing protocols.

3.13.3   Count to Infinity

Count to infinity, which typically happens when a network has slow convergence, is a loop that happens when a link in a network goes down and routers on the network update their routing tables with incorrect hop counts. For example, a loop can occur if the link to Router C goes down. Router B then advertises that the link is down and that it has no route to C. Because Router A has a route to C with a metric of 2, it responds to Router B and sends its link to C. Since two hops is a better route than 16 hops, Router B then updates its table to include a link with metric 3, and the routers continue to announce between the two routers and update their routing entries to C until they reach the number 16, a count to infinity.

To handle the Count to infinity problem, networks use either split horizon or poison reverse. Split-horizon is a route-advertising algorithm that prevents the advertising of routes in the same direction in which they were learned. In other words, the routing protocol differentiates which interface a network route was learned on and then will not advertise that route back out that same interface. This would prevent Router B from sending the updated information it received from Router C back to Router C.

Poison reverse is a process that, used with split horizon, improves RIP convergence over simple split horizon by advertising all network IDs. However, the network IDs learned in a given direction are advertised with a hop count of 16, indicating that the network is unavailable. Therefore, in our example, Router B will initiate route poisoning by entering a table entry for the network that went down by given a value 16. By this poisoning of the route to the downed network, during the next update, it will broadcast to Router A. This way, all routers will eventually be updated that the network went down.

SUMMARY

1. To communicate on LANs, a network card, or network interface card (NIC), which is found as an expansion card or is integrated into the motherboard, uses transceivers (devices that both transmit and receive analog or digital signals).

2. Topology describes the appearance or layout of the network. Depending on how you look at the network, there is the physical topology and the logical topology.

3. The physical topology (part of the physical layer) describes how the network actually appears.

4. The logical topology (part of the data link layer) describes how the data flows through the physical topology or the actual pathway of the data.

5. A bus topology looks like a line. It consists of a single cable along which the data is sent. The two ends of the cable do not meet, and the two ends do not form a ring or loop.

6. Typically with a bus topology network, the two ends of the cable must be terminated.

7. If there is a break anywhere or one system does not pass the data along correctly, the entire network will go down.

8. A ring topology has all devices connected to one another in a closed loop.

9. Traditionally, a break in the ring will cause the entire network to go down and can be difficult to isolate.

10. A star topology is the most popular topology in use. Each network device connects to a central point such as a hub, which acts as a multipoint connector.

11. If a link fails (hub port or cable), the remaining workstations are not affected, as they would be in the bus and ring topology.

12. Another topology is the mesh topology. In this type of topology, every computer is linked to every other computer. While this topology is not very common in LANs, it is common in WANs, where it connects remote sites over telecommunication links.

13. Many technologies that are wireless use a cellular topology, where an area is divided into cells. A broadcast device is located at the center and broadcasts in all directions to form an invisible circle (cell).

14. The hybrid topology scheme combines two of the traditional topologies, usually to create a larger topology.

15. The cabling system used in networks serves as the veins of the network that connect all the computers together and allow them to communicate with each other.

16. Attenuation occurs as the strength of a signal falls off with distance over a transmission medium.

17. Interference occurs when undesirable electromagnetic waves affect the desired signal.

18. A twisted pair consists of two insulated copper wires twisted around each other.

19. Crosstalk results when an electromagnetic signal inducts or transfers from one wire in a twisted pair to the other.

20. Cables in a twisted pair are twisted to reduce crosstalk.

21. UTP cable was categorized by the Electronic Industries Association (EIA) based on the quality and number of twists per unit.

22. Early networks that used UTP typically used Category 3, whereas today's newer high-speed networks typically use Category 5 or Enhanced Category 5 cabling.

23. While your telephone uses a cable with two pairs (four wires) and an RJ-11 connector, a computer network uses a cable with four pairs (eight wires) and an RJ-45 connector.

24. The straight-through cable, which can be used to connect a network card to a hub, has the same sequence of colored wires at both ends of the cable.

25. A crossover cable, which can be used to connect one network card to another network card or a hub to a hub, reverses the transmit and receive wires.

26. Shielded twisted pair is similar to unshielded twisted pair, except that the shielded twisted pair is usually surrounded by a braided shield that serves to reduce both EMI sensitivity and radio emissions.

27. Coaxial cable, sometimes referred to as coax, is a cable that has a center wire surrounded by insulation and then a grounded shield of braided wire (mesh shielding).

28. A BNC connector is a type of connector used with coaxial cables, such as the RG-58 A/U cable used with the 10Base 2 Ethernet system.

29. BNC T-connectors (used with the 10Base 2 system) are female devices for connecting two cables to a network interface card.

30. A BNC barrel connector allows two cables to be connected together.

31. A fiber-optic cable consists of a bundle of glass or plastic threads, each of which is capable of carrying data signals in the form of modulated pulses of light.

32. Fiber-optic cables use several connectors, but the two most popular and recognizable connectors are the straight tip (ST) and subscriber (SC) connectors.

33. The horizontal wiring system has cables that extend from wall outlets throughout the building to the wiring closet or server room.

34. Backbone wiring, which is often designed to handle a higher bandwidth, consists of cables used to interconnect wiring closets, server rooms, and entrance facilities (telephone systems and WAN links from the outside world). Backbone wiring is sometimes referred to as the vertical cabling system.

35. Vertical connections between floors are known as risers.

36. TIA/EIA released their joint 568 Commercial Building Wiring Standard, also known as structured cabling, for uniform, enterprisewide, multivendor cabling systems.

37. Most of the cables used contain halogens, chemicals that give off toxic fumes when they burn.

38. The plenum is the space above the ceiling and below the floor used to circulate air throughout the workplace.

39. A plenum cable is a special cable that gives off little or no toxic fumes when burned. Depending on the building's fire code, the building's wiring will include plenum cabling unless the wiring exists in a special sealed conduit that would prevent the toxic fumes from escaping into the open air.

40. A riser cable is intended for use in the vertical shafts that run between floors.

41. A punch-down block is used to connect several cable runs to each other without going through a hub.

42. To connect the wires to the punch-down block, you would use a punch-down tool.

43. A patch panel is a panel with numerous RJ-45 ports. The wall jacks are connected on the back of the patch panel to the individual RJ-45 ports. You can then use patch cables to connect the port in the front of the patch panel to a computer or a hub. As a result, you can connect multiple computers with a hub located in the wiring closet or server room.

44. Cable ties and basket systems are tools used to help manage the cables anywhere throughout the building.

45. If you have several servers, you should also consider a server rack or rackmount cabinet to hold the servers.

46. A wireless LAN (WLAN) transfers data through the air using radio frequencies instead of cables.

47. A band is a contiguous group of frequencies that are used for a single purpose.

48. Network cards are media-dependent, media-access-method–dependent, and protocol-independent.

49. A media access control (MAC) address is a hardware address identifying a node on the network.

50. When you install a network card, you must bind the protocol stack to the network card and install any additional network software needed.

51. NDIS (network device interface specification, developed by Microsoft) and ODI (open data link interface, developed by Apple and Novell) are sets of standards for protocols and network card drivers to communicate. Both allow multiple protocols to use a single network adapter card.

52. A diskless workstation is a computer that does not have its own disk drive.

53. A hub, which works at the physical OSI layer, is a multiported connection point used to connect network devices via a cable segment.

54. A bridge, which is a layer 2 device, is used to break larger network segments into smaller network segments.

55. A switching hub (sometimes referred to as a switch or a layer 2 switch) is a fast, multiported bridge.

56. A VLAN is a collection of nodes grouped together in a single broadcast domain that is based on something other than physical location.

57. A router, which works at the network ISO layer, is a device that connects two or more LANs. As multiple LANs are connected together, multiple routes are created to get from one LAN to another.

QUESTIONS

1. What basic terminology describes the low-level rules of communication that each network node must use?
   a. Logical protocol
   b. Physical topology
   c. Physical protocol
   d. Logical topology

2. You are a network administrator for the Acme Corporation, and you need to design a network for the corporate office in New York City. You design a network with Category 5 unshielded twisted-pair (CAT 5 UTP) cables arranged in a star topology. Which layer of the Open Systems Interconnection (OSI) model is associated with these design components?
   a. Transport layer
   b. Data link layer
   c. Network layer
   d. Physical layer

3. What is a point-to-point connection?
   a. A link that exists only between two devices
   b. A link that exists between multiple devices
   c. A secure channel on a TCP/IP link
   d. A baseband wireless connection

4. Because each node must be wired to a central point, this physical topology usually requires more cable runs than a topology with serially connected nodes.
   a. Star   b. Bus
   c. Ring   d. Mesh

5. As nodes are added to the network, an interface must be added to each existing node until finally it is impractical to add any more.
   a. Bus   b. Star
   c. Ring   d. Mesh

6. The last node is connected to the first. Each node is connected to two neighboring nodes. These two statements are characteristic of what physical topology?
   a. Star   b. Bus
   c. Mesh   d. Ring

7. Network nodes connected in point-to-point links to a central device are characteristic of what physical topology?
   a. Bus   b. Mesh
   c. Ring   d. Star

8. Nodes connected to central devices in point-to-point links, where the central devices are connected in a serial fashion, are characteristic of what physical topology?
   a. Ring   b. Mesh
   c. Bus   d. Star-bus

9. In which physical topology are all the nodes serially connected to each other?
   a. Bus   b. Star-bus
   c. Star   d. Mesh

10. Which of the following physical topologies has all its nodes interconnected to each other?
   a. Bus   b. Ring
   c. Star   d. Mesh

11. Your network consists of four servers. Each server has three NICs. Server A is connected to servers B, C, and D. Server B is attached to C and D. Server C is attached to server D. No hub is used. All cabling is CAT 3 UTP crossover cables. What is your physical topology?
   a. Star   b. Bus
   c. Ring   d. Mesh

12. Your network consists of a 24-port bridge, two servers, and 15 workstations. Each computer interfaces with the bridge using its own segment of CAT 5 twisted-pair cable, which is terminated at both ends with RJ-45 connectors. What type of physical topology do you have?
    a. Mesh   b. Bus
    c. Ring   d. Star

13. Your network's physical topology is that of a star configuration. From time to time, each workstation will hold a packet called a token, during which time it will transmit data and then pass the token along to the next workstation. What is the logical topology of your network?
    a. Bus   b. Star
    c. Mesh   d. Ring

14. Pat is the network engineer for the Acme Corporation, a nationwide bank. He needs to design a wide area network (WAN) that will connect offices in Dallas, San Francisco, Denver, and Tallahassee. The WAN that Pat designs should provide redundancy to ensure that if a link malfunctions, the critical account data will not be lost. Which WAN topology should Pat design for Acme Corporation?
    a. Ring   b. Bus
    c. Star   d. Mesh

15. Which type of network topology uses terminators?
    a. Star   b. Bus
    c. Ring   d. Mesh

16. The network segment that provides the primary path for data flow is called what?
    a. Primary link
    b. Primary segment
    c. Data link
    d. Backbone

17. Which physical topology typically uses the least amount of cabling?
    a. Star   b. Ring
    c. Bus   d. Mesh

18. A physical topology has these two advantages: It is easy to troubleshoot because it has a central point to isolate faults, and it can be organized in a hierarchical structure. Which physical topology is this?
    a. Ring
    b. Bus
    c. Mesh
    d. Star

19. A physical topology is characterized by the following disadvantage: A single fault will bring down the entire network segment. Which physical topology is this?
    a. Ring   b. Star
    c. Mesh   d. Bus

20. Which of the following is used to connect two pieces of cable on a linear bus topology?
    a. A BNC terminator
    b. A network adapter card
    c. A BNC barrel connector
    d. A medium attachment unit

21. Crosstalk and other types of interference can affect network performance and security. Which type of cable is most susceptible to crosstalk?
    a. RG-58 A/U
    b. STP
    c. RG-58 /U
    d. Category 5 UTP

22. Which type of problem is most likely to be caused by increasing cable lengths?
    a. Attenuation   b. Beaconing
    c. Crosstalk   d. Jitter

23. Twisted-pair cables use different connection hardware than coaxial cables. Which type of connector is commonly used by twisted-pair cables?
    a. BNC   b. DIX
    c. AUI   d. RJ-45

24. When a signal jumps from one wire to an adjacent wire, this is known as _____.
    a. attenuation   b. jitter
    c. crosstalk   d. beaconing

25. Which cable is the one that is most common in networks?
    a. UTP   b. STP
    c. Coaxial   d. Fiber optics

26. When planning a network, you must be concerned with fire. What type of cable emits little or no toxic fumes when it burns?
    a. Polling   b. Plastic cable
    c. Plenum   d. Glass cable

27. Your company has offices in two separate buildings, and each office is networked. You are responsible for connecting the two separate networks. The distance between the two buildings is approximately 450 meters. The current capacity for each network is 10 Mbps. You could use 10Base5 cable to connect the two networks. What is one reason to consider using fiber-optic cable instead?

a. Fiber-optic cable is light and flexible and is easier to install than the heavy and inflexible coaxial cable used by 10Base5.

b. Fiber-optic cable supports higher digital transmission rates and can provide additional capacity for future network expansion without the need to run new cable.

c. Fiber-optic cable is the most widely used network cabling and is usually cheaper than plenum-grade coaxial cable.

d. Fiber-optic cable is better suited for broadband transmission, thus providing a relatively inexpensive way to increase network capacity in the future.

28. You have been given the task of installing cables for an Ethernet network in your office building. The network cable will have to share the existing conduit with telephone cables. Cable segments will be up to 95 meters in length. Which cable is best suited for this installation?

a. Fiber optic

b. Category 3 UTP

c. Category 1 UTP

d. Thicknet coaxial

29. What type of connector is normally used by a thicknet cable for connection to the network adapter card?

a. A BNC T-connector

b. An RJ-45 connector

c. An AUI connector

d. A BNC barrel connector

30. What type of connector is used by thinnet for connection to the network adapter card?

a. An AUI connector

b. A BNC T-connector

c. An RJ-45 connector

d. A BNC barrel connector

31. Plenum cable has which of the following characteristics?

a. It has a lower cost then nonplenum cables.

b. It transmits data faster.

c. It is a military version of the cable, which contains more shielding.

d. It meets fire codes.

32. Which of the following has the highest possible throughput?

a. STP   b. UTP

c. Coax   d. Fiber optic

33. An RJ-45 connector should be wired with _____ pair(s) when used on a Category 5 UTP cable.

a. 1   b. 2

c. 4   d. 8

34. Which network installation tool can be used to connect UTP cable to a 110 block?

a. Cable tester

b. Wire crimper

c. Punch-down tool

d. Media installer

35. Which of the following is not unbound network media?

a. Fiber-optic cable

b. Infrared pulses

c. Radio waves

d. None of the above are unbound network media.

36. Where is the plenum located in a building? (Choose two answers)

a. In the ceilings

b. In the walls

c. In the elevator shaft

d. In the basement

37. What network cabling has four pairs of wires?

a. Coaxial

b. Twisted-pair

c. Fiber optic

d. None of the choices listed

38. What network media uses a single copper conductor covered by an insulating layer, which is in turn covered by a conductive mesh and then finally a tough plastic skin?

a. Fiber optic   b. Twisted pair

c. Stranded pair   d. Coaxial

39. You are put in charge of building a network in a small power plant, which is full of interference, both EMI and RFI. There will be 20 servers and 500 workstations. The physical topology will be star-bus, and the logical topology you've selected is Ethernet. You've drawn the design and found that the longest cable segment you'll need to run is 85 meters. Now that it's time to order the cable, what do you select?

a. Twisted pair   b. Coaxial

c. Fiber optic   d. STP

40. What cable uses a glass thread core?

a. STP   b. Twisted pair

c. Fiber optic   d. Coaxial

41. Some of the NICs in your network have only a single port that accepts an RJ-45 plug. What cable type will attach to these NICs?
a. STP
b. Coaxial
c. Fiber optic
d. UTP

42. Which media connector on a NIC is a D-shell 15-pin female interface? (Select two answers.)
a. Thinnet
b. DIX
c. BNC
d. AUI

43. Which cable connector looks like a large phone plug?
a. BNC
b. DB-15
c. RJ-45
d. RJ-11

44. What is the maximum length allowed for a segment of twisted-pair cabling? (Select two answers.)
a. 425 feet
b. 100 meters
c. 115 meters
d. 328 feet

45. RG-8 and RG-11 coaxial cabling is commonly called what?
a. Cable-net
b. Thinnet
c. Fiber-net
d. Thicknet

46. RG-58 cabling is commonly called what?
a. Thinnet
b. Cable-net
c. Fiber-net
d. Thicknet

47. Your boss wants you to recommend a cable type that will support fast Ethernet. He said the cost to buy the cabling doesn't matter that much. There are no EMI or RFI issues about where the cable is to run. What is the best choice?
a. Coaxial
b. Fiber optic
c. STP
d. UTP

48. You have a NIC with only a single AUI connector. What is needed to connect it to the network?
a. Vampire-tap transceiver
b. RJ-45 plug
c. Twisted-pair cable
d. BNC T-connector

49. What are ST and SC connectors used for? (Select two answers.)
a. Multimode fiber
b. Optiplexed fiber
c. Single-mode fiber
d. Stranded copper cabling

50. Network transmission of data via radio, laser, infrared, or microwave is said to be sent on what?
a. Switching media
b. Bound media
c. Pulse media
d. Unbounded media

51. What physical medium requires the greatest expertise to install?
a. Coaxial
b. Fiber optic
c. Infrared
d. Copper wire

52. Which part of a coaxial cable carries the data?
a. Wire mesh shielding
b. Core
c. Plenum
d. None of the choices listed

53. In an Ethernet network, what device should you use to connect two hubs?
a. A crossover cable
b. A tone generator
c. A hardware loopback
d. A tone locator

54. Pat is not able to access network resources from her computer. When she plugs her Ethernet cable into a coworker's machine, she is able to access the network without any problems. Which of the following network components is causing the problem?
a. The server
b. The cable
c. The router or gateway
d. The network adapter

55. You want to have a 10BaseT segment that is greater than 300 meters. Which device will be required?
a. A repeater
b. A multiplexer
c. An amplifier
d. An RJ-45 connector

56. VLANs are a feature primarily of which type of network device?
a. Hubs
b. Switches
c. Router
d. NICs
e. Cable

57. What is the primary purpose of a VLAN?
a. Testing all router links on your network
b. Simulating a network
c. Segmenting a network inside a switch
d. Connecting a LAN to another LAN

58. Which part of a network interface card identifies whether or not there is basic connectivity from the station to a hub or switch?
a. Link light
b. Port
c. Collision light
d. Jumper

59. You are installing a network interface card (NIC) on your computer. The NIC has a 15-pin attachment unit interface (AUI) connector. Which device do you need to attach to the NIC to connect your computer to a LAN and allow your computer to communicate with other computers on the LAN?

   a. A hub   b. A transceiver
   c. A patch panel   d. A router

60. Due to rapid growth of the company, your supervisor asks you to help him find a good switch that will break up the network into at least three different subnets, each with a different network identifier. What type of switch do you need?

   a. Multiplexing   b. Layer 3
   c. Token passing   d. Layer 2

61. What is the main reason you'd use dynamic routing rather than static routing?

   a. There is much less administrative overhead.
   b. Actually, static routing is the preferred method.
   c. Dynamically routed packets use less bandwidth.
   d. Routers operate faster with dynamic routing.

62. The ABC company and the XYZ company are both in the same building on different floors. One day they decide to get together and merge into a single company, called AXBYCZ. Each company has been running a simple local area network, the first using Token Ring and the second using Ethernet. In order to combine both networks, what device do you use? (Choose two answers.)

   a. Bridge
   b. Gateway
   c. Router
   d. Layer 2 switch

63. You wish to set up a LAN in a star configuration. However, you want each node to have full bandwidth on its port. What device do you need?

   a. Layer 2 switch   b. Router
   c. Active hub   d. Brouter

64. A layer 2 switch is also known as what?

   a. Brouter   b. Router
   c. Repeater   d. Bridge

65. Building tables based on MAC addresses is the job of what device?

   a. Bridge   b. Gateway
   c. Router   d. Switch

66. What switching mode causes the least amount of latency?

   a. Store-and-forward
   b. FragmentFree
   c. Port switching
   d. Cut-through

67. Which devices will forward all network traffic? (Choose all that apply.)

   a. Bridge   b. Gateway
   c. Repeater   d. Router

68. Which of the following is NOT a characteristic of a layer 2 switch?

   a. Builds a MAC address table
   b. Creates multiple collision domains
   c. Operates at the data link layer
   d. Routes packets based on IP address (layer 3)

69. Building tables based on network addresses is the job of what device?

   a. Bridge   b. Gateway
   c. Repeater   d. Router

70. What is another name for a layer 3 switch?

   a. Bridge   b. Brouter
   c. Router   d. Repeater

71. Why should unused protocols be unbound from NICs on the network?

   a. To reduce unnecessary packet traffic.
   b. To improve bandwidth for the protocols that are used.
   c. It is not necessary to unbind them.
   d. To prevent lost packets.

72. A media converter that enables connecting a NIC to various cable types is called a what?

   a. Transceiver   b. Dataplexor
   c. Receiver   d. Multiplexor

73. When multiple protocols are bound to a NIC, the most used protocol on the network should be where in the binding list?

   a. Doesn't matter
   b. Always last
   c. Depends on the protocol
   d. Always first

74. Some NICs have a chip that allows a computer without an OS to boot and attach to the network; also the chip is software-configurable. What sort of chip is this?

   a. EPROM   b. CPU
   c. BIOS   d. EEPROM

75. You want to link two computers together without using a hub. Both have network cards. Which of the following would be best to use?

   a. Serial cable
   b. Coaxial cable
   c. Parallel cable
   d. Crossover cable

76. Suddenly none of the members in your peer-to-peer star topology network can access each other's computers. To find the problem, what would you check first?

   a. Whether a cable terminator has loosened
   b. Whether one of the cables is crimped
   c. Whether one of the computer's NICs has failed
   d. Whether the power light on the hub is off

77. The onboard serial port on your PC is set to use IRQ 4 and I/O port 3f8. Your internal modem is using the standard resources used by Com port 3. You cannot dial out. What is the problem?

   a. Your modem is using IRQ 4, which conflicts with the onboard serial port.
   b. Your modem is using I/O port address 3f8, which conflicts with the onboard serial port.
   c. You cannot configure an internal modem to be used as Com port 3.
   d. A defective modem.

78. A NIC's EEPROM can store all but which of the following?

   a. IRQ
   b. DMA channel
   c. I/O address
   d. Network data

79. You need to prepare a 10-foot run of patch cable that will have two RJ-45 connectors on each end. What tool should you use to put the RJ-45s on?

   a. C-clamp
   b. Flat-head screwdriver
   c. Pulse generator
   d. Wire crimper

80. What are BNC connectors used for?

   a. To connect thinnet Token Ring cabling to network devices
   b. To connect twisted-pair cabling to hubs
   c. To connect thicknet Ethernet cable to network devices
   d. To connect thinnet Ethernet cable to network devices

81. NDIS and ODI are specifications that define interfaces for communication between the data link layer and protocol drivers. Which need do NDIS and ODI fulfill?

   a. The need to dynamically bind a single protocol to multiple MAC drivers
   b. The need to bind multiple protocols to a single network adapter card

   c. The need for monolithic protocols to conform to the OSI model
   d. The need for monolithic protocols to be loaded into the upper memory area

82. ODI is a specification defined by Novell and Apple to simplify driver development. Which of the following statements is true of ODI?

   a. ODI specifies cable and network architecture standards.
   b. ODI specifies network topology standards.
   c. ODI provides support for multiple protocols on a single network adapter card.
   d. ODI provides support for multiple network topologies in a single networking environment.

83. Which of the following tools can be used to check for cable breaks? (Choose all that apply)

   a. Digital voltmeter
   b. Time-domain reflectometer
   c. Advanced cable tester
   d. Multiplexer

84. You connect the leads from a digital multimeter (DMM) to each of the conductors in a twisted pair. Which of the following settings should you use on the DMM to see if there is a short?

   a. Resistance   b. DC voltage
   c. Current   d. Capacitance
   e. AC voltage

85. Which of the following tools can you use to trace the path of a cable through a physical network? (Choose two answers)

   a. A hardware loopback
   b. A tone generator
   c. A protocol analyzer
   d. A tone locator

86. Your company uses a large electric drill press as part of its manufacturing process. Occasionally, the 10BaseT LAN stops responding when the operator uses the drill press. Which of the following could cause this problem? (Choose two answers.)

   a. Electromagnetic interference (EMI)
   b. Power sag
   c. Electrostatic discharge (ESD)
   d. Radiomagnetic interference (RMI)

87. A surge of EMI that could corrupt network data could be caused by what? (Choose two answers)

   a. Fluorescent lighting
   b. Sonic energy wave
   c. Electric motor
   d. Crimped cable

88. Pat uses a Windows 2000 Professional computer that is connected to a 100BaseTX LAN. She attempts to use My Network Places to browse the computers on the LAN, but the Windows 2000 Professional computer displays the error message "Unable to Browse the Network. The Network Is Not Accessible." Pat was able to connect to the network the last time she used the Windows 2000 Professional computer. What should Pat check first to determine the cause of this problem?

a. The link light on her computer's network interface card (NIC)

b. The power switch on the NIC

c. The cable that connects the NIC to the wall outlet in her office

d. The power switch on the hub in the wiring closet that connects the NIC to the rest of the network

89. In addition to testing for cable faults, which of the following can also test for excessive packet collisions, as well as cable voltage?

a. Protocol analyzer

b. TDR

c. SRT

d. Advanced cable tester

90. Which of the following tools can test voltage, resistance, and current, as well as continuity?

a. Advanced cable tester

b. TDR

c. Voltmeter

d. Digital multimeter

91. What is a "fox and hound"?

a. Multiport fluke

b. Packet sniffer

c. Serial/parallel translator

d. Tone generator/locator

92. One of your servers is not accessible by anyone on the LAN. You go to the server and note that no error messages are being displayed. Also, there are no error messages in the event log. What is the next thing you should check?

a. The server's NIC link light

b. The server's network cable

c. The server's port in the wire closet

d. The server's disk drives

93. One day your boss points at an unused RJ-45 jack near the bottom of a wall and asks, "Where does the other end of that cable come out?" What network tool would you use to help answer this question?

a. Fox and hound

b. Wire crimper

c. Some other tool not listed here

d. Media tester

HANDS-ON EXERCISES

Exercise 1: Create a Coaxial Cable

1. Create a 6-foot 10Base2 coaxial cable with two BNC connectors.
2. With a voltmeter, test the BNC cable.
3. If you have a cable tester, test the BNC cable.

Exercise 2: Create an Unshielded Twisted-Pair Cable

1. Create a standard 15-foot Category 5 100BaseT with two RJ-45 connectors.
2. With a voltmeter, test the UTP cable.
3. If you have a cable tester, test the BNC cable.
4. Create a crossover 15-foot Category 5 100BaseT with two RJ-45 connectors.
5. With a voltmeter, test the UTP cable.
6. If you have a cable tester, test the BNC cable.

CHAPTER **4**

Networking Technologies

Topics Covered in this Chapter

Introduction

In chapter 3, you learned about the building blocks of a network. These building blocks, including cables and connection devices, form the network. In this chapter, you will look at the different LAN technologies such as Ethernet, Token Ring, and 802.11 wireless networks.

Objectives

- Specify the main features (speed, access, method, topology, and media) of 802.3 (Ethernet) and 802.5 (Token Ring).
- Specify the characteristics, including speed, length, topology, and cable type, for 802.3 (Ethernet) standards, 10BaseT, 100BaseTX, 10Base2, 10Base5, 100BaseFX, and Gigabit Ethernet.
- Specify the main features and characteristics of 802.11 (wireless LANs).

- Given a troubleshooting scenario involving a network failure, identify the cause of the failure.
- Build a coaxial (thinnet) Ethernet network.
- Build a UTP Ethernet network.
- Explain how to build a Token Ring network.

4.1   ETHERNET

Currently, three LAN technologies are being used. A LAN technology defines topologies, packet structures, and access methods that can be used together on a segment. LAN technologies include Ethernet, Token Ring, and ARCnet. Out of these, **Ethernet** is the most widely used LAN technology today. It offers a good balance of speed, price, reliability, and ease of installation. More than 85 percent of all LAN connections installed are Ethernet. All popular operating systems and applications are Ethernet-compatible, as are upper-layer protocol stacks such as TCP/IP, IPX, and NetBEUI.

The original Ethernet network was developed by Digital, Intel, and Xerox in the early 1970s. This version currently is referred to as Ethernet II. Today, the Ethernet standard is defined by the Institute of Electrical and Electronics Engineers (IEEE) in a specification commonly known as IEEE 802.3. The **IEEE 802.3** specification covers rules for configuring Ethernet LANs, the types of media that can be used, and the way the elements of the network should interact. (See table 4.1.)

Ethernet was traditionally used on light to medium traffic networks, and performs best when a network's data traffic is sent in short bursts. Recently, newer and faster Ethernet standards combined with switches have significantly increased the performance of the network.

Table 4.1   Various Ethernet standards

Ethernet Name	Cable Type	Maximum Speed	Maximum Transmission Distance	Notes
10Base5	Coax	10 Mbps	500 meters	Also called thicknet
10Base2	Coax	10 Mbps	185 meters	Also called thinnet
10BaseT	UTP	10 Mbps	100 meters	
100BaseT	UTP	100 Mbps	100 meters	
100BaseVG	UTP	100 Mbps	213 meters (CAT 5) or 100 meters (CAT 3)	100VG—ANYLAN
100BaseT4	UTP	100 Mbps	100 meters	Requires four pairs of CAT 3, 4, or 5 UTP cable
100BaseTX	UTP or STP	100 Mbps	100 meters	Two pairs of CAT 5 UTP or Type 1 STP
10BaseFL	Fiber (multimode)	10 Mbps	2000 meters	Ethernet over fiber-optic implementation; connectivity between network interface card and fiber-optic hub
100BaseFX	Fiber (multimode)	100 Mbps	2000 meters	100 Mbps Ethernet over fiber-optic implementation
1000BaseTX (Gigabit Ethernet)	UTP	1000 Mbps	100 meters	Uses same connectors as 10BaseT; requires CAT 5 or better
1000BaseSX	Fiber (multimode)	1000 Mbps	260 meters	Uses SC fiber connectors; designed for workstation to hub implementation
1000BaseLX	Fiber (Single mode)	1000 Mbps	550 meters	Uses longer-wavelength laser than 1000BaseSX; typically used for backbone implementation
1000BaseCX	STP Type 1	1000 Mbps	25 meters	Typically used for equipment interconnection such as clusters
1000BaseT	UTP	1000 Mbps	100 meters	Requires CAT 5 or better

Ethernet networks can be configured in either a star topology using unshielded twisted pair connected to a hub or a bus topology using a coaxial cable acting as a backbone. Of these two, UTP cabling is by far the most commonly used. Ethernet cards can have one, two, or possibly all three of the following connectors:

- DIX (Digital Intel Xerox)/AUI connectors support 10Base5 external transceivers.
- BNC (British Naval Connector) connectors support 10Base2 coax cabling.
- RJ-45 connectors support 10BaseT/100BaseTX (UTP) cabling.

When a network device wants to access the network, the access method used is carrier sense multiple access with collision detection (CSMA/CD). When a computer wants to send data over the network, it will listen to see if there is any traffic on the network. If the network is clear, it will then broadcast. Unfortunately, it is possible for two network devices to listen and try to send data at the same time. As a result, a collision occurs and both data packets are corrupted. While this is normal for an Ethernet network, both network devices will wait a different random amount of time and try again. Of course, if the network is heavily congested, the network will have more collisions, which results in more traffic, slowing the entire network even more.

For an efficient Ethernet network, you need to keep collisions to a minimum. To accomplish this, you have to watch your wiring distances and repeater counts (see the 5-4-3 rule shown later in this chapter) in any given path to keep them to a minimum and to optimize your applications and protocols to keep the average frame size up.

Within an Ethernet network, you are limited to a 1024 nodes identified by their MAC/physical addresses, which are burned into the ROM chip on the network card. IEEE assigns the first 3 bytes of the 6-byte address to the network card vendor. The vendor is then responsible for assigning the rest of the address to make sure that the MAC address is unique.

4.1.1   Ethernet Encoding Method

Although many methods are used for encoding the signal, the Manchester Signal Encoding method, which is used on an Ethernet medium, is the most common. When a device driver receives a data packet from the higher-layer protocols such as an IP or IPX packet, the device driver constructs a frame (much like an envelope) with the appropriate Ethernet header information and a frame check sequence at the end for error control. The circuitry on the adapter card then takes the frame and converts it into an electrical signal.

The data is measured as a transition state that occurs in the middle of each bit-time. To represent a binary one, the first half of the bit-time is a low voltage, while the second half of the bit is always the opposite of the first half. To represent a binary zero, the first half of the bit-time is a high voltage and the second half is a low voltage. (See figure 4.1.)

4.1.2   Ethernet Frame Types

When a packet is transmitted, information about the sender, receiver, and upper-layer protocols is attached to the data. The format for the completed packet is called the frame type. There are four Ethernet frame types. To get a good understanding of frame types, think of Ethernet as a language and the four different frame types as the dialects of the language.

Figure 4.1 Ethernet Using Transition State to Encode Data

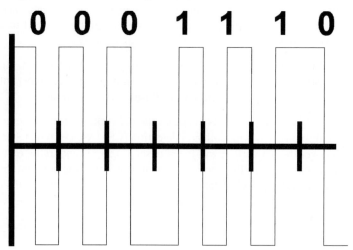

Table 4.2 Ethernet Frame Types

Novell Name	Cisco Name	Ethernet Type
Novell Ethernet_802.3	LLC	Also known as Ethernet-Raw. Default Ethernet frame type of Novell NetWare 3.11 or earlier. Supports IPX.
Novell Ethernet_II	ARPA	Used for IPX and TCP/IP.
IEEE 802.2	Novell	Default Ethernet frame type of Novell NetWare 3.12 or later and Windows. Used for IPX.
IEEE 802.2 SNAP	SNAP	Used for IPX, TCP/IP, and AppleTalk.

The frame types are all similar, and yet there are some differences. The four frame types started with Novell, when people at that company decided they could not wait for the emerging Ethernet standard to be ratified by IEEE. (See table 4.2.)

The Ethernet packet starts with the preamble, which consists of 8 bytes of alternating 1s and 0s, ending in 11. A station on an Ethernet network detects the change in voltage that occurs when another station begins to transmit, and uses the preamble to "lock on" to the sending station's clock signal. When it reads the 11 at the end, it then knows the preamble has ended.

Next, the MAC address of the target (destination) computer, followed by the MAC address of the source of the sending computer, is sent.

NOTE: The address could be represented in a multicast (address sent to multiple computers) message.

Figure 4.2   Ethernet Packet

Data Link Header   LLC Header

Destination Address (6 bytes)	Source Address (6 bytes)	Length (2 bytes)	DSAP (1 byte)	SSAP (1 byte)	CONTROL (1 byte)	**DATA**	FCS (4 bytes)

In addition, most Ethernet adapters can be set into promiscuous mode, where they receive all frames that appear on the LAN whether addressed to them or not. If this poses a security problem, a new generation of smart hub devices can filter out all frames with private destination addresses belonging to another station.

Depending on the Ethernet frame type, a type field or a length field is sent. For Ethernet 802.2, the length field describes the length of the data field, including the LLC and SNAP headers. The data field can be between 64 and 1500 bytes. The LLC header is used to identify the process/protocol that generated the packet and the process/protocol that the packet is intended for. The data field is between 43 and 1497 bytes, which include the TCP/IP or IPX packet. The last 4 bytes that the adapter reads is the frame check sequence or CRC for error control. (See figure 4.2.)

Ethernet_802.3 is an incomplete implementation of the IEEE 802.3 specification, sometimes called Ethernet-Raw. Novell implemented the 802.3 header but not the fields defined in the 802.2 specification. Instead, the IPX packet begins immediately after the 802.3 fields. The Novell frame type Ethernet_802.2 is a complete implementation; i.e., it includes the 802.3 and 802.2 fields. Ethernet_II is the standard frame type for TCP/IP networks. Ethernet_SNAP includes the 802.3, 802.2, and SNAP (Subnetwork Access Protocol) fields.

4.1.3   10Base5 (Thicknet)

The original Ethernet was called **10Base5** or **thicknet.** 10Base5 got its name because it has a 10-Mbps baseband network that can have a cable segment up to 500 meters long. It uses a 50-ohm RG-8 and RG-11 as a backbone cable, and it uses physical and logical bus topology. Most network devices connected to the network use an external transceiver that uses a 4-pair AUI cable (sometimes referred to as a drop cable) via a DIX connector (2-row, 15-pin female connector). Other network devices can connect to the backbone cable by using a vampire tap, which uses teeth that make contact with the inner conductor, or by using a BNC barrel connector. (See figure 4.3.)

When building a 10Base5 network, you must follow these rules:

- The minimum cable distance between transceivers must be 8 feet, or 2.5 meters.
- The AUI cable may be up to 164 feet, or 50 meters.
- You may not go beyond the maximum network segment length of 1640 feet, or 500 meters.
- The entire network cabling scheme cannot exceed 8200 feet, or 2500 meters.
- All unused ends must be terminated with 50-ohm terminating resistors.

Figure 4.3   10Base5 Network

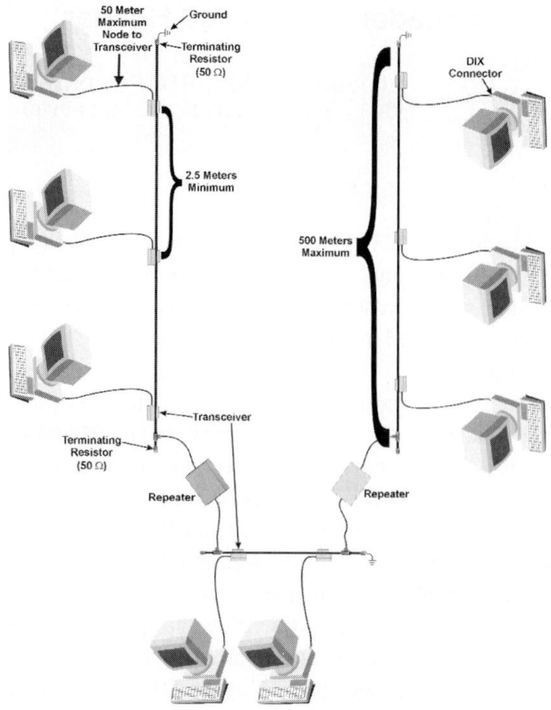

- One end of the terminated network segment must be grounded.
- The maximum number of nodes per network segment including repeaters is 100.
- A signal quality error (SQE) test, which is present on some boards, must be turned off when repeaters are used.

4.1.4   10Base2 (Thinnet)

The **10Base2 (thinnet)** is a simplified version of the 10Base5 network. Its name derives from the fact that it has a 10-Mbps baseband network with a maximum cable segment length of approximately 200 meters (actually 185 meters). Instead of having external transceivers, the transceivers are on the network card, which attaches to the network using a BNC T-connector. The cable used is a 50-ohm RG-58 A/U coaxial-type cable. Different from the 10Base5 network, the 10Base2 does not use a drop cable. (See figure 4.4.)

When building a 10Base2 network, you must follow these rules:

- The minimum cable distance between workstations must be 1.5 feet or, 0.5 meter.
- You may not go beyond the maximum network segment limitation of 607 feet, or 185 meters (not the 200 meters commonly stated).
- The entire network cabling scheme cannot exceed 3035 feet, or 925 meters.

Figure 4.4   10Base2 Network

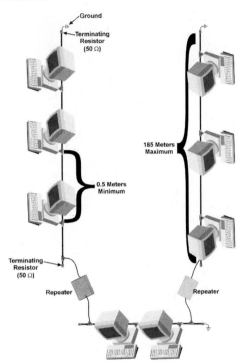

- The maximum number of nodes per network segment is 30 (this includes workstations and repeaters).
- All unused ends must be terminated with 50-ohm terminating resistors.
- A grounded terminator must be used on only one end of the network segment.
- You can connect 10Base5 networks to 10Base2 networks using special boards.

4.1.5   10BaseT (Twisted-Pair Ethernet)

Different from 10Base2 and 10Base5, **10BaseT** uses UTP, which costs less, is smaller, and is easier to work with than coax cable. While Ethernet is a logical bus topology, 10BaseT is a physical star topology; it has the network devices connected to a hub (or switch). It uses a Category 3 (or greater cable) with two pairs of wires (pins 1, 2, 3, and 6) connected with RJ-45 connectors. (See figure 4.5.)

When building a 10BaseT network, you must follow these rules:

- The maximum number of network segments or physical LAN is 1024.
- The minimum unshielded cable segment length is 2 feet, or 0.6 meter.
- The maximum unshielded cable segment length is 328 feet, or 100 meters.
- To avoid EMI, you must route UTP cable no closer than 5 feet to any high-voltage (power) wiring or fluorescent lighting.

Figure 4.5   10BaseT Network

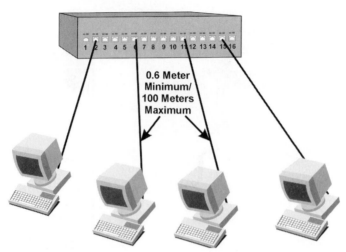

0.6 Meter
Minimum/
100 Meters
Maximum

One advantage of using 10BaseT is that if you need to add another workstation, you just run an additional cable to the hub and plug it in. If there is a break in the cable, the hub has its own intelligence that will route traffic around the defective cable segment. As a result, the entire network is not affected.

NOTE: Since most networks are using Category 5 twisted-pair cabling (or Enhanced Category 5 twisted pair) in a star topology, you should choose the same unless you have an overwhelming reason to choose differently.

4.1.6   100-MBPS Ethernet

Fast Ethernet is an extension of the 10BaseT Ethernet standard that transports data at 100 Mbps and yet still keeps using the CSMA/CD protocol used by 10-Mbps Ethernet. The first type is **100BaseTX,** which uses two pairs of the standard Category 5 UTP. The second type is **100BaseT4,** which runs over existing Category 3 UTP by using all four pairs—three pairs are to transmit data simultaneously, and the fourth pair is used for collision detection.

NOTE: If Category 5 cable is used for backbone for fast Ethernet, it should not exceed 5 meters.

The last one is **100BaseFX,** which operates over multimode fiber-optic cabling.
100VG-ANYLAN was developed by Hewlett-Packard as a 100-Mbps half-duplex transmission that allows 100 Mbps on a four-pair Category 3 cabling system and allows for **voice over IP (VoIP).** Different from the previous Ethernet standards, it uses the Demand Priority Protocol instead of the Ethernet CSMA/CD Protocol. The Demand Priority Protocol allows different nodes and data types to be assigned a priority. Unfortunately,

for a 100VG-ANYLAN network, fast Ethernet has gained a greater market acceptance because it has been standardized by the original IEEE committee.

4.1.7   Gigabit Ethernet

Gigabit Ethernet has been demonstrated to be a viable solution for increased-bandwidth requirements for growing networks, as it is used for high-speed backbones and specialized needs. The Gigabit Ethernet standards include 1000BaseSX (short-wavelength fiber), 1000BaseLX (long-wavelength fiber), 1000BaseCX (short-run copper), and 1000BaseTX (100-meter, four-pair Category 5 UTP).

The Gigabit Ethernet CSMA/CD method has been enhanced in order to maintain a 100-meter collision at gigabit speeds. Without this enhancement, minimum-size Ethernet packets could complete transmission before the transmitting station senses a collision, thereby violating the CSMA/CD method. To resolve this issue, both the minimum CSMA/CD carrier time and the Ethernet slot time have been extended from their present value of 64 bytes to a new value of 512 bytes. Note that the minimum packet length of 64 bytes has not been affected. Packets smaller than 512 bytes have an extra carrier extension. Packets longer than 512 bytes are not extended. These changes, which can impact small-packet performance, have been offset by incorporating a new feature, called packet bursting, into the CSMA/CD algorithm. Packet bursting will allow servers, switches, and other devices to send bursts of small packets in order to fully utilize available bandwidth.

4.1.8   5-4-3 Rule

When creating and expanding an Ethernet network, you must always follow the **5-4-3 rule.** This rule states that an Ethernet network must not exceed five segments connected by four repeaters. Of these segments, only three of them can be populated by computers. This means that the distance of a computer cannot exceed five segments and four repeaters when communicating with other computers on the network. (See figures 4.6 to 4.8.)

Propagation delay is the amount of time that passes between the time a signal is transmitted and the time it is received at the opposite end of the copper or optical cable. As mentioned several times already in this book, when an Ethernet adapter needs to communicate on the wire, it will listen, and if no other adapter is already communicating, it will send its packets on the wire. And of course, if two adapters listen at the same time, both cards will transmit at the same time, resulting in a collision. When you combine Ethernet segments, each segment has a propagation delay. Therefore, when one adapter is communicating, it takes longer for the packets to be transmitted to all the computers on the LAN. If you don't abide by the 5-4-3 rule and you don't follow the maximum cable lengths allowed by the Ethernet specifications, your network will have an increased propagation delay, which will result in a higher number of collisions and thus make the network run less efficiently.

NOTE: You can reduce the number of collisions by using switches instead of hubs since the data packets will only be sent on cables on which the computer exists.

Figure 4.6   5-4-3 Rule for an Ethernet Coaxial Network

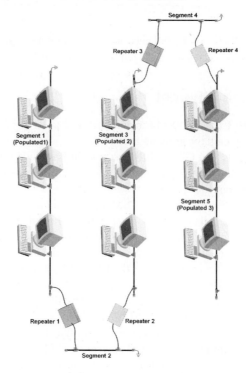

Figure 4.7   5-4-3 Rule for an Ethernet UTP Network

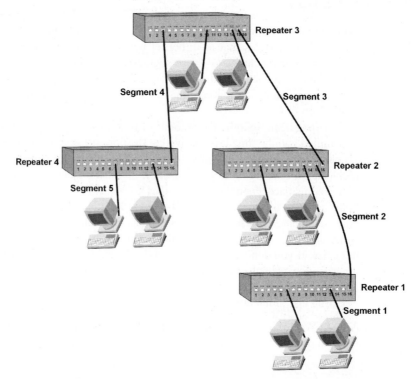

Figure 4.8   5-4-3 Rule for an Ethernet UTP/Coaxial Network

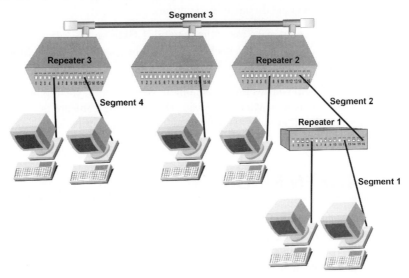

4.2   PLANNING AN ETHERNET NETWORK

Let's take a look at two network examples. The first example consists of a room with approximately 10 computers. This simple network includes purchasing network cards for each computer, one or more hubs, and enough UTP cabling to connect all the network cards. The hub is located somewhere within the room, and the cables are stretched from each computer to the hub. Depending on the function of the network, standard 10-Mbps Ethernet using Category 3 cabling (or better) will probably suffice.

If you need a larger network, it would be a lot more complicated. First, you would place your server or servers in a server room, where the server can be secured and have proper ventilation. Cables are then stretched from wall jacks throughout the building and lead to a patch panel located in the server room or in a wiring closet along with the hub or hubs. The cables in the walls should be eight-wire/four-pair Category 5 or better solid-wire cabling.

NOTE: Depending on the fire code for your area, some or all of the cabling will need to be plenum cabling.

Patch cables are then made using eight-wire/four-pair Category 5 or better stranded cabling. They are used to connect the individual computers to the wall jacks. The port of the patch panel that connects to the wall jack is connected with another patch cable to the hub. The patch panel is a convenient way to connect the network together. In addition, it allows for easy reconfiguration. Again, depending on the function and load of the network, you could do one of two options:

- You could make the entire network 1 Gbps or 100 Mbps.
- You could install a backbone of 1 Gbps connected to 100-Mbps switches to connect the computers.

To help you design and document your network, you might want to use diagramming software such as Microsoft Visio. Diagramming software is a specialized drawing program that allows you to easily draw outlines of your buildings and offices and to specify the location of all cables.

4.3   TOKEN RING

Another major LAN technology in use today is Token Ring. **Token Ring** rules are defined in the IEEE 802.5 specification. As mentioned in the last chapter, the physical topology of Token Ring is a star, but the logical topology is a ring. Therefore, Token Ring is actually implemented in what can best be described as a collapsed ring. In Token Ring LANs, each station is connected to a Token Ring wiring concentrator called a **multistation access unit (MAU)** using a shielded twisted pair or unshielded twisted pair. Like Ethernet hubs, MAUs are usually located in a wiring closet.

Figure 4.9 depicts a typical Token Ring frame. Some of its characteristics are similar to those of the Ethernet frame, but significant differences arise in how the control information is handled. The following list describes the components of the typical Token Ring frame:

- **Start delimiter (SD)**—Signifies the beginning of the packet.
- **Access control (AC)**—Contains information about the priority of the frame.
- **Frame control (FC)**—Defines the type of frame; used in the frame check sequence.
- **Destination address**—Contains the destination node address.
- **Source address**—Contains the address of the originating node.
- **Data**—Contains the data transmitted from the originating node. May also contain routing and management information.
- **Frame check sequence (FCS)**—Used to check the integrity of the frame.
- **End delimiter (ED)**—Indicates the end of the frame.
- **Frame status (FS)**—Indicates whether the destination node recognized and correctly copied the frame or whether the destination node was not available.

The token frame consists of the start delimiter, access control, and end delimiter.

Figure 4.9   A Token Ring Frame

Start Delimiter	Access Control	Frame Control	Destination Address	Source Address	Data	Frame Check Sequence	End Delimiter	Frame Status

The access method used on Token Ring networks is called **token passing.** In token passing, a network device only communicates over the network when it has the token (a special data packet that is generated by the first computer that comes online in a Token Ring network). The token is passed from one station to another around a ring. When a station gets a free token and transmits a packet, it travels in one direction around the ring, passing all the other stations along the way.

Each node acts as a repeater that receives token and data frames from its nearest active upstream neighbor (NAUN). After a frame is processed by the node, the frame is passed or rebroadcast downstream to the next attached node. Each token makes at least one trip around the entire ring. It then returns to the originating node. Workstations that indicate problems send a beacon to identify an address of the potential failure.

If a station has not heard from its upstream neighbor in 7 seconds, it sends a packet down the ring that contains its address and the address of its NAUN. Token Ring can then reconfigure itself to avoid the problem area.

Token Ring network interface cards can run at 4 Mbps or 16 Mbps. The 4-Mbps cards can run only at that data rate. However, 16-Mbps cards can be configured to run at 4 or 16 Mbps. All cards on a given network ring must be running at the same rate. Of course, to run at 4 Mbps, you should use Category 3 UTP or better, whereas 16 Mbps requires Category 4 cable or higher or Type 4 STP cable or higher.

Like Ethernet cards, the node address on each NIC is burned in at the manufacturer on a ROM chip and is unique to each card. For fault tolerance, a maximum of two Token Ring cards can be installed in any node, with each card being defined as the primary or alternate Token Ring card in the machine. The Token Ring card will use two types of connectors: One is equipped with a female (DB-9) 9-pin connector (of which only 4 pins are used), and the second is an RJ-45 connector. MAU's repeaters and most other equipment use a special IBM Type 1 unisex data connector.

NOTE: To connect multiple MAUs, the ring out (RO) port of each MAU must be connected to the ring in (RI) port of the next MAU so that it can complete a larger ring. (See figure 4.10.) In a Token Ring network, you can have up to 33 MAUs chained together.

When building a Token Ring network, you must follow these rules:

- The maximum cable length between a node and a MAU must be 330 feet (100 meters) for Types 1 and 2 cables, 220 feet (66 meters) for Types 6 and 9 cables, and 150 feet (45 meters) for a Category 3 UTP cable.
- Nodes must be separated by a minimum of at least 8 feet (2.5 meters).
- The maximum ring length is 660 feet (200 meters) for Types 1 and 2 cables, 400 feet (120 meters) for a Type 3 cable, 140 feet (45 meters) for a Type 6 cable, and 0.6 mile (1 kilometer) for a fiber-optic segment.
- While IEEE 802.5 specifies a maximum of 250 nodes, if using a small movable Token Ring cable system, IBM STP specifies a maximum of 96 nodes and 12 MAUs (IBM model 8228) using IBM Type 6 cable.
- While IEEE 802.5 specifies a maximum of 250 nodes, if using a large nonmovable Token Ring cable system, you are limited to a maximum of 260 nodes and 33 MAUs (using IBM Types 1 and 2 cables).

Figure 4.10 MAUs Connected Together. Notice That the Cables Attached to the Ring In and the Ring Out Parts are Connected to Form a Larger Ring.

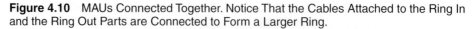
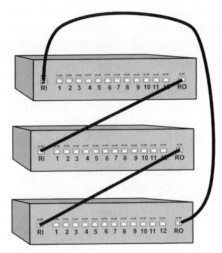

- While IEEE 802.5 specifies a maximum of 250 nodes, if using IBM UTP specifies 72 nodes.
- To run at 4 Mbps, you should use Category 3 UTP or better, whereas 16 Mbps requires Category 4 cable or higher or Type 4 STP cable or higher.
- A media filter is used to convert from Type 1 (shielded twisted-pair) cable to Type 3 (unshielded twisted-pair) cable or the reverse.

4.4   LOCALTALK

The Apple network architecture is called AppleTalk, which has been included in Apple Macintosh computers for several years. By using the computer's built-in network interface, you would usually connect the computers and Apple printers using shielded twisted pair, but could also use unshielded twisted pair and fiber optics. **LocalTalk** is Apple's own network interface and cable system. Since AppleTalk can only operate at 230 Kbps, it is not commonly used. Instead, Macintosh computers will usually use EtherTalk or TokenTalk to communicate with larger and faster networks.

LocalTalk uses a bus topology, where each device in the network is directly connected to the next device in a daisy chain. In addition, for its access method, it uses CSMA/CA. On a Macintosh or Apple IIgs, a small box with a pigtail connector is plugged into the printer port; the pigtail on LocalTalk boxes therefore comes in two variations to support computers with either DIN or 9-pin D printer ports. It is often assumed that the LocalTalk box is only used with a DIN-style printer port, so be careful to purchase the right one for your computer.

The Apple LocalTalk box is equipped with two further DIN sockets so that it can take its place in the daisy chain. The pigtail connector from the LocalTalk box plugs into the printer port, and a length of cable connects the LocalTalk box to the next box on the daisy chain. For the LocalTalk boxes at each end of the chain, only one of the DIN sockets is used. The two ends of the chain should not be joined together. Apple LocalTalk boxes are self-terminating. Therefore, you do not have to provide external terminators.

In addition to the Apple cabling system, LocalTalk is often used with PhoneNet cabling. A PhoneNet box is used in place of the LocalTalk box, which is equipped with RJ-11 telephone-style jack connectors for connecting to the network. PhoneNet also normally uses a bus topology, although some vendors manufacture hubs to support a variety of star topologies. In bus topology, PhoneNet requires a terminator at each end of the chain, and network performance is reduced without the terminators. The terminator is simply an RJ-11 connector with a suitable resistor across two of the pins.

4.5   802.11 WIRELESS STANDARD

The 802.11 standard is the wireless standard that can be compared with the IEEE 802.3 standard for Ethernet for wired LANs. A new standard put out by the Institute of Electrical and Electronics Engineers called 802.11b, or **Wi-Fi,** is making WLAN use faster and easier, and the market is growing quickly. The number of IEEE 802.11b users grew from almost zero in early 2001 to more than 15 million at the end of the 2002. That still isn't much compared with the number of users of cell phones and wired Ethernet, but the growth will likely continue.

4.5.1   802.11 Physical Layer and Architecture

The 802.11 standard defines three physical layers for WLANs: two radio frequency (RF) specifications (direct sequence and frequency hopping spread spectrum) and one infrared (IR). Most WLANs operate in the 2.4-GHz license-free frequency radio band and have throughput rates up to 2 Mbps.

NOTE: The 2.4-GHz band is particularly attractive because it enjoys worldwide allocations for unlicensed operations.

The new 802.11b standard is direct sequence only, and provides throughput rates up to 22 Mbps. The new 802.11a standard will operate in the 5-GHz license-free frequency band and is expected to provide throughput rates up to 54 Mbps in normal mode or 75 Mbps in turbo mode. The newest standard is the 802.11g standard, which has a nominal maximum throughput of 54 Mbps. But because it is using the 2.4-GHz frequency band, its products should be compatible with 802.11b products.

Two operation modes are defined in IEEE 802.11:

■ Ad hoc mode
■ Infrastructure mode

The **ad hoc network,** also referred to as the **independent basic service set (IBSS),** stands alone and is not connected to a base. The wireless stations communicate directly with each other without using an access point or any connection to a wired network. This basic

Figure 4.11   Wireless End Station and Wireless Access Point

topology is useful in order to quickly and easily set up a wireless network anywhere a wireless infrastructure does not exist, such as a hotel room, a convention center, or an airport.

In **infrastructure mode,** also known as the extended service set (ESS), the wireless network consists of at least one **access point (AP)** connected to the wired network infrastructure and a set of wireless end stations. (See figure 4.11.) An access point controls encryption on the network and may bridge or route the wireless traffic to a wired Ethernet network (or the Internet). Access points that act as routers can also assign an IP address to your PCs using DHCP services. APs can be compared with a base station used in cellular networks.

An extended service set consists of two or more basic service sets (BSSs) forming a single subnetwork. Traffic is forwarded from one BSS to another to facilitate movement of wireless stations between BSSs using cellular topology. Almost always the distribution system that connects this network is an Ethernet LAN. Since most corporate WLANs require access to the wired LAN for services (file servers, printers, Internet links), they will operate in infrastructure mode.

One of the requirements of IEEE 802.11 is that it can be used with existing wired networks. 802.11 solved this challenge with the use of a portal. A **portal** is the logical integration between wired LANs and 802.11. It also can serve as the access point to the Distribution System (DS). A distribution system is the means by which an access point communicates with another access point to exchange frames for stations in their respective cell, forward mobile stations as the move from on cell to another, and exchange frames with a wired network. All data going to an 802.11 LAN from an 802.X LAN must pass through a portal. It thus functions as a bridge between wired and wireless.

Today, 802.11a still has some issues to work out, particularly in the area of compatibility. Currently, products aren't backward-compatible with 802.11b products, which clearly dominate the market. And although all 802.11a products use the same chip set, their implementation by each manufacturer differs enough to make them incompatible. Until an interoperability standard is established, 802.11a products from one company may not communicate with those of another.

The Wireless Ethernet Compatibility Alliance (WECA) is an industry consortium that tests for interoperability. Those 802.11b products tests that pass WECA's tests are given the wireless fidelity (Wi-Fi) seal of approval. WECA is working on an 802.11a certification called WiFi5.

4.5.2 Framing

Frame formats are specified for wireless LAN systems by 802.11. Each frame consists of a MAC header, a frame body, and a frame check sequence (FCS). The MAC header consists of seven fields and is 30 bytes long. The fields are frame control, duration, address 1, address 2, address 3, sequence control, and address 4. The frame control field is 2 bytes long and comprises 11 subfields.

The duration/ID field is 2 bytes long. It contains the data on the duration value for each field, and for control frames it carries the associated identity of the transmitting station. The address fields identify the basic service set, the destination address, the source address, and the receiver and transmitter addresses. Each address field is 6 bytes long. The sequence control field is 2 bytes and is split into two subfields: fragment number and sequence number. The fragment number is 4 bits and tells how many fragments the MSDU is broken into. The sequence number field is 12 bits, which indicates the sequence number of the MSDU. The frame body is a variable-length field from 0 to 2312. This is the payload. The frame check sequence is a 32-bit cyclic redundancy check that ensures there are no errors in the frame.

4.5.3 Medium Access Control Protocol

Most wired LANs products use carrier sense multiple access with collision detection (CSMA/CD) as the MAC protocol. Carrier sense means that the station will listen before it transmits. If someone is already transmitting, then the station waits and tries again later. If no one is transmitting, then the station goes ahead and sends what it has. When two stations send at the same time, the transmissions will collide and the information will be lost. This is where collision detection comes into play. The station will listen to ensure that its transmission made it to the destination without collisions. If a collision occurred, then the stations wait and try again later. The time the station waits is determined by a different random amount of time. This technique works great for wired LANs, but wireless topologies can create a problem for CSMA/CD. The problem is the hidden node problem.

The hidden node problem is shown in figure 4.12. Node C cannot hear node A. So if node A is transmitting, node C will not know and may transmit as well. This will result in collisions. The solution to this problem is carrier sense multiple access with collision avoidance, or CSMA/CA. In CSMA/CA, the station listens before it sends. If someone is already transmitting, the station will wait for a random period and try again. If no one is

Figure 4.12   The Hidden Node Problem

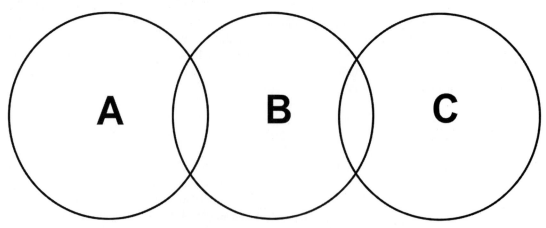

transmitting, then it sends a short message called the "Ready to Send" message (RTS). This message contains the destination address and the duration of the transmission. Other stations now know that they must wait that long before they can transmit. The destination then sends a short message which is the "Clear to Send" message (CTS). This message tells the source that it can send without fear of collisions. Each packet is acknowledged. If an acknowledgment is not received, the MAC layer retransmits the data. This entire sequence is called the four-way handshake, as shown in figure 4.13. This is the protocol that 802.11 chose for the standard.

4.5.4   Security

With traditional wired LANs, you control access to your network by housing it within your office. Tapping into the network remotely is somewhat difficult. But with a WLAN, your network communications are broadcast via radio waves past your office walls, through your neighbor's walls, and out into the parking lot and beyond. Unless you implement proper security, anyone with the right equipment and a little knowledge can see your network traffic.

Minimum steps to take to protect your network should include:

- Change the default network name, ESSID, and the default password needed to sign on to a WLAN on your access points. Each manufacturer's default settings are commonly known by hackers.
- Disable the ESSID broadcast in the access point beacon. By default, access points periodically transmit their ESSID values. Wireless utilities included with Windows XP and as freeware programs can capture these values to present a list of available networks to the user. Disabling this broadcast makes it more difficult for intruders to recognize your network.

Figure 4.13   The Four-way Handshake

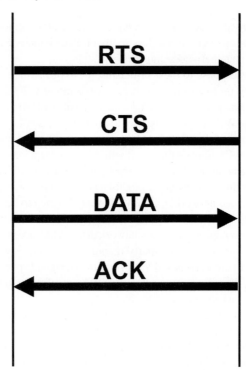

- Enable Wired Equivalent Privacy (WEP). Without encryption, your data is transmitted in readable form. Anyone within radio range using a wireless protocol analyzer or a promiscuous-mode network adapter may capture the data without joining the network. WEP employs RC4 encryption, the same algorithm used for secure online shopping. WEP encryption can generally be found in 64-bit or 128-bit flavors. Of course, use the 128-bit variety if available.
- Change your encryption keys periodically. While the WEP is reasonably strong, the algorithm can be broken in time. The probability of breaking the algorithm is directly related to the length of time that a key is in use. So WEP allows changing of the key to prevent a brute force attack on the algorithm.

  NOTE: WEP can be implemented in hardware or in software. One reason that WEP is optional is because encryption may not be exported from the United States. This allows 802.11 to be a standard outside the United States, albeit without the encryption.

- Enable media access control (MAC) filtering on your access point. Each wireless PC card has a unique identifier known as the MAC address. Many access points let you build a list of MAC addresses that are allowed on the network. Those not listed are denied.

While WEP is probably adequate for most home use, depending on the confidentiality of the data, corporate administrators should add strong encryption to the WLAN. Virtual private networks (VPNs) can be used so that you can also use IPSec or PPTP encryption over the wireless segment. See chapters 5 and 13 for more information on virtual private networks and security.

In response to these criticisms, the Wi-Fi Alliance recently announced a new wireless security protocol that will be available in early 2003. It's called Wi-Fi Protected Access (WPA), and is designed to take the place of WEP and address many of its shortcomings. For starters, WPA requires the user to provide a master key, but this does not become a static encryption key. Instead, the master key is simply a password used as a starting point through which WPA derives the key it will use to encrypt network traffic. Moreover, the key is regularly and automatically changed (and never reused), reducing the likelihood that it will be compromised. The master key also serves as a password by which users can be authenticated and granted network access. WPA was designed to be a software upgrade to WEP, so most existing wireless devices should be upgradeable to WPA via a firmware update. In order to take advantage of WPA, all network devices like access points and clients must be upgraded.

A number of third-party manufacturers have stepped into the breach to offer a new class of hardware-based VPN products designed specifically for wireless networks. In addition, IEEE formed a task group, TGi, specifically to tighten up WLAN security in a nonproprietary systematic format.

TGi proposed a stopgap measure called the Temporary Key Integrity Protocol (TKIP), which is intended to work with existing and legacy hardware. Your WLAN provider may add this feature through a firmware revision. TKIP uses a mechanism called fast-packet rekeying, which changes the encryption keys frequently.

In addition, TGi also ratified the use of **802.1X,** which requires a WLAN client to initiate an authorization request to the access point; it authenticates the client with an Extensible Authentication Protocol (EAP)-compliant RADIUS server. The RADIUS server may authenticate either the user (via passwords) or the machine (by MAC address). In theory, the wireless client is not allowed to join the network until the transaction is complete.

One form of EAP that was created by Cisco and is not supported by Microsoft is PEAP. PEAP, short for Protected Extensible Authentication Protocol, is a form of EAP that is combined with transport-layer security (TLS) to create a TSL-encrypted channel between client and server. To provide mutual authentication, PEAP uses Microsoft Challenge Authentication Protocol (MS-CHAP) version 2. Because the challenge/response packets are sent over a TLS-encrypted channel, the password and the key are not exposed to offline dictionary attacks. See chapter 5 for more information on MS-CHAP and chapter 13 for more information on TLS.

4.6   BLUETOOTH

Bluetooth is a radio frequency (RF) specification for short-range, point-to-multipoint voice and data transfer. It aims to simplify communications among Net devices and between devices and the Internet. It also aims to simplify data synchronization between Net devices and other computers. An advantage of Bluetooth is its similarity to many other specifications already deployed and its borrowing of many features from these specifications. The Bluetooth standard is becoming more and more of a short time network between devices for a small amount of information.

Bluetooth has a present nominal link range of 10 centimeters to 10 meters, which can be extended to 100 meters, with increased transmitting power. Bluetooth operates in the 2.4-GHz industrial-scientific-medical (ISM) band and uses a frequency hop spread-spectrum technology in which packets are transmitted in defined time slots on defined frequencies. A full-duplex information interchange rate of up to 1 Mbps may be achieved in which a time-division duplex (TDD) scheme is used. The second generation of Bluetooth supports up to 2 Mbps.

Work on the Bluetooth specification is progressing and is primarily the responsibility of the Bluetooth Special Interest Group (SIG). This is an industry group consisting of leaders in telecommunications and computing industries. The promoter group within the SIG currently consists of 3Com, Ericsson, IBM, Intel, Lucent Technologies, Microsoft, Motorola, Nokia, and Toshiba.

A Bluetooth system essentially comprises the following four major components:

- **Radio unit**—Consisting of a radio transceiver, which provides the radio link between the Bluetooth devices.
- **Baseband unit**—A hardware consisting of flash memory and a CPU. This interfaces with the radio unit and the host device electronics.
- **Link management software**—A driver software or firmware that enables the application software to interface with the baseband unit and radio unit.
- **Application software**—This implements the user interface and is the application that can run on wireless. For example, this could be chat software that allows two laptop users in a conference hall to talk to each other using wireless technology.

Each device has a unique 48-bit address from the IEEE 802 standard. In addition, a frequency hop scheme allows devices to communicate even in areas with a great deal of electromagnetic interference, and it includes built-in encryption and verification.

One Bluetooth standard is **HomeRF SWAP** (SWAP is short for Shared Wireless Access Protocol). It is designed specifically for wireless networks in homes, in contrast to 802.11, which was created for use in businesses. HomeRF networks are designed to be more affordable to home users than other wireless technologies. HomeRF is based on frequency hopping and using radio frequency waves for the transmission of voice and data with a range of up to 150 feet.

SWAP works together with the PSTN network and the Internet through existing cordless telephone and wireless LAN technologies to enable voice-activated home electronic systems, accessing the Internet from anywhere in the home and forward fax, voice, and email messages. SWAP uses time-division multiple access for interactive data transfer and CSMA/CA for high-speed packet transfer. SWAP operates in the 2400-MHz band at 50 hops per second to provide a data rate between 1 and 2 Mbps.

SUMMARY

1. Ethernet is the most widely used LAN technology today.

2. The Ethernet standard is defined by the Institute of Electrical and Electronics Engineers (IEEE) in a specification commonly known as IEEE 802.3.

3. Ethernet networks can be configured in either a star topology using unshielded twisted pair connected to a hub or a bus topology using a coaxial cable acting as a backbone.

4. When a network device wants to access the network, the access method used is carrier sense multiple access with collision detection (CSMA/CD).

5. The original Ethernet was called 10Base5 or thicknet. 10Base5 got its name because it has a 10-Mbps baseband network that can have a cable segment up to 500 meters long. It uses a 50-ohm RG-8 and RG-11 as a backbone cable.

6. Most network devices connected to the network use an external transceiver that uses a 4-pair AUI cable (sometimes referred to as a drop cable) via a DIX connector (2 row, 15-pin female connector).

7. The 10Base2 (thinnet) is a simplified version of the 10Base5 network. Its name derives from the fact that it has a 10-Mbps baseband network with a maximum cable segment length of approximately 200 meters (actually 185 meters).

8. 10BaseT uses UTP, which costs less, is smaller, and is easier to work with than coax cable. While Ethernet is a logical bus topology, 10BaseT is a physical star topology; it has the network devices connected to a hub (or switch).

9. Fast Ethernet is an extension of the 10BaseT Ethernet standard that transports data at 100 Mbps and yet still keeps using the CSMA/CD protocol used by 10-Mbps Ethernet.

10. 100VG-ANYLAN was developed by Hewlett-Packard as a 100-Mbps half-duplex transmission that allows 100 Mbps on a four-pair Category 3 cabling system and allows for voice of IP (VoIP).

11. Gigabit Ethernet has been demonstrated to be a viable solution for increased-bandwidth requirements for growing networks, as it is used for high-speed backbones and specialized needs.

12. The 5-4-3 rule states that an Ethernet network must not exceed five segments connected by four repeaters. Of these segments, only three of them can be populated by computers.

13. Propagation delay is the amount of time that passes between the time a signal is transmitted and the time it is received at the opposite end of the copper or optical cable.

14. Token Ring rules are defined in the IEEE 802.5 specification.

15. In Token Ring LANs, each station is connected to a Token Ring wiring concentrator called a multistation access unit (MAU) using a shielded twisted pair or unshielded twisted pair.

16. The access method used on Token Ring networks is called token passing.

17. LocalTalk is Apple's own network interface and cable system. Since AppleTalk can only operate at 230 Kbps, it is not commonly used. LocalTalk uses a bus topology, where each device in the network is directly connected to the next device in a daisy chain. In addition, for its access method, it uses CSMA/CA.

18. The 802.11 standard is the wireless standard that can be compared with the IEEE 802.3 standard for Ethernet for wired LANs.

19. The ad hoc network stands alone and is not connected to a base. The wireless stations communicate directly with each other without using an access point or any connection to a wired network.

20. In infrastructure mode, the wireless network consists of at least one access point (AP) connected to the wired network infrastructure and a set of wireless end stations.

21. An access point controls encryption on the network and may bridge or route the wireless traffic to a wired Ethernet network (or the Internet). Access points that act as routers can also assign an IP address to your PCs using DHCP services.

22. A portal is the logical integration between wired LANs and 802.11.

23. Bluetooth is a radio frequency (RF) specification for short-range, point-to-multipoint voice and data transfer.

24. HomeRF SWAP is a Bluetooth standard designed specifically for wireless networks in homes, in contrast to 802.11, which was created for use in businesses.

QUESTIONS

1. A network interface card has only a single DB-9 female connector. What type of network is it used for?
   a. Token Ring
   b. FDDI
   c. Ethernet
   d. None of the choices listed

2. A network interface card has a DB-15 female connector and an RJ-45 connector. This is what type of card?
   a. Token Ring
   b. Ethernet
   c. FDDI
   d. It can be used with all those listed above.

3. Carrier sense multiple access collision detection is a feature of what LAN technology?
   a. FDDI   b. Token Ring
   c. ATM   d. Ethernet

4. Which of the following networks has the greatest bit throughput?
   a. 100VG-ANYLAN
   b. 100BaseTX
   c. 100BaseFX
   d. They are all the same.

5. In a CAN (campus area network) you need to run a network cable between two buildings that are 250 meters apart. You need to make sure that you do not introduce a problem with attenuation, and you need to provide for high-bandwidth capability. You decide to run thicknet coaxial cable. What is the result?
   a. You solve the first problem, but not the second.
   b. You solve the second problem, but not the first.
   c. You solve both problems.
   d. You don't solve either problem.

6. In your network you have five active hubs connected in a serial configuration. You have three of the hubs populated with ten workstations each. How many unique network IDs do you need?
   a. 5   b. 3
   c. 30   d. 1

7. What technology was developed for VoIP?
   a. 100VG-ANYLAN   b. 100BaseFX
   c. 100BaseTX   d. 10Base5

8. What do thicknet and thinnet have in common?
   a. Both support 100-Mbps data transmission speeds.
   b. Both can be run up to 500 meters.
   c. Both support Token Ring.
   d. Both use coaxial cabling.

9. You are troubleshooting a newly installed network to determine why it is running slow. On one side of a 24-port bridge is CAT 5 twisted-pair cabling which is run to the workstations. On the other side, a segment of RG-58 coaxial cabling is run to the network server. All network cards support 100 Mbps, but the network is only running at 10 Mbps. The company's owner will pay you enough to retire if you can get his network up and running at 100 Mbps. What do you do?
   a. Make sure full duplex is selected on the switch.
   b. Replace the RG-58 cable with a CAT 5 cable.
   c. Replace the CAT 5 cables with RG-58 cables.
   d. Replace the RG-58 cable with an RG-59 cable.

10. What is the maximum overall length of a 10-Mb Ethernet network?
    a. 3000 feet   b. 5500 feet
    c. 3000 meters   d. 6000 meters

11. What is the maximum allowable length of a segment of cabling used for a 100BaseFX network? (Select two answers.)
    a. 500 meters
    b. 1.2 miles
    c. 1000 meters
    d. 2000 meters

12. What is the 5-4-3 rule?
    a. 5 repeaters, 4 segments, 3 maximum populated segments
    b. 5 bridges, 4 segments, 3 repeaters
    c. 5 routers, 4 bridges, 3 repeaters
    d. 5 segments, 4 repeaters, 3 maximum populated segments

13. What is the maximum number of stations that a single thinnet segment can handle?
    a. 50   b. 40
    c. 60   d. 30

14. 100BaseTX Ethernet uses how many pairs of wires?
    a. 2   b. 6
    c. 4   d. 8

15. What is the maximum allowable length of a single thinnet cable segment? (Choose two answers)
    a. 185 feet
    b. 601 feet
    c. 601 meters
    d. 185 meters
16. Which technology standard does NOT utilize CSMA/CD?
    a. 100VG-ANYLAN
    b. 10Base2
    c. 100BaseTX
    d. 10Base5
17. What is a BNC T-connector used for?
    a. To connect a workstation to a server
    b. To connect 10Base5 segments
    c. To terminate a coaxial segment
    d. To connect stations using thinnet to the network
18. You need to run a single network cable between two buildings that are 650 meters apart. What cable type must you use?
    a. Thinnet coaxial
    b. Twisted pair
    c. Fiber optic
    d. Thicknet coaxial
19. If a workstation fails, its downstream neighbor sends out a special packet called a beacon. This process is called beaconing. Which technologies use beaconing? (Choose two answers.)
    a. Ethernet
    b. FDDI
    c. ATM
    d. Token Ring
20. What is the specification for running Fast Ethernet over fiber?
    a. 10Base5
    b. 100BaseTX
    c. 100BaseFX
    d. 10BaseT
21. Your boss wants you to recommend an inexpensive cable type and connector that is compatible with the 802.3 IEEE standard and will also support 100-Mbps throughput. What do you suggest she buy?
    a. Category 7 UTP, BNC
    b. STP, RJ-45
    c. Coaxial, BNC
    d. Category 5 UTP, RJ-45
22. Which of the following is completely true of 100BaseTX?
    a. Requires coaxial, 100-Mbps, 500-meter-max cable segment
    b. Requires CAT 5 twisted-pair, 1000-Mbps, 100-meter-max cable segment
    c. Requires coaxial, 100-Mbps, 100-meter-max cable segment
    d. Requires CAT 5 twisted-pair, 100-Mbps, 100-meter-max cable segment

23. Which of the following is completely true of 10Base5?
    a. Requires coaxial, 10-Mbps, 500-meter-max cable segment, 1000 stations per segment
    b. Requires coaxial, 10-Mbps, 185-meter-max cable segment, 500 stations per segment
    c. Requires coaxial, 10-Mbps, 500-meter-max cable segment, 50 stations per segment
    d. Requires coaxial, 10-Mbps, 500-meter-max cable segment, 500 stations per segment
24. A Token Ring hub is also known as what?
    a. VAX
    b. TPU
    c. BTU
    d. MAU
25. Which media connector on a NIC is a D-shell, 15-pin female interface? (Choose two answers)
    a. Thinnet
    b. DIX
    c. BNC
    d. AUI
26. Most 16-Mbps Token Ring networks use what type of cabling?
    a. UTP
    b. Fiber optic
    c. Coaxial
    d. STP
27. Which of the following is completely true of 100BaseFX?
    a. Requires fiber-optic, 100-Mbps, 1-km-max cable segment
    b. Requires fiber-optic, 10-Mbps, 2-km-max cable segment
    c. Requires coaxial, 100-Mbps, 2-km-max cable segment
    d. Requires fiber-optic, 100-Mbps, 2-km-max cable segment
28. Which of the following is completely true of 10Base2?
    a. Requires coaxial, 10-Mbps, 185-meter-max cable segment, 30 stations max per segment
    b. Requires coaxial, 10-Mbps, 500-meter-max cable segment, 30 stations max per segment
    c. Requires CAT 5 twisted-pair, 10-Mbps, 185-meter-max cable segment, 30 stations max per segment
    d. Requires coaxial, 10-Mbps, 185-meter-max cable segment, 50 stations max per segment
29. Which of the following is completely true of 10BaseT?
    a. Requires CAT 5 twisted-pair, 100-Mbps, 100-meter-max cable segment
    b. Requires CAT 3 twisted-pair, 100-Mbps, 100-meter-max cable segment

c. Requires CAT 5 twisted-pair, 10-Mbps-max-speed, 100-meter-max cable segment

d. Requires CAT 3 twisted-pair, 10-Mbps-max-speed, 100-meter-max cable segment

30. Token Ring networks can operate at what speeds? (Choose two answers.)

   a. 24 Mbps   b. 16 Mbps

   c. 100 Mbps   d. 4 Mbps

31. The amount of time required for a packet to travel through a switch is called what?

   a. Latency

   b. Signal oscillation

   c. Attenuation

   d. Frequency

32. What is an AUI?

   a. Attachment user interlink

   b. Application user interface

   c. Application user interlink

   d. Attachment user interface

33. What is the maximum cable length used on a Gigabit Ethernet segment that uses fiber-optic media?

   a. 2 kilometers   b. 5000 feet

   c. 3500 meters   d. 5 kilometers

34. Which two Ethernet standards use fiber-optic cabling? (Choose two answers.)

   a. 100BaseT   b. 1000BaseSX

   c. 10Base2   d. 100BaseFX

35. What is the maximum speed of a gigabit network?

   a. 1 million bits per second

   b. 1 trillion bits per second

   c. 1 billion bits per second

   d. 1 million bytes per second

36. Which two high-speed Ethernet standards are commonly used as backbones? (Choose two answers.)

   a. 10Base5   b. Gigabit

   c. 100BaseT   d. 100BaseFX

37. What is the maximum cable segment length for cable used in 100BaseFX (100 million bits per second)?

   a. 2 miles   b. 1500 feet

   c. 5000 meters   d. 2 kilometers

38. What is the maximum data transfer speed on a 100BaseFX network?

   a. 100 million bits per second

   b. 100 billion bits per second

   c. 1000 million bits per second

   d. 100 kilobits per second

39. Which of the following is an AppleTalk protocol?

   a. MegaTalk   b. MacTalk

   c. RemoteTalk   d. LocalTalk

40. In the 802.3 IEEE standard, what category UTP cable is required to support 10 Mbps?

   a. CAT 4   b. CAT 5

   c. CAT 7   d. CAT 3

41. In the 802.3 IEEE standard, what category UTP cable is required to support 100 Mbps?

   a. CAT 3   b. CAT 5

   c. CAT 4   d. CAT 7

42. A CAT 5 UTP cable extending from your wire closet's patch panel to the workstation's patch panel is exactly 100 meters in length. The patch cable at the workstation is 3 meters long, and the patch cable in the wire closet is 5 meters long. Your network is a 100BaseTX Ethernet network. Which statement below is the most accurate for this configuration?

   a. CAT 5 supports 100BaseTX, so everything is configured properly.

   b. This configuration is correct since the patch cables do not exceed 5 meters in length.

   c. As long as the patch cables are terminated properly, this configuration is correct.

   d. The total cable length exceeds UTP specifications for 100BaseTX.

43. What is an MSAU used as?

   a. As an FDDI ring

   b. As a Token Ring hub

   c. As a WAN link

   d. As an Ethernet hub

44. 100BaseTX requires at a minimum what category level of UTP?

   a. 4   b. 5

   c. 7   d. 3

45. What is propagation delay?

   a. Routing table assimilation delay

   b. Protocol switching time

   c. NIC buffer response time

   d. Period of time for a signal change to traverse the cable

46. Your network consists of five servers, twenty workstations, and one printer. Each device has a thinnet coaxial cable going from it to a transceiver with a vampire tap that is tapped into a single run of 10Base5 cabling. The 10Base5 cable is terminated on both ends with a 50-ohm terminator, and one end of it is grounded. What is your physical topology?

   a. Bus   b. Ring

   c. Star   d. Mesh

47. Your WAN consists of the following: A local area network in Atlanta with a Cisco 2500 router and a local area network in Brisbane with a Cisco 4000 router. What is the least number of unique network IDs you need?
    a. 2   b. 4
    c. 3   d. 5

48. Which is the proper configuration for installing coaxial cable used by Ethernet?
    a. Both ends terminated, both ends grounded
    b. One end terminated, one end grounded
    c. One end terminated, both ends grounded
    d. Both ends terminated, one end grounded

49. You administer a local area network that uses the 10BaseT Ethernet standard. You have connected a Windows 95 computer to your network, but this computer cannot contact the network. You have used a 10-meter cable to connect the network interface card to the wall outlet. A 90-meter drop cable connects the wall outlet to a patch panel in the wiring closet, and a 5-meter cable connects the patch panel to an Ethernet hub. What should you do to allow the Windows 95 computer to contact the network?
    a. Replace the 10-meter cable between the Windows 95 computer and the wall outlet with a 20-meter cable.
    b. Replace the 10-meter cable between the Windows 95 computer and the wall outlet with a 5-meter cable.
    c. Replace the 90-meter UTP cable between the wall outlet and the patch panel with a 90-meter fiber-optic cable.
    d. Replace the 5-meter UTP cable between the patch panel and the Ethernet hub with a 5-meter fiber-optic cable.

50. The best cable choice for linking a few computers in a small office using a bus Ethernet network is cable with an _____ designation.
    a. RG-47   b. RG-58
    c. RG-59   d. RG-62

51. You have a thinnet bus network that has been in use for about a year. You have just added three new client computers to the network. When you test the network after installation, none of the computers on the network can access the server. Which of the following may be preventing the network from functioning? (Choose two answers)
    a. The bus network is not properly terminated.
    b. A client computer on the network has failed.

    c. The new cables added to service the new computers are not compatible with the existing cable type.
    d. The network adapter cards in the new computers are not compatible with the network adapter cards in other computers.

52. What is the distance limitation of a 10BaseT network?
    a. 100 meters
    b. 500 meters
    c. 185 meters
    d. 1 kilometer

53. What type of connector is used by thinnet for connection to the network adapter card?
    a. An AUI connector
    b. A BNC T-connector
    c. An RJ-45 connector
    d. A BNC barrel connector

54. Ethernet uses a special type of data format called frames. Which of the following statements is true of Ethernet frames?
    a. All Ethernet frames are 1518 bytes long.
    b. All Ethernet frames contain preambles that mark the start of the frame.
    c. All Ethernet frames contain starting and ending delimiters.
    d. All Ethernet frames contain CRC fields that store the source and destination addresses.

55. Which type of media access method is used by IBM-based LANs with MAUs?
    a. Token passing
    b. CSMA/CD
    c. Demand priority
    d. CDMS/CA

56. Pat is a network engineer for the Acme Corporation. The architects in the Sacramento branch office want to connect their Windows 98 SE computers and their laser printers in a network for file and printer sharing. They ask Pat to design the network for them. She tells the architects that a 10BaseT Ethernet network is inexpensive and relatively easy to install, and the architects agree to allow Pat to design this type of network. Which physical topology will Pat implement in designing the network for the architects in Sacramento?
    a. Bus   b. Ring
    c. Star   d. Mesh

57. Which of the following are characteristics of a Token Ring network? (Select all correct answers.)
   a. Logical ring
   b. Token passing
   c. Physical star
   d. Demand priority access

58. You have been contracted to build a network for a multimedia development firm that currently uses a 10-Mbps Ethernet network. The company requires a high-bandwidth network for the multimedia team, which constantly views and manipulates large files across the network. The company is expecting moderate growth. You are to come up with a solution to support the high-bandwidth applications and growth potential of this company.

Required result:
■ Increase network bandwidth.

Optional result:
■ Support future growth.
■ Improve server response time.

Proposed solution:
■ Install CAT 5 UTP cable in a star topology with the existing hub.
■ Upgrade the workstation's adapter cards to support 100 Mbps.
■ Increase the amount of RAM in the server.

The proposed solution _____.
   a. achieves the required result and both optional results
   b. does not achieve the required result but achieves both the optional results
   c. does not achieve the required result but does achieve one of the optional results
   d. achieves the required result but cannot achieve either of the optional results
   e. achieves neither the required result nor either of the optional results

59. Suppose the following situation exists: You are the senior network administrator for the Acme Corporation. Pat, the technology officer, has asked you to design a network for a new branch office in Russia. He would like you to use the following specifications in your design.

Required result:
■ The Russian network must use Category 5 unshielded twisted-pair (CAT 5 UTP) cable.

Optional desired results:
■ The Russian network must support a physical star topology.
■ The Russian network must transmit data at 1 gigabit per second (Gbps).

Proposed solution:
■ You design the Russian network using the 100BaseT standard.

Which results does the proposed solution produce?
   a. The proposed solution produces the required result and both of the optional desired results.
   b. The proposed solution produces the required result and only one of the optional desired results.
   c. The proposed solution produces only the required result.
   d. The proposed solution produces only the two optional desired results.

60. Your company is a medium-size firm and leases two buildings. The two buildings are 800 meters apart. Each building has a 100BaseT Ethernet network.

Required result:
■ Connect the two separate networks.

Optional result:
■ The network needs to support a transmission rate of 1000 Mbps.
■ Must be immune to interference.

Proposed solution:
■ Implement fiber-optic Ethernet.

Which statement categorizes the proposed solution?
   a. It achieves the required result and both optional results.
   b. It does not achieve the required result but does achieve both the optional results.
   c. It does not achieve the required result but does achieve one of the optional results.
   d. It achieves the required result but cannot achieve either of the optional results.
   e. It achieves neither the required result nor either of the optional results.

61. Look at the figure and determine the problem with this Ethernet network.

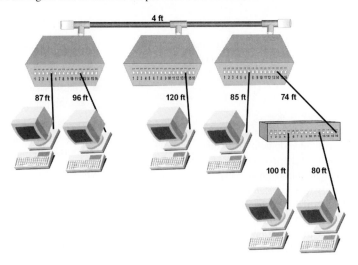

62. Look at the figure and determine the problem with the network.

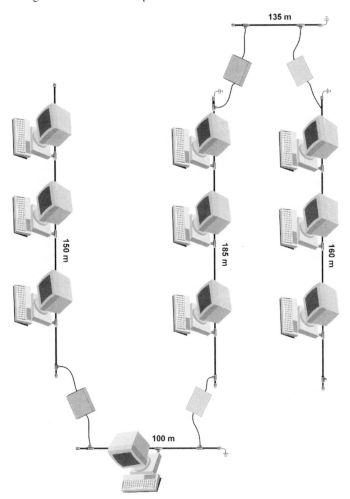

HANDS-ON EXERCISES

Exercise 1: Installing a Windows 2000 Server

NOTE: You can also install Windows 2000 Advanced Server or Windows Server 2003.

1. Reboot the computer with the proper DOS CD-ROM drivers.

   NOTE: You can also boot with a bootable network disk and connect to a network drive to load the Windows 2000 files.

2. Load the Windows 2000 Server CD or go to the network drive/directory where the Windows 2000 installation files are.
3. Change into the I386 directory by using the CD I386 command.
4. Type in WINNT and press the Enter key.
5. When Windows 2000 Setup needs to know where the Windows 2000 files are located, press the Enter key. Windows 2000 will copy the installation files over the C drive. Be patient; this will take a few minutes.
6. When the MS-DOS–based portion of Setup is complete, remove any floppy disks from the A drive and press the Enter key to restart the computer.
7. Windows 2000 will welcome you to Setup. To set up Windows now, press the Enter key.
8. When the license agreement appears, press the F8 key to continue.
9. When Windows shows the partitions, select the C drive (2047 MB partition). This way, the system partition will be the C drive, and the boot partition for Windows 2000 Server will be the C drive. Press the Enter key to install.
10. If needed, convert or format the partition to the NTFS option and press the Enter key. Windows 2000 will copy some more files. The system will reboot again.
11. When the system reboots and starts a graphical interface, it will then automatically detect and install the hardware devices. Next it will ask for the regional settings. Select the appropriate regional settings and click on the Next button.
12. To personalize your software, enter the name and the company that you work for. Click on the Next button.
13. In the next screen, select the Per Server licensing mode and enter 50 connections. Click on the Next button.

    NOTE: To get the most out of this book, you will need to be working with a partner or a second computer. The computer on the left will be designated as computer A, and the computer on the right will be designated as computer B.

14. The next screen shows a random computer name. Change the computer name to Server2000-xxy, where xx represents your two-digit partner number in the class and y represents A if you are computer A or B if you are computer B. If you are not doing this in class, use 01. If you are the first set of partners and you are using the computer on the left, you would use Server2000-01A. If you are the first set of partners and you are using the computer on the right, you would use Server2000-01B. Lastly, enter the password of password in the Administrator password and confirm password text boxes. Click on the Next button.
15. The installation wizard will ask you to add or remove components of Windows 2000. Since there is nothing that you want to add at this time, click on the OK button.
16. If your computer has a modem, a modem dialing information box will appear. Enter your area code and type in the appropriate options for your computer. Click on the Next button.

17. For the date and time settings, enter the proper information and click on the Next button.
18. Windows 2000 will configure the networking settings. Select Custom settings and click on the Next button.
19. The network components chosen by default are client for Microsoft networks, file and printer sharing for Microsoft networks, and Internet Protocol (TCP/IP). Click on the Internet Protocol and click on the Properties button.
20. In the Internet Protocol (TCP/IP) properties dialog box, click on the Use the following IP address option and input the following:

    IP address: 192.168.XXX.1YY
    Subnet mask: 255.255.255.0

    where XXX (1–255) is your room number and YY is the computer number assigned. If you are working at home, use 1 and 01. Therefore, the address would be 192.168.1.101. In addition, if your network has a gateway and a DNS server, specify them. Click on the OK button and click on the Next button.

    NOTE: If the card is not recognized, you are going to have to finish the installation and install the appropriate driver and configure the network card.

21. The next window asks if you want to be a member of a workgroup or a computer domain. For now, select No and click on the Next button. The computer will then copy some files and perform final tasks.
22. When the Windows 2000 Setup wizard is complete, click on the Finish button and the computer will reboot.
23. If the boot menu appears, select the Windows 2000 Server.
24. Log on as the administrator.
25. Open the Device Manager. Either right-click the My Computer or click the Start button, select Settings, select Control Panel, and double-click the System applet. Then click on the Hardware tab and the Device Manager button. Make sure that all drivers loaded properly. If not, load the appropriate drivers.
26. Right-click on the desktop and create a DATA folder.
27. Open the DATA folder. Right-click the empty space of the DATA folder and select New to create a text file. In the text file, type your name in the document.
28. Right-click the DATA folder and select Sharing. Click the Share this folder and click on the OK button.
29. Open the My Network Places and browse to your computer to see the DATA share. You may need to click on Computers Near Me or Entire Network.
30. Browse to your partner's computer to see the DATA share.

Exercise 2: Building an Ethernet Network

1. Look at the hub used in the classroom or at home. Determine the types of lights on the hub and determine if your hub has an uplink port.
2. Connect the cable to the hub and your computer. Notice the status of the link light on the network card and the hub.
3. Disconnect the cable and notice the status of the link light.
4. Reconnect the cable.

CHAPTER 5

WAN and Remote Access Technologies

Topics Covered in this Chapter

Introduction

In this chapter, we look at the wide area network (WAN) technology used to link individual computers or entire networks together, a technology that covers a relatively broad geographic area. We also look at remote access, which allows users to dial into a server/network. These connections are usually made through a telephone company or some other common carrier. While LANs typically are multiple-point connections, WANs typically are made of multiple point-to-point connections.

Objectives

- Identify the purpose, features, and functions of CSU/DSUs, modems, and ISDN adapters.
- Identify the basic characteristics (speed, capacity, and media) of ISDN, ATM, Frame Relay, SONET, SDH, T-1/E-1, T-3/E-3, and OC-x.
- Compare and contrast packet switching and circuit switching.
- Given a troubleshooting scenario involving a small office/home office network failure (e.g., xDSL, cable, wireless, ports), identify the cause of the problem.

- Define the purpose, function, and/or use of the telnet protocols within TCP/IP.
- Define the function of RAS, PPP, PPTP, and ICA protocols and services.
- Given a remote connectivity scenario (IP, IPX, dial-up, authentication, and physical connectivity), configure the connection.
- Given a troubleshooting scenario involving a remote connectivity problem (e.g., authentication failure, protocol configuration, and physical connectivity), identify the cause of the problem.

5.1   WAN DEVICES

Tables 5.1 and 5.2 summarize various WAN and remote access technology. Typically, when talking about the various WAN connection devices, the devices can be divided into the following categories:

- Data terminal equipment (DTE)
- Data circuit-terminating equipment (DCE)

Data terminal equipment (DTE) devices are end systems that communicate across the WAN. They control data flowing to or from a computer. They are usually terminals, PCs, or network hosts that are located on the premises of individual subscribers. **Data circuit-terminating equipment (DCE)** devices are special communication devices that provide the interface between the DTE and the network. Examples include modems and adapters. The purpose of the DCE is to provide clocking and

Table 5.1   WAN and Remote Access Technology

Carrier Technology	Speed	Physical Medium	Connection Type	Comment
Plain old telephone service (POTS)	Up to 56 Kbps	Twisted pair	Circuit switch	Used by home and small business
Asymmetrical digital subscriber line lite (ADSL lite)	Up to 1.544 Mbps downstream Up to 512 Kbps upstream	2 twisted pairs	Circuit switch	Used by home and small business
Asymmetrical digital subscriber line (ADSL)	1.5–8 Mbps downstream Up to 1.544 Mbps upstream	2 twisted pairs	Circuit switch	Used by small to medium business
High bit-rate digital subscriber line (HDSL)	1.544 Mbps full duplex (T-1) 2.048 Mbps full duplex (E-1)	2 pairs of twisted pair	Circuit switch	Used by small to medium business
DS-0 leased line	64 Kbps	1 or 2 pairs of twisted pair	Dedicated point-to-point	The base signal on a channel in the set of digital signal levels
Switched 56	56 Kbps	1 or 2 pairs of UTP	Circuit switch	Used by home and small business
Switched 64	64 Kbps	1 or 2 pairs of UTP	Circuit switch	Used by home and small business
Fractional T-1 leased line	64 Kbps to 1.536 Mbps in 64-Kbps increments	1 or 2 pairs of twisted pair	Dedicated point-to-point	Used by small to medium business
T-1 leased line (DS-1)	1.544 Mbps (24–64-Kbps channels)	2 pairs UTP or UTP or optical fiber	Dedicated point-to-point	Used by medium to large business, ISP to connect to the Internet
T-3 leased line (DS-3)	44.736 Mbps	2 pairs of UTP or optical fiber	Dedicated point-to-point	Used by large business, large ISP to connect to the Internet or as the backbone of the Internet
E-1	2.048 Mbps	Twisted pair, coaxial cable, or optical fiber		32-channel European equivalent of T-1
ISDN—BRI	64 to 128 Kbps	1 or 2 pairs of UTP	Circuit switch	Used by home and small business

Continued

Table 5.1   Continued

Carrier Technology	Speed	Physical Medium	Connection Type	Comment
ISDN—PRI	23–64-Kbps channels plus control channel up to 1.544 Mbps (T-1) or 2.048 (E-1)	2 pairs of UTP	Circuit switch	Used by medium to large business
Frame Relay	56 Kbps to 1.536 Mbps (using T-1) or 2.048 Mbps using E-1	1 or 2 pairs of twisted pair	Packet switch	Popular technology used mostly to connect LANs together
FDDI/FDDI-2	100 Mbps/200 Mbps	Optical fiber	Packet switch	Large, wide-range LAN usually in a large company or a larger ISP
SDH/SONET	51.84 Mbps and up	Optical fiber	Dedicated point-to-point	Used by ATM
SMDS	1.544 to 34 Mbps using T-1 and T-3 lines	1 or 2 pairs of twisted pair	Cell relay	Popular growing technology used mostly to connect LANs together. SMDS is the connectionless component of ATM.
ATM	Up to 622 Mbps	T-1, T-3, E-1, E-3, SDH, and SONET	Cell Relay	The fastest network connection up to date
Cable modems	500 Kbps to 1.5 Mbps or more	Coaxial cable	Leased point-to-point	Used by home and small business

switching services in a network, and the DCE actually transmits data through the WAN. Therefore, the DCE controls data flowing to or from a computer. In practical terms, the DCE is usually a modem, and the DTE is the computer itself, or more precisely, the computer's UART chip. For internal modems, the DCE and DTE are part of the same device.

Another term that is used often when discussing WAN connections is the **channel service unit/data service unit (CSU/DSU).** The DSU is a device that performs all error correction, handshaking, and protective and diagnostic functions for a telecommunications line. The CSU is a device that connects a terminal to a digital line that provides the LAN/WAN connection. Typically, the two devices are packaged as a single unit. You can think of it as a very high-powered and expensive modem.

Table 5.2   Comparison of WAN Technologies

Technology	Speed	Pros	Cons
POTS/V.34	1200 bps to 56 Kbps	• Worldwide network • Proven technology • Readily available • Low cost for small transmissions	• Slow speed • Poor line quality in some areas can slow transmission further • Usage charges
ISDN (BRI)	64 to 128 Kbps	• Fast digital connection • Voice, data, video • Increasing availability	• Usage charges • Not available everywhere • Rates vary widely
Switched 56	56 Kbps	• Digital service	• Uneconomical for low traffic • Bandwidth inadequate for high traffic
T-1	1.544 Mbps	• Mature technology • Widely available • High bandwidth for dedicated connections	• Flat-rate charges • Moving/adding connections
X.25	2400 to 56 Kbps	• Available worldwide • Proven technology • Good if light traffic	• Limited bandwidth
Frame Relay	56 Kbps to 1.544 Mbps	• High bandwidth • Occasional bursts OK	• Connection to LANs requires routers • Not globally available
SMDS	56 Kbps to 44.736 Mbps	• Economical at lower speeds • Available now	• Not available in all metropolitan areas • May never be widely used
ATM	1.5 Mbps to 622 Mbps	• High scalable bandwidth • Voice, data, and video	• Products and services not readily available in all areas

5.2   PUBLIC SWITCHED TELEPHONE NETWORK

The **public switched telephone network (PSTN)** is the international telephone system based on copper wires (UTP cabling) carrying analog voice data. The PSTN, also known as the **plain old telephone service (POTS),** is the standard telephone service that most homes use.

The PSTN is a huge network with multiple paths that link source and destination devices. The PSTN uses circuit switching when you make a call. Therefore, the data is switched to a dedicated path throughout the conversation.

Figure 5.1   Public Switched Telephone Network

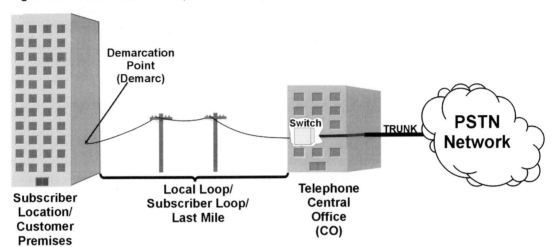

The original concept of the Bell system was a series of PSTN trunks connecting the major U.S. cities. The PSTN originally began with human operators sitting at a switchboard manually routing calls [private branch exchanges (PBXs)]. Today, the PSTN systems still use analog signals from the end node (phone) to the first switch. The switch then converts the analog signal to a digital signal and routes the call on to its destination. Since the digital signal travels on fiber-optic cabling, the signals are switched at high speeds. Once the call is received on the other end, the last switch in the loop converts the signal back to analog, and the call is initiated. The connection will stay active until the call is terminated (the user hangs up). The active circuit enables you to hear the other person almost instantaneously.

The **subscriber loop** or **local loop** is the telephone line that runs from your home or office to the telephone company's central office (CO) or neighborhood switching station (often a small building with no windows). While its cable length can be as long as 20 miles, the local loop is often referred to as the **last mile,** not because of its length, but because it is the slow link in the telecommunications infrastructure as it carries analog signals on a twisted-pair cable. The point where the local loop ends at the customer's premises is called the **demarcation point (demarc).**

NOTE: Unless you have an agreement with the phone company, the phone company is only responsible for maintaining the line from the central office to the demarc. (See figure 5.1.)

The standard home phone communicates over the local loop using analog signals. Therefore, when a PC needs to communicate over the local loop, it must use a modem to convert the PC's digital data to analog signals. The analog lines can only reach a maximum speed of 53 Kbps due to FCC regulations that restrict the power output. Unfortunately, the speed is not guaranteed and is often not reached.

5.3   DIAL-UP TECHNOLOGY

Today, modern networks offer **remote access service (RAS),** which allows users to connect remotely using various protocols and connection types. Current projections call for the number of remote users—telecommuters, road warriors, and other mobile users—to grow from 30 million in 1999 to more than 100 million by the year 2002 as remote LAN and Internet access continue to grow.

Typically, remote access costs anywhere from 63 to 157 percent more to support than its office-bound counterparts. About one-third of these costs are attributed to incremental charges such as equipment costs and access fees. The rest of the costs are devoted to ongoing administration and technical support for the act of keeping remote users connected to the corporate network. Of course, these drawbacks are offset by the reduced real estate costs and increased productivity. But to make remote access a truly sound investment, organizations must find a way to simplify remote desktop management and support while providing the same level of performance and reliability currently available with local solutions.

There are two types of remote access: dial-up networking and virtual private networking. **Dial-up networking** occurs when a remote access client makes a nonpermanent, dial-up connection to a physical port on a remote access server by using the service of a telecommunications provider such as an analog phone, ISDN, or X.25. The best example of dial-up networking is that of a dial-up networking client who dials the phone number of one of the ports/modem of a remote access server.

A **remote access server** is the computer and associated software that is set up to handle users seeking access to a network remotely. Sometimes called a **communication server,** a remote access server usually includes or is associated with a firewall server to ensure security and a router that can forward the remote access request to another part of the corporate network.

5.3.1   The UART

When connecting a computer to a network using dial-up technology, you typically connect using a serial device. The central part of the serial device, including an analog modem, is the **universal asynchronous receiver/transmitter (UART).** The UART is a single IC chip that is the translator between the serial device and the system bus; it's the component that processes, transmits, and receives data.

Inside the computer, data is generated by the processor and moved around in parallel format (system bus). This means that the several bits are being moved from one place to another using several wires. Of course, if you have more wires, you can transport more data. As the data comes from the system bus, the UART converts the data to a serial signal (bits in a single file), adds the start and stop bits, and adds any error control (such as parity). Then the UART converts the digital (+ 5 Vdc TTL) signals to a bipolar signal. The data is then sent to the serial device and out to the serial cable.

As data comes in from the serial port, the UART must reverse the process. First, the UART converts the bipolar signal to a + 5 Vdc TTL digital signal. It then strips the start and stop bits, checks for errors, and strips the error control information. Finally, it con-

Figure 5.2   RJ-11 Connector and Jack

verts the data back into parallel format to the system bus so that the data can be processed by the computer.

Early UARTs (8250 and 16450) used a FIFO (first in, first out) buffer. If the computer did not retrieve the data fast enough, the data would be overwritten by new data coming in through the serial port. Today, newer systems use a 16-byte FIFO-buffered UART, such as the 16550. Since the buffer is larger, the processor only has to process the data every 16 bytes before any data is overwritten. Consequently, the serial communication is faster.

5.3.2   Analog Modems

A **modem (modulator-demodulator)** is a device that enables a computer to transmit data over telephone lines. Since the computer information is stored and processed digitally, and since the telephone lines transmit data using analog waves, the modem converts (modulates) digital signals to analog signals and converts (demodulates) analog signals to digital signals.

A modem can be classified as either an internal modem or an external modem. The internal modem is an expansion card that is plugged into an expansion slot, whereas the external modem can be attached to the computer using a serial port. In either case, the modem has at least one RJ-11 connector to be used to connect the twisted-pair cable between the modem and the wall telephone jack. Some will include a second RJ-11 jack to connect a phone. (See figure 5.2.)

When data needs to be sent through the modem, the modem receives the digital data from the computer. The modem then converts the digital signals into analog signals. The UART of the modem then adds the start and stop bits and any error control bits such as parity. When the data reaches the other modem, the UART strips the start and stop bits, checks for errors, and converts the analog signal to a digital signal, which is then processed by the computer.

As the modem is communicating, it is continuously monitoring the status of the line, the quality of the signals, and the number of errors encountered. When the modems see excessive problems, the modems interrupt the carries and reevaluate the line. Consequently, the speed may be decreased. If the speed is decreased and the errors have decreased, the line will reevaluated again to see if the speed can be increased again.

Of course, when two computers are communicating using a telephone line, both computers must be set to the same number of bits used for data, the same length of the stop bit, and the same type of error control. If not, the receiving computer would misinterpret the data, resulting in errors or garbage. Usually, the calling computer will be configured to the settings of the receiving computer.

Example:

You need to download some technical information that is not available on the Internet but is available on a BBS computer. Therefore, you are given the phone number of the BBS computer and the following information:

9600 baud at 7E1

Of course, the 9600 baud is the speed of the modem on the other end. The 7 indicates 7 bits are used for data, and the E indicates that the computer uses even parity checking (O would be for odd parity). Lastly, the 1 indicates that the computer has 1 stop bit.

Baud rate refers to the modulation rate or the number of times per second that a line changes state. This is not always the same as bits per second (bps). If you connect two serial devices together using direct cables, the baud rate and the bps are the same. Thus, if you are running at 19,200 bps, the line is also changing states 19,200 times per second to represent 19,200 logical 1s and 0s. Much like they do for the data and stop bits, the sending and receiving devices must agree on the baud rate. The speed of a normal serial port is 115.2 kilobits per second.

Baud rates can differ from bps rates when, for example, communicating over telephone lines. The telephone lines are actually limited to a maximum of 2400 baud. However, since they use analog signals and not digital signals, the bps can be increased by encoding more bits into each line change. Consequently, the 9600 bps encodes 4 bits at the same time, 14,400 encodes 6 bits, and 28.8 encodes 12 bits. In addition, the bps can be increased by using data compression.

When installing a modem, the first step is to configure the resources. If it is an internal modem, you would usually choose a free IRQ, COM port, and I/O address. If it is an external modem, you must then configure the serial port. While the serial port is already assigned an IRQ, COM port, and I/O address, the serial port speed usually has to be configured.

Question:

You have a computer running Windows 98 that has two serial ports: COM1 and COM2. COM1 is being used by the mouse, and COM2 is not being used at all. You need to use a proprietary communication software package for your business that will only use COM1 and COM2. Therefore, you purchase an internal modem. How should you configure the modem?

Answer:

If you configure the modem to use COM3, the communication software package will not recognize the modem. If you configure the modem as COM1 or COM2, you have a resource conflict. Since you need the mouse (COM1) for Windows 98, you should configure the modem as COM2/IRQ 3. Of course, for this to work, you will need to disable COM2 using either jumpers, DIP switches, configuration software, or the CMOS Setup program.

The next step is to physically install and connect the modem. The modem is connected to the telephone wall jack using a twisted-pair cable with RJ-11 connectors.

Lastly, you must then load the modem software drivers and install and configure the communication software. If you have Windows, the driver is loaded by using the Add New Hardware icon/applet from the control panel.

With the communication software package, you can usually choose the baud rate (up to the speed of the modem), the number of data bits, the number of stop bits, and the parity method. In addition, many of the packages allow you to establish different parameters for different telephone numbers. Therefore, when you call a particular telephone number, the communication software will automatically use the same parameters that were used before.

While the modem interface is standardized, there are a number of standards and protocols that specify how formatted data is to be transmitted over telephones lines. The International Telecommunications Union (ITU), formerly known as the Comité Consultatif International Téléphonique et Télégraphique (CCITT), has defined many important standards for data communications. Most modems have built-in support for the more common standards. Today, v.90 is the most popular standard for 56K modems.

5.3.3   Integrated Services Digital Network

The **integrated services digital network (ISDN)** is the planned replacement for POTS so that it can provide voice and data communications worldwide using circuit switching while using the same wiring that is currently being used in homes and businesses. Because ISDN is a digital signal from end to end, it is faster and much more dependable with no line noise. ISDN has the ability to deliver multiple simultaneous connections, in any combination of data, voice, video, or fax, over a single line and allows for multiple devices to be attached to the line.

The ISDN uses two types of channels, a B channel and a D channel. The **bearer channels (B channels)** transfer data at a bandwidth of 64 Kbps (kilobits per second) for each channel. The **data channels (D channels)** use a communications language called DSS1 for administrative signaling—it's used, for example, to instruct the carrier to set up or terminate a B-channel call, to ensure that a B channel is available to receive a call, or to provide signaling information for such features as caller identification. Since the D channel is always connected to the ISDN, the call setup time is greatly reduced to 1 to 2 seconds (versus 10 to 40 seconds using an analog modem) as it establishes a circuit.

NOTE: The bandwidth listed here is the uncompressed speed of the ISDN connections. The compressed bandwidth has a current maximum transmission speed of four times the uncompressed speed. Of course, you will probably see a much lower speed, as much of the data flowing across the network is already compressed.

Today, two well-defined standards are used: the **basic rate interface** and the **primary rate interface**. The **basic rate interface (BRI)** defines a digital communications line consisting of three independent channels: two bearer (or B) channels, each carrying 64 kilobytes per second, and one data (or D) channel at 16 kilobits per second. For this reason, the ISDN basic rate interface is often referred to as 2B+D. BRIs were designed to enable customers to use the existing wiring in their home or business. This provided a low-cost solution for customers and is why it is the most basic type of service today intended for small business or home use.

To use BRI services, you must subscribe to ISDN services through a local telephone company or provider. By default, you must be within 18,000 feet (about 3.4 miles) of the telephone company central office. Repeater devices are available for ISDN service to extend this distance, but these devices can be very expensive.

The **primary rate interface (PRI)** is a form of ISDN that includes 23 B channels (30 in Europe) and one 64 KB D channel. PRI service is generally transmitted through a T-1 line (or an E-1 line in Europe).

To connect a computer to an ISDN line, you would use a **terminal adapter.** A terminal adapter is a bit like a regular modem, but whereas a regular modem needs to convert between analog and digital signals, a terminal adapter only needs to pass along digital signals. To connect a router to an ISDN line, you will need to purchase a router with a built-in NT1 or use your router's serial interface connected with an ISDN modem.

The **directory number (DN)** is the 10-digit phone number the telephone company assigns to any analog line. A **Service Profile Identifier (SPID)** includes the DN and an additional identifier used to identify the ISDN device to the telephone network. However, depending on which kind of switch you are connected to and how you are going to use the ISDN service, you may not need a single SPID, or you may need a SPID for each B channel or each device. Unlike with an analog line, a single DN can be used for multiple channels or devices, or up to eight DNs can be assigned to one device. Therefore, a BRI can support up to 64 SPIDs.

NOTE: Most standard BRI installations include only two directory numbers, one for each B channel.

5.3.4   Traditional Remote Access

The typical dial-in session is actually composed of six distinct steps or stages, each of which must be completed successfully before the user gains access to centralized resources. Each step is subject to its own unique set of problems or failures that individually could jeopardize the entire connection. These steps include:

1. Modem connection
2. Dial-out

3. Handshake
4. Authentication
5. IP address negotiation
6. Access to resources

During the modem connection, the modem checks the phone line for a dial tone. If the modem fails, or if the modem cannot detect a dial tone, the connection process fails immediately. A "no dial tone" error can be caused by problems with the phone cable, the phone jack, the modem, the dialer, or any combination of conditions related to these items or devices.

If a dial tone is detected, the remote PC proceeds with the connection process by dialing the remote access number. Any number of events, such as no answer, a busy signal, or no carrier, could cause the connection process to fail.

If the call gets through, the remote (dialing) modem will begin negotiating a common data rate and other transmission parameters with the answering modem. This is referred to as the **handshaking** stage because both modems must agree on the same parameters before continuing. Technology advancements have produced a variety of ways to encode, compress, modulate, and transfer bits over the phone line, adding complexity—and processing time—to the process. These factors make it critical for IT managers to monitor and measure both the final outcome of the negotiation process and the length of time it takes for the negotiation to complete. Connection failures and delays are typically the result of the local and remote modems disagreeing on one or more transmission parameters.

Once authentication is complete, the remote PC is assigned a dynamic IP address for identification purposes. The remote PC must agree to use the specified IP address, or the local system will refuse to establish the link. The remote PC is also assigned a DNS server at this time; if the assignment isn't recorded or is unsuccessful for some reason, attempts to connect will appear as application problems.

Once the previous five stages of the remote connection process have been successfully completed, the dial-in client PC is finally granted access to the corporate network. At this stage, the user can then check email, send messages, copy or transfer files, run application programs, etc. But the connection is still not entirely safe; connectivity problems can occur if the call is abnormally terminated, if the modem speed is too slow, or if the dial-up network has an excessive amount of line noise.

5.3.5   Remote Authentication Dial-in User Service

Adding a remote access point to the centralized corporate network increases the chance of a break-in. Therefore, a secure authentication scheme is required to provide security and protect against remote client impersonation. During the authentication stage, the remote access server collects authentication data from the dial-in client and checks it against its own user database or against a central authentication database server, such as RADIUS.

RADIUS, short for **Remote Authentication Dial-in User Service,** is the industry standard client/server protocol and software that enables remote access servers to communicate with a central server to authenticate dial-in users and authorize their access to the requested system or service. RADIUS allows a company to maintain user profiles in a central database that all remote servers can share. It provides better security, allowing a company to set up a policy that can be applied at a single administered network point. Lastly, since RADIUS has a central server, it also means that it is easier to perform accounting of network usage for billing and for keeping network statistics.

When a user dials into a remote access server, that server communicates with the central RADIUS server to determine if the user is authorized to connect to the LAN. The RADIUS server performs the authentication and responds with an "accept" or a "reject." If the user is accepted, the remote access server routes the user onto the network; if not, the remote access server will terminate the user's connection. If the user's login information—the user name and password—does not match the entry in the remote access server, or if the remote access server isn't able to contact the RADIUS server, the connection will be denied.

NOTE: Besides being deployed in remote access servers, RADIUS can also be deployed on routers and firewalls.

5.3.6   SLIP and PPP

The first protocol used to connect to an IP network using dial-up lines was the **Serial Line Interface Protocol (SLIP).** SLIP is a simple protocol in which you send packets down a serial link delimited with special END characters. SLIP doesn't do a number of desirable things that data link protocols can do. First, it only works with TCP/IP. Therefore, it cannot be used with other protocols such as IPX. It doesn't perform error checking at the OSI data layer, and it doesn't authenticate users who are dialing into an access router. Next, with SLIP, you have to know the IP address assigned to you by your service provider. You also need to know the IP address of the remote system you will be dialing into. If IP addresses are dynamically assigned (which depends on your service provider), your SLIP software needs to be able to pick up the IP assignments automatically or you will have to set them up manually. You will have to configure certain parameters of the device such as the maximum transmission unit (MTU) and the maximum receive unit (MRU) and the use of compression. Lastly, SLIP does not support encrypted passwords, and therefore it transmits passwords in clear text, which is not secure at all. You would typically use SLIP when you are dialing into an older server that does not use Point-to-Point Protocol connections.

The **Point-to-Point Protocol (PPP)** has become the predominant protocol for modem-based access to the Internet. PPP provides full-duplex, bidirectional operations between hosts and can encapsulate multiple network-layer LAN protocols to connect to private networks. PPP is a full-duplex protocol that can be used on various physical media, including twisted-pair or fiber-optic lines or satellite transmission. PPP supports asynchronous serial communication, synchronous serial communication, and ISDN. Furthermore, a multilink version of PPP is also used to access ISDN lines, inverse multiplexing analog phone lines, and high-speed optical lines. PPP is one of the many variants of an early, internationally standardized data link protocol known as the High-Level Data Link Control (HDLC).

To enable PPP to transmit data over a serial point-to-point link, three components are used:

■ **High-Level Data Link Control (HDLC) Protocol**—Encapsulates its data during transmission.

- **Link Control Protocol (LCP)**—Establishes, configures, maintains, and terminates point-to-point links [including Multilink PPP (MP) sessions] and optionally tests link quality prior to data transmission. In addition, user authentication is generally performed by LCP as soon as the link is established.
- **Network Control Protocol (NCP)**—Used to configure the different communications protocols including TCP/IP and IPX, which are allowed to be used simultaneously.

One typical IP-specific function of the NCP is the provision of dynamic IP addresses for remote dial-in users. ISPs commonly employ this PPP feature to assign temporary IP addresses to their modem-based customers.

NOTE: The PPP mechanism is not as complete as DHCP, which can set up a lease period and provide such things as subnet masks, default router addresses, DNS server addresses, domain names, and much more to a newly connected computer.

An ISP or other provider of remote connectivity services can use DHCP as the source of IP addresses that NCP hands out.

There are four distinct phases of negotiation of a PPP connection. Each of these phases must complete successfully before the PPP connection is ready to transfer user data. The four phases are:

1. PPP configuration
2. Authentication
3. Callback (optional)
4. Protocol configuration

PPP configures PPP protocol parameters using the LCP. During the initial LCP phase, each device on both ends of a connection negotiates communication options that are used to send data. Options include PPP parameters, address and control field compression, protocol ID compression, the authentication protocols that are to be used to authenticate the remote access client, and multilink options.

NOTE: An authentication protocol is selected but not implemented until the authentication phase.

After LCP is complete, the authentication protocol agreed upon by the remote access server and the remote access client is implemented. The nature of this traffic is specific to the PPP authentication protocol.

5.3.7   Authentication Protocols

When a client dials into a remote access server, the server will have to verify the client's credentials for authentication by using the client's user account properties and remote access policies to authorize the connection. If authentication and authorization succeed, the server allows a connection.

There are a number of PPP authentication protocols, each of which has advantages and disadvantages in terms of security, usability, and breadth of support. (See table 5.3.)

Table 5.3   Various Security Protocols

Protocols	Security	Description	Use When
Password Authentication Protocol (PAP)	Low	Passwords are sent across the link as unencrypted plain text. Not a secure authentication method.	The client and server cannot negotiate by using a more secure form of validation.
Shiva Password Authentication Protocol (SPAP)	Medium	Sends password across the link in reversibly encrypted form. Not recommended in situations where security is an issue.	Connecting to a Shiva LanRover, or when a Shiva client connects to a Windows 2000–based remote access server.
Challenge Handshake Authentication Protocol (CHAP)	High	Uses MD5 hashing scheme. Client computer sends a hash of the password to the remote access server, which checks the client hash against a hash it generates with the password stored on the remote access server. CHAP does provide protection against server impersonation.	You have clients that are not running Microsoft operating systems.
Microsoft Challenge Handshake Authentication Protocol (MS-CHAP) v1	High	Similar to CHAP. Provides nonreversible, encrypted password authentication. Does not require the password on the server to be in plain text or reversibly encrypted form. It only supports one-way authentication. Therefore, MS-CHAP v1 does not provide protection against remote server impersonation, which means that a client can't determine the authenticity of a remote access server it connects to. It is more secure than CHAP.	You have clients running Windows.
Microsoft Challenge Handshake Authentication Protocol (MS-CHAP) v2	High (most secure)	An improved version of MS-CHAP v1. It provides stronger security for remote access connections and allows for mutual authentication where the client also authenticates the server.	You have clients running Windows.

The **Password Authentication Protocol (PAP)** is the least secure authentication protocol because it uses **clear text** (plain text) passwords. These steps occur when using PAP:

1. The remote access client sends a PAP "Authenticate-Request" message to the remote access server containing the remote access client's user name and clear text password. Clear text, also referred to as plain text, is textual data in ASCII format.

2. The remote access server checks the user name and password. It sends back either a PAP "Authenticate-Acknowledgment" message when the user's credentials are correct or a PAP "Authenticate-No acknowledgment" message when the user's credentials are not correct.

Therefore, the password can easily be read with a protocol analyzer. In addition, PAP offers no protection against replay attacks, remote client impersonation, or remote server impersonation. So to make your remote access server more secure, ensure that PAP is disabled. Another disadvantage of using PAP is that if your password expires, PAP doesn't have the ability to change your password during authentication.

The **Shiva Password Authentication Protocol (SPAP),** Shiva's proprietary version of PAP, offers a bit more security than PAP's plain text password with its reversible encryption mechanism. SPAP is more secure than PAP but less secure than CHAP or MS-CHAP. Someone capturing authentication packets won't be able to read the SPAP password, but this authentication protocol is susceptible to playback attacks (i.e., an intruder records the packets and resends them to gain fraudulent access). Playback attacks are possible because SPAP always uses the same reversible encryption method to send the passwords over the wire. Like PAP, SPAP doesn't have the ability to change your password during the authentication process.

Historically, the **Challenge Handshake Authentication Protocol (CHAP)** is the most common dial-up authentication protocol used. It uses an industry Message Digest 5 (MD5) hashing scheme to encrypt authentication. A **hashing scheme** scrambles information in such a way that it's unique and can't be changed back to the original format.

CHAP doesn't send the actual password over the wire; instead, it uses a three-way challenge-response mechanism with one-way MD5 hashing to provide encrypted authentication. This is what happens:

1. The remote access server sends a CHAP challenge message containing a session ID and an arbitrary challenge string.
2. The remote access client returns a CHAP response message containing the user name in clear text and a hash of the challenge string, session ID, and the client's password using the MD5 one-way hashing algorithm.
3. The remote access server duplicates the hash and compares it with the hash in the CHAP response. If the hashes are the same, the remote access server sends back a CHAP success message. If the hashes are different, a CHAP failure message is sent.

Because standard CHAP clients use the plain text version of the password to create the CHAP challenge response, passwords must be stored on the server to calculate an equivalent response.

Since CHAP uses an arbitrary challenge string per authentication attempt, it protects against replay attacks. However, CHAP does not protect against remote server impersonation. In addition, because the algorithm for calculating CHAP responses is well known, it is very important that passwords be carefully chosen and sufficiently long. CHAP passwords that are common words or names are vulnerable to dictionary attacks if they can be discovered by comparing responses to the CHAP challenge with every entry in a dictionary. Passwords that are not sufficiently long can be discovered by brute force by comparing the CHAP response with sequential trials until a match to the user's response is found.

The **Microsoft Challenge Handshake Authentication Protocol (MS-CHAP)** is Microsoft's proprietary version of CHAP. Unlike PAP and SPAP, it lets you encrypt data that is sent using the Point-to-Point Protocol (PPP) or PPTP connections using Microsoft Point-to-Point Encryption (MPPE). The challenge response is calculated with an MD4 hashed version of the password and the NAS challenge.

NOTE: The two flavors of MS-CHAP (versions 1 and 2) allow for error codes including a "password expired" code and password changes.

1. The remote access server sends an MS-CHAP challenge message containing a session ID and an arbitrary challenge string.
2. The remote client must return the user name and an MD4 hash of the challenge string, the session ID, and the MD4-hashed password.
3. The remote access server duplicates the hash and compares it with the hash in the MS-CHAP response. If the hashes are the same, the remote access server sends back a CHAP success message. If the hashes are different, a CHAP failure message is sent.

A disadvantage of MS-CHAP 1 is that it supports only one-way authentication. Therefore, MS-CHAP v1 does not provide protection against remote server impersonation, which means that a client can't determine the authenticity of a remote access server it connects to.

MS-CHAP v2 provides stronger security for remote access connections and allows for mutual authentication where the client authenticates the server.

1. The remote access server sends an MS-CHAP v2 challenge message to the remote access client; the message consists of a session identifier and an arbitrary challenge string.
2. The remote access client sends an MS-CHAP v2 response that contains the user name, an arbitrary peer challenge string, an MD4 hash of the received challenge string, the peer challenge string, the session identifier, and the MD4 hashed versions of the user's password.
3. The remote access server checks the MS-CHAP v2 response message from the client and sends back an MS-CHAP v2 response message containing an indication of the success or failure of the connection attempt. An authentication response is based on the sent challenge string, the peer challenge string, the client's encrypted response, and the user's password.
4. The remote access client verifies the authentication response and, if it is correct, uses the connection. If the authentication response is not correct, the remote access client terminates the connection.
5. If a user authenticates by using MS-CHAP v2 and attempts to use an expired password, MS-CHAP prompts the user to change the password while connecting to the server. Other authentication protocols do not support this feature, effectively locking out the user who used the expired password.

If you configure your connection to use only MS-CHAP 2 and if the server you're dialing into does not support MS-CHAP 2, the connection will fail. This behavior is different from that of Windows NT, where the remote access servers negotiate a lower-level authentication if possible. In addition, MS-CHAP passwords are stored more securely at the server than CHAP passwords but they have the same vulnerabilities to dictionary and brute force attacks as CHAP's do. When using MS-CHAP, it is important to ensure that passwords are well chosen and long enough that they cannot be calculated readily. Many large customers require passwords to be at least six characters long, with uppercase and lowercase characters and at least one digit.

Since the field of security and authentication is constantly changing, embedding authentication schemes into an operating system is impractical at times. To solve this problem, Microsoft has included support for the **Extensible Authentication Protocol (EAP),** which allows new authentication schemes to be plugged in as needed. Therefore, EAP allows third-party vendors to develop custom authentication schemes such as retina scans, voice recognition, fingerprint identification, smart card, Kerberos, and digital certificates. In addition, EAP offers mutual authentication.

NOTE: Mutual authentication is the process in which both computers authenticate each other so that the client also knows that the server is who it says it is.

Extensible Authentication Protocol Message Digest 5 Challenge Handshake Authentication Protocol (EAP-MD5 CHAP) is a type of EAP that uses the same challenge-handshake protocol as PPP-based CHAP, but the challenges and responses are sent as EAP messages. A typical use for EAP-MD5 CHAP is to authenticate the credentials of remote access clients by using user name and password security systems.

EAP-Transport-Level Security (EAP-TLS) is a type of EAP that is used in certificate-based security environments. If you are using smart cards for remote access authentication, you must use the EAP-TLS authentication method. The EAP-TLS exchange of messages provides mutual authentication, negotiation of the encryption method, and secured private key exchange between the remote access client and the authenticating server. EAP-TLS provides the strongest authentication and key exchange method.

EAP-RADIUS is not a type of EAP, but the passing of EAP messages by an authenticator to a RADIUS server for authentication. The EAP messages sent between the remote access client and remote access server are encapsulated and formatted as RADIUS messages between the remote access server and the RADIUS server.

EAP-RADIUS is used in environments where RADIUS is used as the authentication provider. An advantage of using EAP-RADIUS is that EAP types do not need to be installed at each remote access server, only at the RADIUS server. In a typical use of EAP-RADIUS, a remote access server is configured to use EAP and to use RADIUS as its authentication provider. When a connection is made, the remote access client negotiates the use of EAP with the remote access server. When the client sends an EAP message to the remote access server, the remote access server encapsulates the EAP message as a RADIUS message and sends it to its configured RADIUS server. The RADIUS server processes the EAP message and sends a RADIUS-encapsulated EAP message back to the remote access server. The remote access server then forwards the EAP message to the remote access client. In this configuration, the remote access server is only a pass-through device. All processing of EAP messages occurs at the remote access client and the RADIUS server.

A new member of the EAP protocol family is the **Protected Extensible Authentication Protocol (PEAP).** PEAP uses Transport Level Security (TLS) to create an encrypted channel between an authenticating EAP client, such as a wireless computer, and an EAP authenticator, such as an Internet Authentication Service (IAS), or Remote Authentication Dial-In User Service (RADIUS), server. PEAP does not specify an authentication method, but provides a secure "wrapper" for other EAP authentication protocols, such as EAP-MS-CHAPv2, that operate within the outer TLS encrypted channel provided by PEAP. PEAP is used as an authentication method for 802.11 wireless client computers, but is not supported for virtual private network (VPN) or other remote access clients.

You can select between two built-in EAP types for use within PEAP: EAP-TLS or EAP-MS-CHAPv2. EAP-TLS uses certificates installed in the client computer certificate store or a smart card for user and client computer authentication, and a certificate in the server computer certificate store for server authentication. EAP-MS-CHAPv2 uses credentials (user name and password) for user authentication and a certificate in the server computer certificate store for server authentication.

The unauthenticated access method allows remote access users to log on without checking their credentials. It does not verify the user's name and password. The only user validation performed in the unauthenticated access method is authorization. Enabling unauthenticated access presents security risks that must be carefully considered when deciding whether to enable this authentication method.

5.3.8 PPPOE

With the arrival of low-cost broadband technologies in general and DSL (digital subscriber line) in particular, the number of computer hosts that are permanently connected to the Internet has greatly increased. Computers connected to the Internet via DSL do so through an Ethernet link that has no additional protocols, including protocols for security. Modem dial-up connections, on the other hand, use PPP (Point-to-Point Protocol), which provides secure login and traffic metering among other advanced features. **PPPoE (PPP over Ethernet)** was designed to bring the security and metering benefits of PPP to Ethernet connections such as those used in DSL. DSL is discussed in the next section.

5.4 DIGITAL SUBSCRIBER LINE

A **digital subscriber line (DSL)** is a special communication line that uses sophisticated modulation technology to maximize the amount of data that can be sent over plain twisted-pair copper wiring, which is already carrying phone service to subscribers' homes. Originally, its purpose was to transmit video signals to compete against the cable companies, but it soon found use as a high-speed data connection with the explosion of the Internet. DSL is sometimes expressed as xDSL, because there are various kinds of digital subscriber line technologies including ADSL, R-DSL, HDSL, SDSL, and VDSL.

The best thing about DSL technologies is their ability to transport large amounts of information across existing copper telephone lines. This is possible because DSL modems leverage signal processing techniques that insert and extract more digital data onto analog lines. The key is modulation, a process in which one signal modifies the property of another.

In the case of digital subscriber lines, the modulating message signal from a sending modem alters the high-frequency carrier signal so that a composite wave, called a modulated wave, is formed. Because this high-frequency carrier signal can be modified, a large digital data payload can be carried in the modulated wave over greater distances than on ordinary copper pairs. When the transmission reaches its destination, the modulating message signal is recovered, or demodulated, by the receiving modem.

There are many ways to alter the high-frequency carrier signal that results in a modulated wave. For ADSL, there are two competing modulation schemes: carrierless amplitude phase (CAP) modulation and discrete multitone (DMT) modulation. CAP and DMT use the same fundamental modulation technique—quadrature amplitude modulation (QAM)—but apply it in different ways.

QAM, a bandwidth conservation process routinely used in modems, enables two digital carrier signals to occupy the same transmission bandwidth. With QAM, two independent message signals are used to modulate two carrier signals that have identical frequencies but differ in amplitude and phase. QAM receivers are able to discern whether to use lower or higher numbers of amplitude and phase states to overcome noise and interference on the wire pair.

Carrierless amplitude phase (CAP) modulation is a proprietary modulation of AT&T Paradyne. Income data modulates a single carrier channel that is sent down a telephone line. The carrier is suppressed before transmission and reconstructed at the receiving end. Generating a modulated wave that carries amplitude and phase-state changes is not easy. To overcome this challenge, the CAP version of QAM stores parts of a modulated message signal in memory and then reassembles the parts in the modulated wave. The carrier signal is suppressed before transmission because it contains no information, and it is reassembled at the receiving modem (hence the word *carrierless* in CAP). At start-up, CAP also tests the quality of the access line and implements the most efficient version of QAM to ensure satisfactory performance for individual signal transmissions. CAP is normally FDM-based.

Discrete multitone (DMT) modulation is actually a form of FDM (frequency-division multiplexing). Because high-frequency signals on copper lines suffer more loss in the presence of noise, DMT discretely divides the available frequencies into 256 subchannels, or tones. As with CAP, a test occurs at start-up to determine the carrying capacity of each subchannel. Incoming data is then broken down into a variety of bits and distributed to a specific combination of subchannels based on their ability to carry the transmission. To rise above noise, more data resides in the lower frequencies and less in the upper ones.

To create upstream and downstream channels, ADSL modems divide the phone line's available bandwidth using one of two methods: frequency-division multiplexing or echo cancellation. FDM assigns one band or frequency for upstream data and another band or frequency for downstream data. The downstream path is further divided by time-division multiplexing (TDM) into one or more high-speed channels for data and one or more low-speed channels, one of which is for voice. The upstream path is multiplexed into several low-speed channels.

Echo cancellation, the same technique used by V.32 and V.34 modems, means that the upstream and the downstream signals are sent on the wire at the same frequencies. The advantage of echo cancellation is that the signals are both kept at the lowest possible frequencies (since cable loss and crosstalk noise both increase with frequency), and therefore it achieves greater cable distance for a given data rate. An ADSL receiver will see an incoming signal that is both the incoming signal from the far end and the outgoing signal from the local transmitter. These are mixed together over the same frequency range. In other words, the received signal is composed of not only the signal to be recovered from the far end but also a local echo due to the local transmitter. The local echo must be accurately modeled by DSP circuitry, and then this replica echo is electronically subtracted from the composite incoming signal. If done properly, all that is left behind is the incoming data from the far-end ADSL system. While echo cancellation uses bandwidth more efficiently, the process of modeling the echo is quite complicated and is more costly. Therefore, only a few vendors have implemented it.

Currently, existing POTS data is transmitted over a frequency spectrum that ranges from 0 to 4 kHz. Copper phone lines can actually support frequency ranges much greater than those, and ADSL takes advantage of these ranges by transmitting data in the range between 4 kHz and 2.2 MHz. Therefore, ADSL relies on advanced digital signal processing (DSP) and complex algorithms to compress all the information into the phone line. In addition, ADSL modems correct errors caused by line conditions and attenuation. With this technology, any computer or network can easily become connected to the Internet at speeds comparable to T-1 access for a fraction of the cost, and ADSL can serve as a suitable medium for video streaming and conferencing.

As its name implies, **asymmetrical DSL (ADSL)** transmits an asymmetric data stream, with much more going downstream to the subscriber and much less coming back. The reason for this has less to do with transmission technology than with the cable plant itself. Twisted-pair telephone wires are bundled together in large cables. Fifty pairs to a cable is a typical configuration toward the subscriber, but cables coming out of a central office may have hundreds or even thousands of pairs bundled together. An individual line from a CO to a subscriber is spliced together from many cable sections as they fan out from the central office (Bellcore claims that the average U.S. subscriber line has 22 splices). Alexander Bell invented twisted-pair wiring to minimize crosstalk. Since a small amount of crosstalk does occur, the amount of crosstalk increases as the frequencies and the length of the line increase. Therefore, if you have symmetric signals in many pairs within the same cable, the crosstalk significantly limits the data rate and the length of the line.

Since most people download information such as web pages and files, the amount of information downloaded is far greater than the amount of information that a user uploads or transfers to other computers. This asymmetry, combined with "always-on" access (which eliminates call setup), makes ADSL ideal for Internet/intranet surfing, video-on-demand, and remote LAN access.

ADSL modems usually include a POTS (plain old telephone service) splitter, which enables simultaneous access to voice telephony and high-speed data access. Some vendors provide active POTS splitters, which enable simultaneous telephone and data access. However, if the power fails or the modem fails with an active POTS splitter, then the telephone fails. A passive POTS splitter, on the other hand, maintains lifeline telephone access even if the modem fails (due to a power outage, for example), since the telephone is not powered by external electricity. Telephone access in the case of a passive POTS splitter is a regular analog voice channel, the same as customers currently receive in their homes.

Downstream, ADSL supports speeds between 1.5 and 8 Mbps, while upstream, the rate is between 640 Kbps and 1.544 Mbps. ADSL can provide 1.544-Mbps transmission rates at distances of up to 18,000 feet over one wire pair. Optimal speeds of 6 to 8 Mbps can be achieved at distances of 10,000 to 12,000 feet using standard 24-gauge wire.

If you order DSL for your home, it is most likely ADSL lite. ADSL lite, also known as g.lite, is a low-cost, easy-to-install version of ADSL specifically designed for the consumer marketplace. ADSL lite is a lower-speed version of ADSL (up to 1.544 Mbps downstream, up to 512 Kbps upstream) that will eliminate the need for the telephone company to install and maintain a premises-based POTS splitter. ADSL lite is also supposed to work over longer distances than full-rate ADSL, making it more widely available to mass market consumers. It will support both data and voice and provide an evolution path to full-rate ADSL.

To connect to an ADSL line, you install an ADSL modem via an expansion card or use a USB port or external stand-alone device connected to a network card. If you are using the expansion card or USB device, you would assign and configure TCP/IP. If you are using the external stand-alone device, you would assign and configure TCP/IP for the network card.

At home and in small offices, it is very common to connect to the Internet by ordering a DSL line and using a DSL router or using some other form to have several computers share the DSL line. The DSL router will allow multiple computers to connect to the router, and the router will act as a DHCP server for the clients connected to it.

5.5   CABLE MODEMS

Cable systems were originally designed to deliver broadcast television signals efficiently to subscribers' homes. But they can also be used to connect to the Internet. To ensure that consumers could obtain cable service with the same TV sets they use to receive over-the-air broadcast TV signals, cable operators re-create a portion of the over-the-air radio frequency (RF) spectrum within a sealed coaxial cable line.

Traditional coaxial cable systems typically operate with 330 or 450 MHz of capacity, whereas modern hybrid fiber/coax (HFC) systems are expanded to 750 MHz or more. Logically, downstream video programming signals begin around 50 MHz, the equivalent of channel 2 for over-the-air television signals. Each standard television channel occupies 6 MHz of the RF spectrum. Thus a traditional cable system with 400 MHz of downstream bandwidth can carry the equivalent of 60 analog TV channels, and a modern HFC system with 700 MHz of downstream bandwidth has the capacity for about 110 channels.

A **hybrid fiber coaxial (HFC) network** is a telecommunication technology in which optical fiber cable and coaxial cable are used in different portions of a network to carry broadband content (such as video, data, and voice). Using HFC, a local CATV company installs fiber-optic cable from the cable head-end (distribution center) to serving nodes located close to business and residential users and then from these nodes they use coaxial cable to connect to individual businesses and homes. An advantage of HFC is that some of the characteristics of fiber-optic cable (high bandwidth and low noise and interference susceptibility) can be brought close to the user without having to replace the existing coaxial cable that is installed all the way to the home and business. Both cable TV and telephone companies are using HFC in new and upgraded networks and, in some cases, sharing the same infrastructure to carry both video and voice conversations in the same system.

While regular cable uses analog signals, digital cable (as you can tell from its name) uses digital signals. Digital signals can be compressed much more than analog signals. Digital cable can give you 200 to 300 channels of the same bandwidth as analog cable.

To deliver data services over a cable network, one television channel in the 50- to 750-MHz range is typically allocated for downstream traffic to homes, and another channel in the 5- to 42-MHz band is used to carry upstream signals. A cable modem termination system (CMTS) communicates through these channels with **cable modems** located in subscriber homes to create a virtual local area network connection. Most cable modems are external devices that connect to a personal computer (PC) through a standard 10BaseT Ethernet card and twisted-pair wiring, through external Universal Serial Bus (USB) modems and internal PCI modem cards.

A single downstream 6-MHz television channel may support up to 27 Mbps of downstream data throughput from the cable using 64 QAM (quadrature amplitude modulation) transmission technology. Speeds can be boosted to 36 Mbps using 256 QAM. Upstream channels may deliver 500 Kbps to 10 Mbps from homes using 16 QAM or QPSK (quadrature phase shift key) modulation techniques, depending on the amount of spectrum allocated for service. This upstream and downstream bandwidth is shared by the active data subscribers connected to a given cable network segment, typically 500 to 2000 homes on a modern HFC network.

An individual cable modem subscriber may experience access speeds from 500 Kbps to 1.5 Mbps or more, depending on the network architecture and traffic load—a blazing performance compared with dial-up alternatives. However, when surfing the web, performance can be affected by Internet backbone congestion. In addition to speed, cable modems offer another key benefit: constant connectivity. Because cable modems use connectionless technology, much like in an office LAN, a subscriber's PC is always online with the network. That means there's no need to dial in to begin a session, so users do not have to worry about receiving busy signals. Additionally, going online does not tie up their telephone line.

Similar to DSL routers, cable modems can be replaced with the cable routers. The cable router will allow multiple computers to connect to the router, and the router will act as a DHCP server for the clients connected to it.

5.6 LEASED DIGITAL LINES

The **T-carrier system** was introduced by the Bell System in the United States in the 1960s. It was the first successful system that converted the analog voice signal to a digital bit stream. While the T-carrier system was originally designed to carry voice calls between telephone company central offices, today it is used to transfer voice, data, and video signals between different sites and to connect to the Internet.

NOTE: E-carrier is the digital system used in Europe.

The T-carrier and E-carrier systems are entire digital systems that consist of permanent dedicated point-to-point connections. These digital systems are based on 64-Kbps channels (DS-0 channels), where each voice transmission is assigned a channel.

NOTE: DS stands for "digital signal."

In North America and Japan, you would typically find a T-1 line that has 24 64-Kbps channels for a bandwidth of 1.544 Mbps and T-3 line that has 672 64-Kbps channels for a bandwidth of 44.736 Mbps. In Europe, you will find E-1 lines with 32 64-Kbps chan-

Table 5.4   Digital Carrier Systems

Digital Signal Designator	T-Carrier	E-Carrier	Data Rate	DS-0 Multiple
DS-0	—	—	64 Kbps	1
DS-1	T-1	—	1.544 Mbps	24
—	—	E-1	2.048 Mbps	32
DS-1C	—	—	3.152 Mbps	48
DS-2	T-2	—	6.312 Mbps	96
–	—	E-2	8.448 Mbps	128
—	—	E-3	34.368 Mbps	512
DS-3	T-3	—	44.736 Mbps	672
—	—	E-4	139.264 Mbps	2048
DS-4/NA	—	—	139.264 Mbps	2176
DS-4	—	—	274.176 Mbps	4032
—	—	E-5	565.148 Mbps	8192

nels for a bandwidth of 2.048 Mbps and an E-3 line with 512 64-Kbps channel for a bandwidth of 34.368 Mbps. T-1 lines are popular leased line options for businesses connecting to the Internet and for Internet service providers (ISPs) connecting to the Internet backbone. T-3 connections make up the Internet backbone and are used by larger ISPs to connect to the backbone. (See table 5.4.)

If your company is not ready for a full T-1 line, your company can lease a fractional T-1 line—this means your company will use only a portion of the 24 channels. Since the hardware for using a full T-1 line already exists, you just have to call the carrier to increase the number of channels. Therefore, a fractional T-1 line leaves room for growth.

Traditionally, the T-1 system uses two unshielded twisted-pair copper wires to provide full-duplex capability (one pair to receive and one pair to send). Today, the T-carrier system can also use coaxial cable, optical fiber, digital microwave, and other media. (See table 5.5.)

While analog signals can travel up to 18,000 feet from the subscriber to the central office or switching station, T-1 lines require cleaner lines for the faster speeds. Therefore, repeaters are placed about every 6000 feet or so.

To complete the T-carrier connection, you must also use a CSU/DSU and possibly a multiplexer and router. The CSU/DSU physically and electrically terminates the connection to the telephone company, controls timing and synchronization of the signal, and handles line coding and framing.

Table 5.5   Different Media Used in T-Carrier System

Media	T-Carrier Capacity
Twisted pair	Up to 1 T-1 circuit
Coaxial cable	Up to 4 T-1 circuits
Microwave	Up to 8 T-3 circuits
Fiber optics	Up to 24 T-3 circuits

Figure 5.3   A Multiplexer on a T-1 Line

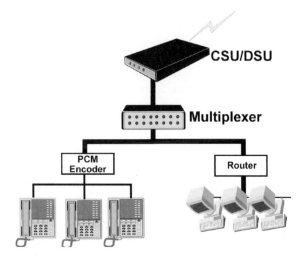

NOTE: By law, all T-1s require that a CSU be connected between your DTE and the T-1 line to act as a surge protector and to monitor the line.

A multiplexer is used to split the circuit into the various channels, and some can also provide PCM encoding for analog devices such as an analog phone.

NOTE: Remember that most business systems use digital phone systems.

The router can connect an Ethernet LAN to a WAN. A router would generally provide one or more RJ-45 connections for the Ethernet LAN or LANs and will include a serial connection for the CSU/DSU or include a built-in CSU/DSU. (See figure 5.3.)

5.7   FRAME RELAY

As mentioned in chapter 2, packet switching breaks the messages into small parts called packets and sends each packet individually onto the network. Since the packets have their own source and destination addresses, each packet can travel a different path than the

Figure 5.4   Dedicated Point-to-point Circuits Versus Frame Relay

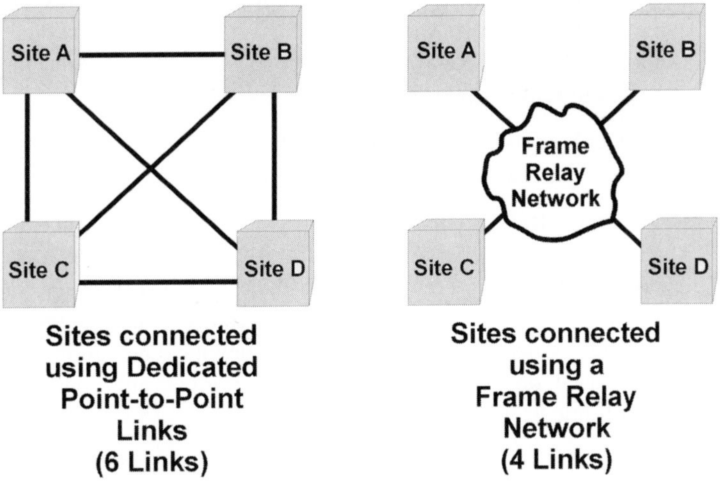

other packets travel. When the packets get to their destination, they must be reassembled. Two well-known examples of packet switching networks are X.25 and Frame Relay.

When a subscriber accesses a WAN network such as X.25, Frame Relay, or ATM, the leased network is sometimes referred to as a cloud. A **cloud** represents a logical network with multiple pathways as a black box. The subscribers that connect to the cloud don't worry about the details inside the cloud. Instead, the only thing the subscribers need to know is that they connect at one edge of the cloud and the data is received at the other edge of the cloud.

Frame Relay is a direct descendant of X.25 and is defined by the International Telecommunications Union (ITU-T) Q.922 and Q.933 standards. **Frame Relay** is a packet-switching protocol designed to use high-speed digital backbone links to support modern protocols that provide for error handling and flow control for connecting devices on a wide area network. As noted above, packet-switching protocols have messages divided into packets, transmitted individually with possible different routes to their destination, and recompiled in their original form at their destination. Frame Relay only defines the physical and data link layers of the OSI model, which allows for greater flexibility of upper-layer protocols to be run across Frame Relay. Frame Relay networks are typically found using a fractal T-1 line or a full T-1 line, but they can also be found on a T-3 line. In addition, Frame Relays can run on fiber-optic cable, twisted-pair copper wire, or digital microwave.

The most common use of a Frame Relay network is to connect individual LANs together. Data terminal equipment (DTE) is user terminal equipment such as PCs and routers, that creates information for transmissions. It typically costs less to use a Frame Relay network than dedicated point-to-point circuits since customers can use multiple virtual circuits on a single physical circuit. For example, if you look at figure 5.4, you can see that it would take six dedicated point-to-point links to connect four sites. With a Frame Relay network, it would only take four links to connect the same sites. If you have five sites, it would take 10 links using dedicated point-to-point links versus five links for a Frame Relay network.

To connect the DTEs to the frame relay network, you use a **Frame Relay Access Device (FRAD),** sometimes referred to as a Frame Relay Assembler/Dissembler. It multiplexes and formats traffic for entering a Frame Relay network. A FRAD is a device that allows non-Frame Relay devices to connect to a Frame Relay network. FRADs can be included in routers and bridges or can be stand-alone.

NOTE: You also may need a CSU/DSU to connect to common links such as a T-1 line. These CSU/DSUs may be included with the FRAD.

Frame Relay routers are the most versatile of these since they can handle traffic from other WAN protocols, reroute traffic if a connection goes down, and provide flow and congestion control. Frame Relay bridges, which are used to connect a branch office to a hub, are basically low-cost unintelligent routers. Stand-alone FRADs are designed to aggregate (gather) and convert data, but they have no routing capabilities. They are typically used on sites that already have bridges and routers or are used for sending mainframe traffic.

NOTE: These distinctions between the devices are blurring as vendors combine their functions into a single device.

Initially, Frame Relay gained acceptance as a means to provide end users with a solution for LAN-to-LAN connections and other data connectivity requirements. Due to advances in areas such as digital signal processing and faster backbone links within the Frame Relay network, end users are beginning to see viable methods being developed that incorporate nondata traffic such as voice or video over the Frame Relay network.

Frame Relay networks are statistically time-division multiplexed. This means, instead of assigning an equal number of time slots among the network devices, time slots are assigned dynamically so that empty time slots are used by busy network devices rather than being wasted. As a result, statistically time-division multiplexing uses the bandwidth more efficiently, particularly for bursty data traffic among clients and servers and for inter-LAN links. Lastly, Frame Relay uses variable-length packets to make efficient and flexible transfers.

5.8   FDDI

Fiber Distributed Data Interface (FDDI) is a MAN protocol that provides data transport at 100 Mbps (a much higher data rate than that for standard Ethernet or Token Ring) and can support up to 500 stations on a single network. Originally, FDDI networks required fiber-optic cable, but today they can also run on UTP.

NOTE: The copper part adopted into the FDDI standard is known as Copper Distributed Data Interface (CDDI).

Of course, fiber-optic cable offers much greater distances than UTP cable. Traditionally, FDDI was used as a backbone to connect several department LANs together within a single building or to link several building LANs together in a campus environment.

FDDI communicates all of its information using symbols. Symbols are 4 bits encoded in a 5-bit sequence. To measure a bit in FDDI as expressed by the change of state of the light on the other side, the station will take a sample of the light coming from the other machine approximately every 8 nanoseconds. The light will be either on or off. If it has changed since the last sample, that translates into a bit of 1. If the light has not changed since the last sample, the bit is a zero.

Like Token Ring, FDDI is a token passing ring using a physical star topology. While a Token Ring node can transmit only a single frame when it gets the token, an FDDI node can transmit as many frames as can be generated within a predetermined time before it has to give up the token.

FDDI can be implemented in two basic ways: as a dual-attached ring and as a concentrator-based ring. In the dual-attached scenario, stations are connected directly one to another. FDDI's dual counter-rotating ring design provides a fail-safe in case a node goes down. If any node fails, the ring wraps around the failed node. However, one limitation of the dual counter-rotating ring design is that if two nodes fail, the ring is broken in two places, effectively creating two separate rings. Nodes on one ring are then isolated from nodes on the other ring. External optical bypass devices can solve this problem, but their use is limited because of FDDI optical power requirements.

FDDI has a maximum total ring length of 100 km or 60 miles (or 200 km or 120 miles in wrap state—rings collapse to single ring). The maximum length of 2 km or 1.24 miles between stations without repeaters.

Another way around this problem is to use concentrators to build networks. Concentrators are devices with multiple ports into which FDDI nodes connect. FDDI concentrators function like Ethernet hubs or Token Ring multiple access units (MAUs). Nodes are single-attached to the concentrator, which isolates failures occurring at those end stations. With a concentrator, nodes can be powered on and off without disrupting ring integrity. Since concentrators make FDDI networks more reliable, most FDDI networks are now built with concentrators.

An extension to FDDI, called FDDI-2, supports the transmission of voice and video information as well as data. Another variation of FDDI, called FDDI Full-Duplex Technology (FFDT), uses the same network infrastructure but can potentially support data rates up to 200 Mbps.

5.9   SYNCHRONOUS DIGITAL HIERARCHY AND SYNCHRONOUS OPTICAL NETWORK

Synchronous Digital Hierarchy and Synchronous Optical Network are transmission technology standards for synchronous data transmission over fiber-optic cables. They provide a high-speed transfer of data, video, and types of information across great distances without regard to the specific services and applications they support. Service providers that have to aggregate (combine) multiple T-1s to provide connections across the country or around the world typically use them. **Synchronous Digital Hierarchy (SDH)** is an international standard. The International Telecommunications Union, formerly known as

Table 5.6 SONET/SDH Digital Hierarchy

Optical Level	SDH Equivalent	Electrical Level	Line Rate, Mbps	Payload Rate, Mbps	Overhead Rate, Mbps
OC-1	—	STS-1	51.840	50.112	1.728
OC-3	STM-1	STS-3	155.520	150.336	5.184
OC-9	STM-3	STS-9	466.560	451.008	15.552
OC-12	STM-4	STS-12	622.080	601.344	20.736
OC-18	STM-6	STS-18	933.120	902.016	31.104
OC-24	STM-8	STS-24	1,244.160	1,202.688	41.472
OC-36	STM-13	STS-36	1,866.240	1,804.032	62.208
OC-48	STM-16	STS-48	2,488.320	2,405.376	82.944
OC-96	STM-32	STS-96	4,976.640	4,810.752	165.888
OC-192	STM-64	STS-192	9,953.280	9,621.504	331.776
OC-768	STM-256	STS-768	39,813.12	38,486.01	1,327.10

CCITT, coordinates the development of the SDH standard. **SONET** (short for **Synchronous Optical Network**) is the North American equivalent of SDH, which is published by the American National Standards Institute (ANSI).

SONET and SDH do not apply directly to switches. In fact, there is no such thing as a SONET or SDH switch. Instead, the terms are used to specify the interfaces between switches that are linked by fiber-optic cable. In regard to the switches themselves, an ATM switch (or a similar switch such as FDDI, ISDN, or SMDS) is used. ATM, which we will discuss later in the chapter, can support a variety of other services including Frame Relay and voice. SDH and SONET map to the physical layer of the OSI model.

5.9.1 Building Blocks of SDH and SONET

The basic foundation of SONET consists of groups of DS-0 signals (64 Kbits/sec) that are multiplexed to create a 51.84-Mbps signal, which is also known as STS-1 (synchronous transport signal). STS-1 is an electrical signal rate that corresponds to the optical carrier line rate of OC-1, SONET's building block.

NOTE: OC-1 has enough bandwidth to support 28 T-1 links or one T-3 link.

Higher rates of transmission are a multiple of this basic rate. So, for example, STS-3 is three times the basic rate, or 155.52 Mbps. OC-12 is 12 times the basic rate, or 622 Mbps. The rate for OC-192 would be 9.95 Gbps. The SDH standard is based on the STM-1, which is equivalent to OC-3. Therefore, SDH uses multipliers of 155.52 Mbps. (See table 5.6.)

Figure 5.5   SDH/SONET Networks Use Dual Rings to Provide Redundant Pathways

Carrier SDH/SONET Network

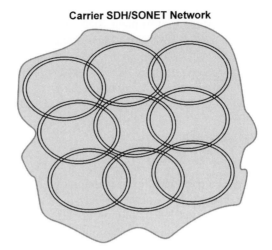

In essence, SONET and SDH are the same technology. There are minor differences in header information, payload size, and framing. But at 155 Mbps and above, the two are completely interoperable.

Just because the basic level of SONET starts at approximately 51 Mbps doesn't mean that lower-bit-rate asynchronous signals are ignored. The basic STS-1 frame contains 810 DS-0s, 783 of which are used for sending data (including slower asynchronous signals) and 27 of which are overhead. The overhead in this case is information concerning framing, errors, operations, and format identification.

Signals with speeds below STS-1, such as DS-1 and the European E-1 (2.048 Mbps), can be accommodated by dividing the STS-1 payload into smaller segments that are known as virtual tributaries (VTs). The lower-data-rate signals are combined with overhead information, which leads to the creation of synchronous payload envelopes (SPEs). SPEs allow these signals to be transported at high speeds without compromising integrity. Each VT on an STS-1 signal includes its own overhead information and exists as a distinct segment within the signal.

5.9.2   SDH and SONET Lines and Rings

Local SDH and SONET services are sold in two forms: point-to-point dedicated lines and dual fiber rings. Both deliver high speed, but only dual fiber rings guarantee automatic rerouting around outages. Long distance SDH and SONET connections employ multiple rings within the public network. Some are even using dual-ring pairs (four redundant rings on one circuit). (See figure 5.5.)

When using dual rings, if one circuit is broken, the traffic reverses and flows in the opposite direction on the same ring—avoiding the cut altogether. If the ring is broken in two places, the traffic is automatically rerouted onto the second circuit. This takes a few milliseconds longer than merely reversing direction, but it still happens nearly instantaneously.

Long distance point-to-point SDH and SONET services are similar to standard leased lines. A company buys a connection between two points, but its traffic is sent over multiple rings within the public network. Although the customer isn't buying rings, the customer gets all the benefits associated with rings within a long distance portion of the carrier network. Customers who need high reliability end to end should use local access rings.

5.10 CELL RELAY

Cell Relay is a data transmission technology based on transmitting data in relatively small, fixed-size packets or **cells.** Each cell contains only basic path information that allows switching devices to route the cell quickly. Cell Relay systems can reliably carry live video and audio because cells of fixed size arrive in a more predictable way than systems with packets or frames of varying size. Examples of Cell Relay are SMDS and ATM.

5.10.1 Switched Multimegabit Data Service

Switched Multimegabit Data Service (SMDS) is a high-speed, Cell Relay, wide area network (WAN) service designed for LAN interconnection through the public telephone network. SMDS can use fiber- or copper-based media. While it mostly supports speeds between 1.544 Mbps and 34 Mbps, it has been extended to support lower and higher bandwidth to broaden its target market. While SMDS and Frame Relay are both very desirable ways of gaining Internet access for your business, Frame Relay is popular for speeds of T-1 and below and SMDS is popular for speeds above T-1 up to T-3. An SMDS circuit is committed at its specified speed. Bursting of the circuit to full bandwidth is never required.

Different from Frame Relay and ATM, SMDS is a connectionless service. Connectionless service means that there is no predefined path or virtual circuit set up between devices. Instead, SMDS simply sends the traffic into the network for delivery to any destination on the network. With no need for a predefined path between devices, data can travel over the least congested routes in an SMDS network. As a result, it provides faster transmission for the networks bursty data transmissions and greater flexibility to add or drop network sites. SMDS currently supports only data, and is being positioned as the connectionless part of ATM services.

An SMDS digital service unit (DSU) or channel service unit (CSU) takes frames that can be up to 7168 bytes long (large enough to encapsulate entire IEEE 802.3, IEEE 802.5, and FDDI frames) from a router and breaks them up into 53-byte cells. The cells are then passed to a carrier switch. Each cell has 44 bytes of payload or data and 9 bytes for addressing, error correction, reassembly of cells, and other control features. The switch then reads addresses and forwards cells one by one over any available path to the desired end point. SMDS addresses ensure that the cells arrive in the right order.

To connect to an SMDS network, you will communicate with a common carrier to provide T-1 connection. To connect to SMDS, you would use a CSU/DSU (or some other connection device) and router. Often, the CSU/DSU and router can be found packaged together.

5.10.2   Introduction to ATM

Asynchronous Transfer Mode (ATM) is both a LAN and a WAN technology, which is generally implemented as a backbone technology. It is a cell-switching and multiplexing technology that combines the benefits of circuit switching and packet switching. The small, constant cell size allows ATM equipment to transmit video, audio, and computer data. Current implementations of ATM support data transfer rates of from 25 to 622 Mbps. The ATM standards allow it to operate in virtually every transmission medium including T-1, T-3, E-1, E-3, SDH, and SONET.

Because of its asynchronous nature, ATM is more efficient than synchronous technologies, such as time-division multiplexing. With TDM, users are assigned to time slots, and no other station can be sent in that time slot. If a station has a lot of data to send, it can send only when its time slot comes up, even if all other time slots are empty. If, however, a station has nothing to transmit when its time slot comes up, the time slot gets sent empty and is wasted. Because ATM is asynchronous, time slots are available on demand.

The ATM cell has a size of 53 bytes. This cell contains 48 bytes of payload or data and 5 bytes to hold the control and routing information. A fixed cell was chosen for two reasons. First, the switching devices necessary to build the ATM network need to use very fast silicon chips to perform the cell switching at such high speeds. These chips can run much faster if the cells do not vary, but instead remain a fixed, known value. By using a payload length of 48 bytes for data, ATM offers a compromise between a larger cell size (such as 64 bytes) optimized for data and a smaller cell size (such as 32 bytes) optimized for voice.

The second reason is that ATM is designed to carry multimedia communications, composed of data, voice, and video. Unlike data, however, voice and video require a highly predictable cell arrival time in order for sound and video to appear natural. These cells need to arrive one after the other in a steady stream, with no late cells. If the cells varied in size, a very long cell might block the timely arrival of a shorter cell. Therefore, a constant cell size was chosen.

An ATM network is made up of ATM switches (sometimes referred to as network-node interfaces, or NNI) and ATM end points (sometimes referred to as user-network interfaces, or UNI), consisting of PCs with ATM adapter cards, routers, bridges, CSU/DSUs, and video coder-decoders. When a packet initially enters the ATM network, the packet is segmented into ATM cells. This process is known as cell segmentation.

5.11   USING ROUTERS WITH WAN LINKS

As you recall, routers connect different LANs together to form a bigger network. Each connection on the router will have a unique network ID. Typically when you connect WAN links to a router, each LAN will have its own network ID, and the WAN links will have a unique network ID. (See figure 5.6.)

Figure 5.6   A WAN Connecting Several LANs Together. Each LAN Will Have Its Own Unique ID, and the WAN Will Have Its Own Unique ID

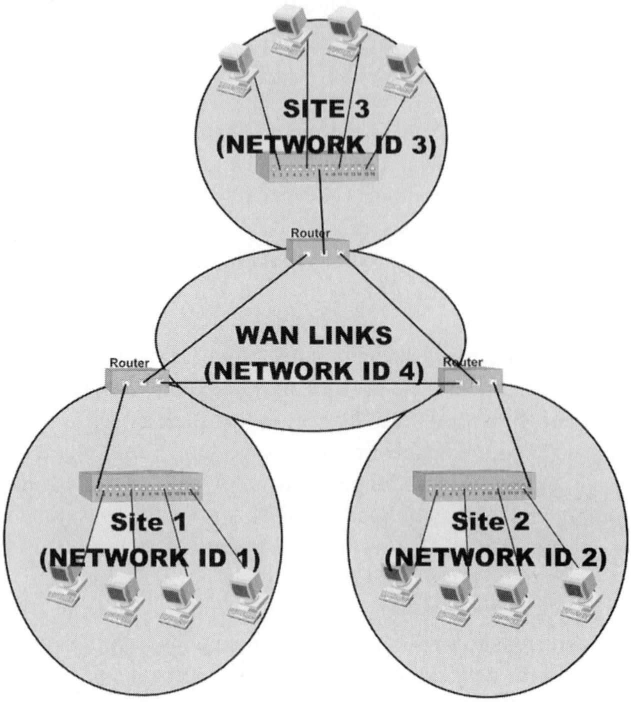

5.12   VIRTUAL PRIVATE NETWORKING

A **Virtual private networking (VPN)** is a network made of secured, point-to-point connections across a private network or a public network such as the Internet. In other words, a VPN connects the components of one network using another network.

The basic technology that defines a VPN is tunneling. **Tunneling** is a method of transferring data packets over the Internet or other public network, providing the security and features formerly available only on private networks. A tunneling protocol encapsulates the data packet in a header that provides routing information to enable the encapsulated payload to securely traverse the network. The entire process of encapsulation and transmission of packets is called tunneling, and the logical connection through which the packets travel is known as a tunnel.

Tunneling is similar to sending a letter from one building to another building, both of which belong to the same corporation but are located in two different cities. You take your letter and send it through the corporation main services. When the letter gets to the mail room, it will be sent on, using the U.S. mail. The U.S. mail then delivers it to the second building, after which the letter is sent through the corporate mail to the correct office.

Remote users who can gain access to their networks via the Internet are probably using **Point-to-Point Tunneling Protocol (PPTP).** PPTP, developed by Microsoft and Ascend Communications, uses the Internet as the connection between remote users and a local network, as well as between local networks. It is easy to see that the PPTP is the most popular implementation of the virtual private network and is an inexpensive way to create wide area networks (WANs) with PSTN, ISDN, and X.25 connections. PPTP wraps various protocols inside an IP datagram, an IPX datagram, or a NetBEUI frame. This lets the protocols travel through an IP network tunnel, without user intervention. In addition, PPTP saves companies the need to build proprietary and dedicated network connections for their remote users and instead lets them use the Internet as the conduit.

PPTP is based on the Point-to-Point Protocol. The difference between PPP and PPTP is that PPTP allows access using the Internet as the connection medium, rather than requiring a direct connection between the user and the network. In other words, instead of having to dial up the corporate network directly, a remote user could log in to a local Internet service provider, and PPTP will make the connection from that provider to the corporate network's Internet connection. From there, the connection continues into the corporate network the same as if the user dialed in directly. The process can be broken down this way:

1. The remote client makes a point-to-point connection to the front-end processor via a modem.
2. The front-end processor connects to the remote access server, establishing a secure "tunnel" connection over the Internet. This connection then functions as the network backbone.
3. The remote access server handles the account management and supports data encryption through IP, IPX, or NetBEUI protocols.

Microsoft Point-to-Point Encryption (MPPE) is a data encryption method that is used to encrypt data on PPTP connections. MPPE uses the RSA RC4 stream cipher to encrypt data, and MPPE can use a 40-bit, 56-bit, or 128-bit encryption key. Windows NT 4.0 only supports 40-bit MPPE encryption, so Windows 2000 and Windows .NET MPPE provide 40-bit encryption for backward compatibility with down-level clients. The level of MPPE encryption that will be used on a virtual private network (VPN) connection is negotiated as the link is being established.

PPTP has drawbacks. PPTP does not offer the strongest encryption technology and authentication features available. Therefore, for highly secure transmission over the Internet, most corporations tend to prefer IPSec.

5.13   TERMINAL EMULATION

Telecommunication network (telnet) is a virtual terminal protocol (**terminal emulation**) allowing a user to log on to another TCP/IP host to access network resources. Such a computer is frequently called a host computer, while the client is called a dumb terminal. With telnet, you log on as a regular user with whatever privileges you may have been granted to the specific applications and data on that computer. You can then enter commands through the telnet program, and they will be executed as if you were entering them directly on the server console. This enables you to control the server and communicate with other servers on the network. Telnet is a common way to remotely control web servers. The default port for telnet is TCP port 23.

A popular terminal program/brand of terminal from DEC was Visual Terminal 100 (vt100). It was widely used by companies and universities that ran Berkeley UNIX on their VAXes (also from DEC). Most communication packets support vt100.

DOS, Windows, and Linux computers can start a telnet session by executing the telnet command at the command prompt, which can be used with or without a computer name. If no computer name is used, telnet provides command mode and provides a prompt to the user. After a connection is established, telnet enters input mode.

The telnet command can be used to test TCP connections. You can telnet to any TCP port to see if it is responding, which is especially useful when checking SMTP and HTTP ports. The syntax for the telnet command is:

```
telnet hostname port
```

whereas the host name could be the host name or the IP address. If you do not specify the port number, it will default to port 23. For example, to verify that HTTP is working on a computer, you can type in the following command:

```
telnet acme.com 80
```

If you connect to an SMTP server, you can further test your email server by actually sending an email message from the telnet session.

5.14   INDEPENDENT COMPUTING ARCHITECTURE PROTOCOL

Although the term *thin client* has frequently been used to refer to software, it is increasingly being used to refer to computers, such as network computers and Net PCs, that are designed to serve as the clients for client/server architectures. In this latter sense, a **thin client** is a computer that is between a dumb terminal and a PC. A thin client is designed to be especially small so that the bulk of the data processing occurs on the server.

A **fat client** is a client that performs the bulk of the data processing operations. The data itself is stored on the server. Although a fat client also refers to software, we use the term here to apply to a network computer that has relatively strong processing abilities.

A relatively new technology for remote access consists of the Citrix WinFrame (or MetaFrame) products (including the Windows Terminal Server), which uses the **Independent Computing Architecture (ICA)** Protocol. The Citrix WinFrame allows multiple computers to take control of a virtual computer and use it as if it were a desktop (thin client). The ICA Protocol is the protocol that the Citrix or Terminal Server client uses to commu-

Figure 5.7   The Windows 2000 Terminal Server Allows You to Attach to and Run Programs on a Server from Another Computer

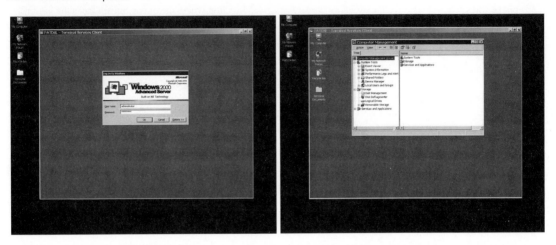

nicate with the server. This protocol sends screenshots, mouse movements, and so forth between the client and server. This way you can connect to a server and run configuration and administration programs just as if you were sitting at the server. (See figure 5.7.)

SUMMARY

1. Data terminal equipment (DTE) devices are end systems that communicate across the WAN.

2. Data circuit-terminating equipment (DCE) devices are special communication devices that provide the interface between the DTE and the network.

3. In a channel service unit/data service unit (CSU/DSU), the DSU is a device that performs protective and diagnostic functions for a telecommunications line. The CSU is a device that connects a terminal to a digital line.

4. The public switched telephone network (PSTN) is the international telephone system based on copper wires (UTP cabling) carrying analog voice data.

5. The PSTN, also known as the plain old telephone service (POTS), is the standard telephone service that most homes use.

6. The subscriber loop or local loop is the telephone line that runs from your home or office to the telephone company's central office (CO) or neighborhood switching station (often a small building with no windows).

7. The local loop is often referred to as the last mile, not because of its length, but because it is the slow link in the telecommunications infrastructure as it carries analog signals on a twisted-pair cable.

8. The point where the local loop ends at the customer's premises is called the demarcation point (demarc).

9. Modern networks offer remote access service (RAS), which allows users to connect remotely using various protocols and connection types.

10. Dial-up networking occurs when a remote access client makes a nonpermanent, dial-up connection to a physical port on a remote access server by using the service of a telecommunications provider such as an analog phone, ISDN, or X.25.

11. The central part of the serial device, including an analog modem, is the universal asynchronous receiver/transmitter (UART). The UART is a single IC chip that is the translator between the serial device and the system bus; it's the component that processes, transmits, and receives data.

12. A modem (modulator-demodulator) is a device that enables a computer to transmit data over telephone lines.

13. Baud rate refers to the modulation rate or the number of times per second that a line changes state.

14. The integrated services digital network (ISDN) is the planned replacement for POTS so that it can provide voice and data communications worldwide using circuit switching while using the same wiring that is currently being used in homes and businesses.

15. The ISDN uses two types of channels, a B channel and a D channel. The bearer channels (B channels) transfer data at a bandwidth of 64 Kbps (kilobits per second) for each channel.

16. The data channels (D channels) use a communications language called DSS1 for administrative signaling—it's used, for example, to instruct the carrier to set up or terminate a B-channel call, to ensure that a B channel is available to receive a call, or to provide signaling information for such features as caller identification.

17. The basic rate interface (BRI) defines a digital communications line consisting of three independent channels: two bearer (or B) channels, each carrying 64 kilobytes per second, and one data (or D) channel at 16 kilobits per second.

18. The first protocol used to connect to an IP network using dial-up lines was the Serial Line Interface Protocol (SLIP). SLIP is a simple protocol in which you send packets down a serial link delimited with special END characters.

19. The Point-to-Point Protocol (PPP) has become the predominant protocol for modem-based access to the Internet. PPP provides full-duplex, bidirectional operations between hosts and can encapsulate multiple network-layer LAN protocols to connect to private networks.

20. PPP supports asynchronous serial communication, synchronous serial communication, and ISDN.

21. When a client dials into a remote access server, the server will have to verify the client's credentials for authentication by using the client's user account properties and remote access policies to authorize the connection. If authentication and authorization succeed, the server allows a connection.

22. The Password Authentication Protocol (PAP) is the least secure authentication protocol because it uses clear text (plain text) passwords.

23. MS-CHAP v2 provides stronger security for remote access connections and allows for mutual authentication where the client authenticates the server.

24. PPPoE (PPP over Ethernet) was designed to bring the security and metering benefits of PPP to Ethernet connections such as those used in DSL.

25. A digital subscriber line (DSL) is a special communication line that uses sophisticated modulation technology to maximize the amount of data that can be sent over plain twisted-pair copper wiring, which is already carrying phone service to subscribers' homes.

26. Asymmetrical DSL (ADSL) transmits an asymmetric data stream, with much more going downstream to the subscriber and much less coming back.

27. Cable systems, originally designed to deliver broadcast television signals efficiently to subscribers' homes, can also be used to connect to the Internet.

28. The T-carrier system was the first successful system that converted analog voice signals to a digital bit stream.

29. In North America and Japan, you would typically find a T-1 line that has 24 64-Kbps channels for a bandwidth of 1.544 Mbps, and a T-3 line that has 672 64-Kbps channels for a bandwidth of 44.736 Mbps.

30. In Europe, you will find E-1 lines with 32 64-Kbps channels for a bandwidth of 2.048 Mbps and an E-3 line with 512 64-Kbps channel for a bandwidth of 34.368 Mbps.

31. Packet switching breaks the messages into small parts called packets and sends each packet individually onto the network.

32. A cloud represents a logical network with multiple pathways as a black box. The subscribers that connect to the cloud don't worry about the details inside the cloud. Instead, the only thing the subscribers need to know is that they connect at one edge of the cloud and the data is received at the other edge of the cloud.

33. Frame Relay is a packet-switching protocol designed to use high-speed digital backbone links to support modern protocols that provide for error handling and flow control for connecting devices on a wide area network.

34. Synchronous Digital Hierarchy (SDH) and Synchronous Optical Network (SONET) are transmission technology standards for synchronous data transmission over fiber-optic cables.

35. OC-1 has enough bandwidth to support 28 T-1 links or one T-3 link.
36. Cell Relay is a data transmission technology based on transmitting data in relatively small, fixed-size packets or cells.
37. Switched Multimegabit Data Service (SMDS) is a high-speed, Cell Relay, wide area network (WAN) service designed for LAN interconnection through the public telephone network.
38. Asynchronous Transfer Mode (ATM) is both a LAN and a WAN technology, which is generally implemented as a backbone technology. It is a cell-switching and multiplexing technology that combines the benefits of circuit switching and packet switching.

QUESTIONS

1. Of the following choices, which remote access connection technology requires the use of a modem?
   a. ISDN
   b. PSTN
   c. POTS
   d. T-1
2. Pat wants to connect his home computer to a remote access connection that will provide data transmission rates of up to 1.544 Mbps. Which service must Pat use?
   a. ISDN with two B channels
   b. PSTN
   c. Frame Relay
   d. T-1
3. Pat wants to connect his Windows 98 computer to an Internet service provider (ISP). He wants to use a remote access technology that provides 128 Kbps of bandwidth in two 64-Kbps channels. He also wants to send his email messages on a separate 16-Kbps channel. Which remote access technology should Pat use to connect his Windows 98 computer to an ISP?
   a. ISDN
   b. PSTN
   c. T-1
   d. T-3
4. You are trying to determine whether to use an integrated services digital network (ISDN) service or a public switched telephone network (PSTN) service to connect your home computer to the LAN at your office. What benefit is offered by ISDN but not by PSTN? (Select two answers)
   a. Faster connection rates
   b. Less expensive
   c. Widely available
   d. More bandwidth
5. Name two advantages of xDSL technology. (Select two answers.)
   a. Inexpensive and uses commonly available modems
   b. Provides fast digital service and "always-on" Internet access
   c. Provides high-speed digital access and is inexpensive
   d. Runs on fiber-optic cables and increased bandwidth
   e. Widely available in rural areas for a modest cost
6. What is the major roadblock to providing full digital service over POTS?
   a. It is too expensive for the telephone companies to implement.
   b. The "last mile" of telephone cable is still copper.
   c. Competition isn't strong enough to warrant full digital service.
   d. Fiber-optic lines cannot be brought out to rural areas.
7. The ISDN D channel is responsible for

   _____.
   a. providing redundancy
   b. carrying voice communications
   c. voice grade transmission
   d. providing control signals
   e. provides flow control
8. PSTN is an acronym for what?
   a. Partial switched telephone network
   b. Public switched transmission network
   c. Partial switched transmission network
   d. Public switched telephone network
9. What is another representation for a standard basic rate interface ISDN line?
   a. 64B2+16D
   b. B2+D
   c. 2B+D
   d. 2B64+D16

10. Which of the following WAN technologies can provide subscribers with bandwidth as needed?
    a. Frame Relay
    b. X.25
    c. ISDN
    d. T-1

11. SONET systems are _____ technology.
    a. twisted-pair, copper-based
    b. thinnet cabling
    c. fiber-optic
    d. wireless

12. SONET's base signal (STS-1) operates at a bit rate of _____.
    a. 64 kbps
    b. 1.544 Mbps
    c. 51.840 Mbps
    d. 155.520 Mbps

13. _____ is the standard for North America, while _____ is the standard for the rest of the world.
    a. SONET, SDH
    b. SDH, SONET
    c. ATM, SONET
    d. ATM, SDH

14. Which network component is used in conjunction with a router to provide access to a T-1 circuit?
    a. Gateway
    b. T-1 modem
    c. CSU/DSU
    d. Switch

15. What do FDDI and 100BaseFX have in common?
    a. Both use fiber-optic media to transmit data.
    b. Both are Ethernet technologies.
    c. Both use CSMA/CD as a media access method.
    d. Both use Token Passing as a media access method.

16. Which logical topology can be used in both a LAN and a WAN?
    a. Ethernet
    b. X.25
    c. Token Ring
    d. ATM

17. A device commonly called an ISDN modem connects computers to ISDN networks. What is the technically correct name for this device?
    a. Terminal adapter
    b. Digital transceiver
    c. Digital/analog modem
    d. Duplex transmitter

18. In an FDDI-based 100-Mbps LAN, if one ring fails, what will be the result?
    a. The other ring will take over at 50 Mbps.
    b. The other ring will take over at 10 Mbps.

    c. The other ring will take over at 100 Mbps.
    d. The network will go down.

19. Which is the most popular standard for 56K modems?
    a. V.32bis
    b. V.42
    c. V.90
    d. V.32

20. Which UART chip do you need to support 56K transport speeds and above?
    a. 8550
    b. 16550
    c. 85500
    d. 65550

21. What is the maximum data throughput that a typical PC serial port can support?
    a. 19,200 bps
    b. 65,536 bps
    c. 115,000 bps
    d. 56 kbps

22. Which of the following remote access protocols sends application screenshots from a server to a client?
    a. TCP/IP
    b. RAS
    c. IPX
    d. ICA

23. You are in charge of setting up a connection to the Internet for a large law firm. Money is no object. Which of the following would you use as your connection method?
    a. T-3
    b. T-1
    c. POTS
    d. OC-3

24. What is a VPN?
    a. Virtual private network
    b. Virtual primary network
    c. Visual point-to-point network
    d. Verifiable particle-stream network

25. What are the two service levels of ISDN? (Select two answers.)
    a. SRI
    b. PRI
    c. GRI
    d. BRI

26. Without using a repeater, what is the maximum cable length allowed in the rings of an FDDI network? (Select two answers.)
    a. 7500 feet
    b. 2000 meters
    c. 1500 meters
    d. 1.2 miles

27. What logical WAN topology operates up to 1.544 Mbps?
    a. SONET
    b. X.25
    c. ATM
    d. Frame Relay

28. What is PPP?
    a. Point-to-Point Protocol
    b. Private-Point Protocol
    c. Positive-Point Protocol
    d. Public-Point Protocol

29. What is SLIP?
    a. Single Line Interface Protocol
    b. Simple Link Internet Protocol
    c. Serial Link Interface Protocol
    d. Serial Line Internet Protocol
30. Leasing individual channels of a T-1 line is called what type of T-1?
    a. Partial rate   b. Serial
    c. Narrow bandwidth   d. Fractional
31. What is POTS?
    a. Pass-over transport service
    b. Plain old telephone service
    c. Positive online transport sensor
    d. Positive online transmission service
32. What is the maximum throughput of a T-1 line?
    a. 128 Kbps   b. 1.544 Mbps
    c. 64 Kbps   d. 56 Kbps
33. FDDI supports which physical network topologies? (Choose two answers.)
    a. Bus   b. Dual ring
    c. Mesh   d. Star
34. Network broadcasts can be blocked by which device?
    a. Bridge   b. Repeater
    c. Hub   d. Router
35. What media access method does FDDI employ?
    a. Token passing
    b. Polling
    c. CSMA/CD
    d. Contention
36. What network medium is used in an FDDI ring?
    a. Laser   b. Copper
    c. Infrared   d. Fiber optic
37. What is RAS?
    a. Remote access scan
    b. Random access sensor
    c. Rapid access service
    d. Remote access service
38. At what speed does an FDDI ring typically operate at?
    a. 500 Mbps   b. 100 Mbps
    c. 1000 Mbps   d. 10 Mbps
39. What is the function of a CSU?
    a. Provides LAN/WAN interface in a CSU/DSU
    b. Collision sensor unit interface provider
    c. Cable select unit interface provider
    d. Provides carrier signals in unison

40. What is the function of a DSU?
    a. Provides for synchronous communications over a LAN
    b. Provides for analog transmission over a WAN
    c. Provides for asynchronous communications over a LAN
    d. Provides for error correction and handshaking
41. The process of two modems communicating with each other to determine what protocol they will use is called what?
    a. Handshaking
    b. Digital synchronization
    c. Vectoring
    d. Serial replicating
42. SDH is the international equivalent of what data transmission standard?
    a. Token Ring   b. Ethernet
    c. AppleTalk   d. SONET
43. What utility can enable a Windows 95 Workstation to connect to a mainframe server?
    a. Link net   b. Subnet
    c. Portnet   d. Telnet
44. Frame Relay is what type of network?
    a. Link routing
    b. Circuit switching
    c. OSI linking
    d. Packet switching
45. ATM can transmit at what speed over fiber-optic media?
    a. 100 Mbps
    b. 100 Kbps
    c. 1.2 Mbps
    d. 155 Mbps
46. A UART is used by what device?
    a. Modem   b. Router
    c. Switch   d. Bridge
47. Frame Relay can use which of the following media?
    a. Fiber-optic cable
    b. Twisted-pair copper wire
    c. Digital microwave
    d. All of the choices can be used.
48. Which of the following is the Internet standard for the serial transmission of IP packets over asynchronous and synchronous lines?
    a. LPK   b. SLIP
    c. LKIP   d. PPP

49. Which device converts analog signals from a POTS line to digital signals?

   a. Switch   b. Router

   c. Modem   d. T-1 hub

50. What is the maximum access speed of Frame Relay? (Select two answers.)

   a. 100 Gbps

   b. 2445 Mbps

   c. 1.544 Mbps when running over a T-1 line

   d. 45 Mbps when running over a T-3 line

51. Which is NOT true of PPTP?

   a. It's an implementation of PPP over TCP.

   b. It's a VPN solution.

   c. It's a Point-to-Point Tunneling Protocol.

   d. All of the choices are true.

52. Your computer is using standard serial ports. IRQ 4 is therefore claimed by which serial or COM ports?

   a. 3   b. 2

   c. 4   d. 1

53. A dumb terminal is not able to send data to its mainframe server. What utility can you run from a UNIX or Microsoft workstation to see if the problem is with the server?

   a. Usenet   b. Sysnet

   c. Net-view   d. Telnet

54. Which UART chip do you need to support 56K transport speeds and above?

   a. 8550   b. 16550

   c. 85500   d. 65550

55. One of your users brings a laptop to work and proceeds to unplug the line from her office phone and plug it into her laptop computer to dial in to the Internet to check her private email. She can't dial out. What is the most likely reason?

   a. She did not dial a "9" first.

   b. The office phone line is digital and not analog.

   c. The office phone line does not have a trunk to the Internet.

   d. Network admin has blocked IP access on the office phone line.

56. Your Windows 95 or 98 computer at home needs to remotely dial in to a server at your office which is running a remote access server. What virtual device do you need to install on your computer to allow it to connect to your remote access server?

   a. Modem

   b. Dial-up adapter

   c. Router

   d. NIC

57. You notice your fellow tech setting COM1 to use I/O address 3F8. Which statement most accurately describes this configuration?

   a. COM1 is set up as standard.

   b. COM1 will function only if you disable COM2.

   c. COM1 won't function at all.

   d. COM1 is set up as nonstandard, but it will work fine.

58. You notice your fellow tech setting COM2 to use I/O address 3E8. Which statement most accurately describes this configuration?

   a. COM2 will function only if COM3 is disabled.

   b. COM2 is set up as standard.

   c. COM2 is nonstandard but will work fine.

   d. COM2 won't work at all.

59. What two advantages does modem communication on a PSTN/POTS network have over ISDN? (Select two answers.)

   a. Greater bandwidth

   b. Less costly

   c. Higher network throughput

   d. More widely available

60. Your computer is using standard serial (COM) ports. Therefore, IRQ 3 will be claimed by which COM port? (Choose all that apply)

   a. 3   b. 2

   c. 4   d. 1

61. Your Windows 98 workstation has a 56K modem installed. TCP/IP is also installed. You have an Internet service provider, and you have a local access number with which to connect to your ISP. You configure your TCP/IP properties to automatically obtain an IP address from your ISP. You dial in fine, but you cannot access any web pages. Which is the most likely cause?

   a. You did not configure your web browser with an IP address.

   b. Your 56K modem cannot support full-duplex transmissions.

   c. You need the IPX protocol installed to view web pages.

   d. Your dial-up connection is configured to use SLIP and not PPP.

62. Which protocol supports PPTP?

   a. NetBEUI   b. DecNet

   c. UNIX   d. TCP/IP

63. Your company asks you to set up a secure VPN between its Atlanta and Chicago offices. You are also required to use the Internet to send data between offices. Each office is using a different Internet service provider. You propose to set up PPTP between each office and its ISP. Your proposal accomplishes which of the following?
- a. The second requirement
- b. Both requirements
- c. The first requirement
- d. Neither requirement

64. ISA and PCI modems require all but which of the following settings?
- a. I/O port address
- b. DMA channel
- c. IRQ number
- d. COM port number

65. What logical LAN topology supports a physical dual-ring network?
- a. SONET
- b. ATM
- c. X.25
- d. FDDI

66. Which device must use a different network address off each of its ports?
- a. Hub
- b. Bridge
- c. Repeater
- d. Router

67. Which of the following functions does PPP LCP perform? (Choose all that apply.)
- a. Establish link connection
- b. Configure link connection
- c. Terminate link connection
- d. Test link connection

68. Which of these are authentication protocols for PPP? (Choose all that apply.)
- a. CHAP
- b. LCP
- c. MS-CHAP
- d. PAP
- e. NCP

69. What is RADIUS used for?
- a. Dial-up authentication
- b. WAN-based encryption
- c. Remote access compression algorithm
- d. Dynamic network configuration

70. Which well-known port does telnet use?
- a. 23
- b. 110
- c. 80
- d. 119

71. When running terminal services for Windows, what protocol does terminal services use?
- a. ICA
- b. SMTP
- c. Telnet
- d. WINS

72. Which of the following are characteristics of the Point-to-Point Protocol? (Choose all that apply.)
- a. It can transport data over serial links.
- b. It can create a virtual private network between two computers that are connected to the Internet.
- c. It can retrieve TCP/IP configuration data from a Dynamic Host Configuration Protocol server computer.
- d. It uses the Link Control Protocol to communicate between computers using PPP.

73. Which of the following allows a computer to remotely dial into a network and use the connection as if the computer were part of the internal network?
- a. IAS
- b. CAS
- c. XAS
- d. Remote access server

PART II

PROTOCOLS

CHAPTER 6

TCP/IP

Topics Covered in this Chapter

Introduction

Up to now, we have discussed the physical part of the network, but we haven't talked in depth about the protocols that make the network run. If you recall from chapter 1, protocols are the rules or standards that allow the computer to connect to one another and enable computers and peripheral devices to exchange information with as little error as possible. It is essential that the administrator understand these protocols in order to configure and troubleshoot the network. Of course, of all these, TCP/IP is the most common protocol suite being used today.

Objectives

- Explain TCP/IP in terms of routing, addressing schemes, interoperability, and naming conventions.
- Define the purpose, function, and/or use of IP, TCP, UDP, ARP, ICMP, and NTP within TCP/IP.
- Identify the purpose of the NAT network services.
- Identify the differences between public and private networks.
- Define the function of TCP/UDP ports.
- Identify well-known ports.
- Identify IP addresses (IPv4 and IPv6) and their default subnet masks.
- Identify the purpose of subnetting and default gateways.
- Given a troubleshooting scenario, select the appropriate TCP/IP utility (tracert, ping, ARP, netstat, nbtstat, IPCONFIG, ifconfig, winipcfg, and nslookup).
- Given output from a diagnostic utility (tracert, ping, IPCONFIG, etc.), identify the utility and interpret the output.
- List the protocols that can connect to a TCP/IP network.
- Given an IP address and subnet mask, determine the network address and host address.
- Given a network situation, determine a subnet mask that can be used for the network.
- List and describe the ways to translate from NetBIOS names to IP addresses and from host names to IP addresses.
- Explain how Classless Interdomain Routing differs from Classful IP.

6.1   TCP/IP AND THE INTERNET

The Internet that you know today began as an experiment funded by the U.S. Department of Defense (DoD) to interconnect DoD-funded research sites in the United States and to provide vendor-independent communications. In December 1968, the Advanced Research Projects Agency (ARPA) awarded a grant to design and deploy a packet-switching network (messages divided into packets, transmitted individually, and recompiled into the original message). In September 1969, the first node of the ARPAnet was installed at UCLA. By 1971, the ARPAnet spanned the continental United States, and it had connections to Europe by 1973.

Over time, the initial protocols used to connect the hosts together proved incapable of keeping up with the growing network traffic load. Therefore, a new TCP/IP suite was proposed and implemented. By 1983, the popularity of TCP/IP grew as it was included in the communications kernel for the University of California's UNIX implementation, 4.2BSD (Berkeley Software Distribution) UNIX. Today, TCP/IP is the primary protocol used on the Internet and is supported by Microsoft Windows, Novell NetWare, UNIX, and Linux.

The standards for TCP/IP are published in a series of documents called **Request for Comments (RFC).** An RFC can be submitted by anyone. Eventually, if it gains enough interest, it may evolve into an Internet standard. Each RFC is designated by an RFC number. Once published, an RFC never changes. Modifications to an original RFC are assigned a new RFC number.

RFCs are classified as one of the following: approved Internet standards, proposed Internet standards (circulated in draft form for review), Internet best practices, or For Your Information (FYI) documents. You should always follow approved Internet standards.

For more information on RFCs, visit the following websites:

http://www.rfc-editor.org/

http://www.cis.ohio-state.edu/hypertext/information/rfc.html

6.2   TCP/IP SUITE

With the TCP/IP suite, TCP/IP does not worry about how the hosts (computers or any other network connections) connect to the network. Instead, TCP/IP was designed to operate over nearly any underlying local or wide area network. (See figure 6.1.) This would include:

- LAN protocols: Ethernet, Token Ring, and ARCnet networks
- WAN protocols: ATM, Frame Relay, and X.25
- Serial line protocols: Serial Line Internet Protocol (SLIP) and Point-to-Point Protocol (PPP)

When you send or receive data, the data is divided into little chunks called packets. Each of these packets contains both the sender's TCP/IP address and the receiver's TCP/IP address. When the packet needs to go to another computer on another network, the packet is sent to a gateway computer, usually a router. The gateway understands the networks that it is connected to directly. The gateway computer reads the destination address to determine which direction the packet needs to be forwarded. It then forwards the packet to an adjacent gateway. The packet is then forwarded from gateway to gateway until it gets to the network that the destination host belongs to. The last gateway forwards the packet directly to the computer whose address is specified. (See figure 6.2.)

The lowest protocol within the TCP/IP suite is the **Internet Protocol (IP),** which is primarily responsible for addressing and routing packets between hosts. IP is a connectionless protocol, which means that there is no established connection between the end points that are communicating. Each packet (also known as a datagram) that travels through

Figure 6.1   TCP/IP Suite

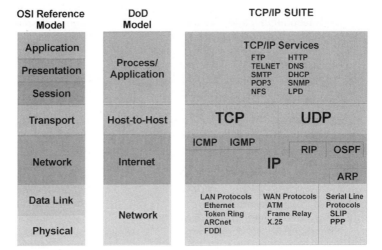

OSI Reference Model	DoD Model	TCP/IP SUITE

Internet Protocols

These kinds of protocols route data packets between different hosts or networks.

- **Internet Protocol (IP)**–Connectionless protocol primarily responsible for addressing and routing packets between hosts. (RFC 791)
- **Address Resolution Protocol (ARP)**– Used to obtain hardware addresses (MAC addresses) of hosts located on the same physical network. (RFC 826)
- **Internet Control Message Protocol (ICMP)**–Sends messages and reports errors regarding the delivery of a packet. (RFC 792)
- **Internet Group Management Protocol (IGMP)**–Used by IP hosts to report host group membership to local multicast routers. (RFC 1112)
- **Router Information Protocol (RIP)**– Distance-Vector Route Discovery Protocol where the entire routing table is periodically sent to the other routers. (RFC 1723)
- **Open Shortest Path First (OSPF)**–Link State Route Discovery Protocol where each router periodically advertises itself to other routers. (RFCs 1245, 1246, 1247, and 1253)

Host-to-Host Protocols

Host-to-host protocols maintain data integrity and set up reliable, end-to-end communication between hosts.

- **Transmission Control Protocol (TCP)**–Provides connection-oriented, reliable communications for applications that typically transfer large amounts of data at one time or that require an acknowledgment for data received. (RFC 793)
- **User Datagram Protocol (UDP)**–Provides connectionless communications and does not guarantee that packets will be delivered. Applications that use UDP typically transfer small amounts of data. Reliable delivery is the responsibility of the application. (RFC 768)

Process/Application Protocols

Process/application protocols act as the interface for the user. They provide applications that transfer data between hosts.

- **File Transfer Protocol (FTP)**–Allows a user to transfer files between local and remote host computers. (RFC 959)

Continued

Figure 6.1 TCP/IP Suite Continued

- **Telecommunication Network (telnet)**–a Virtual Terminal Protocol (terminal emulation) allowing a user to log on to another TCP/IP host to access network resources. (RFC 854)
- **Simple Mail Transfer Protocol (SMTP)**–The standard protocol for the exchange of electronic mail over the Internet. It is used between email servers on the Internet or to allow an email client to send mail to a server. (RFCs 821 and 822)
- **Post Office Protocol (POP)**–Defines a simple interface between a user's mail client software and email server. It is used to download mail from the server to the client and allows the users to manage their mailboxes. (RFC 1460)
- **Network File System (NFS)**–Provides transparent remote access to shared files across networks. (RFC 1094)
- **HyperText Transfer Protocol (HTTP)**–Serves as the basis for exchange over the World Wide Web (WWW). WWW

pages are written in the Hypertext Markup Language (HTML), an ASCII-based, platform-independent formatting language. (RFCs 1945 and 1866)
- **Domain Name System (DNS)**–Defines the structure of Internet names and their association with IP addresses. (RFCs 1034 and 1035)
- **Dynamic Host Configuration Protocol (DHCP)**–Used to automatically assign TCP/IP addresses and other related information to clients. (RFC 2131)
- **Simple Network Management Protocol (SNMP)**–Defines procedures and management information databases for managing TCP/IP-based network devices. (RFCs 1157 and 1441)
- **Line Printer Daemon (LPD)**–Provides printing on a TCP/IP network.
- **Network Time Protocol (NTP)**–An Internet standard protocol that assures accurate synchronization to the millisecond of computer clock times in a network of computers. (RFC 1305)

Figure 6.2 TCP/IP Packet

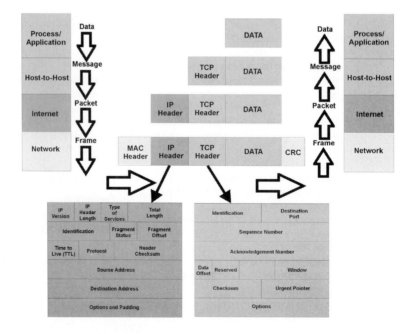

the Internet is treated as an independent unit of data. Therefore, the packets are not affected by other data packets. In addition, IP does not guarantee any deliveries. Thus, packets can get lost, delivered out of sequence, or delayed. Instead, IP must rely on TCP to determine that the data arrived successfully at its destination and to retransmit the data if it did not.

When a packet is received from the TCP, the IP protocol inserts its own header in the datagram. The main contents of the IP header are the source and destination addresses, the protocol numbers, and a checksum.

The protocols that works on top of IP are TCP and UDP. The **Transmission Control Protocol (TCP)** is a reliable, connection-oriented delivery service that breaks the data into manageable packets, wraps them with the information needed to route them to their destination, and then reassembles the pieces at the receiving end of the communication link. TCP establishes a virtual connection between the two hosts or computers so that they can send messages back and forth for a period of time. A virtual connection appears always to be connected, but in reality it is made of many packets being sent back and forth independently.

The most important information in the header includes the source and destination port numbers, a sequence number for the datagram, and a checksum. The source port number and destination port number ensure that the data is sent back and forth to the correct process (or program) running on each computer. The sequence number allows the datagram to be rebuilt in the correct order in the receiving computer, and the checksum allows the protocol to check whether the data sent is the same as the data received.

TCP has two other important functions. First, TCP uses acknowledgments to verify that the data was received by the other host. If an acknowledgment is not sent, the data is resent. In addition, since the data packets can be delivered out of order, TCP must put the packets back in the correct order.

TCP packets are sent in a sliding window. Sliding windows are used so that more data can be sent when network traffic is low (larger window), and less data can be sent when network traffic is high (smaller window). The transport layer communicates with its peer layer (transport) on the other machine to determine window size as data is being sent.

Another transport-layer protocol is the **User Datagram Protocol (UDP).** Unlike TCP, which uses acknowledgments to ensure data delivery, UDP does not. Therefore, UDP is considered unreliable, "best-effort" delivery. Since it is considered unreliable, UDP is used for protocols that transmit small amounts of data at one time or for broadcasts (packets sent to everyone).

NOTE: Unreliable doesn't mean that the packets will not get delivered; it is just that there is no guarantee or check to make sure that they get to their destination.

6.3   IPv4 ADDRESSING

The current version of the IP protocol is IPv4. Each connection on a TCP/IP address (logical address) is called a **host** (a computer or other network device that is connected to a TCP/IP network) and is assigned a unique **IP address** by the IP protocol. A host is any network interface, including each network interface card, or a network printer that connects directly onto the network. The format of the IP address is four 8-bit numbers (octet) divided by a period (.). Each number can be zero to 255. For example, a TCP/IP address could be 131.107.3.1 or 2.0.0.1.

IP addresses will be manually assigned and configured (**static IP addresses**) or dynamically assigned and configured by a DHCP server (**dynamic IP addresses**). Since the address is used to identify the computer, no two connections can use the same IP address. Otherwise, one or both of the computers would not be able to communicate and you will usually see a message stating "IP address conflict."

When communication occurs on a TCP/IP network, communications can be classified as unicast, multicast or anycast. **Unicast** is when a single sender communicates with a single receiver over the network. The opposite of unicast is **multicast,** which is communication between a single sender and multiple receivers, and **anycast,** which is communication between any sender and the nearest of a group of receivers in a network. Broadcasts are packets that are sent to every computer on the network or a subnet.

NOTE: Broadcast are not normally forwarded by routers.

When connecting to the Internet, network numbers are assigned to a corporation or business. The number system is divided into classes and therefore is known as a classful network. If the first number is between 1 and 126 (the first bit is a 0), the network is a class A. If the first number is between 128 and 191 (the first two bits are 1 0), the network is a class B. (You may notice that the number 127 is absent here. Its function is explained later in the chapter.) If the first number is between 192 and 223 (the first three bits are 1 1 0), the network is a class C. (See table 6.1.)

Because Internet addresses must be unique and because address space on the Internet is limited, there is a need for some organization to control and allocate address number blocks. IP number management was formerly a responsibility of the Internet Assigned Numbers Authority (IANA), which contracted with Network Solutions Inc. for the actual services. If you go to the following website, you will see exactly what blocks are assigned to what purposes:

http://www.iana.org/assignments/ipv4-address-space

Some of the class A blocks are staked out by corporations or government departments.

In December 1997, IANA turned this responsibility over to the following organizations to manage the world's Internet address assignment and allocation:

- **American Registry for Internet Numbers (ARIN)**—North and South America, the Caribbean, and sub-Saharan Africa
- **Réseaux IP Européens Network Coordination Centre (RIPE NCC)**—Europe, the Middle East, and parts of Africa and Asia
- **Asia Pacific Network Information Centre (APNIC)**—Asia Pacific region

Domain name management is still the separate responsibility of Network Solutions and a number of other registrars accredited by the Internet Corporation for Assigned Names and Numbers (ICANN).

Two additional classes should be mentioned. They are used for special functions only and are not commonly assigned to individual hosts. Class D addresses begin with a value between 224 and 239 and are used for IP multicasting. Multicasting is sending a single

Table 6.1   Standard IP Classes for the IP Address of w.x.y.z

Class Type	First Octet	Network Number	Host Number	Default Subnet Mask	Comments
A	1–126	w	x.y.z	255.0.0.0	Supports 16 million hosts on each of 126 networks
B	128–191	w.x	y.z	255.255.0.0	Supports 65,000 hosts on each of 16,000 networks
C	192–223	w.x.y	z	255.255.255.0	Supports 254 hosts on each of 2 million networks

data packet to multiple hosts. Class E addresses begin with a value between 240 and 255 and are reserved for experimental use.

Since TCP/IP addresses are growing scarce for the Internet, a series of addresses have been reserved to be used by **private networks** (networks not connected to the Internet). They are:

Class A—10.x.x.x (1 class A addresses)
Class B—Between 172.16.x.x and 172.31.x.x (16 class B addresses)
Class C—Between 192.168.0.x and 192.168.255.x (256 class C addresses)

If you are not connected to the Internet or are using a proxy server, it is recommended that you use private addresses to prevent a renumbering of your internetwork when you eventually connect to the Internet.

The TCP/IP address is broken down into a network number (sometimes referred to as a **network prefix**) and a host number. The network number identifies the entire network, and the host number identifies the computer or connection on the specified network. If it is a class A network, the first octet describes the network number while the last three octets describe the host address. If it is a class B network, the first two octets describe the network number while the last two octets describe the host address. If it is a class C network, the first three octets describe the network number while the last octet describes the host number. (See figure 6.3.)

Example 1:

You have the following network address:

131.107.20.4

Since 131 is between 128 and 191, the address is a class B network. In the address, 131.107 identifies the network, and 20.4 identifies the host or computer on the 131.107 network.

Figure 6.3   IP Network With Addresses and Subnet Masks. Notice the Multihomed Computer (Computer With Two Network Cards Connected to Two Subnets)

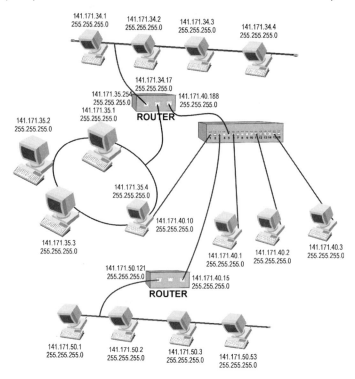

Example 2:

You have the following network address:

208.234.23.4

Since 208 is between 192 and 223, the address is a class C network. In this case, 208.234.23 identifies the network, and 4 identifies the host or computer on the 208.234.23 network.

NOTE: Several address values are reserved and/or have special meaning. The network number 127 is used for loopback testing, and the specific host address 127.0.0.1 refers to the local host or the actual host or computer you are currently using.

Usually when you define TCP/IP for a network connection, you would also specify a subnet mask. The subnet mask is used to define which bits describe the network number and which bits describe the host address. The default subnet mask for a class A network is 255.0.0.0. If you convert this to a binary equivalent, you would have 11111111.00000000.00000000.00000000, showing that the first 8 bits (first octet), marked with 1s, is used to define the network address: the last 24 bits, marked with 0s, are used to define the host address. The default subnet mask for a class B network is 255.255.0.0 (11111111.11111111.00000000.00000000), while the default subnet mask for a class C network is 255.255.255.0 (11111111.11111111.11111111.00000000).

If an individual network is connected to another network and you must communicate with a computer on the other network, you must also define the **default gateway,** which specifies the local address of a router. If the default gateway is not specified, you will not be able to communicate with computers on other networks.

NOTE: If the local area network is connected to two or more networks, you only have to specify one gateway. This is because when a data packet is sent, the sending computer will first determine if the data packet needs to go to a local computer or a computer on another network. If the data packet is meant to be sent on a computer on another network, the sending computer will forward the data packet to the router. The router will then determine the best direction that the data packet must go to get to its destination. Occasionally, it will have to go back through the network to get to another gateway. (See figure 6.4.)

Broadcasts are used to reach all devices on a network or subnetwork. The broadcast address is used when a machine wants to send the same packet to all devices on the network. They are two types of broadcasts, an all-nets broadcast and a subnet broadcast. Of course, while broadcast are not forwarded by most routers, broadcasts can take up valuable bandwidth and processing power in the receiving devices. All-nets broadcasts packets are addressed to 255.255.255.255 in the IP header. Literally, the packets are addressed to all networks. Subnet broadcasts contain the subnet address in the broadcast packet and all 1s for the host address, and are aimed at all computers within the subnet. To get your broadcast address, you set the device or host portion of the IP address to all 1s. Therefore, if you have the IP address 129.23.123.2 (A class C with a mask of 255.255.255.0), your broadcast address will be 129.23.123.255. When the device or host portion is all 0s it will give your network address, which will be 129.23.123.0.

Figure 6.4   IP Addresses With Default Gateways That Point to the Nearest Port on the Router

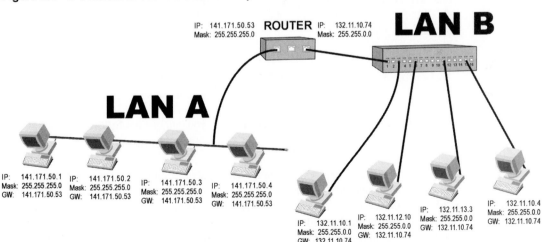

6.4   SUBNETTING THE NETWORK

The subnet mask can be changed to take a large network and break it into several small networks called subnets. This allows an organization to assign a distinct subnetwork number for each of its internal networks, which means that the organization can deploy additional subnets without needing to obtain a new network number from the Internet.

As noted earlier, the **subnet mask** is used to define which bits represent the network address (including the subnet number) and which bits represent the host address. For a subnet, the network prefix, subnet number, and subnet mask must be the same for all computers. (See figure 6.5.)

Example 3:

Your network is assigned a network number of 161.13. Since it is a class B network, it already has a default subnet mask of 255.255.0.0. Therefore, the TCP/IP address of any network interface card that belongs to this network must begin with 161.13. Since it is a class B network, it uses 16 bits to define the host address, which allows the network to have up to 65,534 computers. I don't know of any single local area network that has 65,534 computers.

The network administrator could take this large network and divide it into several smaller subnets. For example, if 161.13 defines the entire corporation network (network prefix), the third octet could be used to define a site network or individual local area network (subnet number), while the last octet is used to define the host address (host number).

Therefore, if you have three individual LANs, you can use the third octet to define the three LANs. The first building would have a network address of 161.13.1, the second building would have a network address of 161.13.2, and the third building would have a network address of 161.13.3. Since the last octet is used to define the host number and 8 bits are used to define the host number, there can be 254 hosts for each local area network. Of course, to let the network know that the 161.13 network is subnetted, the mask would have to be changed from 255.255.0.0 to 255.255.255.0 (11111111.11111111.11111111.00000000) to indicate that the first 24 bits represent the network address.

To calculate the maximum number of subnets, you can use the following equation:

$$\text{Number of subnets} = 2^{\text{number of masked bits}} - 2$$

To calculate the maximum number of hosts, you can use the following equation:

$$\text{Number of hosts} = 2^{\text{number of unmasked bits}} - 2$$

The -2 is used because you cannot use a network number, subnet number, or host number of all 0s and all 1s. For example, if you had an address of 131.107.3.4, the network address would be 131.107 and the host address would be 3.4. If you send a packet to the 131.107.0.0 (the host address is all zeros), you are sending the packet to the network itself, not to the individual computer on the network. But if you send it to 131.107.255.255

Figure 6.5 By Changing the Default Subnet Mask, You Split the Host Number Into a Subnet Number and a Host Number

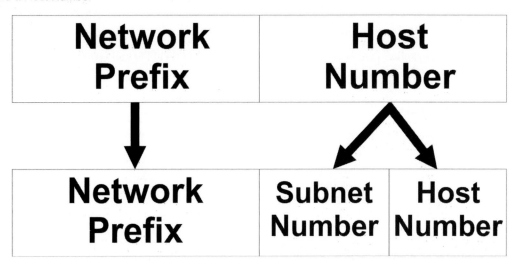

(the host address is all 1s), then you are doing a broadcast to all the computers on network 131.107. If you use the address of 0.0.3.4 (the network number is all 0s), the computer assumes 3.4 is on the current or local network.

Example 4:

If you look at example 3, you'll see that 8 bits define the subnet/site number and 8 bits define the host number. Therefore,

Number of subnets = $2^{\text{number of masked bits}} - 2 = 2^8 - 2 = 254$ subnets or sites
Number of hosts = $2^{\text{number of unmasked bits}} - 2 = 2^8 - 2 = 254$ hosts for each subnet

Example 5:

Your network is assigned an address of 207.182.15. You choose to subnet your network into several smaller networks. The largest network you have will have 25 computers. How many bits can you mask so that you can have the largest number of subnets or sites?

To start with, you only have the 8 bits used for the host address to play with. The easiest way is to use the formula, which calculates the number of hosts. To determine how many unmasked bits will allow 25 or more computers, you would use the following calculations:

Number of hosts = $2^{\text{number of unmasked bits}} - 2 = 2^1 - 2 = 0$
Number of hosts = $2^{\text{number of unmasked bits}} - 2 = 2^2 - 2 = 2$
Number of hosts = $2^{\text{number of unmasked bits}} - 2 = 2^3 - 2 = 6$
Number of hosts = $2^{\text{number of unmasked bits}} - 2 = 2^4 - 2 = 14$
Number of hosts = $2^{\text{number of unmasked bits}} - 2 = 2^5 - 2 = 30$

Since you use 5 of the 8 bits for the host number, that leaves 3 bits that are masked, which gives:

Number of subnets $= 2^{\text{number of masked bits}} - 2 = 2^3 - 2 = 6$ subnets

The subnet mask would be 11111111.11111111.11111111.11100000, which is equivalent to 255.255.255.224.

Example 6:

Using the network discussed in example 5, find the range of TCP/IP addresses for the second subnet.

The network number is 207.182.15 with a subnet mask of 255.255.255.224, which describes the first 3 bits of the fourth octet is used for the subnet number. The possible subnet numbers (in binary) are:

~~0 0 0~~
0 0 1
0 1 0
0 1 1
1 0 0
1 0 1
1 1 0
~~1 1 1~~

Of course, you cannot use 0 0 0 or 1 1 1, thus leaving six possible subnets.

By using binary counting, the second subnet is defined by 0 1 0. Since there are 5 bits left, the 5 bits can range from 0 0 0 0 1 (remember that they can't be all 0s) to 1 1 1 1 0 (remember that they can't be all 1s). Therefore, the last octet will be 0 1 0 0 0 0 0 1 to 0 1 0 1 1 1 1 0. If you translate these to decimal numbers, the last octet will be 65 to 94. So the address range is 207.182.15.65 to 207.182.15.94 with a subnet mask of 255.255.255.224.

Example 7:

You are the IT administrator for the Acme Manufacturing Company. Your company designs and produces widgets. It is your responsibility to plan out and implement the IP addresses for your company. You currently have 35 sites throughout the country, and there are plans to add 5 more sites within the next 3 years. Currently, the largest site has 275 people, but it may add another 25 to 75 employees at that site over the next 2 years depending on the market. In addition, the corporate building has 10 web servers and 15 corporate servers that also need IP addresses. ARIN has assigned 182.24 to your corporation.

1. How would you subnet your network so that you can get the maximum number of people per site?
2. What is the number of sites your network can have with this configuration?

3. What is the maximum number of people your network can have with this configuration?

4. What would the subnet mask be for your corporation?

5. The primary site (which is also the largest site) is the corporate office. If you assign this site as your first subnet, what would the extended network prefix be for the corporate office site?

6. What would the range of addresses be for this site?

7. What is the broadcast address for this site?

Let's take the first question and determine how you would subnet your network to get the maximum number of people per site.

You know that you have a network number of 182.24 (10110110.00011000), assigned by ARIN. Since the network number starts with 182, you know that it is a class B network with a default subnet mask of 255.255.0.0 (11111111.11111111.00000000. 00000000). This tells you that the first 16 bits are locked for you and the last 16 bits are yours to assign. You also know that you will have 40 (35 + 5) sites to plan for, with the maximum number of 375 (275 + 75 + 10 + 15) hosts. Since you want to maximize the number of hosts, let's figure out how many bits it would take to assign the 40 sites.

$$\text{Number of subnets} = 2^{\text{number of masked bits}} - 2 = 2^1 - 2 = 0$$
$$\text{Number of subnets} = 2^{\text{number of masked bits}} - 2 = 2^2 - 2 = 2$$
$$\text{Number of subnets} = 2^{\text{number of masked bits}} - 2 = 2^3 - 2 = 6$$
$$\text{Number of subnets} = 2^{\text{number of masked bits}} - 2 = 2^4 - 2 = 14$$
$$\text{Number of subnets} = 2^{\text{number of masked bits}} - 2 = 2^5 - 2 = 30$$
$$\text{Number of subnets} = 2^{\text{number of masked bits}} - 2 = 2^6 - 2 = 62$$

Therefore, it would take 6 bits to define the subnets. As you can see, if you use 6 bits, you can actually grow to 62 sites (question 2). This leaves 10 bits (16 − 6) to use to define your hosts. You would use the following equation:

$$\text{Number of hosts} = 2^{\text{number of masked bits}} - 2 = 2^{10} - 2 = 1022$$

Thus you can have up to 1022 hosts per site (question 3). Note that if this number was smaller than what you needed, you would have to come up with another solution so that you could get enough addresses for each site. Your options would be:

■ Assign multiple subnets per site.

■ Acquire a class A network from ARIN (highly unlikely).

■ Use a proxy server or a network address translation solution so that you can use one set of IP addresses for internal traffic (private network) and a second set of addresses for external traffic (IP addresses assigned by ARIN).

Since you are going to use 6 bits to define the subnets, the new subnet mask (question 4) would be:

11111111.11111111.**111111**00.00000000

This is equivalent to:

255.255.252.0

which would have to be assigned to every host on every subnet.

Looking at the 6 bits that define our subnets, you will number your subnets as shown below:

Site Number (Decimal)	Site Number (Binary)	Extended Network Number (Binary)	Extended Network Number (Decimal)
0*	000000	~~10110110.00011000.000000XX.XXXXXXXX~~	~~182.24.0.0~~
1	000001	**10110110.00011000.000001**XX.XXXXXXXX	182.24.4.0
2	000010	**10110110.00011000.000010**XX.XXXXXXXX	182.24.8.0
3	000011	**10110110.00011000.000011**XX.XXXXXXXX	182.24.12.0
4	000100	**10110110.00011000.000100**XX.XXXXXXXX	182.24.16.0
5	000101	**10110110.00011000.000101**XX.XXXXXXXX	182.24.20.0
6	000110	**10110110.00011000.000110**XX.XXXXXXXX	182.24.24.0
7	000111	**10110110.00011000.000111**XX.XXXXXXXX	182.24.28.0
8	001000	**10110110.00011000.001000**XX.XXXXXXXX	182.24.32.0
.	.	.	.
.	.	.	.
.	.	.	.
62	111110	**10110110.00011000.111110**XX.XXXXXXXX	182.24.248.0
63*	~~111111~~	~~10110110.00011000.111111XX.XXXXXXXX~~	~~182.24.252.0~~

NOTE: To calculate the extended network number in decimal, assign 0s to the host bits.

*Remember that the subnet number cannot be all 0s or all 1s.

Every host on your first site (the corporate site) will start with the following bits:

10110110.00011000.000001XX.XXXXXXXX

To calculate the extended network number in decimal, assign 0s to the host bits. Therefore, the extended network number is:

10110110.00011000.000001**00.00000000**

which gives an extended network number of 182.24.4.0 (question 5).

NOTE: This is the reason that you cannot use all 0s for the host since it is used to define the network numbers.

Since the Xs define the hosts (10110110.0001100.000001**XX.XXXXXXXX**) on that site. If you remember, you have 10 bits to assign to the host. The first host will be assigned:

10110110.00011000.00000**100.00000001**

which is equivalent to:

182.24.4.1

The last host at this site would be:

10110110.00011000.00000**111.11111110**

which is equivalent to:

182.24.7.254

So as you can see, the range of addresses for the first subnet is 182.24.4.1 to 182.24.7.254 (question 6).

Host Number (Binary)	Host Number (Decimal)	IP Address (Binary)	IP Address (Decimal)
~~00.00000000~~*	~~0.0~~	~~10110110.00011000.00000100.00000000~~	~~182.24.4.0~~
00.00000001	0.1	10110110.00011000.00000**100.00000001**	182.24.4.1
00.00000010	0.2	10110110.00011000.00000**100.00000010**	182.24.4.2
00.00000011	0.3	10110110.00011000.00000**100.00000011**	182.24.4.3
00.00000100	0.4	10110110.00011000.00000**100.00000100**	182.24.4.4
00.00000101	0.5	10110110.00011000.00000**100.00000101**	182.24.4.5
00.00000110	0.6	10110110.00011000.00000**100.00000110**	182.24.4.6
00.00000111	0.7	10110110.00011000.00000**100.00000111**	182.24.4.7
00.00001000	0.8	10110110.00011000.00000**100.00001000**	182.24.4.8
00.00001001	0.9	10110110.00011000.00000**100.00001001**	182.24.4.9
.	.	.	.
.	.	.	.
.	.	.	.
11.11111110	3.254	10110110.00011000.00000**111.11111110**	182.24.7.254
~~11.11111111~~*	~~3.255~~	~~10110110.00011000.00000111.11111111~~	~~182.24.7.255~~

*Remember that the subnet number cannot be all 0s or all 1s. The broadcast address is calculated by assigning all host bits as a 1. This is the reason we cannot use all 1s for the host bits assigned to a host address.

The broadcast address (question 7) for the entire network would be:

10110110.00011000.1**1111111.11111111**

which is equivalent to 182.24.255.255.

The broadcast address for the corporate subnet (site 1) would be:

10110110.00011000.000001**11.11111111**

which is equivalent to 182.24.7.255.

6.5   VARIABLE LENGTH SUBNET MASK AND CLASSLESS INTER-DOMAIN ROUTING

One of the major problems with the earlier limitations of IP addresses is that once the mask was selected across a given network prefix, locks the organization into a fixed number of fixed-sized subnets. For example, assume that a network administrator decided to configure the 130.5.0.0 network with a default mask of 255.255.0.0 with a subnet mask of 255.255.252.0 (6 bits used for the subnet number). This permits 8 subnets ($2^{number\ masked\ bits}$ = 8), each of which supports a maximum of 1,022 hosts ($2^{number\ unmasked\ bits} - 2 = 2^{10} - 2$). Unfortunately, if you have some subnets that are small (maybe 20 or 30 hosts), you would waste approximately 1,000 IP host addresses for each small subnet deployed.

In 1987, TCP/IP was modified with RFC 1009, which specified how a subnetted network could use more than one subnet mask. When an IP network is assigned more than one subnet mask, it is considered a network with **variable length subnet masks (VLSM)** since the extended network prefixes have different lengths. The advantages of having more than one subnet mask assigned to a given IP network number are:

■ Multiple subnet masks permit more efficient use of an organization's assigned IP address space.
■ Multiple subnet masks permit route aggregation which can significantly reduce the amount of routing information at the "backbone" level within an organization's routing domain.

As the class B network IDs would be depleted and for most organizations, a class C network does not contain enough host IDs to provide a flexible subnetting scheme for an organization, IP addresses are no longer given out under the class A, B, or C designation. Instead a method called Classless Internetwork Domain Routing (or CIDR, which is usually pronounced "cider") is used.

Routers that support CIDR do not make assumptions on the first 3-bits of the address; they rely on the prefix length information. The prefix length is described as slash x (/x), where x represents the number of network bits.

NOTE: Allocation can also be specified using the traditional dotted-decimal mask notation.

Under the IPv4 addressing, a class C address (for example, 198.23.27.32) uses a subnet mask of 255.255.255.0 (which is equivalent to 11111111.11111111.11111111.00000000)—the first 24 bits are the network bits, and the last 8 bits are the host bits. In the CIDR addressing, the same address would be indicated as 198.23.27.32/24. The number 24 indicates that the first 24 bits are the network bits.

Since CIDR eliminates the traditional concept of class A, class B, and class C network addresses, CIDR supports the deployment of arbitrarily sized networks rather than the standard 8-bit, 16-bit, or 24-bit network numbers associated with classful addressing. Therefore, with CIDR, the addresses that were wasted for the class A and class B networks are reclaimed and redistributed.

Prefixes are viewed as bitwise contiguous blocks of IP address space. For example, all prefixes with a /20 prefix represent the same amount of address spaces ($2^{12} - 2$, or 4094 host addresses). Examples for the traditional class A, class B, and class C are:

Traditional class A	10.23.64.0/20	**00001010.00010111.0100**0000.00000000
Traditional class B	140.5.0.0/20	**10001100.00000101.0000**0000.00000000
Traditional class C	200.7.128.0/20	**11001000.00000111.1000**0000.00000000

Table 6.2 provides information about the most commonly deployed CIDR address blocks.

Much like VLSM, CIDR allows you to take a single network ID and subdivide its network into smaller segments, depending upon its requirements. While VLSM divides the addresses on the private network, CIDR divides the addresses by the Internet Registry, a high-level ISP, a mid-level ISP, a low-level ISP, or a private organization.

Like VLSM, another important benefit of CIDR is that it plays an important role in controlling the growth of the Internet's routing tables. The reduction of routing information requires that the Internet be divided into addressing domains. Within a domain, detailed information is available about all the networks that reside in the domain. Outside of an addressing domain, only the common network prefix is advertised. This allows a single routing table entry to specify a route to many individual network addresses.

Supernetting is the process of combining multiple IP address ranges into a single IP network, such as combining several class C networks. For example, rather than allocating a class B network ID to an organization that has up to 2000 hosts, the InterNIC allocates a range of eight class C network IDs. Each class C network ID accommodates 254 hosts, for a total of 2032 host IDs.

Although this technique helps conserve class B network IDs, it creates a new problem. Using conventional routing techniques, the routers on the Internet now must have eight class C network ID entries in their routing tables to route IP packets to the organization. To prevent Internet routers from becoming overwhelmed with routes, CIDR is used to collapse multiple-network ID entries into a single entry corresponding to all of the class C network IDs allocated to that organization.

Table 6.2   CIDR Address Blocks

CIDR Prefix Length	Dotted Decimal	Number of Individual Addresses	Number of Classful Networks
/13	255.248.0.0	512K	8 class B or 2048 class C
/14	255.252.0.0	256K	4 class B or 1024 class C
/15	255.254.0.0	128K	2 class B or 512 class C
/16	255.255.0.0	64K	1 class B or 256 class C
/17	255.255.128.0	32K	128 class C
/18	255.255.192.0	16K	64 class C
/19	255.255.224.0	8K	32 class C
/20	255.255.240.0	4K	16 class C
/21	255.255.248.0	2K	8 class C
/22	255.255.252.0	1K	4 class C
/23	255.255.254.0	510	2 class C
/24	255.255.255.0	254	1 class C
/25	255.255.255.128	126	1/2 class C
/26	255.255.255.192	62	1/4 class C
/27	255.255.255.224	30	1/8 class C

Example 8:

For example, consider the following block of contiguous 32-bit addresses (192.32.0.0 through 192.32.7.0 in decimal notation). The supernet address (network address) for this block is 192.32.0.0 (11000000 00100000 00000000 00000000), the 21 upper bits (/21) shared by the 32-bit addresses. The mask for the supernet address in this example is 255.255.248.0 (11111111 11111111 11111000 00000000).

```
192.32.0.0
192.32.1.0
192.32.2.0
192.32.3.0    /24           →              192.32.0.0    /21
192.32.4.0
192.32.5.0
192.32.6.0
192.32.7.0
```

Example 9:

To list the individual network numbers defined by the CIDR block 198.35.64.0/20, first express the CIDR block in binary format:

198.35.64.0/20 **11000110.00100011.0100**0000.00000000

The /20 mask is 4 bits shorter than the natural mask for a traditional /24. This means that the CIDR block identifies a block of 16 (or 2^4) consecutive /24 network numbers.

The range of /24 network numbers defined by the CIDR block 198.35.68.0/24 includes:

Net #0:   11000110.00100011.0100**0000**.xxxxxxxx 198.35.64.0/20
Net #1:   11001000.00111000.0100**0001**.xxxxxxxx 198.35.65.0/20
Net #2:   11001000.00111000.0100**0010**.xxxxxxxx 198.35.66.0/20
Net #3:   11001000.00111000.0100**0011**.xxxxxxxx 198.35.67.0/20
Net #4:   11001000.00111000.0100**0100**.xxxxxxxx 198.35.68.0/20
Net #5:   11001000.00111000.0100**0101**.xxxxxxxx 198.35.69.0/20
Net #6:   11001000.00111000.0100**0110**.xxxxxxxx 198.35.70.0/20
Net #7:   11001000.00111000.0100**0111**.xxxxxxxx 198.35.71.0/20
Net #8:   11001000.00111000.0100**1000**.xxxxxxxx 198.35.72.0/20
Net #9:   11001000.00111000.0100**1001**.xxxxxxxx 198.35.73.0/20
Net #10: 11001000.00111000.0100**1010**.xxxxxxxx 198.35.74.0/20
Net #11: 11001000.00111000.0100**1011**.xxxxxxxx 198.35.75.0/20
Net #12: 11001000.00111000.0100**1100**.xxxxxxxx 198.35.76.0/20
Net #13: 11001000.00111000.0100**1101**.xxxxxxxx 198.35.77.0/20
Net #14: 11001000.00111000.0100**1110**.xxxxxxxx 198.35.78.0/20
Net #15: 11001000.00111000.0100**1111**.xxxxxxxx 198.35.79.0/20

NOTE: Different from classful networks, CIDR allows subnet numbers of all 0s and all 1s.

Thus, if you aggregate the 16 IP /24 network addresses to the highest degree, you would get 198.35.68.0.

6.6   IPv6

Since TCP/IP and the Internet became popular, the Internet has grown and continues to grow at an exponential rate. At this pace, it is easy to see that the Internet will eventually run out of network numbers. Therefore, a new IP, currently called IPv6, previously called IP Next Generation (IPng), has been developed.

The IPv6 header includes a simplified header format that reduces the processing requirements and includes fields so that the packets can be identified for real-time traffic used in multimedia presentations in order that the routers can handle real-time traffic differently (quality of service). (See figure 6.6.)

Figure 6.6 IPv6 IP Header

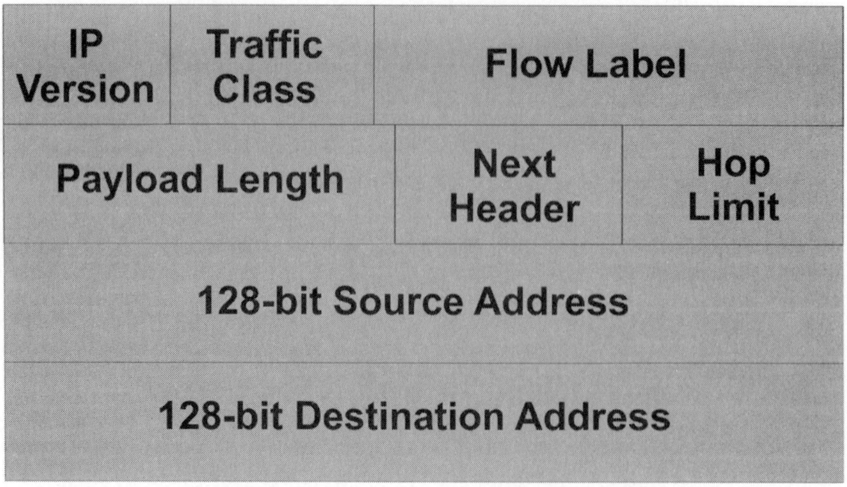

IP Version	Traffic Class	Flow Label	
Payload Length		Next Header	Hop Limit
128-bit Source Address			
128-bit Destination Address			

In addition, IPv6 introduces the extension header, which is described by a value in the IPv6 next header field. Routers can view the next header value and then independently and quickly decide if the extension header holds useful information. The extension headers carry much of the information that contributed to the large size of the IPv4 header. This information supports authentication, data integrity, and confidentiality, which should eliminate a significant class of network attacks, including host masquerading attacks.

IPv6 has security built into it. While it is not bulletproof security, it is strong enough to resist many of the common crippling problems that plagued IPv4. In IPv4, IPSec was optional; in IPv6, it is mandatory. IPSec, short for IP Security, is a set of protocols to support a secure exchange of packets at the IP level. In an IPv6 packet's authentication header are fields that specify IPv6 parameters.

IPv4 is based on 32-bit-wide addresses that allow a little over 4 billion hosts. IPv6 uses 128 bits for its addresses, which can have up to 3.4×10^{38} hosts. This means that IPv6 can handle all of today's IP-based machines without using Network Address Translation, allow for future growth, and also handle IP addresses for upcoming mobile devices like PDAs and cell phones.

When writing IPv6 addresses, they are usually divided into groups of 16 bits written as four hex digits, and the groups are separated by colons. An example is:

```
FE80:0000:0000:0000:02A0:D2FF:FEA5:E9F5
```

Leading zeros within a group can be omitted. Therefore, the above address can be abbreviated as:

```
FE80:0:2a0:d2ff:fea5:e9f5
```

You can also drop any single grouping of zero octets (as in the previous number) between numbers as long as you replace them with a double colon (::) and they are complete octets. You cannot use the zero compression rule to drop more than one grouping of zero octets. For example, the previous number can be further abbreviated as:

```
fe80::2a0:d2ff:fea5:e9f5
```

Because there must always be a certain number of bytes in the address, IPv6 can intelligently determine where the zeros are through this scheme.

To make addresses manageable, they are split in two parts: the bits that identify the network a machine is on and the bits that identify a machine on a network or subnetwork. The bits are known as net bits and host bits, and in both IPv4 and IPv6, the netbits are the left, or most significant, bits of an IP number; and the host bits are the right, or least significant, bits.

In IPv4, the border is drawn with the aid of the netmask, which can be used to mask all net/host bits. In CIDR routing, the borders between net and host bits stopped being 8-bit boundaries, and started to use the /x designation such as /64 to indicate 64 bits as network bits. The same scheme is used in IPv6.

IPv6 addresses can come with prefixes, which replace the "subnet mask" convention from IPv4. Prefixes are shown with a slash:

```
2180:FC::/48
```

In this example, the /48 is a routing prefix. A /64 would be a subnet prefix.

Usually, though, you would have an address such as:

```
FE80:0000:0000:0000:02A0:D2FF:FEA5:E9F5/64
```

which tells you that the first (leftmost) 64 bits of the address used here are used as the network address, and the last (rightmost) 64 bits are used to identify the machine on the network.

With 128 bits available for addressing in IPv6, the scheme commonly used is the same, only the fields are wider. Providers usually assign /48 networks, which leaves 16 bits for a subnetting and 64 host bits. This means that for your corporation or organization, you can have up to 65,536 subnets, and each subnet could have 18,446,744,073,709,551,616 hosts.

The idea behind having fixed-width, 64-bit-wide host identifiers is that they aren't assigned manually as in IPv4. Instead, IPv6 host addresses are recommended to be built from the so-called EUI64 addresses. The EUI64 addresses are of course built on 64-bit-wide identifiers and are derived from the MAC address of the network interface card. If the MAC address is:

```
01:23:45:67:89:ab
```

a FF:FE is inserted in the middle of the MAC to become:

```
01:23:45:ff:fe:67:89:ab
```

Therefore, the host bits of the IPv6 address would be:

```
:0123:45ff:fe67:89ab
```

These host bits can now be used to automatically assign IPv6 addresses to hosts, which supports autoconfiguration of v6 hosts. All that is needed to get a complete v6 IP number is the net and subnet bits, which can be assigned automatically.

An IPv4-mapped IPv6 address is the address of an IPv4-only node represented as an IPv6 address. The IPv4 address is stored in the low-order 32 bits. The high-order 96 bits bear the prefix 0:0:0:0:0:0. The address of any IPv4-only node may be mapped into the IPv6 address space by prepending the prefix 0:0:0:0:0:0 to its IPv4 address.

NOTE: The IPv4-mapped IPv6 addresses always identify IPv4-only nodes; they never identify IPv6/IPv4 or IPv6-only nodes.

An IPv4-compatible IPv6 address is an address, assigned to an IPv6 node, that can be used in both IPv6 and IPv4 packets. An IPv4-compatible IPv6 address holds an IPv4 address in the low-order 32 bits. The high-order 96 bits bear the prefix 0:0:0:0:0:ffff. The entire 128-bit address can be used when sending IPv6 packets. The low-order 32 bits can be used when sending IPv4 packets.

NOTE: The addresses can never identify IPv4-only nodes.

As with IPv4, IPv6 has several addresses that are reserved for special uses. The IPv6 address ::/0 is the default address for a host (like 0.0.0.0 in IPv4). The address ::1/128 (0:0:0:0:0:0:0:1) is reserved for the local loopback (like 127.0.0.1 in IPv4).

To support autoconfiguration, the Neighbor Discovery Protocol is used to discover local nodes, routers, and link-layer addresses and to maintain reachability information based on them. The aim of autoconfiguration is to make connecting machines to the network as easy as possible. In many cases it actually allows you to plug your machine into the network and start using it right away.

Each interface will have at least one globally unique unicast address and one link-local address. The link-local address will be used to provide addressing on a single link for the purpose of address autoconfiguration, neighbor discovery, and internal routing. The unicast address will be used to establish hierarchical boundaries for the operation of routing protocols.

The long-term plans for IPv6 are probably going to stretch out over the next 10 to 15 years. While, for instance, Windows XP and Linux have support for it, this support is nowhere near complete. The stacks for OSes are still being tested and revised. Getting IPv6 to the average person on the desktop is going to be a long and involved procedure.

6.7   TCP/IP PORTS AND SOCKETS

Every time a TCP/IP host communicates with another TCP/IP host, it will use the IP address and port number to identify the host and service/program running on the host. A TCP/IP port number is a logical connection placed for client programs to specify a particular server program running on a computer on the network defined at the transport layer. (See figure 6.7.) The source port number identifies the application that sent the data, and the destination port number identifies the application that receives the data. Port numbers range from 0 to 65536. Additionally, there are two types of ports: TCP and UDP, which are based on their respective protocols.

Figure 6.7   TCP/IP Ports

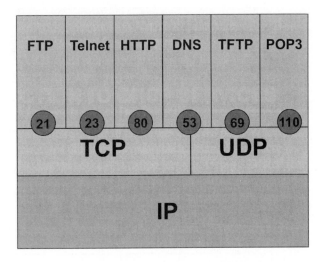

Today, the very existence of ports and their numbers is typically transparent to the users of the network, as many ports are standardized. Thus, a remote computer will know which port it should connect to for a specific service. IANA divides port numbers into three groups:

- **Well-known ports**—These are the most commonly used TCP/IP ports. These ports are in the range of 0 through 1023. These ports can be used only by system processes or privileged programs. Well-known ports are TCP ports, but are usually registered to UDP as well. Some of the well-known protocols are shown in table 6.3.
- **Registered ports**—These are the ports in the range of 1024 through 49251. On most systems, user programs use registered ports to create and control logical connections between proprietary programs.
- **Dynamic (private) ports**—These ports are in the range of 49152 through 65535. These ports are unregistered and can be used dynamically for private connections. In a multiuser system, a program can define a port on the fly if more than one user requires access to the same service at the same time.

For example, when using your browser to view a web page, the default port to indicate the HTTP service is identified as port 80, and the FTP service is identified as port 21. Other application processes are given port numbers dynamically for each connection so that a single computer can run several services. When a packet is delivered and processed, TCP (connection-based services) or UDP (connectionless-based services) will read the port number and forward the request to the appropriate program.

A socket identifies a single-network process in terms of the entire Internet. An application creates a socket by specifying three items: the IP address of the host, the type of service, and the port the application is using.

NOTE: For Windows, it is the Windows socket (WinSock) that provides the interface between the network program or service and the Windows environment.

Table 6.3   Popular TCP/IP Services and Their Default Assigned Port Numbers

Network Program/Service	Default Assigned Port Number
FTP data transfer	TCP port 20
FTP control	TCP port 21
Telnet	TCP port 23
Simple Mail Transfer Protocol (SMTP)	TCP port 25
Domain Name System (DNS)	UDP port 53
HTTP	TCP/UDP port 80
POP3	TCP port 110
Network News Transport Protocol (NNTP)	TCP port 119
NetBIOS Name Service	UDP port 137
NetBIOS Session Service	UDP port 139
Simple Network Management Protocol (SNMP)	UDP port 161
Secure HTTP	TCP/UDP port 443

For a complete list of registered well-known port numbers, go the following website: http://www.isi.edu/in-notes/rfc1700.txt

By default, when using a web page browser, port 80 is the default port. But if a website is using a different port, you can override that default by specifying the URL followed by a colon and the port number. For example, to specify port 2232, you would use: http://www.acme.com:2232

6.8   ROUTING IN TCP/IP

TCP/IP hosts use a routing table to maintain knowledge about other IP networks and hosts. As discussed earlier in this chapter, networks and hosts are identified with an IP address, and corresponding subnet masks specify which bits of the address are the network address and which bits are the host address.

When a computer prepares to send an IP datagram, it inserts its own source IP address and the destination IP address of the recipient into the IP header. The computer then examines the destination IP address, compares it with a locally maintained IP routing table, and takes appropriate action based on what it finds. The computer does one of three things:

- It passes the datagram up to a protocol layer above IP on the local host.
- It forwards the datagram through one of its attached network interfaces.
- It discards the datagram.

IP searches the routing table for the route that is the closest match to the destination IP address (either a route that matches the IP address of the destination host or a route that matches the IP address of the destination network). If a matching route is not found, IP discards the datagram.

To determine if a packet is to be sent to a host on the local network or to determine if the packet needs to be a remote network via the router, TCP/IP will complete a couple of calculations. When TCP/IP is initialized on a host, the host's IP address is ANDed (logical bitwise AND) with its subnet mask at a bit-by-bit level. Before a packet is sent, the destination IP address is ANDed with the same subnet mask. If both results match, IP knows that the packet belongs to a host on the local network. If the results don't match, the packet is sent to the IP address of an IP router so that it can be forwarded to the host.

When ANDing an IP address and subnet mask, each bit in the IP address is compared with the corresponding bit in the subnet mask. If both bits are 1s, the resulting bit is 1. If there is any other combination, the resulting bit is 0.

NOTE: By ANDing the address and subnet mask, you are essentially isolating the network addresses.

Example 10:

You have a host 145.17.202.56 with a subnet mask of 255.255.224.0 with data that needs to be delivered to 145.17.198.75 with a subnet mask of 255.255.224.0. Are the two hosts on one local network, or are they located on two remote networks?

To figure this out, you need to first convert the addresses and subnet mask to binary:

145.17.202.56	10010001.00010001.11001010.00111000
145.17.198.75	10010001.00010001.11000110.01001011
255.255.224.0	11111111.11111111.11100000.00000000

Next, AND the source address and the subnet mask:

145.17.202.56	10010001.00010001.11001010.00111000
255.255.224.0	*11111111.11111111.11100000.00000000*
	10010001.00010001.11000000.00000000

AND the destination address and the subnet mask:

145.17.198.75	10010001.00010001.11000110.01001011
255.255.224.0	*11111111.11111111.11100000.00000000*
	10010001.00010001.11000000.00000000

If you compare the two results, you will find that they are the same. Therefore, the packet will be sent directly to the host and not the router. If they were different, the packet would be sent to the router.

6.8.1   Routing Information Protocol

The Routing Information Protocol (RIP) is designed for exchanging routing information within a small to medium-size network. The biggest advantage of RIP is that it is extremely simple to configure and deploy.

RIP uses a single routing metric of hop counts (the number of routers) to measure the distance between the source and a destination network. Each hop in a path from source to destination is assigned a hop-count value, which is typically 1. When a router receives a routing update that contains a new or changed destination network entry, the router adds 1 to the metric value indicated in the update and enters the network in the routing table. The IP address of the sender is used as the next hop.

RIP prevents routing loops from continuing indefinitely by implementing a limit on the number of hops allowed in a path from the source to a destination. The maximum number of hops in a path is 15. If a router receives a routing update that contains a new or changed entry, and if increasing the metric value by 1 causes the metric to be infinity (in this case, 16), the network destination is considered unreachable. Of course, this means RIP is unable to scale to large or very large internetworks.

Since RIP is a distance-vector protocol, then as internetworks grow larger in size, the periodic announcements by each RIP router can cause excessive traffic. Another disadvantage of RIP is its high convergence time. When the network topology changes, it may take several minutes before the RIP routers reconfigure themselves to the new network topology. While the network reconfigures itself, routing loops may form that result in lost or undeliverable data. To help prevent routing loops, RIP implements split horizon.

To overcome some of RIP's shortcomings, RIP Version 2 (RIP II) was introduced. RIP II provides the following features:

- You can use a password for authentication by specifying a key that is used to authenticate routing information to the router.
- RIP II includes the subnet mask in the routing information and supports variable-length subnets. Variable-length subnet masks can be associated with each destination, allowing an increase in the number of hosts or subnets that are possible on your network.
- The routing table can contain information about the IP address of the router that should be used to reach each destination. This helps prevent packets from being forwarded through extra routers on the system.
- Multicast packets only speak to RIP II routers and are used to reduce the load on hosts not listening to RIP II packets. The IP multicast address for RIP II packets is 224.0.0.9.

RIP II is supported on most routers and end nodes. Make sure that the mode you configure is compatible with all implementations of RIP on your network.

6.8.2   OSPF

OSPF, short for Open Shortest Path First, is a link-state routing protocol used in medium-size and large networks that calculates routing table entries by constructing a shortest-path tree. It is considered superior to RIP for the following reasons:

- It is a more efficient protocol than RIP and does not have the restrictive 16 hop-count problem, which causes data to be dropped after the 16th hop. An OSPF network can have an accumulated path cost of 65,535, which enables you to construct very large networks (within the maximum time-to-live value of 255) and assign a wide range of costs. An OSPF network is best suited for a large infrastructure with more than 50 networks.

- OSPF networks can detect changes to the network and calculate new routes quickly (rapid convergence). The period of convergence is brief and involves minimal overhead. The count-to-infinity problem does not occur in OSPF internetworks.

- Less network traffic is generated by OSPF than by RIP. RIP requires each router to broadcast its entire database every 30 seconds. OSPF routers only broadcast link-state information when it changes.

- Link-state advertisements (LSAs) include subnet mask information about networks. You can assign a different subnet mask for each segment of the network (variable-length subnetting). This increases the number of subnets and hosts that are possible for a single network address.

The shortest-path-first (SPF) routing algorithm is the basis for OSPF operations. When an SPF router is started, it initializes its routing protocol data structures and then waits for indications from lower-layer protocols that its interfaces are functional.

After a router is assured that its interfaces are functioning, it uses the OSPF Hello Protocol to acquire neighbors, which are routers with interfaces to a common network. The router sends hello packets to its neighbors and receives their hello packets. In addition to helping acquire neighbors, hello packets also act as keep-alives to let routers know that other routers are still functional.

Adjacency is a relationship formed between selected neighboring routers for the purpose of exchanging routing information. When the link-state databases of two neighboring routers are synchronized, the routers are said to be adjacent. Not every pair of neighboring routers becomes adjacent. From the topological database generated from these hello packets, each router calculates a shortest-path tree, with itself as root. The shortest-path tree, in turn, yields a routing table. To maintain the table, the established adjacencies are compared with the hello messages sent out so that the router can quickly detect failed routers and the network's topology can be altered appropriately.

6.8.3   Other Routing Protocols

The Enhanced Interior Gateway Routing Protocol (EIGRP) for IP, IPX, and AppleTalk was developed in the mid-1980s by Cisco Systems. EIGRP combines the advantages of link-state protocols with those of distance-vector protocols. It has a fast convergence time and a low network overhead. In addition, it is easier to configure than OSPF. EIGRP maintains separate routing tables for IP, IPX, and AppleTalk protocols but forwards routing update information by using a single protocol. While it can accommodate very large and heterogeneous networks, it is only supported by Cisco routers.

The Border Gateway Protocol (BGP) for IP is the routing protocol of Internet backbones. Since the Internet has grown so rapidly, BGP was developed to handle 100,000 routes and to route traffic efficiently through hundreds of Internet backbones. BGP is a vector-distance protocol, but unlike traditional vector-distance protocols such as RIP where there is a single metric, BGP determines a preference order by applying a function mapping each path to a preference value and selects the path with the highest value.

6.8.4   ICMP

Another protocol worth mentioning is the Internet Control Message Protocol (ICMP), which works at the Network layer and is used by IP for many different services. ICMP manages IP by reporting on the IP status information. It transmits "destination unreachable" messages to the source device if the target device cannot be located. It can also provide notification that a router's buffer has become full, indicating that the router is congested and it is forced to drop packets. Additionally, the ICMP protocol notifies a transmitting device that a better route to a destination exists. Both the ping and traceroute utilities use ICMP message to identify whether a destination device is reachable and to track a packet as it is routed from a source to a destination device. As the IP processes the datagrams generated by ICMP, ICMP is not directly apparent to the application user.

6.9   NETWORK ADDRESS TRANSLATION AND PROXY SERVERS

Many corporations will connect their corporate network (private network) to the Internet (public network). A **private network** is a network where only authorized users have access to the data while a **public network,** everyone connected has access to the data. In a public network, since the IP hosts are directly accessible from the Internet, the network addresses has to be registered with IANA. In a private network, the IP addresses are assigned by the administrator. Of course, it is recommended that you use the addresses reserved for private addresses (see section 7.3). If you connect a private network to a public network, you increase the possibility for security break-ins. This is the reason that firewalls are implemented to protect a private network from unauthorized users on a public network.

Another difference between a public and private network is that a public network sits in front of a firewall, and does not enjoy its protection. A private network sits behind the firewall and does. To combat certain types of security problems, a number of firewall products are available, which are placed between the user and the Internet to verify all traffic before allowing it to pass through.

A firewall is what keeps intruders out of your network. Just as a real firewall will protect one side of a building from fire on the other side, a network firewall acts as a barrier to network traffic on one side to protect the network on the other. This means, for example, that no unauthorized user would be allowed to access the company's file or email server.

A firewall can be configured with rules to control which packets will be accepted into the private network, and which can pass out of it. It reads the headers of every packet, in- or out-bound, and compares that information with its settings. Packets that do not comply are dropped. The problem with firewall solutions is that they are expensive and difficult to set up and maintain, putting them out of reach for home and small business users.

NOTE: Recently, personal firewalls have become very popular and are available in Windows XP and Linux to protect individual computers that surf the Internet. Firewalls are explained more in chapter 13.

Because IP addresses are a scarce resource, most Internet service providers (ISPs) will only allocate one address to a single customer. In a majority of cases this address is assigned dynamically, so every time a client connects to the ISP, a different address will be provided. Big companies can buy more addresses, but for small businesses and home users, the cost of doing so is prohibitive. Because such users are given only one IP address, they can have only one computer connected to the Internet at one time. **Network Address Translation (NAT)** is a method of connecting multiple computers to the Internet (or any other IP network) using one IP address. With a NAT gateway running on this single computer, it is possible to share that single address with multiple local computers and connect them all at the same time. The outside world is unaware of this division and thinks that only one computer is connected.

NAT automatically provides firewall-style protection without any special setup. The basic purpose of NAT is to multiplex traffic from the internal network (private network) and present it to the Internet (public network) as if it were coming from a single computer having only one IP address. The TCP/IP suite includes a multiplexing facility so that any computer can maintain multiple simultaneous connections with a remote computer. For example, an internal client can connect to an outside FTP server, but an outside client will not be able to connect to an internal FTP server because it would have to originate the connection, and NAT will not allow that. It is still possible to make some internal servers available to the outside world via inbound mapping, which maps certain well-known TCP ports (e.g., 21 for FTP) to specific internal addresses, thus making services such as FTP or the web available in a controlled way.

To multiplex several connections to a single destination, client computers label all packets with unique port numbers. Each IP packet starts with a header containing the source and destination addresses and port numbers. This combination of numbers completely defines a single TCP/IP connection. The addresses specify the two machines at each end, and the two port numbers ensure that each connection between this pair of machines can be uniquely identified.

Each separate connection is originated from a unique source port number in the client, and all reply packets from the remote server for this connection contain the same number as their destination port, so that the client can relate them to their correct connection. In this way, for example, it is possible for a web browser to ask a web server for several images at once and to know how to put all the parts of all the responses back together.

A modern NAT gateway must change the source address on every outgoing packet to be its single public address. It therefore also renumbers the source ports to be unique, so that it can keep track of each client connection. The NAT gateway uses a port mapping table to remember how it renumbered the ports for each client's outgoing packets. The port mapping table relates the client's real local IP address and source port plus its translated source port number to a destination address and port. The NAT gateway can therefore reverse the process for returning packets and route them back to the correct clients.

When any remote server responds to a NAT client, incoming packets arriving at the NAT gateway will all have the same destination address, but the destination port number will be the unique source port number that was assigned by the NAT. The NAT gateway looks in its port mapping table to determine which "real" client address and port number a packet is destined for, and replaces these numbers before passing the packet on to the local client.

This process is completely dynamic. When a packet is received from an internal client, NAT looks for the matching source address and port in the port mapping table. If the entry is not found, a new one is created, and a new mapping port is allocated to the client:

- Incoming packet received on non-NAT port.
- Look for source address and port in the mapping table.
- If found, replace source port with previously allocated mapping port.
- If not found, allocate a new mapping port.
- Replace source address with NAT address, and replace source port with mapping port.

Packets received on the NAT port undergo a reverse translation process:

- Incoming packet received on NAT port.
- Look up destination port number in port mapping table.
- If found, replace destination address and port with entries from the mapping table.
- If not found, the packet is not for us and should be rejected.

Many higher-level TCP/IP protocols embed client addressing information in the packets. For example, during an "active" FTP transfer the client informs the server of its IP address and port number and then waits for the server to open a connection to that address. NAT has to monitor these packets and modify them on the fly to replace the client's IP address (which is on the internal network) with the NAT address. Since this changes the length of the packet, the TCP sequence/acknowledge numbers must be modified as well.

A **proxy** is any device that acts on behalf of another. The term is most often used to denote web proxying. A web proxy acts as a "halfway" web server; network clients make requests to the proxy, which then makes requests on their behalf to the appropriate web server. Proxy technology is often seen as an alternative way to provide shared access to a single Internet connection.

The other aspect of proxy servers is to improve performance. This capability is usually called proxy server caching. In simplest terms, the proxy server analyzes user requests and determines which, if any, should have the content stored temporarily for immediate access. A typical corporate example would be a company's home page located on a remote server. Many employees may visit this page several times a day. Since this page is requested repeatedly, the proxy server would cache it for immediate delivery to the web browser. Cache management is a big part of many proxy servers, and it is important to consider how easily the cache can be tuned and for whom it provides the most benefit.

Unlike NAT, web proxying is not a transparent operation. It must be explicitly supported by its clients. Due to early adoption of web proxying, most browsers, including Internet Explorer and Netscape Communicator, have built-in support for proxies, but to use proxy, you must normally be configured on each client machine, and can be changed by the naive or malicious user.

A proxy server operates above the TCP level and uses the machine's built-in protocol stack. For each web request from a client, a TCP connection has to be established between the client and the proxy machine, and another connection has to be established between the proxy machine and the remote web server. This puts a lot of strain on the proxy server machine; in fact, since web pages are becoming more and more complicated, the proxy itself may become a bottleneck on the network. This contrasts with a NAT, which operates on a packet level and requires much less processing for each connection.

6.10   ARP AND MAC ADDRESS RESOLUTION

Early IP implementations ran on hosts commonly interconnected by Ethernet local area networks (LANs). Every transmission on these LANs contained the MAC address of the source and destination nodes. Since there is no structure to identify different networks, routing could not be performed.

When a host needs to send a data packet to another host on the same network, the sender application must know both the IP and MAC addresses of the intended receiver; this is because the destination IP address is placed in the IP packet and the destination MAC address is placed in the LAN's protocol frame (such as Ethernet or Token Ring). If the destination host is on another network, the sender will look instead for the MAC address of the default gateway or router. (See figure 6.8.)

Unfortunately, the sender's IP process may not know the MAC address of the intended receiver on the same network. Therefore, ARP—the Address Resolution Protocol (RFC 826)—provides a mechanism so that a host can learn a receiver's MAC address when knowing only the IP address.

Anytime a computer needs to communicate with a local computer, it will first look in the ARP cache in memory to see if it already knows the MAC address of a computer with the specified IP address. If the address isn't in the ARP cache, the computer will try to discover the MAC address by broadcasting an ARP request packet. The station on the LAN recognizes its own IP address, which then sends an ARP response with its own MAC address. Then both the sender of the ARP reply and the original ARP requester record each other's IP address and media access control address as an entry in ARP cache for future reference. (See figure 6.9.)

If a computer needs to communicate with another computer that is located on another network, it will do the same, except it will send the packet to the local router. Therefore, it will search for the MAC address of the local port of the router, or it will send a broadcast looking for the address of the local port of the router.

Figure 6.8 Showing the IP Addresses (Logical Addresses) That Are Mapped to MAC Addresses (Physical Addresses)

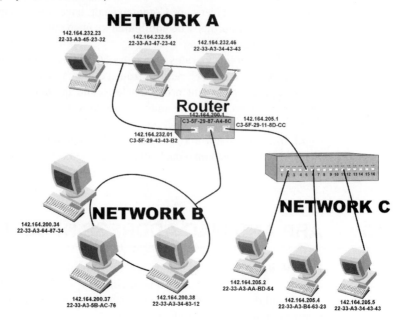

Figure 6.9 Name and Address Resolution Done on an IP Network

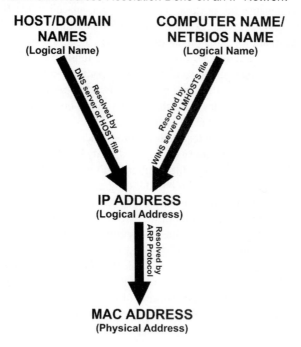

The **Reverse Address Resolution Protocol (RARP)** is a TCP/IP that permits a physical address, such as an Ethernet address, to be translated into an IP address. Hosts such as diskless workstations often only know their hardware interface address, or MAC address, when booted, but not their IP address. They must discover their IP address from an external source, usually a RARP server. RARP is defined in RFC 903.

6.11   NAVIGATING A TCP/IP NETWORK

Fully Qualified Domain Names (FQDN), sometimes referred to as just domain names, are used to identify computers on a TCP/IP network. Examples include MICROSOFT. COM and EDUCATION.NOVELL.COM.

While IP addresses are 32 bits (or 128 bits for IPv6) in length, most users do not memorize the numeric addresses of the hosts to which they attach. Instead, people are more comfortable with host names. Most IP hosts, then, have both a numeric IP address and a host name. And while the host name is convenient for people, the name must be translated back to a numeric address for routing purposes. This is done either by using a HOSTS file or by using a DNS server.

The HOSTS file is a text file that lists the IP address followed by the host name. Each entry should be kept on an individual line. In addition, the IP address should be placed in the first column followed by the corresponding host name. A # symbol is used as a comment or REM statement. This means that anything after the # symbol is ignored. (See figure 6.10.)

If a computer is using the host table shown in figure 6.10, then if rhino.acme.com is entered into a browser such as Internet Explorer or Netscape Navigator, the computer will find the equivalent address of 102.54.94.97 to connect to it.

NOTE: The HOSTS file is kept in the /ETC directory on most UNIX/Linux machines and in the WINDOWS directory in Windows 9X machines and *%systemroot%*\SYSTEM32\ DRIVERS\ETC directory in Windows NT, Windows 2000, Windows XP, and Windows 2003 Server.

Another way to translate the Fully Qualified Domain Name to the IP address is to use a **Domain Name System (DNS)** server. DNS is a distributed database (the database is contained in multiple servers) containing host name and IP address information for all do-

Figure 6.10   Sample HOST File

```
102.54.94.97     rhino.acme.com          # source server

38.25.63.10      x.acme.com              # x client host

127.0.0.1        localhost
```

Figure 6.11   Sample LMHOSTS File

```
102.54.94.97   rhino        #PRE   #DOM:networking #File Server

182.102.93.122 MISSERVER    #PRE                   #MIS Server

122.107.9.10   SalesServer                         #Sales Server

131.107.7.29   DBServer                            #Database Server

191.131.54.73  TrainServ                           #Training Server
```

mains on the Internet. For every domain, there is a single authoritative name server that contains all DNS-related information about the domain.

For example, you type in a web address of Microsoft.com in your browser. Your computer will then communicate with your local area network's DNS server. If the DNS server does not know the address of Microsoft.com, another DNS server will be asked. This will continue until the DNS server finds the address of Microsoft.com or it determines that the host name is not listed and sends back a reply with a "No DNS Entry" message.

When you share a directory, drive, or printer on PCs running Microsoft Windows or Linux machines running Samba, you would access these resources by using the Universal Naming Convention (UNC) (also known as Uniform Naming Convention) to specify the location of the resources. UNC uses the following format:

```
\\computer_name\shared-resource-pathname
```

So to access the shared directory called data on the server1 computer, you would type the following:

```
\\server1\data
```

The computer name can actually be the IP address of the PC or the NetBIOS name. Of course, if you use the NetBIOS name, something will be needed to translate the NetBIOS name to the IP address. It could broadcast onto the network asking for the IP address of the computer. Therefore, you would have to connect the TCP/IP address with the computer name (NetBIOS name). Microsoft networks can use a LMHOSTS file (see figure 6.11) or a WINS server.

A **Windows Internet Naming Service (WINS)** server contains a database of IP addresses and NetBIOS (computer names) that updates dynamically. For clients to access the WINS server, the clients must know the address of the WINS server. Therefore, the WINS server needs to have a static address—one that does not change. When the client accesses the WINS server, the client doesn't do a broadcast. Instead, the client sends a message directly to the WINS server. When the WINS server gets the request, it knows which computer sent the request and so it can reply directly to the originating IP address. The WINS database stores the information and makes it available to the other WINS clients.

When a WINS client starts up, it registers its name, IP address, and type of services within the WINS server's database. Since WINS was only made for Windows operating systems, other network devices and services (such as a network printer and UNIX machines) cannot register with a WINS server. Therefore, these addresses would have to be added manually.

6.12   TROUBLESHOOTING A TCP/IP NETWORK

Several utilities can be used to test and troubleshoot a TCP/IP network. Yet, no matter which utility you use, when you troubleshoot TCP/IP types of problems, you should use the following systematic approach:

1. Check the configuration.
2. Ping 127.0.0.1 (the loopback address).
3. Ping the IP address of the computer.
4. Ping the IP address of the default gateway (router).
5. Ping the IP address of the remote host.

The first thing you need to do when troubleshooting an apparent TCP/IP problem is to check your TCP/IP configuration, specifically, your IP address, subnet mask, default gateway, DNS server, and WINS server. If the subnet mask is wrong, you may not be able to communicate with machines on the same subnet or remote subnets. If the default gateway is wrong, you will not be able to connect to any computer on a remote subnet. If the DNS server is wrong, you will not be able to perform name resolution, and you will not be able to surf the Internet. Specifying the wrong address of a WINS server will limit your using computer BIOS names when using UNC names.

To verify the TCP/IP configuration in Microsoft Windows, you would use either the IP configuration program (WINIPCFG.EXE) command (available in Windows 9X) or the IPCONFIG.EXE (available in Windows 98, Windows NT, Windows 2000, Windows XP, and Windows Server 2003). To verify your TCP/IP in Linux, you can use ifconfig and route commands. See chapters 10 to 12 for more information on various network operating systems and their specific TCP/IP utilities.

6.12.1   The Ping Command

The ping command sends packets to a host computer and receives a report on their round-trip time. (See figure 6.12.) For example, you can ping an IP address by typing the following command at a command prompt:

```
ping 127.0.0.1
ping 137.23.34.112
```

The ping command can also be used to ping a host/computer by NetBIOS (computer) name or host/DNS name. Some examples would include:

```
ping FS1
ping WWW.MICROSOFT.COM
```

Figure 6.12   Ping Command

```
C:\>ping 132.233.150.4

Pinging 132.233.150.4 with 32 bytes of data:

Reply from 132.233.150.4: bytes= 32 time< 10ms TTL= 128
Reply from 132.233.150.4: bytes= 32 time< 10ms TTL= 128
Reply from 132.233.150.4: bytes= 32 time< 10ms TTL= 128
Reply from 132.233.150.4: bytes= 32 time< 10ms TTL= 128

Ping statistics for 132.233.150.4:
    Packets: Sent = 4, Received = 4, Lost = 0 (0% loss),
Approximate round-trip times in milliseconds:
    Minimum = 0ms, Maximum = 0ms, Average = 0ms
```

Suppose you can ping by address but not by name, it will tell you that the TCP/IP is running fine but the name resolution is not working properly. Therefore, you must check the LMHOSTS file and the WINS server to resolve computer names, and you must check the HOSTS file and the DNS server to resolve domain names.

If the time takes up to 200 milliseconds, the time is considered very good. If the time is between 200 and 500 milliseconds, the time is considered marginal. And if the time is over 500 milliseconds, the time is unacceptable. A "Request timed out" message indicates total failure, as shown in figure 6.13.

For the current configuration, pinging the loopback address (ping 127.0.0.1) and pinging the IP address of your computer will verify that the TCP/IP is properly functioning on your PC. By pinging the IP address of the default gateway or router, as well as other local IP computers, you determine if the computer is communicating on the local network. If it cannot connect to the gateway or any other local computer, either you are not connected properly, or the IP is misconfigured (the IP address, IP subnet mask, or gateway address). If you cannot connect to the gateway but you can connect to other local computers, check your IP address, IP subnet mask, and gateway address. Also check to see if the gateway is functioning by using the ping command at the gateway to connect to your computer and other local computers on your network, as well as pinging the other network connections on the gateway/router or pinging computers on other networks. If you cannot ping another local computer but you can ping the gateway, most likely the other computer is having problems and you need to restart this procedure at that computer. If you can ping

Figure 6.13   Ping Command Showing Total Failure

```
C:\>ping 132.233.150.2


Pinging 132.233.150.2 with 32 bytes of data:


Request timed out.
Request timed out.
Request timed out.
Request timed out.


Ping statistics for 132.233.150.2:
    Packets: Sent = 4, Received = 0, Lost = 4 (100% loss),
Approximate round-trip times in milliseconds:
    Minimum = 0ms, Maximum = 0ms, Average = 0ms
```

the gateway but you cannot ping a computer on another gateway, you need to check the routers and pathways between the two computers by using the ping or tracert commands. The tracert command is shown next.

NOTE: Some servers and/or firewalls will be configured to block ICMP packets. Therefore, when you try to ping these computers or use the tracert command on these computers, nothing is returned—although you can contact the server in other ways such as using your browser to access a mapped drive. One such example would be Microsoft.com.

6.12.2   The Tracert Command

Another useful command is the tracert command (Linux uses the traceroute command), which sends out a packet of information to each hop (gateway/router) individually. Therefore, the tracert command can help determine where the break is in a network. (See figure 6.14 and table 6.4.)

In addition to ping and tracert, newer versions of Microsoft offer a command called pathping. Pathping is a command that combines ping and tracert into one command.

6.12.3   The ARP Utility

The ARP utility is primarily useful for resolving duplicate IP addresses. For example, your workstation receives its IP address from a DHCP server, but it accidentally receives

Figure 6.14   Tracert Command

```
C:\>tracert www.novell.com


Tracing route to www.novell.com [137.65.2.11]

over a maximum of 30 hops:


  1    97 ms     92 ms    107 ms   tnt3-e1.scrm01.pbi.net [206.171.130.74]

  2    96 ms     98 ms    118 ms   core1-e3-3.scrm01.pbi.net [206.171.130.77]

  3    96 ms     95 ms    120 ms   edge1-fa0-0-0.scrm01.pbi.net [206.13.31.8]

  4    96 ms    102 ms     96 ms   sfra1sr1-5-0.ca.us.ibm.net [165.87.225.10]

  5   105 ms    108 ms    114 ms   f1-0-0.sjc-bb1.cerf.net [134.24.88.55]

  6   107 ms    112 ms    106 ms   atm8-0-155M.sjc-bb3.cerf.net [134.24.29.38]

  7   106 ms    110 ms    120 ms   pos1-1-155M.sfo-bb3.cerf.net [134.24.32.89]

  8   109 ms    108 ms    110 ms   pos3-0-0-155M.sfo-bb1.cerf.net [134.24.29.202]

  9   122 ms    105 ms    115 ms   atm8-0.sac-bb1.cerf.net [134.24.29.86]

 10   121 ms    120 ms    117 ms   atm3-0.slc-bb1.cerf.net [134.24.29.90]

 11   123 ms    131 ms    130 ms   novell-gw.slc-bb1.cerf.net [134.24.116.54]

 12     *          *          *    Request timed out.

 13   133 ms    139 ms    855 ms   www.novell.com [137.65.2.11]


Trace complete.
```

the same address as another workstation. When you try to ping it, you get no response. Your workstation is trying to determine the MAC address, and it cannot do so because two machines are reporting that they have the same IP address. To solve this problem, you can use the ARP utility to view your local ARP table and see which TCP/IP address is resolved to which MAC address. To display the entire current ARP table, use the ARP command with the -a switch. (See figure 6.15.)

NOTE: You can also use the IPCONFIG/ALL (Windows) or ifconfig (Linux) command if you need to identify the MAC address on the network interface on your current machine.

In addition to displaying the ARP table, you can use the ARP utility to manipulate it. To add static entries to the ARP table, use the ARP command with the -s switch. These entries stay in the ARP table until the machine is rebooted. A static entry hard-wires a specific IP address to a specific MAC address so that when a packet needs to be sent to that

Table 6.4   Tracert Options

-d	In the event a name resolution method is not available for remote hosts, you can specify the -d option to prohibit the utility from trying to resolve host names as it runs. If you don't use this option, tracert will still function, but it will run very slowly as it tries to resolve these names.
-h	By specifying the -h option, you can specify the maximum number of hops to trace a route to.
Timeout_value	The timeout value is used to adjust the timeout value. The value determines the amount of time in milliseconds the program will wait for a response before moving on. If you raise this value and the remote devices are responding, but they were not responding before, this may indicate a bandwidth problem.
-j	Known as lose source routing, tracert -j < router name> < local computer> allows tracert to follow the path to the router specified and return to your computer.

Figure 6.15   Using the ARP Command

```
C:\>arp -a


Interface: 192.168.1.100 --- 0x2
   Internet Address        Physical Address        Type
   192.168.1.254           00-00-89-2d-40-da       dynamic
   192.168.1.223           00-a0-b1-2d-32-45       dynamic
   199.223.164.5           00-a2-c0-c3-c2-14       static
```

IP address, it is sent automatically to the MAC address. The syntax for this command would be:

```
arp -s IP_Address MAC_Address
```

An example of using this command would be:

```
arp -s 199.223.164.5 00-a2-c0-c3-c2-14
```

If you want to delete entries from the ARP table, you can either wait until the dynamic entries time out or use the -d switch with the IP address of the static entry you would like to delete. An example would be:

```
arp -d 199.223.164.5
```

Figure 6.16   The Netstat Command Without Any Parameters

```
C:\Documents and Settings\Pat>netstat
Active Connections

  Proto  Local Address          Foreign Address        State
  TCP    pregan:3001            pregan:3497            ESTABLISHED
  TCP    pregan:3497            ftp.redhat.com:ftp     ESTABLISHED
  TCP    pregan:3499            ftp.redhat.com:ftp     ESTABLISHED
  TCP    pregan:4275            ftp.redhat.com:ftp-data ESTABLISHED
  TCP    pregan:4445            200-207-217-21.dsl.telesp.net.br:1214  TIME_WAIT

  TCP    pregan:4446            ads.web.aol.com:http   TIME_WAIT
  TCP    pregan:4447            ads.web.aol.com:http   TIME_WAIT
  TCP    pregan:4448            ads.web.aol.com:http   TIME_WAIT
  TCP    pregan:4449            ads.web.aol.com:http   TIME_WAIT
  TCP    pregan:4450            192.168.124.101:1214   SYN_SENT
  TCP    pregan:4451            192.168.150.102:1214   SYN_SENT
  TCP    pregan:4649            cs34.msg.sc5.yahoo.com:telnet  ESTABLISHED
  TCP    pregan:4669            64.12.29.24:5190       ESTABLISHED
  TCP    pregan:4715            msgr-ns42.msgr.hotmail.com:1863  ESTABLISHED
  TCP    pregan:4889            24.244.138.36:1214     ESTABLISHED
  TCP    pregan:4925            64.12.25.7:5190        ESTABLISHED
  TCP    pregan:4926            64.12.27.196:5190      ESTABLISHED

C:\Documents and Settings\Pat>netstat -n

Active Connections

  Proto  Local Address          Foreign Address        State
  TCP    127.0.0.1:3001         192.168.1.100:3497     ESTABLISHED
  TCP    192.168.1.100:3497     63.240.14.62:21        ESTABLISHED
  TCP    192.168.1.100:3499     63.240.14.62:21        ESTABLISHED
  TCP    192.168.1.100:4275     63.240.14.62:20        ESTABLISHED
  TCP    192.168.1.100:4445     200.207.217.21:1214    TIME_WAIT
  TCP    192.168.1.100:4446     205.188.165.57:80      TIME_WAIT
  TCP    192.168.1.100:4447     205.188.165.57:80      TIME_WAIT
  TCP    192.168.1.100:4448     205.188.165.89:80      TIME_WAIT
  TCP    192.168.1.100:4449     205.188.165.89:80      TIME_WAIT
  TCP    192.168.1.100:4453     192.168.150.102:1214   SYN_SENT
  TCP    192.168.1.100:4453     4.62.189.73:1214       TIME_WAIT
  TCP    192.168.1.100:4454     152.163.226.121:80     TIME_WAIT
  TCP    192.168.1.100:4455     152.163.226.121:80     TIME_WAIT
  TCP    192.168.1.100:4456     152.163.226.153:80     TIME_WAIT
  TCP    192.168.1.100:4457     152.163.226.153:80     TIME_WAIT
  TCP    192.168.1.100:4458     192.168.1.2:1214       SYN_SENT
  TCP    192.168.1.100:4459     4.62.189.73:1214       TIME_WAIT
  TCP    192.168.1.100:4649     216.136.227.168:23     ESTABLISHED
  TCP    192.168.1.100:4669     64.12.29.24:5190       ESTABLISHED
  TCP    192.168.1.100:4715     64.4.13.71:1863        ESTABLISHED
  TCP    192.168.1.100:4889     24.244.138.36:1214     ESTABLISHED
  TCP    192.168.1.100:4925     64.12.25.7:5190        ESTABLISHED
  TCP    192.168.1.100:4926     64.12.27.196:5190      ESTABLISHED
```

6.12.4   The Netstat Utility

The netstat command is a great way to see the TCP/IP connections, both inbound and outbound, on your machine. You can also use it to view the packet statistics, such as how many packets have been sent and received and the number of errors. Novell NetWare uses the MONITOR.NLM utility.

When netstat is used without any options, it produces output that shows all the outbound TCP/IP connections. (See figure 6.16.) The netstat utility, used without any options, is particularly useful in determining the status of outbound web connections.

NOTE: If you use -n, addresses and port numbers are converted to names.

The netstat -a command displays all connections, and netstat -r displays the route table plus active connections. The netstat -e command displays Ethernet statistics, and netstat -s displays per-protocol statistics. (See figure 6.17.)

On occasion, you may need to have netstat occur every few seconds. Try placing a number after the netstat -e command, like so:

```
netstat -e 15
```

The command executes, waits the number of seconds specified by the number (in this example, 15), and then repeats until you press the Ctrl+ C command.

Figure 6.17   The Netstat Command With the -s and -e Options

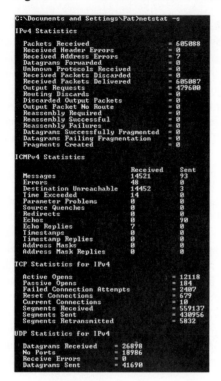

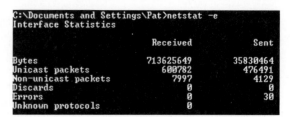

6.12.5   The nslookup Utility

The nslookup utility, which comes with most versions of Microsoft Windows (Windows NT, Windows 2000, Windows XP, and Windows Server 2003) and Linux, allows you to query a name server (DNS server) and quickly find out which name resolves to which IP address. nslookup has two modes, interactive and noninteractive. Use the interactive mode when you require more than one piece of data. To run the interactive mode, at the command prompt, type nslookup. If you type a question mark (**?**), nslookup shows all available commands. To exit the interactive mode, type exit.

Use the noninteractive mode when you require a single piece of data. Type the nslookup command at the command prompt with the proper parameters, and the data is returned. It does this by resolving a host name into an IP address or an IP address into a domain name. To use nslookup, simply provide the address you want to resolve as a command-line argument. For example, if you want to find out the address for mydesk. domain.cxm, you would use:

```
nslookup mydesk.acme.com
```

6.12.6   The nbtstat Utility

You can use the nbtstat command to track NetBIOS over TCP/IP statistics, show the details of incoming and outgoing packets over TCP/IP connections, and resolve NetBIOS names. NetBIOS name resolution is primarily on Microsoft Windows systems. To display a basic description of nbtstat and its associated options, type nbtstat at the command prompt. The -a switch displays a remote machine's NetBIOS name table, which is a list of all the NetBIOS names that a particular machine knows about. If you use the -A switch instead of the -a option, you would then specify an IP address instead of a computer name. For example, you could look at the NetBIOS name table of the server1 computer by typing the following command:

```
nbtstat -a server1
```

The nbtstat -c command displays the local NetBIOS name cache on the workstation on which it is run. The -n switch is used to display the local NetBIOS name table on a Windows station. The -r switch displays the statistics that show how many NetBIOS names have been resolved to TCP/IP addresses. The -R allows you to purge and reload the remote cache name table. This can come in handy when you get a bad name in the NetBIOS name cache. The -S option displays both workstation and server sessions, but it lists remote addresses by IP address only. The -s produces the same output as nbtstat -S, except that it tries to resolve remote host IP addresses into host names.

As you can do with the netstat command, you can place a number at the end of the nbtstat command to indicate that the command should execute once every so many seconds as specified by the number. The screen will continue displaying the statistics until you press Ctrl+ C.

SUMMARY

1. Today, TCP/IP is the primary protocol used on the Internet and is supported by Microsoft Windows, Novell NetWare, UNIX, and Linux.

2. IP is a connectionless protocol, which means that there is no established connection between the end points that are communicating.

3. The Transmission Control Protocol (TCP) is a reliable, connection-oriented delivery service that breaks the data into manageable packets, wraps them with the information needed to route them to their destination, and then reassembles the pieces at the receiving end of the communication link.

4. Unlike TCP, which uses acknowledgments to ensure data delivery, the User Datagram Protocol (UDP) does not.

5. Each connection on a TCP/IP address (logical address) is called a host (a computer or other network device that is connected to a TCP/IP network) and is assigned a unique IP address.

6. The format of the IP address is four 8-bit numbers (octet) divided by a period (.). Each number can be zero to 255.

7. Since TCP/IP addresses are growing scarce for the Internet, a series of addresses have been reserved to be used by private networks (networks not connected to the Internet).

8. The subnet mask is used to define which bits describe the network number and (including the subnet number) which bits describe the host address.

9. The subnet mask can be changed to take a large network and break it into small networks called subnets.

10. Classless Inter-domain Routing (or CIDR, which is usually pronounced "cider") networks are described as slash x networks, where x represents the number of network bits in the IP address range. When an organization is assigned a single network ID, CIDR can be used to subdivide the network into smaller segments, depending upon the requirements of each subnet.

11. Since TCP/IP and the Internet became popular, the Internet has grown and continues to grow at an exponential rate. It will eventually run out of network numbers. As a result, the IPv6 was created. IPv6 uses 128 bits for its addresses.

12. A TCP/IP port number is a logical connection placed for client programs to specify a particular server program running on a computer on the network defined at the transport layer.

13. The Routing Information Protocol (RIP) is designed for exchanging routing information within a small to medium-size network. The biggest advantage of RIP, which is a distance-vector protocol, is that it is extremely simple to configure and deploy.

14. OSPF, short for Open Shortest Path First, is a link-state routing protocol used in medium-size and large networks that calculates routing table entries by constructing a shortest-path tree.

15. Network Address Translation (NAT) is a method of connecting multiple computers to the Internet (or any other IP network) using one IP address.

16. A web proxy acts as a "halfway" web server; network clients make requests to the proxy, which then makes requests on their behalf to the appropriate web server.

17. Proxy servers also use a cache to store recently accessed web pages. Therefore, when the same page is accessed, the proxy server can provide the page without going out to the Internet.

18. Fully Qualified Domain Names (FQDN), sometimes referred to as just domain names, are used to identify computers on a TCP/IP network.

19. DNS servers and host files are used to resolve host and domain names to IP addresses.

20. A Windows Internet Naming Service (WINS) server and LMHOSTS are used to resolve Net-BIOS computer names to IP addresses.

21. To troubleshoot TCP/IP problems, you can use several utilities, including ipconfig, ping, and tracert.

QUESTIONS

1. The protocol used by the Internet is _____.
   a. TCP/IP   b. NetBEUI
   c. IPX   d. AppleTalk

2. To connect to a TCP/IP network, you must configure which of the following? (Choose two answers.)
   a. The TCP/IP address
   b. The IPX address
   c. The DNS server address
   d. The gateway
   e. The subnet mask

3. To connect to a TCP/IP network that contains several subnets, which of the following must be configured? (Select all that apply.)
   a. The TCP/IP address
   b. The IPX address
   c. The DNS server address
   d. The gateway
   e. The subnet mask

4. What is the default subnet mask for a class B network?
   a. 255.0.0.0
   b. 255.255.255.0
   c. 127.0.0.1
   d. 255.255.0.0
   e. 255.255.255.255

5. What does the 127.0.0.1 address represent?
   a. A broadcast address
   b. A loopback address
   c. A network address
   d. A subnet address

6. _____ is the process of taking a large network and dividing it into smaller networks.
   a. Subnetting
   b. Gatewaying
   c. Broadcasting
   d. Hosting

7. Which of the following must you consider when deciding which subnet mask you should apply to a TCP/IP network? (Choose all that apply.)
   a. The IP address class
   b. The types of computers used on the network
   c. The number of subnets
   d. The potential for network growth

8. Your company is assigned the network address 150.50.0.0. You need to create seven subnets on the network. A router on one of the subnets will connect the network to the Internet. All computers on the network will need access to the Internet. What is the correct subnet mask for the network?
   a. 0.0.0.0
   b. 255.255.0.0
   c. 255.255.240.0
   d. The subnet mask assigned by InterNIC

9. A company with the network ID 209.168.19.0 occupies four floors of a building. You create a subnet for each floor. You want to allow for the largest possible number of host IDs on each subnet. Which subnet mask should you choose?
   a. 255.255.255.192
   b. 255.255.255.240
   c. 255.255.255.224
   d. 255.255.255.248

10. The Acme Corporation has been assigned the network ID 134.114.0.0. The corporation's eight departments require one subnet each. However, each department may grow to over 2500 hosts. Which subnet mask should you apply?
    a. 255.255.192.0
    b. 255.255.240.0
    c. 255.255.224.0
    d. 255.255.248.0

11. All IP addresses are eventually resolved to network interface card addresses. Which of the following is used to map an IP address to a network interface card address?
    a. WINS   b. DNS
    c. DHCP   d. ARP

12. The enhanced version of IP (IPv4) is:
    a. IPv6   b. IPv5
    c. IPX   d. IPING

13. There are several UNIX computers and mainframes on your Windows NT network. You want to standardize the network protocol used on all computers and provide access to the Internet. Which protocol would you choose for your network?
    a. TCP/IP
    b. NetBEUI
    c. NWLink
    d. IPX/SPX

14. What is the default assigned ports for a web server browser?
    a. 21   b. 119
    c. 80   d. 139

15. What two types of names can be resolved in a Windows TCP/IP platform? (Select two answers.)
    a. Host
    b. Network name
    c. IPX computer
    d. NetBIOS name

16. What is used to resolve host names such as MICROSOFT.COM? (Select two answers.)
    a. DNS server
    b. LMHOSTS files
    c. HOSTS files
    d. WINS server

17. You are troubleshooting a workstation and find the following information:

    IP address: 131.123.140.14
    Subnet mask: 255.255.255.0
    Default gateway: 131.123.140.200
    DNS servers: 130.13.18.3, 140.1.14.240
    Router's IP address: 131.123.140.1

    The workstation cannot get to the Internet. Why is this? (Choose two answers.)
    a. The subnet mask is invalid for this network.
    b. There can only be one IP address for DNS servers.
    c. The IP address of the DNS should match the default gateway address.
    d. The IP address of the workstation is wrong.
    e. The default gateway should match the router's address.

18. You have static IP addresses assigned to your workstation. If two workstations are assigned the same IP address, what would happen as far as communication with these two workstations?
    a. The second workstation will take over communication when it boots.
    b. The first workstation to boot and log in will communicate.

c. Both workstations will be okay.

d. Neither workstation will be able to communicate on the network.

19. Which of the following uses 15-character names to identify computers on a network?

a. TCP/IP
b. NetBIOS
c. IPX/SPX
d. AppleTalk

20. What file is used to resolve a host name to an IP address?

a. LMHOSTS
b. HOSTS.SAM
c. HOSTS
d. LMHOSTS.SAM

21. What is the purpose of a WINS server in a network?

a. To keep a database of hosts names and corresponding IP addresses

b. To keep communication flowing on a TCP/IP network by assigning IP addresses dynamically to each workstation as it logs in

c. To keep a database of NetBIOS names and corresponding IP addresses

d. To authenticate each workstation as it logs in and determine the MAC address and corresponding IP address of each workstation

22. What protocol provides reliable, connection-based delivery?

a. TCP
b. IP
c. UDP
d. ARP

23. Several users are complaining that they cannot access one of your Windows NT file servers, which has an IP address that is accessible from the Internet. When you get paged, you are not in the server room. Instead you are at another company in a friend's office that only has a UNIX workstation available, which also has Internet access. What can you do to see if the file server is still functioning on the network?

a. Use PING from the UNIX workstation.
b. Use ARP from the UNIX workstation.
c. Use WINS from the UNIX workstation.
d. Use DNS from the UNIX workstation.

24. What tool can you use to see the path taken from a Windows NT system to another network host?

a. PING
b. WINIPCFG
c. TRACERT
d. IPCONFIG
e. SNMP

25. Which bits of a subnet mask correspond to the network ID?

a. The 1s
b. The 0s
c. Both
d. Neither

26. You are troubleshooting the network connectivity for one of the client systems at your company, but you do not know the IP address for the system. You are at the client system, and you want to make sure the system can communicate with itself using TCP/IP. How can you accomplish this?

a. Ping 127.0.0.1
b. Ping 255.255.0.0
c. Ping 255.255.255.255
d. Ping 0.0.0.0

27. How are the network ID and the host ID for an IP address determined?

a. Subnet mask
b. Range mask
c. Unicast mask
d. Multicast mask

28. What TCP/IP addresses are available for multicast transmissions?

a. 128.0.0.0 to 191.255.0.0
b. 224.0.0.0 to 239.255.255.255
c. 192.0.0.0 to 223.255.255.0
d. 240.0.0.0 to 247.255.255.255

29. At one of your company's remote locations, you have decided to segment your class B address down, since the location has three buildings and each building contains no more than 175 unique hosts. You want to make each building its own subnet, and you want to utilize your address space the best way possible. Which subnet mask meets your needs in this situation?

a. 255.0.0.0
b. 255.255.255.0
c. 255.255.0.0
d. 255.255.255.240

30. You have successfully obtained a class C subnet for your company. What is the default subnet mask?

a. 255.255.255.0
b. 255.0.0.0
c. 255.255.0.0
d. 255.255.240.0

31. A Windows 2000 Server computer named SERVER1 resides on a remote subnet. Pat cannot ping SERVER1 using its IP address. But he can successfully ping his default gateway address and the addresses of other computers on the remote subnet. What is the most likely cause of the problem?

a. Pat's computer is set up with an incorrect default gateway address.

b. Pat's computer is set up with an incorrect subnet mask.

c. SERVER1 is not WINS-enabled.

d. The IP configuration on SERVER1 is incorrect.

e. The LMHOSTS file on Pat's computer has no entry for SERVER1.

32. Pat is planning to set up an intranet web server at his company. Employees will be able to access the web server using the server's host name. Which of the following services should Pat install on his intranet to provide name resolution?

a. DHCP

b. DNS

c. FTP

d. WINS

33. Pat fails repeatedly to access a Windows 2000 Server computer on a remote subnet. Using Network Monitor to troubleshoot the problem, he finds that every time he tries to connect to the server, his workstation broadcasts an ARP request for the IP address of the remote address of the remote Windows 2000 Server computer. No other users on the TCP/IP network have trouble accessing the server. What is the most likely cause of Pat's problem?

a. The workstation is not set up to use DNS.

b. The workstation is not set up to use WINS.

c. The workstation is set up with a duplicate IP address.

d. The workstation is set up with an incorrect subnet mask.

34. Pat connects to a remote UNIX computer using its IP address. He knows that TCP/IP is properly installed on this computer. How should he check to see if the router is working correctly?

a. Ping 127.0.0.1.

b. Ping the local server.

c. Ping the far side of the router.

d. Ping the near side of the router.

35. A user with a Windows 2000 Professional computer complains that she cannot connect to any other computers on her network. The network uses both a DHCP server and a DNS server. Seated at her workstation, you ping 127.0.0.1 and fail to get a response. What is the most probable cause of the problem?

a. TCP/IP is not properly installed on the workstation.

b. The default gateway address on the workstation is incorrect.

c. The subnet mask on the workstation is incorrect.

d. The workstation is not set up for DHCP.

e. The workstation is not set up for DNS.

36. Which switch would you use if you want to prohibit the tracert utility from resolving names as it runs?

a. -r

b. -h

c. -n

d. -d

37. What value determines the amount of time in milliseconds the program will wait for a response before moving on?

a. Hops

b. Wait

c. Time-out

d. Time to live

38. Which of the following entries in a HOSTS file residing on a Windows 2000 Server computer will connect to the UNIX server SERVER1?

a. 132.132.11.53 #SERVER1 #corporate server

b. 132.132.11.53 SRV1 #corporate server

c. 132.132.11.53 SERVER1 #corporate server

d. 132.132.11 #SERVER1 #corporate server

39. IP version 6 (IPv6) uses how many bits in its addressing scheme?

a. 16

b. 32

c. 64

d. 128

40. An HTTP proxy cache server will increase the apparent performance of what activity?

a. Bank interaction

b. Web browsing

c. Networking login and authentication

d. Connection refresh

41. When you are using CIDR, how do you specify 26 network bits?

a. 255.255.255.192

b. /26

c. /25

d. -N 25

42. The class B address range for the first octet is _____.

a. 1–127

b. 1–128

c. 128–191

d. 129–223

e. 224–255

43. What does a subnet mask separate?

a. Network ID and host ID

b. Host IDs

c. Workgroups and domains

d. All of the above

44. HTTP usually connects to a web server on port number _____.

a. 21

b. 25

c. 80

d. 121

45. What is the default subnet mask for a class C address?

a. 255.0.0.0

b. 255.255.255.0

c. 255.255.255.240

d. 255.255.255.255

46. Network Address Translation, or NAT, is found in
_____.

   a. Windows 95

   b. NIC protocol driver

   c. Windows NT

   d. Routers

47. Which protocol is considered connection-oriented?

   a. DLC

   b. NetBEUI

   c. TCP

   d. UDP

48. FQDN is an acronym for _____.

   a. Fully Qualified Division Name

   b. Fully Qualified DNS Name

   c. Fully Qualified Dynamic Name

   d. Fully Qualified Domain Name

49. What delimiter separates domain spaces?

   a. : (colon)

   b. # (pound)

   c. . (period)

   d. / (forward slash)

   e. \ (backward slash)

50. Which of the following is not a feature of a proxy
server?

   a. It can reduce Internet traffic requests.

   b. It can assist with security.

   c. It can reduce user wait time for a request.

   d. It can convert a nonroutable protocol to a
routable protocol.

51. Which Windows 9X or Windows NT utility can
display NetBIOS over TCP/IP statistics?

   a. nbtstat   b. netstat

   c. Tracert   d. ARP

   e. IPCONFIG   f. WINS

52. Which ping commands will verify that your local
TCP/IP interface is working? (Choose all that
apply.)

   a. Ping 0.0.0.0.

   b. Ping localhost.

   c. Ping self.

   d. Ping 127.0.0.1.

   e. Ping iphost.

   f. Ping 255.255.255.255.

53. Which nbtstat utility switch will purge and reload
the remote NetBIOS name table cache?

   a. -r   b. /r

   c. -p   d. -R

   e. /R   f. -P

54. You are a network administrator. A user calls you
complaining that the performance of the intranet
web server is sluggish. When you try to ping the
server, it takes several seconds for the server to respond. You suspect that the problem is related to a
router. Which workstation utility could you use to
find out which router is causing the problem?

   a. netstat   b. nbtstat

   c. Ping   d. Tracert

   e. IPCONFIG   f. ARP

55. Which utility will display a list of all routers
that a packet passes through on the way to an IP
destination?

   a. netstat   b. nbtstat

   c. Ping   d. Tracert

   e. IPCONFIG   f. ARP

56. Which ARP command can you use to display the
current ARP entries?

   a. ARP   b. ARP -A

   c. ARP -a   d. ARP /A

   e. ARP /a   f. ARP -C

57. Which netstat switch will enable you to view the
ICMP packets your workstation has sent and
received?

   a. -a   b. -r

   c. -s   d. -I

58. You administer a local area network for a large
company. You have connected the LAN to the Internet. How would you go about broadcasting only
one IP address to the Internet and allowing all the
computers on the LAN to connect to the Internet?

   a. By configuring the computers on the LAN to
use the same IP address

   b. By configuring the computers on the LAN to
use a default gateway

   c. By configuring the computers on the LAN to
use an IP proxy

   d. By installing a firewall between the LAN and
the Internet

59. You administer a local area network. You have
added several users to the LAN, and you notice
that Internet access has become considerably
slower. What can you use to improve the speed of
Internet access from the LAN?

   a. A firewall

   b. An IP proxy

   c. Hypertext Transfer Protocol (HTTP)

   d. File Transfer Protocol (FTP)

60. How does a UNIX computer use port numbers?
  a. To determine which users have access to resources on the network
  b. To locate hosts on the network
  c. To establish communications between network interface cards (NICs)
  d. To determine which TCP/IP service receives data packets

61. What character or character combination would you place at the beginning of a line in a HOSTS file to comment out the line?
  a. ;   b. /*
  c. #   d. //

62. Which of the following are characteristics of a HOSTS file?
  a. It is static.
  b. It resolves TCP/IP host names to IP addresses.
  c. It resolves NetBIOS names to IP addresses.
  d. It resolves IP addresses to MAC addresses.

63. You administer a network with four subnets connected with three routers. Subnet A is connected to subnet B, subnet B is connected to subnet C, and subnet C is connected to subnet D. You use the tracert utility on COMPUTER_A on subnet A to trace the route of a packet as it travels to COMPUTER_D on subnet D. If the packet travels from COMPUTER_A to COMPUTER_D successfully, how many hops will be reported by the tracert utility?
  a. 1   b. 3
  c. 2   d. 4

64. You are using a Windows NT Workstation 4.0 computer on a Novell NetWare 5.0 network. You attempt to contact a computer with the TCP/IP host name Rover, but you cannot contact this computer by using its TCP/IP host name. You know that Rover's IP address is 192.168.0.1, and you use it to successfully ping the Rover computer. What is the most likely cause of this problem?
  a. The Domain Name System (DNS) is incorrectly configured.
  b. The LMHOSTS file is incorrectly configured.
  c. The Windows Internet Naming Service (WINS) is incorrectly configured.
  d. The IP address of the remote computer is incorrectly configured.

65. Pat oversees a network with five subnets: subnet 1, subnet 2, subnet 3, subnet 4, subnet 5. A Windows 2000 computer user on subnet 4 complains that although he can connect to local servers, he cannot connect to a Windows 2000 Server computer on subnet 1. Other users on subnet 4 have no trouble connecting to the subnet 1 server. Pat runs IPCONFIG on the workstation and notes the following output:

```
Ethernet adapter Local Area
Connection:
Connection-specific DNS suffix:
IP address. . . . . . . . . .
: 132.132.223.19
Subnet mask . . . . . . . . .
: 255.255.224.0
Default gateway . . . . . . .
: 132.132.224.1
```

Why can't the workstation connect to the remote server?
  a. The default gateway is incorrect.
  b. The IP address must fall between 132.132.224.1 and 132.132.255.254.
  c. The subnet mask must be 255.255.192.0.
  d. The subnet mask must be 255.255.255.224.

66. Packets sent via UDP are considered to be what type of communication?
  a. Connection-oriented
  b. Connectionless-oriented
  c. Session-oriented
  d. Application-oriented

67. Pat is the network admin in charge of configuring TCP/IP for his company. The company's ISP has assigned a block of IP addresses from 215.45.96.0 to 215.95.36.0. He doesn't want to implement a single address space, because the company may open up offices in other states, which would require setting up different networks (each network would need its own unique network ID). Which of the following would allow for multiple IP networks using the address range above?
  a. Segmenting   b. Superplexing
  c. Cross-linking   d. Subnetting

68. Which of the following is an example of an IPv6 subnet mask?
  a. 255.255.255.0
  b. A119BBC01E48

c. 0AC9:0101:09A8:BBCD

d. 0AC9:0101:09A8:BBCD:921F:AE3C:
   1968:C00C

69. In configuring a TCP/IP client, which of the following is not an entry option?

a. Default gateway

b. Subnet mask

c. MAC address

d. IP address

70. How many address classes are there for TCP/IP?

  a. 5   b. 4

  c. 6   d. 3

71. You need to give your workstations the ability to dynamically request an IP address along with DNS and WINS server addresses as well as the subnet mask and default gateway. Which network service should you install?

  a. DNS   b. WINS

  c. HGP   d. DHCP

72. Which TCP port does SMTP use?

  a. 21   b. 25

  c. 23   d. 53

73. Which TCP port does telnet use?

  a. 25   b. 23

  c. 119   d. 21

74. Which type of proxy hides internal IP addresses from the public Internet?

  a. HTTP   b. DNS

  c. FTP   d. NAT

75. TCP packets are sent in what type of window?

a. Sliding

b. Expanding

c. Slipping

d. Diminishing

76. Pat, one of your users, can't log onto the network. You go to his computer, and after checking the cable connection, you obtain his IP address using the IPCONFIG utility. Next you use the nbtstat utility to see what MAC address is mapped to the IP address. You then look at the label on his network card and see that a different MAC address is listed. What is the most likely cause of the problem?

a. IP port conflict

b. MAC address conflict

c. ARP conflict

d. IP address conflict

77. You ping a server by its domain name in order to get its IP address, but nothing is returned. You know the server is up because you have a network drive mapped to it. What could be the cause of the problem?

a. The server's IP address is duplicated on the network.

b. You didn't use the -p switch with the ping.

c. The ping is turned off on the server.

d. ICMP packets are being blocked at the router.

78. You are a network engineer implementing a network consisting of approximately 150 Windows 2000 Workstations and 5 Windows 2000 Servers. You have multiple sites located in the same city that are connected through a PSTN. Your manager wants you to ensure fast communication by implementing the proper protocol. Each site has a router that is already configured by the telephone company.

Required result:

■ Use a routable protocol to ensure connectivity between sites.

Optional results:

■ Ensure communication between the Windows 2000 Servers.

■ Log in to all the servers without maintaining multiple logins.

Proposed solution:

■ Implement TCP/IP as your protocol and use the same user names on all servers, but use different passwords for security purposes.

How would you characterize the proposed solution?

a. It achieves the required result and both optional results.

b. It does not achieve the required result but does achieve both the optional results.

c. It does not achieve the required result but does achieve one of the optional results.

d. It achieves the required result but cannot achieve either of the optional results.

e. It achieves neither the required result nor either of the optional results.

79. You are a network engineer implementing a network consisting of approximately 150 Windows NT Workstations and 5 Windows NT Servers. You have multiple sites located in the same city that are connected through a PSTN. Your manager wants you to ensure fast communication by implementing the proper protocol. Each site has a router that is already configured by the telephone company. The workstation is running Windows 98/SE.

Required result:
■ Use a routable protocol to ensure connectivity between sites.

Optional results:
■ Ensure communication between the Windows NT Servers.
■ Log in to all the servers without maintaining multiple logins.

Proposed solution:
■ Implement TCP/IP as your protocol and use the same user names and passwords on all servers.

How would you characterize the proposed solution?
a. It achieves the required result and both optional results.
b. It does not achieve the required result but does achieve both the optional results.
c. It does not achieve the required result but does achieve one of the optional results.
d. It achieves the required result but cannot achieve either of the optional results.
e. It achieves neither the required result nor either of the optional results.

80. Look at the figure. The first number of each host is the IP address, the second number is the subnet mask, and the last number is the default gateway. What is the problem with the TCP/IP network?

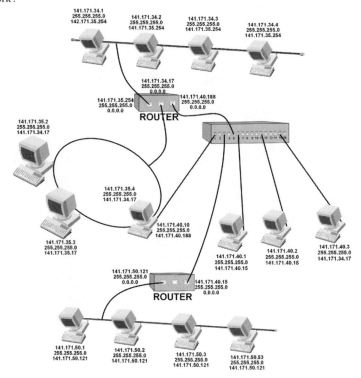

81. Look at the figure. The first number of each host is the IP address, the second number is the subnet mask, and the last number is the default gateway. There are two problems with the TCP/IP network. What are they?

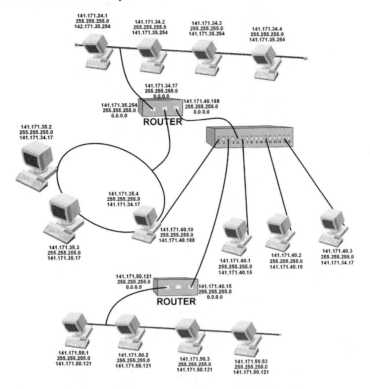

HANDS-ON EXERCISE

Exercise 1: Using the TCP/IP Network

1. Start a DOS prompt by clicking on the Start button, selecting the Programs option, selecting the Accessories option, and selecting the Command prompt option.
2. Execute the IPCONFIG command and record the following settings:

   IP address
   Subnet mask
   Default gateway

3. Use the IPCONFIG /ALL command and record the following:

   MAC address
   WINS server (if any)
   DNS server (if any)
   If it is DHCP-enabled or not

4. Ping the loopback address of 127.0.0.1.
5. Ping your IP address.
6. Ping your instructor's computer.
7. Ping your partner's computer.
8. Use the tracert command to your instructor's computer.
9. If you are connected to the Internet, use the tracert command to Novell.com.
10. If you have a router on your network, ping your gateway or local router connection.
11. Your network has a DHCP network. Right-click My Network Places and select the Properties option.
12. Click on the Internet Protocol (TCP/IP) and then click on the Properties button.
13. In the Internet Protocol (TCP/IP) Properties dialog box, select Obtain an IP address automatically. Click on the OK button.
14. Execute the IPCONFIG command at the command prompt and record the following:

    IP address
    Subnet mask
    Default gateway
    WINS server (if any)
    DNS server (if any)
    If it is DHCP-enabled or not

15. Ping the loopback address of 127.0.0.1.
16. Ping your IP address.
17. Ping your instructor's computer.
18. Ping your partner's computer.
19. If you have a router on your network, ping your gateway or local router connection.
20. At the command prompt, execute the IPCONFIG /RELEASE to remove your values specified by a DHCP server.
21. At the command prompt, execute the IPCONFIG command and compare the recorded values from step 2.
22. At the command prompt, execute the IPCONFIG /RENEW command.
23. At the command prompt, execute the IPCONFIG command and compare the recorded values from step 2.
24. Have your instructor stop the DHCP server (or DHCP service).
25. At the command prompt, execute the IPCONFIG /RELEASE command followed by the IPCONFIG/RENEW.

    Note: There will be a pause while Windows 2000 attempts to locate a DHCP server.

26. At the command prompt, execute the IPCONFIG prompt. Record the address and try to determine where this address came from.
27. Try to ping your instructor's computer and the local gateway. You should not be able to ping these host addresses.
28. After your partner has acquired an Automatic Private IP address, try to ping each other. Since these addresses are on the same network (physically and logically), it should work.
29. Go back into the TCP/IP dialog box and enter the static addresses that you recorded in steps 2 and 3.
30. Test your network by pinging your partner's computer, instructor's computer, and gateway.
31. Disconnect the network cable from the back of the computer.
32. Look at the taskbar in the notification area (near the clock) and notice the red X.
33. From the command prompt, type in IPCONFIG.

34. Go into the Network and Dial-up Connection dialog box by right-clicking My Network Places and selecting Properties. Notice the red X.
35. Connect the cable back into the network card. Notice that the red X in both places goes away. In addition, notice that the icon disappears altogether from the notification area.
36. Right-click Local Area Connection and select Properties.
37. Select the Show icon in the taskbar when connected. Click on the OK button. Close the Local Area Connection dialog box.
38. Go to the notification area and notice the new icon representing the network connection.
39. Without clicking on the new icon, move the mouse pointer onto the icon. Without moving the mouse, notice the information given.
40. Double-click the icon to bring up the Local Area Connection dialog box.
41. From you the command prompt, execute the `ARP -A`.
42. Close all windows.

CHAPTER 7

IPX, NetBIOS, and AppleTalk Protocols

Topics Covered in this Chapter

Introduction

In the last chapter, TCP/IP was discussed. In this chapter, the other common protocol suites will be discussed, including IPX, NetBIOS, and AppleTalk. IPX is the protocol associated with Novell NetWare, and AppleTalk is associated with Apple Macintosh computers. NetBIOS is a protocol that was developed for Windows.

Objectives

- Explain the IPX/SPX, NetBEUI, and AppleTalk protocols in terms of routing, addressing schemes, interoperability, and naming conventions.
- Explain the difference between IPX and SPX.
- Diagnose and troubleshoot a given IPX problem.

- Explain how NetBEUI and NetBIOS relate to each other and how they can relate to TCP/IP and IPX protocols.
- Describe what the DLC Protocol is typically used for.

7.1   IPX PROTOCOL SUITE

When Novell NetWare was introduced, it was designed to be a server platform for local area and wide area networks. To that end, the designers devised a protocol stack that were very efficient over local area networks and that would also work on wide area networks. The protocol stack was known as the NetWare Protocol Suite, the Internetwork Packet eXchange/Sequenced Packet eXchange, and the IPX/SPX. IPX is the fastest routable network protocol suite available. (See figure 7.1.)

Similar to connecting to the TCP/IP network, you connect to an IPX network by using one of several interfaces, such as Ethernet, Token Ring, or ARCnet. For a Novell NetWare network, the software used to connect to an IPX network must first load a Multiple Link Interface Driver (MLID); an MLID is basically the driver to control the network card. Different from previous software interfaces used by early versions of IPX and other protocols, the MLID complies with the Novell Open Data-Link Interface (ODI) architecture that allows a network driver to communicate with multiple protocol stacks such as TCP/IP and IPX.

Since the MLID supports multiple protocols, software is needed to identify the packets that come through the network card and route the packet to the proper protocol stack. This software is known as the link support layer (LSL).

NWLink is Microsoft's NDIS-compliant, 32-bit implementation of the Internetwork Packet Exchange (IPX), Sequenced Packet Exchange (SPX), and NetBIOS protocols used in Novell networks. NWLink is a standard network protocol that supports routing and can

Figure 7.1   IPX Protocol

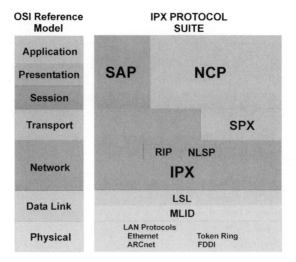

support NetWare client/server applications, where NetWare-aware sockets-based applications communicate with IPX/SPX sockets-based applications.

NOTE: You must use this protocol on Microsoft Windows if you want to use Gateway Service for NetWare or Client Service for NetWare to connect to NetWare servers.

7.1.1   IPX and SPX

The **Internetwork Packet Exchange (IPX)** is a networking protocol used to interconnect networks. It is a connectionless protocol and therefore doesn't require a connection to set up before packets are sent to a destination.

Another protocol that works with the IPX Protocol is the **Sequenced Packet Exchange (SPX).** Different from the IPX Protocol, the SPX Protocol uses packet acknowledgments to guarantee delivery of packets. SPX provides segmentation, reassembly, and segment sequencing of data streams that are too large to fit into the frame size. SPX uses virtual circuits, referred to as connections. These connections are assigned specific connection identifiers as defined in the SPX header. Multiple-connection IDs can be attached to a single socket.

7.1.2   SAP and NCP Protocols

The **Service Advertising Packet (SAP)** is used to advertise the services of all known servers on the network, including file servers, print servers, and so on. Servers periodically broadcast their service information while listening for SAPs on the network and storing the service information. Clients then access the service information table when the clients need to access a network service.

When a user tries to log into a Novell NetWare server using the IPX Protocol, the user's computer will broadcast a Get Nearest Server (GNS) request from a client. Only a local server or router can respond to a GNS request. Since most networks would have a NetWare server on each network segment, the server would respond to the GNS request. If you do not have a Novell NetWare server on the segment and since broadcasts are typically not forwarded by routers, then the router would respond to GNS requests from that segment. Although routers do not forward SAP broadcasts, routers build their own SAP tables and broadcast them at scheduled intervals. By default, the interval is 60 seconds. Each router in the network eventually learns about all the other routers and NetWare servers in the network.

The NetWare Core Protocol consists of the majority of network services offered by a Novell NetWare server. This would include file services, print services, Novell Directory Services (NDS), and message services just to name a few.

7.1.3   IPX Addressing

When assigning addresses within the IPX network, there are two types of addresses: the internal and external IPX network numbers. The internal IPX network number, an 8-digit (4-byte) hexadecimal number, is used to identify a server. Unlike TCP/IP, the clients are not assigned internal network numbers. Instead, they use the external IPX address and the MAC address.

NOTE: In addition to a station address, routers are given an internal IPX address. This address uniquely identifies a router to the rest of a network.

The external IPX address is used to identify the network. Therefore, all servers and routers on the same physical network must be assigned the same external IPX number. When assigning both the internal and external network numbers, the numbers must be unique for each other. These include numbers assigned to the network or to the individual servers. (See figure 7.2.)

Much like TCP/IP, IPX also uses socket numbers to identify which network software the packet is sent to. Some examples of these are shown in table 7.1.

7.1.4   Frame Types

Throughout the years, there have been several versions of the IPX packet. On Ethernet networks, the standard frame type for NetWare 2.2 and NetWare 3.11 is 802.3. Starting with NetWare 3.12, the default frame type was changed to 802.2. (See table 7.2.) One of the biggest problems when connected to an IPX network is using the wrong frame type. For example, if one machine is using 802.2 and the other machine is using 802.3, the two machines cannot communicate with each other using IPX. Fortunately, most of these machines can support multiple frame types if necessary.

NOTE: Each subnet must use the same external addresses and frame type.

7.1.5   Routing in IPX Protocol

Most routers that route TCP/IP traffic can also route IPX traffic (although they may require additional software or configuration). IPX/SPX is a routable protocol stack because it has routing protocols designed into it. The routing protocols for IPX/SPX are RIP and NSLP.

Figure 7.2 IPX Network

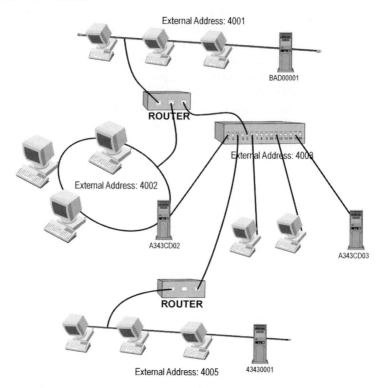

Table 7.1 Common IPX Network Services

Port Number	Network Service
0x0451	NetWare Core Protocol (NCP)
0x0452	Service Advertising Protocol (SAP)
0x0453	Routing Information Protocol (RIP)
0x0455	Novell NetBIOS
0x9001	NetWare Link Services Protocol (NLSP)

IPX RIP is very similar to the RIP in TCP/IP in that RIP for IPX is the distance-vector routing protocol for IPX. Similarly, NSLP is the link-state routing protocol for IPX/SPX. Both work similarly to their TCP/IP counterparts. RIP uses broadcasts of the entire IPX routing tables to keep all IPX routers updated. And just like OSPF, NSLP sends out only the changes to the routing tables and then only to a select group of network addresses.

Table 7.2   Frame Types

Topology	Supported Frame Type
Ethernet	Ethernet II, 802.3, 802.2, and Subnetwork Access Protocol (SNAP), which defaults to 802.2
Token Ring	802.5 and SNAP
Fiber Distributed Data Interface (FDDI)	802.2 and 802.3

Figure 7.3   IPXROUTE CONFIG Command

```
NWLink IPX Routing and Source Routing Control Program v2.00



Num   Name                        Network    Node          Frame

======================================================================

1.    IpxLoopbackAdapter          1234cdef   000000000002  [802.2]

2.    Local Area Connection       00000000   00c0f056bb98  [802.2]

3.    Local Area Connection 2     00000000   00a0c9e73b65  [802.2]

4.    NDISWANIPX                  00000000   7ab920524153  [EthII] -



Legend

======

- down wan line
```

7.1.6   Troubleshooting IPX Networks

To help troubleshoot the IPX Protocol in Windows 2000, Windows XP, and Windows Server 2003 use the IPXROUTE command, which can determine computer settings and perform diagnostic tests to resolve communication problems. If you type IPXROUTE CONFIG at the prompt, the screen will display the current IPX status, including the network number, media access control (MAC) address, interface name, and frame type. If you use the IPXROUTE RIPOUT network_number, it will use RIP to determine if there is connectivity to a specific network. The network_number is the external IPX number representing the network. (See figure 7.3.)

7.2   NETBIOS AND NETBEUI

The Network Basic Input/Output System (NetBIOS) is a common program that runs on most Microsoft networks. Originally created for IBM for its early PC networks, NetBIOS was adopted by Microsoft and has since become a de facto industry standard.

NetBIOS is a session-level interface used by applications to communicate with NetBIOS-compliant transports such as NetBEUI, IPX, or TCP/IP. It is responsible for establishing logical names (computer names) on the network, establishing a logical connection between the two computers, and supporting reliable data transfer between computers that have established a session.

7.2.1   SMB Protocol

Once a logical connection is established, computers can then exchange data in the form of NetBIOS requests or in the form of a **Server Message Block (SMB).** The SMB Protocol, which was jointly developed by Microsoft, Intel, and IBM, defines a series of commands used to pass information between networked computers. Clients that are connected to a network using NetBIOS over TCP/IP, NetBEUI, or IPX/SPX can send SMB commands.

NOTE: Microsoft refers to NetBIOS over TCP/IP as NBT.

SMB can be broken into four message types: session control, file, printer, and message. The session control consists of commands that start and end a redirector connection to a shared resource at the server. The file SMB messages are used by the redirector to gain access to files at the server. The printer SMB messages are used by the redirector to send data to a print queue at a server and to get status information about the print queue. The message SMB type allows an application to send messages to and receive messages from another workstation. When you share a drive, directory, or printer on a Microsoft Windows computer, you are using the SMB Protocol.

7.2.2   Common Internet File System

The **Common Internet File System (CIFS)** is a standard protocol that lets programs make requests for files and services on remote computers on the Internet. CIFS is a public or open variation of the Server Message Block Protocol used by Microsoft Windows and is currently used by Linux with Samba, Novell NetWare 6.0, and AppleTalkIP. A client program makes a request of a server program (usually in another computer) to gain access to a file or to pass a message to a program that runs in the server computer. The server takes the requested action and returns a response. This allows Microsoft Windows computers to access these files on Linux and NetWare 6.0 in the same way as accessing files on another Windows computer, and it allows Linux computers to access the shares on the Microsoft Windows computers.

Figure 7.4   Sample LMHOSTS File

```
102.54.94.97    rhino         #PRE   #DOM:networking #File Server

182.102.93.122  MISSERVER     #PRE                   #MIS Server

122.107.9.10    SalesServer                          #Sales Server

131.107.7.29    DBServer                             #Database Server

191.131.54.73   TrainServ                            #Training Server
```

7.2.3   Accessing a SMB/CIFS Share

When using a SMB/CIFS Protocol to share a directory, drive, or printer, these resources are accessed using the Universal Naming Convention (UNC) also known as Uniform Naming Convention:

    \\computer_name\shared-resource-pathname

The computer_name could be a NetBIOS name or an IP address.

7.2.4   Name Resolution with the NetBIOS Protocol

To navigate a SMB network, the computers are identified with computer names (NetBIOS names), no matter what protocol (TCP/IP, IPX, or NetBEUI) you are using. The Net-BIOS/computer names are up to 15 characters long. To connect the TCP/IP address with the computer name (NetBIOS name), SMB networks can use an LMHOSTS file or a WINS server.

An LMHOSTS file is a text file similar to a HOSTS file. Each entry should be kept on an individual line. The IP address should be placed in the first column followed by the corresponding computer name. The address and the computer name should be separated by at least one space or tab. Again, like the HOSTS file, the # character is generally used to denote the start of a comment. (See figure 7.4 and table 7.3.)

A **WINS** (**Windows Internet Naming Service**) server determines the IP address with the computer's NetBIOS name. WINS uses a distributed database that is automatically updated with the names of computers available and the IP address assigned to each one.

NOTE: UNIX machines do not use NetBIOS names. Therefore, if you want to use name resolution for these machines, you must use HOSTS files/DNS servers.

7.2.5   NetBEUI Protocol

NetBEUI is short for NetBIOS Enhanced User Interface, which provides the transport and network layers for the NetBIOS Protocol, usually found on Microsoft networks. NetBEUI

Table 7.3   LMHOSTS Special Entries

Predefined Keywords	Description
#PRE	Defines which entries should be initially preloaded as permanent entries in the name cache. Preloaded entries reduce network broadcasts, because names are resolved from cache rather than from broadcast or by querying the LMHOSTS file. Entries with a #PRE tag are loaded automatically when TCP/IP initializes, or manually by typing NBTSTAT –R at a command prompt.
#DOM: [*domain_name*]	Facilitates domain activity, such as logon validation over a router, account synchronization, and browsing.
#NOFNR	Avoids using NetBIOS-directed name queries for older LAN Manager UNIX systems.
#INCLUDE	Loads and searches NetBIOS entries in a separate file from the default LMHOSTS file. Typically the #INCLUDE file is a centrally located shared LMHOSTS file. Refer to the #INCLUDE file by using a Universal Naming Convention (UNC) path. Any computer referred to in the UNC path must also have a name-to-IP address mapping in the LMHOSTS file. This enables Windows 2000 to resolve a computer's NetBIOS name when it reads the #INCLUDE file.
#BEGIN_ALTERNATE **#END_ALTERNATE**	Defines a redundant list of alternate locations for LMHOSTS files. When you include multiple #INCLUDE lines between #BEGIN_ALTERNATE and #END_ALTERNATE, Windows 2000 includes the first file that it can locate in the list.
#MH	Adds multiple entries for a multihomed computer. A multihomed computer is a computer that contains the IP addresses that belong to different networks.

Note: The NetBIOS name cache and file are always read sequentially. Add the most frequently accessed computers to the top of the list. Add the #PRE-tagged entries near the bottom, because they are loaded when TCP/IP initializes and are not accessed again.

is a protocol used to transport data packets between two nodes. While NetBEUI is smaller than TCP/IP or IPX and is extremely quick, it will only send packets within the same network. Unfortunately, NetBEUI is not a routable protocol, which means that it cannot send a packet to a computer on another network.

NOTE:

- If you need to communicate between two networks, you should use TCP/IP or IPX on all servers and client computers.
- You can use the NetBEUI Protocol on the servers and then use NetBEUI on client boot disks so that you can perform downloads and installations over the network.

7.3   APPLETALK

Apple Macintosh computers include networking software that uses a protocol known as AppleTalk. The cabling system known as LocalTalk was already discussed in chapter 4. LocalTalk is a very simple and elegant protocol in that the computer takes care of most of the configuration. You simply plug LocalTalk in and it works. Eventually, Apple users wanted a faster version and developed AppleTalk version 2 with support for EtherTalk (AppleTalk over an Ethernet network).

7.3.1   AppleTalk Subprotocols

The AppleTalk subprotocols that are significant for node-to-node communications are:

- **AppleShare**—Provides file sharing services, print queuing services, password access to files or folders, and user accounting information
- **AppleTalk Filing Protocol (AFP)**—Provides transparent access to files on both local and remote systems
- **AppleTalk Session Protocol (ASP)**—Establishes and maintains connections between nodes and servers
- **AppleTalk Transaction Protocol (ATP)**—Ensures reliable delivery of data by checking connections between nodes, checking packet sequence, and retransmitting any data packets that become lost
- **Name Binding Protocol (NBP)**—Translates human-readable node names into numeric AppleTalk addresses
- **Routing Table Maintenance Protocol (RTMP)**—Maintains a routing table of AppleTalk zones and their networks, and uses ZIP (see below) to manage data in the routing table
- **Zone Information Protocol (ZIP)**—Updates zone information maps that tie zones to their networks for routing purposes
- **Datagram Delivery Protocol (DDP)**—Assigns an AppleTalk node's address upon start-up and manages addressing for communications between AppleTalk nodes

Traditional file sharing for Apple computers use the AppleTalk Protocol to communicate with an AppleShare server. The AppleTalk Protocol can be used over many network cabling systems, including LocalTalk, Ethernet, and Token Ring.

NOTE: TCP/IP can run over LocalTalk networks, but there are many limitations.

7.3.2   AppleTalk Addressing

There are two types of routers—seed and nonseed routers. A seed router has the net numbers programmed into it and acts as a source of information for other routers connected to that wiring system. A nonseed router learns the net number for the wiring system to which it is attached by communicating with the other routers on the system. Normally you only need to define the net number for one side of the router. It finds out the other net number(s) from the seed router or other routers on the network.

Each station on an AppleTalk network uses an address that is 24 bits long. Of those bits, 16 are given to the network, and each network can support 254 nodes. Each network segment can be given either a single 16-bit network number or a range of 16-bit network numbers. If a network is assigned a range of numbers, that network is considered an Extended AppleTalk network because it can support more than 254 nodes. The node address is automatically assigned by the computer itself or can be dynamically assigned by a seed router. If more than 254 nodes are needed, then multiple addresses can be assigned to a single network by using a cable range. The cable range is the network number or numbers used by the end nodes connected to the transmission media.

Larger networks can be divided into AppleTalk zones containing printers and computers. Zones are used to split a network into logical subdivisions; in a company, it might be convenient to create three zones called Accounting, Production, and Sales. A zone can be used to describe computers in a single physical location on a network, or it can describe computers from the same department that are in entirely different locations. To manage zones, you will need a dedicated server or hardware router.

NOTE: Although you can have multiple zones on an AppleTalk network, an AppleTalk node can belong to only one zone.

AppleTalk wasn't originally designed to be routed over a WAN, but with the release of AppleTalk version 2, Apple included routing functionality with the introduction of the Routing Table Maintenance Protocol (RTMP). RTMP is a distance-vector routing protocol, like RIP for both IP and IPX.

AppleTalk uses the Name Binding Protocol (NBP) to associate the name of the computer with its network address. As every station broadcasts its name when it comes up on a network, the AppleTalk router on the network caches the name and responds to the NBP request. When a node requests a name resolution, the local router will answer with information it has obtained from this cache. If an AppleTalk network doesn't have a router, each node will perform both NBP requests and NBP responses.

The only computer that comes with AppleTalk installed by default is the Macintosh. Most Windows operating systems are able to use the AppleTalk Protocol but require that additional software be installed.

7.4   DLC PROTOCOL

The Data Link Control (DLC) is a special, nonroutable protocol that enables computers running Windows to communicate with computers running a DLC Protocol stack such as IBM mainframes, IBM AS/400 computers, and Hewlett-Packard network printers that are connected directly to the network using older HP JetDirect network cards.

NOTE: When using the DLC Protocol to connect to a network printer, the DLC Protocol must be installed and running on the print server for the printer. The protocol does not need to be installed on the computers that send print jobs to the print server.

SUMMARY

1. When Novell NetWare was introduced, it was designed to be a server platform for a local area and wide area networks.

2. Novell designed the NetWare Protocol suite (also known as the Internetwork Packet eXchange/Sequenced Packet eXchange, or the IPX/SPX) to be very efficient over local area networks and that would also work on wide area networks.

3. IPX is the fastest routable network protocol suite available.

4. The Internetwork Packet Exchange (IPX) is a networking protocol used to interconnect networks. It is a connectionless protocol and therefore, doesn't require a connection to set up before packets are sent to a destination.

5. The SPX protocol uses packet acknowledgments to guarantee delivery of packets.

6. The Service Advertising Packet (SAP) is used to advertise the services of all known servers on the network, including file servers, print server and so on.

7. The NetWare Core Protocol (NCP) consists of the majority of network services offered by a Novell NetWare server.

8. When assigning addresses within the IPX network, there are two types of address, the internal and external IPX network numbers.

9. The internal IPX network number, an 8-digit (4-byte) hexadecimal number, is used to identify a server.

10. Unlike TCP/IP, the IPX clients are not assigned internal network numbers. Instead, they use the external IPX address and the MAC address.

11. The external IPX address is used to identify the network.

12. Throughout the years, there were several versions of the IPX packet.

13. On Ethernet networks, the standard frame type for NetWare 2.2 and NetWare 3.11 is 802.3.

14. Starting with NetWare 3.12, the default frame type was changed to 802.2.

15. The Network Basic Input/Output System (NetBIOS) is a common program that runs on most Microsoft networks.

16. The Server Message Block (SMB) protocol, which was jointly developed by Microsoft, Intel and IBM, defines a series of commands used to pass information between networked computers. Clients connected to a network using NetBIOS over TCP/IP, NetBEUI or IPX/SPX can send SMB commands.

17. Common Internet File System (CIFS) is a standard protocol that lets programs make requests for files and services on remote computers on the Internet. CIFS is a public or open variation of the Server Message Block Protocol used by Microsoft Windows and is currently used by Linux with Samba, Novell Netware 6.0 and AppleTalkIP.

18. When using SMB/CIFS protocol to share a directory, drive or printer, these resources are accessed using the Uniform Naming Convention (UNC).

19. To connect the TCP/IP address with the computer name (NetBIOS name), SMB networks can use a LMHOSTS file or a WINS server.

20. A WINS (Windows Internet Naming Service) server determines the IP address from computer's NetBIOS name.

21. NetBEUI is short for NetBIOS Enhanced User Interface, which provides the transport and network layers for the NetBIOS protocol, usually found on Microsoft networks.

22. NetBEUI is not a routable protocol, which means that it cannot send a packet to a computer on another network.

23. Apple Macintosh computers include networking software, which uses a protocol known as AppleTalk.

24. LocalTalk is a very simple and elegant protocol in that the computer takes care of most of the configuration. You simply plug it in and it works.

25. Traditional filesharing for Apple computers used the AppleTalk protocol to communicate with an AppleShare server.

26. The Data Link Control (DLC) is a special, non-routable protocol that enables computers running Windows to communicate with computers running DLC protocol stack such as IBM mainframes, IBM AS/400 computers and Hewlett-Packard network printers that connected directly to the network using older HP JetDirect network cards.

QUESTIONS

1. Which of the following are transport-layer protocols? (Choose all that apply.)
   a. TCP
   b. SPX
   c. IP
   d. IPX

2. Which of the following protocols is an NDIS-compliant version of the Internetwork Packet Exchange Protocol?
   a. IP
   b. NetBEUI
   c. NWLink
   d. SMB

3. One computer on your IPX/SPX network cannot connect to the network. What is the most likely cause of the problem?
   a. Incorrect frame type
   b. Upper memory area conflicts
   c. Protocol mismatch
   d. Faulty connectivity devices

4. There is a frame type mismatch on your Novell NetWare network. It is only affecting one computer. Which of the following needs to be reconfigured?
   a. The frame type on the client machine
   b. The frame type on the server machine
   c. The frame binding setting on the client machine
   d. The frame binding setting on the server machine

5. Which of the following protocols are network-layer protocols? (Choose all that apply.)
   a. IP
   b. IPX
   c. TCP
   d. SPX

6. Your company has a network that contains several NetWare 4.11 servers and uses IPX as the routing protocol. Each of the network segments has at least one NetWare server on it. Which of the following responds to a Get Nearest Server (GNS) request from a client?
   a. A NetWare server on the same segment
   b. A NetWare server on any segment
   c. A Cisco router on the same segment
   d. A Cisco router on any segment

7. Which protocol is native to Novell NetWare version 4.11 and guarantees packet delivery?
   a. SPX
   b. IP
   c. IPX
   d. NetBEUI

8. The IPX/SPX Protocol is typically used in NetWare networks. It is a routable protocol. Which of the following is a valid IPX network address?
   a. 64-32-16-0
   b. 115-255-64
   c. 1110110101
   d. 0ac05d9e

9. What is NWLink?
   a. Macintosh protocol
   b. UNIX protocol
   c. Microsoft's implementation of Novell's IPX/SPX
   d. Printer-only protocol

10. Which of the following is NOT part of the IPX/SPX Protocol suite?
    a. SAP - Service Advertising Protocol
    b. NCP - NetWare Core Protocol
    c. NISP - NetWare Link Services
    d. RIP - Routing Information Protocol
    e. All of the choices are part of IPX/SPX.

11. What service does Sequenced Packet Exchange (SPX) provide?
    a. Routing
    b. Connectionless packet delivery
    c. Connection to the Internet
    d. Retransmission of undelivered packets

12. Routers interconnect networks and provide filtering functions. Which of the following protocols can be used with a router?
    a. IPX/SPX
    b. NetBEUI
    c. TCP/IP
    d. AppleTalk

13. What is the primary protocol used with Novell NetWare 3.12?
    a. TCP/IP
    b. IPX/SPX
    c. NetBEUI
    d. DLC

14. Transport protocols provide for communication sessions between computers. Which of the following is a transport protocol?
    a. SNMP
    b. IP
    c. NetBEUI
    d. IPX

15. NetBEUI is the extended user interface of NetBIOS. Which of the following statements is true of NetBEUI?
    a. NetBEUI is a small, fast, and efficient transport-layer protocol used primarily with Microsoft networks.
    b. NetBEUI is a small, fast, and efficient session-layer protocol, but it is limited to Novell NetWare networks.

c. NetBEUI does not support routing.

d. NetBEUI is a relatively large protocol, which can cause problems in MS-DOS—based clients.

16. Which nonroutable protocol fits in at both the network and transport layers of the OSI model?

a. TCP/IP

b. IPX/SPX

c. AppleTalk

d. NetBEUI

17. What command can you use to help troubleshoot an IPX network in Windows 2000?

a. Ping   b. ipx

c. ipxconf   d. ipxroute

18. You are using a computer with NetBEUI. Unfortunately, you can't connect to a computer located on another LAN, although you can connect to the computers that are on the same LAN. What is the problem?

a. You are using the wrong MAC address.

b. You are using the wrong NetBEUI name.

c. You are using the wrong gateway address.

d. NetBEUI is a nonroutable protocol.

19. What protocol allows you to share directories, drives, and printers on a Windows machine so that other Windows users can access those shared resources?

a. SMB   b. WINS

c. DHCP   d. FTP

e. DNS   f. HTTP

20. What protocol is available for users on non-Windows operating systems to share directories, drives, and printers?

a. WINS   b. DHCP

c. FTP   d. DNS

e. CIFS   f. HTTP

21. Which of the following uses 15-character names to identify computers on a network?

a. TCP/IP   b. NetBIOS

c. IPX/SPX   d. AppleTalk

22. A peer-to-peer Microsoft network with no more than 200 hosts, no routers, and no need to connect to the Internet would be best served by what protocol?

a. TCP/IP   b. NetBEUI

c. DLC   d. IPX/SPX

23. Look at the figure. What is the problem with the IPX network?

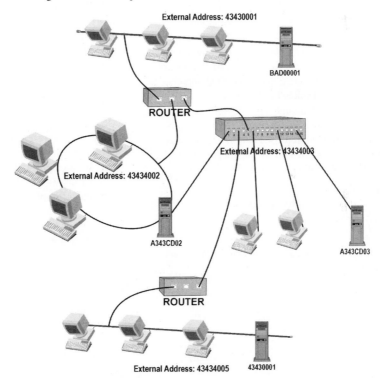

24. A business owner wants you to set up a small work-group network for her. She's heard that NetBEUI is a small, fast protocol and that's what she wants to use. She also wants all her workstations to have access to the Internet. What do you tell her?

   a. NetBEUI can only be used on a client-server network.

   b. NetBEUI cannot be used on the Internet.

   c. You'll need a NetBEUI-capable router to connect to the Internet.

   d. Sounds good. I'll get started right away.

25. Pat administers an Ethernet network that has 50 Windows NT Workstation 4.0 computers, 5 Windows NT Server 4.0 computers, and 25 Windows 98 computers. The Windows NT workstation 4.0 computers and 4 of the 5 Windows NT Server 4.0 computers use TCP/IP. Subnet A has 25 Windows NT Workstation 4.0 computers and 2 Windows NT Server 4.0 computers. Subnet B has 25 Windows NT Workstation 4.0 computers and 2 Windows NT Server 4.0 computers. The Windows 98 computers and the remaining Windows NT Server 4.0 computers use NetBEUI and reside in subnet C. What device could Pat install on subnet B to allow computers on this subnet to send and receive messages from subnets A and C?

   a. A passive hub   b. A brouter

   c. A router   d. An active hub

26. You are a network engineer implementing a network consisting of approximately 150 Windows 2000 Workstations and 5 Windows 2000 Servers. You have multiple sites located in the same city that are connected through a PSTN. Your manager wants you to ensure fast communication by implementing the proper protocol. Each site has a router that is already configured by the telephone company. The workstation is running Windows 98.

Required result:

■ Use a routable protocol to ensure connectivity between sites.

Optional results:

■ Ensure communication between the Windows 2000 Servers.

■ Log in to all the servers without maintaining multiple logins.

Proposed solution:

■ Implement NetBEUI as your protocol and use the same user names on all servers, but use different passwords for security purposes.

How would you characterize the proposed solution?

   a. It achieves the required result and both optional results.

   b. It does not achieve the required result but does achieve both the optional results.

   c. It does not achieve the required result but does achieve one of the optional results.

   d. It achieves the required result but cannot achieve either of the optional results.

   e. It achieves neither the required result nor either of the optional results.

27. What is the name of the distance-vector routing protocol for AppleTalk?

   a. RIP   b. OSPF

   c. NLSP   d. RTMP

28. Windows NT services for Macintosh enables NT servers to communicate with Macintosh computers using what protocol?

   a. Proto-talk   b. AppleTalk

   c. DEC   d. Mac-talk

29. How can you enable communications between your UNIX server and your Macintosh workstations?

   a. Install unix-comm on your workstation.

   b. It cannot be done.

   c. Install AppleTalk on both server and workstations.

   d. Install AppleTalk on your UNIX server.

30. You want your NetWare server to communicate with a Macintosh workstation. How do you do this?

   a. Install AppleTalk on both server and workstation.

   b. Install AppleTalk on the Macintosh workstation.

   c. Install AppleTalk on the NetWare server.

   d. It cannot be done.

HANDS-ON EXERCISE

Exercise 1: Using IPX Protocol

1. Bring up the Network and Dial-up Connection window.
2. Double-click on the Local Area Connection and click on the Properties button.
3. Click on the Install button. In the Select Network Component Type dialog box, click the Protocol option and click on the Add button. In the Select Network Protocol dialog box, click NWLink IPX/SPX/NetBIOS Compatible Transport Protocol and then click OK.
4. In the Local Area Connection Properties dialog box, click the NWLink IPX/SPX/NetBIOS Compatible Transport Protocol and then click Properties. Notice which type of frame detection is selected by default.
5. Close the Local Area Connection Properties dialog box.
6. In the Network and Dial-up Connections window, open the Advanced menu and select the Advanced Settings option.
7. In the Advanced Settings dialog box, under Client for Microsoft Networks, unbind TCP/IP by clearing the Internet Protocol (TCP/IP) check box. Click the OK button and close the Network and Dial-up Connections window.
8. Restart the computer.
9. Open a Command prompt window.
10. At the prompt, type in IPXROUTE CONFIG and press the Enter key. Notice the frame type, network address, and MAC address.
11. Close the Command prompt window.
12. Use My Network Places to access your partner's share that you created in chapter 4.
13. Open the Network and Dial-up Connections window.
14. Double-click the Local Area Connection and click on the Properties button.
15. In the Local Area Connection Properties dialog box, click NWLink IPX/SPX/NetBIOS Compatible Transport Protocol and click the Uninstall button. Click on the Yes button to Uninstall NWLink. Reboot your computer if necessary.
16. Open the Network and Dial-up Connections window. Then open the Advanced menu and select the Advanced Settings option. In the Advanced Settings dialog box, under Client for Microsoft Networks, unbind TCP/IP by clearing the Internet Protocol (TCP/IP) check box. Click the OK button and close the Network and Dial-up Connections window.

CHAPTER 8

Network Services

Topics Covered in this Chapter

Introduction

Before users can start navigating your network, you will need some type of name resolution. In chapter 4, you learned that a DNS and a WINS server are used to resolve names and return IP addresses. In this chapter you will learn to install and configure these types of servers. In addition, you will also install and configure DHCP servers so that you can automatically assign IP addresses and other IP parameters to your clients.

Objectives

- Define the purpose, function, and/or use of the FTP, TFTP, HTTP, HTTPS, SMTP, POP3, and IMAP4 protocols within TCP/IP.
- Identify the purpose of the DHCP/BOOTP, DNS, WINS, and SNMP network services.
- Given a scenario, predict the impact of modifying, adding, or removing network services (e.g., DHCP, DNS, and WINS) on network resources and users.

- Given a name, specify which method of name resolution would be used to resolve the name to an IP address.
- Given a Fully Qualified Domain Name, identify each component of the name.
- Explain how SNMP agents and managers communicate with each other.

8.1 INTRODUCTION TO DOMAIN NAME SYSTEM NAME RESOLUTION

There are two services that can be used to find and provide IP addresses. If a computer knows the destination's NetBIOS name, it can send the NetBIOS name to a WINS and the WINS will send back the corresponding IP address of the destination. If a computer knows the Fully Qualified Domain Name (FQDN), it can send that to a DNS service and the DNS service will return the corresponding IP address. Most modern OSs and programs can use FQDNs.

A **Domain Name Service (DNS),** also known as Domain Name System, is a hierarchical client/server-based distributed database management system that translates Internet domain names such as MICROSOFT.COM to an IP address. As mentioned in previous chapters, it is used because domain names are easier to remember than IP addresses. DNS clients are called **resolvers,** and DNS servers are called **name servers.**

A DNS can be thought of as its own little network. If one DNS server doesn't know how to translate a particular domain name, it will ask another DNS server. DNS is most commonly associated with the Internet, but private networks can also use DNS to resolve computer names and to locate computers within their local networks without being connected to the Internet.

Table 8.1   RFC Documents That Describe DNS

RFC	Title
974	Mail Routing and the Domain System
1034	Domain Names—Concepts and Facilities
1035	Domain Names—Implementation and Specification
1123	Requirements for Internet Hosts—Application and Support
1886	DNS Extensions to Support IP Version 6
1912	Common DNS Operational and Configuration Errors
1995	Incremental Zone Transfer in DNS
1996	A Mechanism for Prompt DNS Notification of Zone Changes
2136	Dynamic Updates in the Domain Name System (DNS UPDATE)
2181	Clarifications to the DNS Specification
2182	Selection and Operation of Secondary DNS Servers
2219	Use of DNS Aliases for Network Services
2308	Negative Caching of DNS Queries (DNS NCACHE)

The first TCP/IP networks used **HOSTS** files to translate from domain names (such as MICROSOFT.COM, ACME.COM, or MIT.EDU) to IP addresses. Since some networks grew so fast, manually updating and distributing the HOSTS file was not very effective. For the Internet (the largest network that uses DNS), there is not a single organization that is responsible for keeping the DNS updated. Instead, it is a distributed database that exists on many different name servers around the world, with no one server storing all the information. Because of this, DNS allows for almost unlimited growth.

The most popular implementation of the DNS Protocol is the Berkeley Internet Name Domain (BIND), which was developed for UC Berkeley's BSD UNIX operating system. The primary specifications for DNS are defined in Requests for Comments (RFCs) 1034 and 1035. DNS uses either UDP port 53 or TCP port 53 as the underlying protocol. (See table 8.1.) Windows NT family servers, Linux, and Novell NetWare include all the necessary software to operate as a DNS server.

8.1.1   Domain Name Space

The **DNS name space** describes the hierarchical structure of the DNS database as an inverted logical tree structure. Each node on the tree is a partition of the name space called

Figure 8.1   DNS Name Space

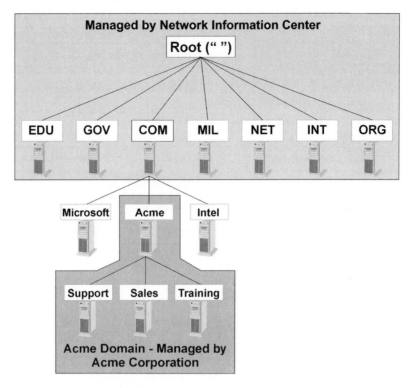

a domain. Domains can be further partitioned at node points within the domain into subdomains. The names of the domain and subdomains can be up to 63 characters long. (See figure 8.1.)

The top of the tree is known as the **root domain.** It is sometimes shown as a period (.) or as empty quotation marks (""), indicating a null value. Immediately below the root domain, you will find the **top-level domains.** The top-level domains indicate a country, region, or type of organization. Three-letter codes indicate the type of organization. For example, COM indicates commercial (business), and EDU stands for educational institution. The codes are listed in table 8.2.

Two-letter codes indicate countries, which follow International Standard 3166. For example, CA stands for Canada, AU for Australia, FR for France, and UK for United Kingdom. For a list of two-letter codes, go to:

http://www.din.de/gremien/nas/nabd/iso3166ma/codlstp1/index.html

The **second-level domain names** are variable-length names registered to an individual or organization for use on the Internet. These names are almost always based on the appropriate top-level domain, depending on the type of organization or geographic location where a name is used.

Table 8.2   Top-Level Domain Codes Indicating the Type of Organization

Traditional Top-Level Domains		New Top-Level Domains	
Code	**Meaning**	**Code**	**Meaning**
COM	Commercial	**AERO**	Airline-related services
EDU	Educational	**BIZ**	Businesses
GOV	Government	**COOP**	Cooperatives
INT	International organization	**INFO**	Websites providing information
MIL	Military	**MUSEUM**	Museums
NET	Network-related	**NAME**	Personal websites and email addresses
ORG	Miscellaneous organization	**PRO**	Professionals such as doctors and lawyers

Examples:

Domain Name	**Second-Level Domain Name**
MICROSOFT.COM	Microsoft Corporation
CISCO.COM	Cisco Corporation
MIT.EDU	MIT University
ED.GOV	United States Department of Education
ARMY.MIL	United States Army
W3.ORG	World Wide Web Consortium
NATO.INT	North Atlantic Treaty Organization
PM.GOV.AU	Prime Minister of Australia

The second-level domain names must be registered by the authorized party. For example, for years, Network Solutions Inc. ran a government-sanctioned monopoly on registrations for .COM, .NET, and .ORG domain names. But as the U.S. government handed the control of the Internet to an international body, several companies now handle the registration of these three-letter codes.

NOTE: Since most of the common top-level domain names are already taken, some countries such as Tonga (TO) and Tuvalu (TV) are selling their domain name. Therefore, some commercial and user sites may be using one of these two-letter codes. This is especially true with the TV domain name since it is easily linked to television.

Subdomain names are additional names that an organization can create that are derived from the registered second-level domain name. The subdomain allows an organization to divide a domain into a department or geographic location, making the partitions of the domain name space more manageable. A subdomain must have a contiguous domain name space. This means that the domain name of a zone (child domain) is the name of that zone added to the name of the domain, or parent domain.

A **host name** is a name assigned to a specific computer within a domain or subdomain by an administrator to identify the TCP/IP host. Multiple host names can be associated with the same IP address, although only one host name can be assigned to a computer. If the DNS is seen as a tree, the host name represents the leaf or object of the tree. Much like a subdomain, it is the leftmost label of the DNS domain name. The host name can then be used in place of an IP address such as the ping or other TCP/IP utilities. The total length of an FQDN cannot exceed 255 characters.

NOTE: The host name does not have to be the same as the NetBIOS (computer) name.

Registering Your Domain

The quickest and easiest way to register your domain name is through your ISP. If you cannot do that, you need to use the following steps:

1. Find the company to register the domain name. For domain names with the COM, ORG, and NET domain suffixes, you can go to the www.netsol.com, www.register.com or www.verisign websites.
2. Pick a name and make sure that it is available. On these websites, you can search for a name without any obligation. Of course, you should pick a name that indicates your company or business. Unfortunately, that may not be an easy task since most easily identifiable domain names are already taken.
3. Arrange your domain DNS either through your ISP or by setting up your own DNS server.
4. Complete the registration form at the company with which you are registering the domain name. Follow that company's instructions. During the entire registration period, keep checking for small errors and omissions since these can delay the processing of the registration form. The registration form will usually need three contacts: admin, technical, and billing.

NOTE: After you submit your registration form, you will usually get an acknowledgment that will contain a tracking number. Make sure to keep that tracking number for future inquiries. You will receive a message telling you the date your domain will be activated in the root name servers. Be aware that it will take a few hours to a few days after the activation date since it takes time for the distributed database to be replicated throughout the Internet with the added address.

NOTE: There will be a fee to activate the domain name.

A **Fully Qualified Domain Name (FQDN)** describes the exact position of a host (computer) within the domain hierarchy, and it is considered to be complete. When used in a DNS domain name, it is stated by a trailing period (.) to designate that the name of the host is located off the root or highest level of the domain hierarchy.

Example:

> SERVER1.SALES.ACME.COM
>
> COM indicates a commercial business.
> ACME is the name of the domain name.
> SALES is the name of the subdomain.
> SERVER1 is the name of the server located within the SALES subdomain.

When you create a domain name space, consider the following domain guidelines and standard naming conventions:

- To minimize the level of administrative tasks, limit the number of domain levels.
- If possible, each subdomain should have a unique name throughout the entire domain to ensure the subdomain name is unique.
- Use simple, yet meaningful, names so that domain names are easy to remember and navigate.
- Use standard DNS characters including A–Z, a–z, 0–9, and hyphen and Unicode characters, which include the additional characters needed for foreign languages such as French, German, and Spanish.
- The names of the domain and subdomains can be up to 63 characters long.
- The total length of an FQDN cannot exceed 255 characters.

8.1.2   DNS Zones

A **DNS zone** is a portion of the DNS name space whose database records exist and are managed in a particular DNS database file. (See figure 8.2 and table 8.3.) Each zone is based on a specific domain node, which is also referred to as the zone's root domain. It is the authority source for that node. Zone files do not necessarily contain the complete DNS branch since a subdomain may be its own zone.

NOTE: If subdomains are added below the domain, the subdomains can be part of the same zone or belong to another zone.

The computer that maintains the master list for a zone is the primary name server for that zone, which is considered the authority for that zone. A DNS server might be configured to manage one or more zones.

Question:

> Why would you take a domain and then partition it into several subdomains (zones) so that they can be controlled by separate DNS name servers?

Answer:

> By breaking up domains across multiple zone files, the zone files can be stored, distributed, and replicated to other DNS servers. This allows you:
>
> - To delegate management of a domain/subdomain as it relates to its location or department within the organization

Figure 8.2   Sample Zone Database Files

```
;
;   Database file acme.com.dns for acme.com zone.
;      Zone version:  6
;
@                        IN  SOA server1.acme.com.
administrator.acme.com. (
                            6              ; serial number
                            900            ; refresh
                            600            ; retry
                            86400          ; expire
                            3600         ) ; minimum TTL
;
;   Zone NS records
;
@                        NS server1.acme.com.
;
;   Zone records
;
server2000-01a        A 132.132.60.45
server1               A 132.132.20.1
testserver            A 132.132.20.20
```

■ To divide one large zone into smaller zones so that the traffic load can be distributed among multiple servers for better DNS name resolution performance and for fault tolerance

Of course, when choosing how to structure zones, you should use a DNS structure that reflects the structure of your organization.

Most BIND implementations have two types of zones that you can configure: a standard primary zone and a standard secondary zone. Windows 2000 and Windows Server 2003 also use a third type called the Active Directory integrated zone. (See table 8.4.)

Table 8.3 Common Resource Record Types Used in DNS Database Files

Resource Record	Purpose
SOA (start of authority)	Identifies the name server that is the authoritative source of information for data within a domain. An SOA record is created automatically when you create a new zone. A primary server for a given zone lists itself in the SOA record to show that it's the source for this zone. The first record in the zone database file must be the SOA record.
NS (name server)	Provides a list of name servers that are assigned to a domain.
A (host address)	Provides a host name to an Internet Protocol (IP) version 4 32-bit address. For more information, see RFC 1035.
PTR (pointer)	Resolves an IP address to a host name (reverse mapping). For more information, see RFC 1035.
CNAME (canonical name)—alias	Creates an alias or alternate DNS domain name for a specified host name. The most common or popular use of an alias is to provide a permanent DNS aliased domain name for generic name resolution of server-based names such as www.acme.com and ftp.acme.com to more than one computer or IP address used in a web server. This way, you can assign acme.com to one server, www.acme.com to a second server, and ftp.acme.com to a third server. If you do use the same server for all three entries and you decide to split the service to separate services, you just have to change the CNAME resource record to point to the new server.
SRV (service)	Locates servers that are hosting a particular service. *Note:* SRV records are new in Windows 2000 DNS Server services. This record enables you to maintain a list of servers for a well-known server port and transport the protocol type ordered by preference for a DNS domain name. For more information, see the Internet draft "A DNS RR for Specifying the Location of Services (DNS SRV)."
MX (mail exchanger)	Identifies which mail exchanger to contact for a specified domain and in what order to use each mail host.

Table 8.4 Zone Types

Zone Type	Description
Standard primary	The master copy of a new zone.
Standard secondary	A replica of an existing zone. Standard secondary zones are read-only.
Active Directory integrated	A zone that is stored in Active Directory (Windows 2000/Windows 2003 only). Updates of the zone are performed during Active Directory replication.

The **primary name server** is a name server that stores and maintains the zone file locally. Changes to a zone, such as adding domains or hosts, are done by changing files at the primary name server. A **secondary name server** gets the data for its zone from another name server, either a primary name server or another secondary name server. The process of obtaining this zone information across the network is referred to as a zone transfer. Zone transfers occur over TCP port 53. The **Active Directory integrated zone** has the zone defined using the Active Directory, not the zone files.

The source of the zone information for a secondary name server is referred to as a master name server. A master name server can be either a primary or a secondary name server for the requested zone.

Question:

Why do you want to have a secondary name server?

Answer:

There are three reasons. The first reason is for fault tolerance. You should have at least two DNS name servers serving each zone. In each client's configuration, both name servers would be listed. If the first name server listed cannot be contacted, the client will then contact the second name server. The second reason would be to divide the load among different name servers so that the performance for name resolution will be increased. Lastly, the DNS servers could be used to service computers that are located in remote locations so that they would not have to use a slow WAN link.

8.1.3   Name Resolution

When a client computer needs an IP address, the client computer will send a name query to the DNS client service (resolver) located on the client. The DNS client service will then check the locally cached information and local HOSTS file (if present).

NOTE: The cache area is not the cache inside the processor but an area of memory in RAM set aside to hold DNS entries. The entries in the cache come from the preloaded entries of the hosts file or from previous answered responses. When previous queries are cached, the data is kept for a preset time period known as the time to live (TTL).

If the query does not match an entry in the cache, the resolution process continues with the client querying a DNS server to resolve the name.

When the resolver queries a DNS server, it will perform a recursive query. A **recursive query** asks the DNS server to respond with the requested data or with an error message stating that the requested data doesn't exist or that the domain name specified doesn't exist.

NOTE: The name server does not refer the query to another name server unless it is configured as a forwarder, in which case it will forward the DNS request as a recursive query.

When the DNS server receives a request, it will check to see if the host name is located in its own zone database file, in which it is an authority. If it is not listed in the zone file, it will then check the cache area. From then on, the DNS server use iterative queries to resolve the name. An **iterative query** gives the best answer it currently has back as a response. The best answer will be the address being sought or an address of a server that would have a better idea of its address.

Example:

A program needs to find the address of SERVER.SUPPORT.ACME.COM. The program will send the request to the resolver. The resolver will look in its cache area and its HOSTS file. If the address has not been found, the resolver will forward the query to its preferred DNS name server.

The DNS server will check to see if it knows the address of SERVER. SUPPORT. ACME.COM. If it doesn't know, it will then ask the root server if it knows the address of SERVER.SUPPORT.ACME.COM. Since the root server does not know the address of SERVER.SUPPORT.ACME.COM, it will respond with the best answer by providing the address of the COM root server. The DNS server will then ask the COM root server for the address of SERVER.SUPPORT.ACME.COM. The COM root server will respond with its known best answer of the ACME.COM name server address. The DNS server will query the ACME.COM name server for the address of SERVER. SUPPORT.ACME.COM. The DNS server will then respond to the resolver with the address of the SUPPORT.ACME.COM. The DNS server will query the SUPPORT. ACME.COM name server for the address of SERVER.SUPPORT.ACME.COM and will respond by sending back the address.

The preferred server will then respond to the client by giving the correct address. The program will then use the IP address to communicate its needs.

As you can see by this example, the resolver queried its DNS server using a recursive query, while the DNS server queried other DNS servers using iterative queries. (See figure 8.3.)

Although the recursive query process can be resource-intensive, it has some performance advantages for the DNS server. When the recursion process is complete, the information is cached by the server. The cache information will be used again to help speed the answering of subsequent queries. Over time, the cache information can grow to occupy a significant portion of the server memory resources.

Typically, the process of domain name resolution occurs very quickly. Occasionally, it may be delayed. If the delay is too long, the browser will come back and say that the domain does not exist even though you know that it does. This is because your computer got tired of waiting and timed out. Yet when you try again, there is a good chance it will work because the authoritative server has had enough time to reply and the name server has stored the information in its cache.

DNS servers can be configured to send all recursive queries to a selected list of servers, known as **forwarders.** Servers used in the list of forwarders provide recursive

Figure 8.3   A DNS Client Will Use a Recursive Query With the Preferred Server to Find an IP Address, While the Preferred Server Will Typically Use an Iterative Query to Discover the IP Address

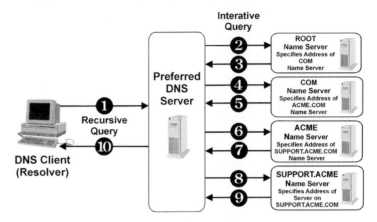

lookup for any queries that a DNS server receives that it cannot answer based on its own zone records. During the forwarding process, a DNS server configured to use forwarders essentially behaves as a DNS client to its forwarders. Typically forwarders are used on re-mote DNS servers that use a slow link to access the Internet.

DNS servers use a mechanism called **round-robin** or **load sharing** to share and dis-tribute loads for network resources. Round-robin rotates the order of resource records data returned in a query answer in which multiple resource records exist of the same resource record type for a queried DNS domain name. Since the client is required to try the first IP address listed, a DNS server configured to perform round-robin rotates the order of the A resource records when answering client requests.

8.1.4   Reverse Queries

Occasionally, the resolver may perform a **reverse query;** in this case, a resolver knows an IP address and wants to know the host name. To prevent from searching all the do-mains for an inverse query, a special domain called in-addr.arpa was created. Nodes in the in-addr.arpa domain are named after the numbers in the IP address. Since IP addresses get more specific from left to right, the order of IP addresses is reversed when building the in-addr.arpa domain. Once the in-addr.arpa domain is built, special resource records called pointer records (PTRs) are added to associate the IP addresses with the correspon-ding host names. To find a host name for an IP address, the resolver would query the DNS server for a PTR for the address.in-addr.arpa. For example, to find the IP address of 1.2.3.4, the resolver would query the DNS server for a PTR of 4.3.2.1.in-addr.arpa. See figure 8.4 for a sample reverse zone database file.

Figure 8.4 Sample Reverse Zone Database File

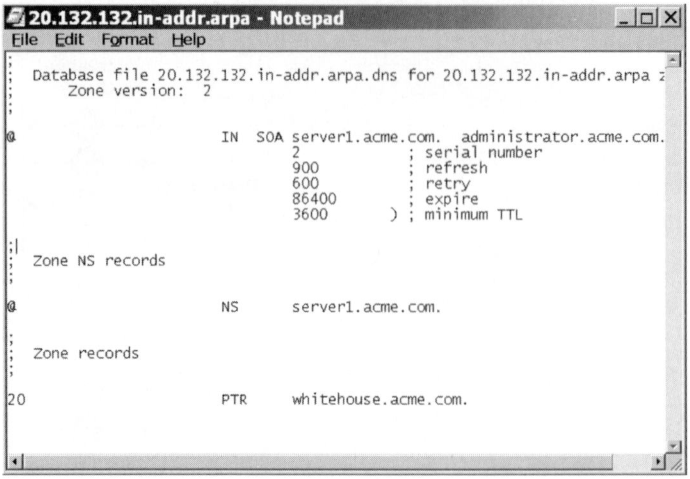

20.132.132.in-addr.arpa - Notepad

File Edit Format Help

```
;
;  Database file 20.132.132.in-addr.arpa.dns for 20.132.132.in-addr.arpa z
;      Zone version:  2
;

@                        IN  SOA server1.acme.com.  administrator.acme.com.
                                 2            ; serial number
                                 900          ; refresh
                                 600          ; retry
                                 86400        ; expire
                                 3600       ) ; minimum TTL

;
;  Zone NS records
;

@                        NS      server1.acme.com.

;
;  Zone records
;

20                       PTR     whitehouse.acme.com.
```

In addition to reverse lookups, some DNS servers support an inverse query. Just as it would for the reverse lookup, a client making an inverse query provides the IP address and requests an FQDN. Instead of using the in-addr.arpa domain to find the answer, the DNS server will check its own zone for the answer. If the answer is not in the zone, it will return an error message. Because inverse queries are not very thorough, they are not used often.

8.1.5 Dynamic DNS

Since DNS has become the primary naming resolution tool, every computer that will hold a network service such as file or print sharing would have to be registered. A large network—one that has a lot of computers that use DHCP to obtain a new IP address—needs to dynamically register and update the DNS servers' resource records. Windows 2000, Windows Server 2003, newer versions of Linux, and Novell NetWare 5.1 and higher provide client and server support for the use of dynamic updates, as described in RFC 2136.

8.1.6 Troubleshooting Name Resolution

Problem:

A DNS client received a "Name not found" error message.

Solution:

First, make sure that the DNS client computer has a valid IP configuration for the network. Verify that the TCP/IP configuration settings for the client computer are correct, including the IP address, subnet mask, gateway, and those settings that are used for DNS name resolution using WINIPCFG, IPCONFIG /ALL, or another appropriate command. If you are not sure what the IP address is for the preferred DNS server, you can observe it by using the IPCONFIG command.

Next, make sure the DNS server is running properly by trying to ping it. If you cannot ping the server, you need to determine if the problem is the client, the DNS server, or some link in between the two. Remember that the problem can be caused by the DNS server's TCP/IP configuration, the DNS server's network cards, the cable connected to the DNS server, or the routes to the DNS server.

The DNS server the client is using does not have authority for the failed name and cannot locate the authoritative server for this name. If the server is the authority, make sure that the record is there. Lastly, make sure that the DNS server is still running.

8.2   INTRODUCTION TO WINDOWS INTERNET NAME SERVICE

If you try to access a computer using its NetBIOS (computer) name, such as using a UNC name to specify a network resource, the computer needs to determine what the IP address is. Initially, it would broadcast onto the network asking for the IP address of the computer. Unfortunately, the broadcast usually doesn't go across routers, which means that computers on other subnets don't get resolved. In addition, if you have a lot of computers doing these types of broadcast, the broadcasts will slow the performance of the network.

Other methods to resolve NetBIOS names would be either to use LMHOSTS files or to access a WINS server. Just as it would be for HOSTS files, it would be extremely difficult to maintain and distribute the LMHOSTS files to all the computers in a large network, especially since addresses may change frequently under a network using DHCP servers.

8.2.1   Using a WINS Server

A **WINS server** contains a database of IP addresses and NetBIOS (computer) names that update dynamically. For clients to access the WINS server, the clients must know the address of the WINS server. Therefore, the WINS server needs to have a static address—one that does not change. When the client accesses the WINS server, the client doesn't do a broadcast; rather, the client sends a message directly to the WINS server. When the WINS server gets the request, it knows which computer sent the request, and it can reply directly to the originating IP address. The WINS database stores the information and makes it available to the other WINS clients.

NOTE: The WINS database is located at *systemroot*\System32\WINS\Wins.mdb.

The WINS registration generates little excessive network traffic because it doesn't use broadcast.

NOTE: The WINS Protocol is based on and is compatible with the protocols defined for NetBIOS Name Server (NBNS) in RFCs 1001 and 1002.

Table 8.5 NetBIOS Network Services

NetBIOS Name Suffix	Network Service/Resource Identifier
\\computer_name[00h]	Workstation service
\\computer_name[03h]	Messenger service
\\computer_name[06h]	RAS
\\computer_name[20h]	Server service
\\computer_name[21h]	RAS client service (on a RAS client)
\\computer_name[BEh]	Network monitoring agent service
\\domain_name[1Bh]	The PDC in its role as the domain master browser
\\domain_name[1Dh]	The master browser for each subnet
\\domain_name[1Ch]	The domain controllers (up to 25 IP addresses) within the domain

When a WINS client starts up, it registers its name, IP address, and type of services within the WINS server's database. The type of service is designated by a hexadecimal value, which is placed at the end of the name. For example, when starting a Windows 2000 computer called Server2, it will register three mappings, including Server2[00h] (workstation), Server2[03h] (messenger), and Server2[20h] (file server).

NOTE: The NetBIOS can only be up to 15 characters, not counting the hexadecimal value, which represents the service. See table 8.5 for the list of NetBIOS network services.

Since WINS was only made for Windows operating systems, other network devices and services (such as a network printer and UNIX machines) cannot register with a WINS. Therefore, these addresses would have to be added manually.

Names that are held in the WINS database are given a time to live, or renewal interval, during name registration. A name must be refreshed before this interval ends or the name will be released from the database. Names are refreshed by sending a name refresh request to the WINS server by the WINS client. Windows clients will attempt a refresh at one-half of the renewal interval and will keep trying till they contact the WINS server or the time has expired. NetBIOS names are explicitly released when the client performs a proper shutdown or remains silent when the name is not refreshed within the renewal interval.

When a client node registers a name that already exists in the WINS database and that has a different IP address, the WINS server must determine if the name with the old IP address still exists. Therefore, the WINS server will send a name query request to the old IP address. If the old address responds with a positive name query response, the WINS server will reject the new registration with a negative name registration response. If the old address does not respond to the name query, the new registration is accepted.

Table 8.6   NetBIOS Resolution Nodes

Node Types	Windows Registry Value	Description
B (broadcast) node	1	A computer doing B-node name resolution relies on broadcasts to convert names into IP addresses. B-node name resolution is not the best option on larger networks because the broadcast will load the network and will usually not go through routers.
		Note: Microsoft really uses B-node name resolution, which will check the LMHOSTS file after doing a broadcast.
P (point-to-point) node	2	A computer doing P-node name resolution uses a NetBIOS Name Server (NBNS)/WINS server to look up NetBIOS names to get IP addresses. All systems must know the IP address of the NBNS. The main drawback of P-node name resolution is that if the NBNS cannot be accessed, there will be no way to resolve names and thus no way to access other systems on the network by using NetBIOS names.
M (mixed) node	4	An M-node computer first tries a broadcast to resolve a name. If that attempt fails, the computer looks up the name in a NetBIOS name server. In other words, an M-node computer first acts as a B-node one, and if that fails, it tries to act as a P-node one. M-node has the advantage over P-node in that if the NBNS is unavailable, systems on the local subnet can still be accessed through B-node resolution. M-node is typically not the best choice for larger networks because it uses B-node and thus results in broadcasts. However, when you have a large network that consists of smaller subnetworks connected via slow WAN links, M-node is a preferred method since it will reduce the amount of communication across the slow links.
H (hybrid) node (default)	8	Finally, an H-node computer first does a P-node lookup. If that fails, the computer does a broadcast. In either case, the NetBIOS name resolution will try the LMHOSTS file after trying a broadcast and/or WINS server.

8.2.2   WINS Clients

A WINS client can be configured to use one of four NetBIOS name resolution methods. They include B (broadcast) node, P (point-to-point) node, M (mixed) node, and H (hybrid) node. (See table 8.6.) In any case, when trying to resolve a computer name, the client will always check its own local NetBIOS name cache. Just like the DNS name cache, it remembers names and addresses of computers that it recently communicated with. By default, when a system is configured to use WINS for its name resolution, it adheres to H node for name registration. By using DHCP servers or by using the registry, you can force the client into one of these nodes.

Example:

> If you are using an H (hybrid) node, it will first check to see if the name is the local machine name. It will then check the NetBIOS cache area for remote names. Resolved names will remain in the cache area for 10 minutes. If the name hasn't been resolved yet, the H node will try the WINS server followed by a broadcast. Lastly, it looks in the LMHOSTS file if the system has one, followed by using the HOSTS file and the DNS server if it was configured.

8.2.3   WINS Server Replication

To provide fault tolerance, it is recommended to have more than one WINS server with the same WINS database. By having more than one WINS server, a WINS client can go to the second WINS server when the first one is unavailable. To make sure that all the WINS servers have the same information, you must replicate the information from one WINS server to the others. These servers are known as replication partners.

A **WINS replication partner** can be added and configured as either a pull partner, a push partner, or a push/pull partner. The push/pull partner type is the default configuration and is the type recommended for use in most cases. A **pull partner** is a WINS server that requests new database entries from its partner. The pull occurs at configured time intervals or in response to an update notification from a push partner. A **push partner** is a WINS server that sends update notification messages. The update notification occurs after a configurable number of changes to the WINS database.

Since pull partners configure at certain time intervals, you should use a pull partner across slow links. For example, you could have it replicate every 24 hours beginning at 12 at night. Therefore, the replication will occur when traffic is at a minimum. A push parent should be used with servers connected across fast links because push replication occurs when a particular number of updated WINS entries is reached.

NOTE: If you need certain changes to replicate immediately, you can force the WINS servers to replicate using a WINS console.

You can configure a WINS server to automatically configure other WINS server computers as its replication partners by using periodic multicasts to announce their presence. These announcements are sent as IGMP messages for the multicast group address of 224.0.1.24 (the well-known multicast IP address reserved for WINS server use). With this automatic partner configuration, other WINS servers are discovered when they join the network and are added as replication partners.

8.2.4   WINS Proxy Agent

A **WINS proxy agent** is a WINS-enabled computer configured to act on behalf of other host computers that cannot directly use WINS. WINS proxies help resolve NetBIOS name queries for computers located on a subnet where there is no WINS server. The proxies do this by hearing broadcasts on the subnet of the proxy agent and forwarding those responses directly to a WINS server. This keeps the broadcasts local and yet gets responses from a WINS server without using the P node. Since most WINS proxies are only useful or necessary on networks that include NetBIOS broadcast-only (B-node) clients, WINS proxies are typically not needed for most networks.

8.3   INTRODUCTION TO DHCP SERVICES

A DHCP server maintains a list of IP addresses called a pool. When a user needs an IP address, the server removes the address from the pool and issues it to the user for a limited amount of time. Issuing an address is called leasing. Using a DHCP server to issue an address is more reliable and requires less labor than setting every computer manually, and you can get by with fewer IP addresses because computers not on the network are not using IP addresses.

Created for diskless workstations, the **Bootstrap Protocol (BOOTP)** enables a booting host to configure itself dynamically. (See table 8.7.) **DHCP,** which stands for **Dynamic Host Configuration Protocol,** is an extension of BOOTP. It is used to automatically configure a host during bootup on a TCP/IP network and to change settings while the host is attached. There are many parameters that you can automatically set with the DHCP server. Some of the more common parameters include:

- IP address
- Subnet mask
- Gateway (router) address
- Address of DNS servers
- Address of WINS servers
- WINS client mode

Table 8.7   Common RFC for DHCP

RFC Number	Description
RFC 2131	Dynamic Host Configuration Protocol
RFC 2132	DHCP Options and BOOTP Vendor Extensions
RFC 951	The Bootstrap Protocol (BOOTP)
RFC 1534	Interoperation between DHCP and BOOTP
RFC 1542	Clarifications and Extensions for the Bootstrap Protocol
RFC 2136	Dynamic Updates in the Domain Name System (DNS UPDATE)
RFC 2241	DHCP Options for Novell Directory Services
RFC 2242	Netware/IP Domain Name and Information

8.3.1   The DHCP Requests

A host computer that is configured to get a DHCP address sends a DHCPDISCOVER message on the local IP subnet to find the DHCP server or servers. The client doesn't know the address or addresses of the DHCP server, so it uses an IP broadcast address for the DHCPDISCOVER message. All available DHCP servers respond with a DHCPOFFER message. If more than one server is available, the client usually selects the first server to respond, but no rule specifies which server the client has to use. Regardless of how many servers respond, the client broadcasts a DHCPREQUEST message that identifies which server the client will use and implicitly informs all other servers that the client won't use them. The selected server responds to the client with a DHCPPACK message that contains the assigned IP address, any other network parameter assignments, and the lease or amount of time for which the DHCP server assigns the client the IP address. The client sends messages to UDP port 67 on the DHCP server, and the server sends messages to UDP port 68 on the DHCP client.

To provide fault tolerance, you can assign two or more DHCP servers for every subnet. With two DHCP servers, if one server is unavailable, the other server can take its place and continue to lease new addresses or renew existing clients. When setting up the two DHCP servers, the scopes of the two servers should include scopes with different addresses to avoid both servers handing out the same address to two different computers. Before installing a DHCP server, you must make sure that the TCP/IP is installed and a static IP address is assigned to the DHCP server.

If you have a client that must always use the same address, you can reserve the address by using a client reservation. The DHCP server will then assign the reserved address to the computer with the specified MAC address. If multiple DHCP servers are configured with a scope that covers the range of the reserved IP address, the client reservation must be made and duplicated at each of these DHCP servers. Otherwise, the reserved client computer can receive a different IP address, depending on which DHCP responds.

8.3.2   DHCP Relay Agent

A **DHCP relay agent** is a computer that relays DHCP and BOOTP messages between clients and servers on different subnets. This way, you can have a single DHCP server handle several subnets without the DHCP server being connected directly to those subnets. You should enable the DHCP relay agent on those router interfaces that are attached to subnets that contain network client computers and do not contain a DHCP server. When you want to use DHCP to configure TCP/IP information on computers that reside on remote subnets from the DHCP server, you must ensure either that the routers can function as BOOTP/DHCP relay agents or that a server on each remote subnet has a BOOTP/DHCP relay agent installed. A router that has a BOOTP relay agent installed is RFC 1542–compliant.

8.4   SIMPLE NETWORK MANAGEMENT PROTOCOL

The **Simple Network Management Protocol (SNMP)** has become the de facto standard for internetwork management. It allows configuring remote devices, monitoring network performance, detecting network faults, detecting inappropriate access, and auditing network usage. Remote devices include hubs, bridges, routers, and servers.

NOTE: SNMP uses UDP.

SNMP contains two primary elements: a manager and agents. The **SNMP manager** is the console through which the network administrator performs network management functions. The SNMP works by sending messages called protocol data units (PDUs) to different parts of a network. Agents store data about themselves in a **management information base (MIB).** The manager, which is the console through which the network administrator performs network management functions, will request the information from the MIB. The **SNMP agent** returns the appropriate information to the manager.

A **trap** is an unsolicited message sent by an SNMP agent to an SNMP management system when the agent detects that a certain type of event has occurred locally on the managed host. The SNMP manager console receives a trap message known as a trap destination. For example, a trap message might be sent when a system restarts or when a router link goes down.

Each SNMP management host and agent belongs to an SNMP community. An **SNMP community** is a collection of hosts grouped together for administrative purposes. Deciding what computers should belong to the same community is generally, but not always, determined by the physical proximity of the computers. Communities are identified by the names you assign to them.

While most networks use a user name and a password for authentication, SNMP messages are only authenticated by the community name. Although a host can belong to several communities at the same time, an SNMP agent does not accept requests from a management system in a community that is not on its list of acceptable community names. If the community name is incorrect, the agents send an "authentication failure" trap to its trap destination. Therefore, it is the responsibility of the administrator to set hard-to-guess community names.

After the SNMP authenticates a message, the request is evaluated against the agent's list of access permissions for that community. The types of permissions that can be granted to a community are shown in table 8.8.

Community names are transmitted as clear text, that is, without encryption. Because unencrypted transmissions are vulnerable to attacks by hackers with network analysis software, the use of SNMP community names represents a potential security risk. However, Windows 2000 and Windows Server 2003 IP Security can be configured to help protect SNMP messages from these attacks.

The SNMP service requires the configuration of at least one default community name. The name Public is generally used as the community name because it is the common name that is universally accepted in all SNMP implementations. You can delete or change the default community name or add multiple community names. If no community names are defined, the SNMP agent will deny all incoming SNMP requests.

When an SNMP agent receives a message, the community name contained in the packet is verified against the agent's list of acceptable community names. After the name is determined to be acceptable, the request is evaluated against the agent's list of access

Table 8.8   Community Permissions

Permission	Description
None	The SNMP agent does not process the request. When the agent receives an SNMP message from a management system in this community, it discards the request and generates an authentication trap.
Notify	This is currently identical to the permission of "none."
Read-only	The agent does not process SET requests from this community. It processes only GET, GET-NEXT, and GET-BULK requests. The agent discards SET requests from manager systems in this community and generates an authentication trap.
Read/create	The SNMP agent processes or creates all requests from this community. It processes SET, GET, GET-NEXT, and GET-BULK requests, including SET requests that require the addition of a new object to a MIB table.
Read/write	Currently identical to "read/create."

permissions for that community. The types of permissions that can be granted to a community include the following:

There are two versions of SNMP. SNMP 1 reports only whether a device is functioning properly. The industry has attempted to define a new set of protocols called SNMP 2 that would provide additional information, but the standardization efforts have not been successful. Instead, network managers have turned to a related technology called RMON that provides more detailed information about network usage.

8.5   INTRODUCTION TO WEB PAGES AND WEB SERVERS

The terms *Internet* and the *World Wide Web* are often used interchangeably, but they have different meanings. As was pointed out earlier, the **Internet** refers to the huge global WAN. Until the early 1990s, to use the computer to communicate with another computer via the Internet required a good deal of knowledge and the ability to understand and use some fairly unfriendly commands. The **World Wide Web (WWW)** was created in 1992, and refers to the means of organizing, presenting, and accessing information over the Internet.

To access the web, a user would use the following technologies:

1. HTML
2. Web server
3. Web browser
4. HTTP and FTP

Web pages are written using the **HyperText Markup Language (HTML).** This language is pretty simple and is implemented as special ASCII tags or codes that you embed within your document to give the browser a general idea of how the information should be displayed. The browsers understand the standard HTML tags, although they may display the same document a little differently. If you want your documents to be accessible by people using different browsers, you should stick with the standard tags. The HTML standard is still actively evolving, so new tags are constantly becoming available to support new browser features.

A **web server** is a computer equipped with the server software that uses Internet protocols such as HyperText Transfer Protocol (HTTP) and File Transfer Protocol (FTP) to respond to web client requests on a TCP/IP network via web browsers. One server can service a large number of clients. Several free server programs are available on the Internet. Most web browsers are built to process two basic types of requests: file server and database server requests. New features are always being added to provide additional support for new technology. A web server acting as a file server simply accepts a request for a document, validates the request, and sends the requested files back to the browser. In addition, the browser can act as a front-end tool or interface to collect data and feed it into a database or script. The database can be located either on the same server as the web server or on a different server. When the database responds with the results, it will then send the results back to the browser.

The **web browser** is the client program/software that you run on your local machine to gain access to a web server. It receives the HTML commands, interprets the HTML, and displays the results. It is strictly a user-interface/document presentation tool. It knows nothing about the application it is attached to and only knows how to take the information from the server and present it to the user. It is also able to capture data entry made into a form and get the information back to the server for processing. The most common browsers are Microsoft Internet Explorer and Netscape Communicator. Both of these tools are available for little or no charge on the Internet.

The application protocol that makes the web work is **HyperText Transfer Protocol, or HTTP.** Whereas HTML is the language used to write web pages, HTTP is the protocol that web browsers and web servers use to communicate with each other over the Internet. It is an application-level protocol because it sits on top of the TCP layer in the protocol stack and is used by specific applications to talk to one another. In this case the applications are web browsers and web servers.

HTTP is a text-based protocol. Clients (web browsers) send requests to web servers for web elements such as web pages and images. Most protocols are connection-oriented—two computers communicating with each other keep the connection open over the Internet. Although HTTP uses the connection-oriented service of TCP (a connection-oriented data transmission service guarantees that packets will arrive at their destination in the order they were sent and that they will arrive error-free), HTTP does not. Before an HTTP request can be made by a client, a new connection must be made to the server. After the request is serviced by a server, the connection between client and server across the Internet is disconnected. A new connection must be made for each request.

Currently, most web browsers and servers support HTTP 1.1. One of the main features of HTTP 1.1 is that it supports persistent connections. This means that once a browser connects to a web server, the browser can receive multiple files through the same connection. This should improve performance by as much as 20 percent.

HTTP is called a stateless protocol because each command is executed independently, without any knowledge of the commands that came before it. This is the main reason that

Figure 8.5 Typical URL

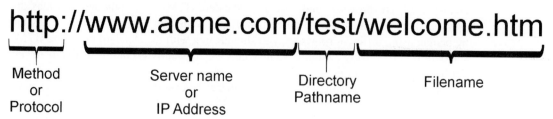

Method or Protocol   Server name or IP Address   Directory Pathname   Filename

it is difficult to implement websites that react intelligently to user input. This shortcoming of HTTP is being addressed in a number of new technologies, including ActiveX, Java, JavaScript, and cookies.

Two other common protocols are the **File Transfer Protocol (FTP)** and the **Trivial File Transfer Protocol (TFTP),** which are used on the Internet to send files. FTP is the more popular of the two. It uses TCP, the connection-oriented service of the TCP/IP suite. As noted before, a connection-oriented data transmission service guarantees that packets will arrive at their destination in the order they were sent and that they will arrive error-free.

TFTP uses UDP to send and receive data packets. UDP is the connectionless service provided by the TCP/IP suite. A connectionless service sends data packets but does not guarantee that those packets will arrive in the order they were sent or that they will arrive error-free. You would typically use TFTP when you need to send small files over a reliable link. Cisco routers use TFTP to backup and update their configuration files.

Typically when accessing a web page, you will specify the web page location by using a **Uniform Resource Locator,** or **URL.** The first part of the address indicates what protocol to use, and the second part specifies the IP address or the domain name where the resource is located. The last part indicates the folder and file name. (See figure 8.5.) Here are two examples:

ftp://ftp.acme.com/files/run.exe
http://www.acme.com/index.html

When you type a URL into a web browser, this is what happens:

1. If the URL contains a domain name, the browser first connects to a DNS server and retrieves the corresponding IP address for the web server.
2. The web browser connects to the web server and sends an HTTP request (via the protocol stack) for the desired web page.
3. The web server receives the request and checks for the desired page. If the page exists, the web server sends it. If the server cannot find the requested page, it will send an HTTP 404 error message. (404 means "page not found," as anyone who has surfed the web probably knows.)
4. The web browser receives the page back, and the connection is closed.
5. The browser then parses through the page and looks for other page elements it needs to complete the web page. These usually include images, applets, etc.
6. For each element needed, the browser makes additional connections and HTTP requests to the server.

7. When the browser has finished loading all images, applets, etc., the page will be completely loaded in the browser window.

A **gopher** is a system that predates the World Wide Web for organizing and displaying files on Internet servers. A gopher server presents its contents as a hierarchically structured list of files. You don't really hear much about gophers today since they have been basically replaced by search engines.

Since the Internet is so vast, consisting of countless web pages, it is often difficult to know the URL that a person wants. Therefore, the World Wide Web has several search engines that can be used to locate desired documents. A **search engine** is a program that searches documents for specified keywords and returns a list of the documents where the keywords were found. Typically, a search engine works by sending out a spider to fetch as many documents as possible. Another program, called an indexer, then reads these documents and creates an index based on the words contained in each document. Each search engine uses a proprietary algorithm to create its indexes such that, ideally, only meaningful results are returned for each query. Popular search engines include google.com, yahoo.com, lycos.com, and altavista.com.

For more information on HTTP, go to the W3C Architecture Domain website located at:

http://www.w3.org/Protocols/

8.6 INTRODUCTION TO HTML LANGUAGE

To create or modify a web page, you would use a text editor such as DOS's EDIT or Windows Notepad. The text file will begin with the < HTML> tag and end with the < /HTML> tag. Web browsers can identify an HTML file by the file extension, and many browsers will display pages that don't carry the < HTML> tag. When viewing your document with your own browser, you might be able to get away without the tag, but you can't be sure that everyone who wants to view your page will be using the same browser you use. It's best to stick with good programming habits and to avoid the temptation to take shortcuts.

The HTML document is divided into two: the head and the body. The head begins with the < H> tag and ends with the < /H> tag, while the body begins with < BODY> and ends with < /BODY>. While these tags are not case-sensitive, I would recommend putting them in uppercase so that they will be easy to identify.

NOTE: Most HTML tags are used in pairs, with an opening tag and an ending tag to delineate which text you want handled in a particular way. Ending tags are the same as opening tags, but with the addition of a forward slash (/).

8.6.1 The Head

The head includes the title, which is defined with the < TITLE> < /TITLE> tags. The information enclosed within the title tags will be displayed by your browser at the top of the web page in the title bar. Beside the title, the head can also include META tags, which are

Figure 8.6   HTML Web Page That Defines the Title and Body

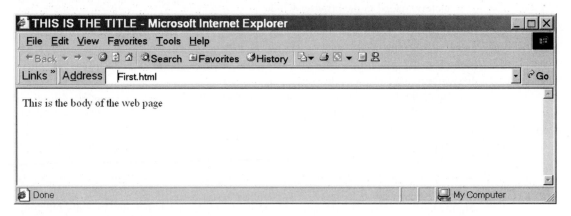

```
<HTML>

<HEAD>

<TITLE>THIS IS THE TITLE</TITLE>

<META NAME="KEYWORD"CONTENTS="HOT,IMPORTANT">

<META NAME="DESCRIPTION"CONTENTS="This is a sample website">

</HEAD>


<BODY>

This is the body of the web page

</BODY>

</HTML>
```

used for presenting information about the document. This can be valuable if you want to list the contents of the page in a series of keywords or a description that can be targeted by search engines.

NOTE: The <META> tag is a singlet with no closing tag. (See figure 8.6.)

8.6.2   The Body

Everything that shows up on the web page itself is found in the body of the HTML document. The HTML equivalent to hitting the Return key in a word processing program is the use of the paragraph <P> </P> tags or the break tag <BR>. The paragraph tags <P> </P>

produce the same results as hitting the Return key twice, creating a blank line as well as a break. The break tag <BR> is like hitting the Return key once.

To center the text on the screen, you would use the <CENTER> </CENTER> tag. One excellent way to organize materials on a web page is through the use of a list. There are two types of lists: ordered lists and unordered lists. Ordered lists are usually numbered and are often used when items are listed in some kind of sequence. Unordered lists are lists that are not numbered and are usually bulleted. <OL> </OL> are the tags used to create an ordered list, while <UL> </UL> are used to indent the text. To make the list into a bulleted list or a numbered list, you would add the list tag <LI> in front of each item on the list. A <BLOCKQUOTE> </BLOCKQUOTE> is usually used when you want to set off a quotation or any paragraph by indenting on the left and the right.

Standard text in a browser will usually appear as 12-point text. The font that will appear depends upon which font is the default font for the browser. To vary the size of the text on the web page, you can use the heading tab <H> </H>. The heading tab always includes a number 1 through 6, which identifies the size of the text. <H1> </H1> is the largest, and <H6> </H6> is the smallest.

The font tags <FONT> </FONT> can also be used to change the size of the text, specify the font, and change the color of the text. Unlike the <H> </H> tags, the font tags do not force a break after the text that they enclose. By itself, the font tag doesn't do anything. It must be given "attributes" for size, color, or font. An attribute is an option that can be used within the tags. An attribute is usually followed by a "value," indicating a specific choice within the range of choices for that attribute. All the font attributes can be nested within a single set of font tags.

To change the size of text, enclose it with the font tags. After the word FONT in the opening tag, leave a space and include SIZE= N where N is any number between 1 and 7. The number 1 is the smallest text size and 7 is the largest. The default text size is about 3 on the font scale.

To change the font, include a <FONT FACE= "name of font"> inside the font tags. For example, to change the font to Arial, you would use <FONT> <FONT FACE= "Arial"> text</FONT>. Of course, you must have the Arial font on your computer for this to work.

Changing the color of the text is a little bit more complicated than changing the size and font of the text. Inside the font tags, the color is specified with the < FONT COLOR= "rrggbb">, where rr represents two hexadecimal numbers that specify the amount of red. The gg of course specifies the amount of green, and bb specifies the amount of blue. If you want the text to be red, you would specify FF0000, where FF is the hexadecimal equivalent to 255. If you want black text, you would use 000000; and for white, you would use FFFFFF. To change the background color of the page, use the < BODY BG COLOR= "rrggbb"> inside the body tags. To change the default text color, you would use < BODY TEXT= "rrggbb"> < /BODY>

Many colors can be indicated by entering the name of the color rather than its hexadecimal number value. Some samples of colors include black, navy, silver, blue, maroon, purple, red, fuchsia, green, teal, lime, aqua, olive, gray, yellow, and white. (See figure 8.7.)

Figure 8.7 HTML Web Page With a Centered Text, Ordered List, and Colors

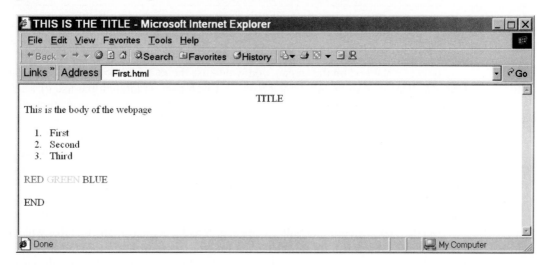

```
<HTML>

<HEAD>
   <TITLE> THIS IS THE TITLE</TITLE>

   <META NAME="KEYWORD" CONTENTS="HOT,IMPORTANT">
   <META NAME="DESCRIPTION" CONTENTS="This is a sample website">
</HEAD>

<BODY>
   <CENTER>TITLE</CENTER>
   This is the body of the web page
   <OL>
     <LI>First
     <LI>Second
     <LI>Third
   </OL>

   <FONT COLOR="FF0000">RED
   <FONT COLOR="00FF00">GREEN
   <FONT COLOR="0000FF">BLUE
   <BR><BR>
   <FONT COLOR="000000">END
</BODY>
</HTML>
```

8.6.3   **Adding Hyperlinks**

Links on a web page are most often achieved through the use of hypertext. Hypertext is text that is connected, or "hyperlinked," to other information on the World Wide Web. When you click on a hyperlink you will be taken to the information that the link is anchored to. Hypertext is colored and underlined to set it off from the rest of the text. The default color for hypertext is blue, but it can be changed.

Essential to creating hypertext links are the anchor tags < A> < /A>. The anchor tags enclose text that will become the hypertext link. For example, if you want to create a hypertext link to the Acme Toy Company, you would start by enclosing the text with the anchor tags. By themselves, the anchor tags don't do anything; they must contain information that indicates where the Acme home page is located. You need to add an attribute to the tag that has the address. This attribute is called the hypertext reference, or HREF. For example, to have the Acme Toy Company link to the Acme home page, you would use:

<A HREF= "http://www.acme.com\list.htm"> Acme Toy Company< /A>

When links are made between files that recode on the same server and in the same directory, only a partial address or URL is required. The partial address is called a relative URL address. For example, if you have a file called LIST.HTM in the same directory, you can then use:

<A HREF= "list.htm"> Acme Toy Company< /A>

8.6.4   **Adding Pictures to Your Web Page**

It is easy to see why you would want to add pictures to your web page. To add a picture (either a GIF or JPG file, you would use the <IMG> tag, which is another singlet. Inside the <IMG> tag, and preceded by a space, you would include the attribute SRC, referring to the source or location of the image. The SRC attribute is followed by an equal sign (=) and the URL, in quotations, for the location of the graphic file. For example, to add a picture of the LOGO.JPG, you would use:

<IMG SRC= "LOGO.JPG">

or

<IMG SRC= "PICTURES\LOGO.JPG">

if the LOGO.JPG is located in the pictures directory, which is located underneath the directory of the current web page that you are viewing or using.

This is assuming that the LOGO.JPG is located in the same directory as the current page. One thing I need to point out is that the picture is not embedded into the web page. Instead, it is a separate file that is shown by a link. Therefore, if you are copying a web page, you will need to also copy the individual pictures.

Hyperlinks are not limited to text; they can also be images. The following shows how to display the LOGO.JPG that is linked to the ACME.COM home page.

<A HREF= "http://www.acme.com"> <IMGSRC= "LOGO.JPG"> </A>

Table 8.9   Input Tag

Attribute	Description
Type	Specifies the type of input. Choices include text, check boxes, and radio (Radio buttons). The default type is TEXT.
Value	The value attribute inserts a label on the form to indicate the input.
Size	The size attribute specifies the size of the input.
Maximum length	The maximum length attribute specifies the maximum length of the input.

8.6.5   Forms and Common Gateway Interface

The common gateway interface (CGI) is a standard for interfacing external applications with web servers. A plain HTML web page is a static document that doesn't change. A CGI program, on the other hand, is executed in real time, so that it can output dynamic information. For example, let's say you want to "hook up" your server database to the World Wide Web, to allow people from all over the world to query it. Basically, you need to create a CGI program that the PC with the web page will execute to transmit information to the database engine, receive the results back, and display them to the client.

When a web server executes a CGI program, it typically begins with displaying an HTML form on the screen. The form provides space for specific data items in the form of text boxes, pull-down boxes, check boxes, radio buttons, and so on. The form typically has a Submit button that a user can click on to submit the entry, and some will have a Clear button to put the data values back to their default values. The FORM tag in the HTML web page specifies a form to fill out. More than one fill-out form can be included in a single document, but forms cannot be nested (one inside another).

To create a form, you would use the </FORM> </FORM> tags. Inside the form, you will identify inputs using the <input> tag. The <input> tag has several inputs including the type, value, size, and maximum length. (See table 8.9.)

To create a Submit Query button, you would use <INPUT TYPE= "SUBMIT">, which will submit the data into the specified database. To create a Reset button, which will reset all the values back to their default values, you would use <INPUT TYPE= "RESET">. (See figure 8.8.)

The method attribute, which is entered into the < FORM> tag, specifies what to do with the data that is entered into the form. The < FORM METHOD= POST> will send the information to the database. For example:

<FORM ACTION= "http://www.acme.com/script" METHOD= POST>

PERL, which is short for Practical Extraction and Report Language, is a programming language especially designed for processing text. Because of its strong text processing abilities, PERL has become one of the most popular languages for writing CGI scripts. PERL is an interpretive language, which makes it easy to build and test simple programs.

Figure 8.8   HTML Web Page Acting as a Form

```
<html>

<FORM action="http://www.acme.com/script/"
method="GET">

   Name:<INPUT TYPE="TEXT" NAME="Name"
   Size="35"><BR>
   <BR>

   E-mail:<INPUT TYPE="TEXT" NAME="E-mail"
   SIZE="35"><BR>
   <BR><BR>

   Sex:<BR>
   Male<INPUT TYPE="RADIO" NAME="sex">
   Female<Input Type="RADIO":NAME="sex">
   <BR><BR>

   Check if you want to be added to our mailing list:
   <Input Type="checkbox" NAME="mailinglist">
   <BR><BR><BR>

   <INPUT TYPE="SUBMIT">
   <INPUT TYPE="RESET">
</FORM>

</html>
```

Continued

Figure 8.8   Continued

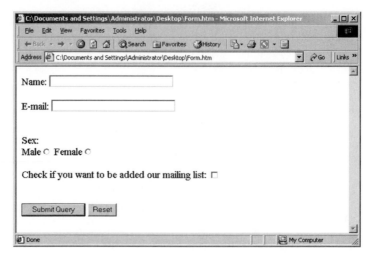

8.6.6   Dynamic HTML and Active Server Pages

For the most part, HTML creates static and unanimated documents. Once a web page is loaded into a browser window, it doesn't change in content or in form without being reloaded. This truly limits HTML's potential as an interactive multimedia format. To make the web a more compelling medium, HTML must provide web authors with the ability to dynamically update content, change the appearance of content, and hide, show, and animate content.

Active Server Pages (ASPs) is a specification for a dynamically created web page with a .ASP extension that utilizes ActiveX scripting. When a browser requests an ASP, the web server generates a page with HTML code and sends it to the browser. So ASPs are similar to CGI scripts, but they enable Visual Basic programmers to work with familiar tools.

Control over page layout is one of HTML's most obvious limitations. Even with today's third-generation HTML specifications, web authors do not have the ability to lay out pages with the pixel-level accuracy available to the traditional desktop publishing author. A standard HTML document cannot specify that text and images be located at exact coordinates or on top of each other, or even that the text be displayed in a particular point size. Since the beginning of HTML's existence, web authors have struggled to get browsers to display HTML content exactly the way they want it.

Dynamic HTML (DHTML) allows content to be displayed with more design flexibility and accuracy through the use of cascading style sheets (CSS). Using CSS, a standard from the World Wide Web Consortium (W3C), web authors can define fonts, margins, and line spacing for different parts of an HTML document. In addition to these stylistic improvements, CSS allows the absolute positioning of content by specifying x,y coordinates (and even a z index, which allows different elements to overlap).

The Microsoft implementation of DHTML also adds built-in multimedia and data objects (treated as properties of cascading style sheets) that can be controlled through scripting languages, allowing for stereo sound, on-the-fly image manipulation, and even access to server-side databases.

8.6.7   Extensible Markup Language

XML is an acronym for **eXtensible Markup Language.** Whereas HTML programs are written using tags that are predefined in the language standards, XML is more of a meta-language, which uses tags to markup the contents of your document. By being a meta-language, it allows users to create their own collection of tags, elements, and attributes as needed and in so doing to accurately describe the physical contents of a document. Therefore, it lets you design your own customized markup languages for limitless different types of documents. You will see more use of XML on web pages, especially as Microsoft's implements its .NET initiative and more applications will support the XML format such as Open Office and future versions of Microsoft Office.

8.6.8   ActiveX, VBScript, and JavaScript

An ActiveX control can be automatically downloaded and executed by a web browser. ActiveX is not a programming language, but rather a set of rules for how applications should share information. Programmers can develop ActiveX controls in a variety of languages, including C, C++, Visual Basic, and Java.

An ActiveX control is similar to a Java applet. Unlike Java applets, however, ActiveX controls have full access to the Windows operating system. This gives them much more power than Java applets, but with this power comes a certain risk that the applet may damage software or data on your machine. To control this risk, Microsoft developed a registration system so that browsers can identify and authenticate an ActiveX control before downloading it. Another difference between Java applets and ActiveX controls is that Java applets can be written to run on all platforms, whereas ActiveX controls are currently limited to Windows environments.

Related to ActiveX is a scripting language called VBScript that enables web authors to embed interactive elements in HTML documents. Just as JavaScript is similar to Java, so VBScript is similar to Visual Basic. Currently, Microsoft Internet Explorer supports Java, JavaScript, and ActiveX, whereas Netscape's Navigator browsers support only Java and JavaScript (though plug-ins can enable support of VBScript and ActiveX).

VBScript, short for Visual Basic Scripting Edition, is a scripting language developed by Microsoft and supported by Microsoft's Internet Explorer web browser. VBScript is based on the Visual Basic programming language, but it is much simpler. It enables web authors to include interactive controls, such as buttons and scroll bars, on their web pages.

JavaScript is a programmable API that allows cross-platform scripting of events, objects, and actions. It allows the page designer to access events such as start-ups, exits, and users' mouse clicks. JavaScript extends the programmatic capabilities of Netscape Navigator (and to a slightly lesser extent, Microsoft's Internet Explorer) to a wide range of authors, and it's easy enough for anyone who can compose HTML.

8.6.9   Web Page Editors

Today, many web pages are created using a web page editing program such as Microsoft Front Page. With these programs, you can create a web page much like you could using Microsoft Word or Microsoft PowerPoint. In fact, both these programs also have the capability to save as an HTML format. Of course, these programs don't have the full capability of creating a web page as do the web page editors.

8.6.10   Web Servers

For its web server, Microsoft offers the Internet Information Server (IIS), while Linux uses Apache for its web server. Typically, when you install a web server, the web server will use a directory that will be used for a home directory for the website. Therefore, when you access a website, you are accessing the web pages within that directory.

Today, when you type in a URL in a browser, you typically specify the protocol, the host name, and the file name (possibly with a file name path). For example, you may type:

http://www.acme.com/welcome.htm

and it will run the welcome.htm page from the home directory.

Often, you will access the home page of a website even if you do not specify a home page. For example, if you put the following URL in a browser:

http://www.acme.com

it will execute the index.htm automatically. That is because these web servers have a default page that will load when a web page is not specified.

Normally, a domain name such as WWW.ACME.COM refers to a single computer host. You can, however, configure a single server to appear as different hosts. You might, for example, create virtual servers, one for each department in your organization, or you can be an ISP that is hosting several hosts for various clients. Each of the servers would be recognized by a different IP address in your DNS server, but all would be hosted on the same server. They are called virtual servers because they all are hosted on the same physical server.

8.6.11   HTTP Error Codes

HTTP status codes are returned by web servers to indicate the status of a request. The status code is a three-digit code indicating the particular response. The first digit of this code identifies the class of the status code. The remaining two digits correspond to the specific condition within the response class. Table 8.10 lists all status codes defined for the HTTP/1.1 draft specification outlined in IETF RFC 2068.

The 4xx errors indicate to a client that the client has sent bad data or a malformed request to the server. Client errors are generally issued by the web server when a client tries to gain access to a protected area using a bad user name and password. For example, a 403 indicates that the requested resource is forbidden, which generally means you don't have the privileges needed to access that page. A 404 indicates that a host server responded to your browser but it cannot find the web page on the server. This usually means that the web page was moved or deleted or that you typed in the wrong URL. (See table 8.10.)

The 5xx errors indicate a server error, which indicates that the client's request couldn't be successfully processed due to some internal error in the web server. These error codes may indicate something is seriously wrong with the web server. (See table 8.11.)

Table 8.10 4xx Browser Errors

400	Bad request	
401	Unauthorized	
	401.1	Logon failed
	401.2	Logon failed due to server configuration
	401.3	Unauthorized due to ACL on resource
	401.4	Authorization failed by filter
	401.5	Authorization failed by ISAPI/CGI application
402	Payment required	
403	Forbidden	
	403.1	Execute access forbidden
	403.2	Read access forbidden
	403.3	Write access forbidden
	403.4	SSL required
	403.5	SSL 128 required
	403.6	IP address rejected
	403.7	Client certificate required
	403.8	Site access denied
	403.9	Too many users
	403.10	Invalid configuration
	403.11	Password change
	403.12	Mapper denied access
	403.13	Client certificate revoked
	403.14	Directory listing denied
	403.15	Client access licenses exceeded
	403.16	Client certificate untrusted or invalid
	403.17	Client certificate has expired or is not yet valid
404	Not found	
405	Method not allowed	
406	Not acceptable	
407	Proxy authentication required	
408	Request timeout	
409	Conflict	
410	Gone	
411	Length required	
412	Precondition failed	
413	Request—entity too long	
414	Request—URL too long	
415	Unsupported media type	

Table 8.11   5xx Browser Errors

500	Internal server error–An internal server error has caused the server to abort your request. This is an error condition that may also indicate a misconfiguration with the web server. However, the most common reason for 500 server errors is when you try to execute a script that has syntax errors.
501	Not implemented–This code is generated by a web server when the client requests a service that is not implemented on the server. Typically, not-implemented codes are returned when a client attempts to post data to a non-CGI (i.e., the form action tag refers to a nonexecutable file).
502	Bad gateway–The server, when acting as a proxy, issues this response when it receives a bad response from an upstream or support server.
503	Service unavailable–The web server is too busy processing current requests to listen to a new client. This error represents a serious problem with the web server (normally solved with a reboot). Try again later.
504	Gateway timeout–Gateway timeouts are normally issued by proxy servers when an upstream or support server doesn't respond to a request in a timely fashion.
505	HTTP version not supported–The server issues this status code when a client tries to talk using an HTTP that the server doesn't support or is configured to ignore.

8.7   EMAIL OVERVIEW

Electronic mail (email) is the transmission of messages over communications networks. The messages can be notes entered from the keyboard or electronic files stored on disk. Most mainframes, minicomputers, and computer networks have an email system. Some electronic mail systems are confined to a single computer system or network, but others have gateways to other computer systems, enabling users to send electronic mail anywhere in the world. Companies that are fully computerized make extensive use of email because it is fast, flexible, and reliable. In recent years, the use of email has exploded.

Most email systems include a rudimentary text editor for composing messages, but many allow you to edit your messages using any editor you want. You then send the message to the recipient by specifying the recipient's address. You can also send the same message to several users at once.

All online services and Internet service providers (ISPs) offer email, and most also support gateways so that you can exchange mail with users of other systems. Usually, it takes only a few seconds or minutes for mail to arrive at its destination. This is a particularly effective way to communicate with a group because you can broadcast a message or document to everyone in the group at once.

8.7.1   Email and the Internet

No matter which email package you use or which service you use to host the email, your email travels the same road that all Internet-based information such as web page downloads do. That is, your email traverses the Internet backbone. The sender creates an email message on an application. The client system is known as a user agent, or UA. When the user sends the message, it is transmitted to the user's Internet mail server.

Once the message reaches the Internet mail server, it enters the Internet's message transfer system, or MTS. The MTS relies on other Internet mail servers to act as **message transfer agents (MTAs),** which relay the message toward the receiving UA. Once an MTA passes the message to the recipient's Internet mail server, the receiving UA can access the message.

RFC 822 defines the standard format for email messages, treating an email message as having two parts: an envelope and its contents. The envelope contains information needed to transmit and deliver an email message to its destination. The contents are the message that the sender wants delivered to the recipient.

The envelope contains the email address of the sender, the email address of the receiver, and a delivery mode, which in our case states that the message is to be sent to a recipient's mailbox. We can divide the contents of the message into two parts: a header and a body. The header is a required part of the message format, and the sending UA automatically includes it at the top of the message; the user does not input this information. The receiving UA may reformat the header information or delete it entirely to make the message easier for the recipient to read.

The header contains detailed information about who sent the message, which MTA received the message, and how the message got from the sending point to the receiving point. In addition, the header displays the date of the message, the times at which the different MTAs received the message, and the unique ID of the message.

The body of the message contains the actual text the sender typed and is separated from the header by a "null" line. RFC 822 doesn't define the message body, as it can be anything the user enters as long as it is ASCII text.

A standard email address usually follows this form:

<mailbox ID>@<domain name>

The mailbox ID is the name of an individual mailbox on a local machine. The domain name is the name of a valid domain registered in the domain name service (DNS). The DNS servers are key to Internet email because they allow MTAs to find the machine specified in the recipient's email address.

8.7.2   Simple Mail Transfer Protocol

The transmission of an email message through the Internet relies on the **Simple Mail Transfer Protocol,** which is defined in RFC 821. SMTP specifies the way a UA establishes a connection with an MTA and the way it transmits its email message. MTAs also use SMTP to relay the email from MTA to MTA, until it reaches the appropriate MTA for delivery to the receiving UA.

The interactions that happen between two machines, whether a UA to an MTA or an MTA to another MTA, have similar processes and follow a basic call-and-response procedure. The main difference between a UA-to-MTA transaction and an MTA-to-MTA transaction is that with the latter, the sending MTA must locate a receiving MTA.

To do this, the sending MTA contacts the DNS to look up the domain name specified in the recipient email address. The DNS may return the IP address of the domain name—in which case the sending MTA tries to establish a mail connection to the host at that domain—or the DNS may return a set of mail-relaying records that contain the domain names of intermediate MTAs that can act as relays to the recipient. In this case, the sending MTA tries to establish a mail connection to the first host listed in the mail-relaying record.

When an MTA is sending to an MTA, the sending MTA chooses a receiving MTA, which may be the final destination of the message or an intermediate MTA that will relay the message to another MTA. Next, the sending MTA requests a TCP connection to the receiving MTA. The receiving MTA responds with a server ID and a status report, which indicates whether or not it is available for the mail transaction. If it isn't, the transaction is over; the sending MTA can try again later or attempt another route. If the receiving MTA is free to handle a session, it will accept the TCP connection.

The sending MTA then sends a hello command followed by its domain name information to the receiving MTA, which responds with a greeting. Next, the sending MTA sends a mail from command that identifies the email address from which the message originated, as well as a list of the MTAs that the message has passed through. This information is also known as a return path. If the receiving MTA can accept mail from that address, it responds with an OK reply.

The sending MTA then sends a Rcpt To command, which identifies the email address of the recipient. If the receiving MTA can accept mail for that recipient [it may perform a DNS lookup to verify this, particularly using the Mail Exchange (MX) records], it responds with an OK reply. If not, it rejects that recipient. (An email message may be addressed to more than one recipient, in which case this process is repeated for each recipient address.)

Once the receiving MTA identifies the recipient's address, the sending MTA sends the Data command. The receiving MTA accepts the command by responding with OK. It then considers all succeeding lines of data to be the message text. Once the sending MTA gets an OK reply, it starts sending the message. The sending MTA signals the end of the message by transmitting a line that contains only a period (.).

When the receiving MTA receives the signal for the end of the message, it replies with an OK to signal its acceptance of the message. If, for some reason, the receiving MTA can't process the message, it will signal the sending MTA with a failure code. After the message has been sent to the receiving MTA and the sending MTA gets an OK reply, the sending MTA can either start another message transfer or use the Quit command to end the session.

Once the receiving MTA accepts the message, it reverses its role and becomes a sending MTA, contacting the MTA next in line for the relay of the message. The process stops once the message reaches the Internet mail server that services the recipient specified in the Rcpt To email address.

If at any point along the way an MTA can't deliver the email, it generates an error report, also known as an undeliverable mail notification. The MTA uses MTAs identified in the return path to relay the error report to the original sender.

8.7.3   MIME

Initially, the Internet email system was limited to simple text messages because SMTP, the protocol used to transport mail across the Internet, could carry only 7-bit ASCII text. In the United States, the main ASCII standard used for email is US-ASCII. This version of ASCII offers only a basic set of characters—128 characters in all, each represented as a 7-bit binary number. This ASCII set was designed to cover the English alphabet, including uppercase and lowercase letters and the numbers 0 through 9, as well as some other characters.

If you have used email today, you know that email is not just simple text anymore. You can use rich text with italics, boldface, bullets, and other types of enriched formatting. You can even embed graphics in documents. But because SMTP was developed to handle only basic text messages in a 7-bit format—as laid out by RFC 822, which defines the standard for Internet text messaging—these more sophisticated data formats can't be sent via email. The problem with the 128-character set is that it cannot accommodate rich text. In addition, you cannot use foreign language characters that aren't represented in US-ASCII. Therefore, the 7-bit format required by SMPT prevents users from sending other types of data via email—for example, the 8-bit binary data found in many executable files and in files created by applications such as Microsoft Word. For email to handle diverse data such as text, word processing documents, and images, users could share all sorts of data without having to ship disks or make any actual real-time network connections to copy or download files.

To work around the limitation of SMPT's 7-bit ASCII format, today's email uses **Multipurpose Internet Mail Extensions (MIME).** MIME makes the data appear as standard email messages to the Internet's SMTP servers regardless of the data it contains, although the SMTP didn't have to change the way it handles such data. In other words, the solution didn't require that all Internet mail servers be upgraded to a new version of SMTP. In effect, the transport system remained untouched.

As defined in RFC 2045, MIME provides three main enhancements to standard email. First, with MIME, email can contain text that goes beyond basic US-ASCII, including various keystrokes such as different line and page breaks, foreign language characters, and enriched text. Second, users can attach different types of data to their email, including such files as executables, spreadsheets, audio, and images. And third, users can create a single email message that contains multiple parts, and each part can be in a different data format. For example, you could compose a single email message that consists of a plain text message, an image file, and a binary-based document, such as a Word file.

RFC 2045 defines seven types of email content that MIME can package and pass across the Internet: text, image, audio, video, application, message, and multipart. Each of these data types can come in a few different formats, or subtypes. (MIME also augments types and subtypes with certain parameters, which are specified in RFC 2045. However, these parameters contain too much detail to cover in this overview of MIME.)

Obviously, the text type of email content supports messages carrying text. However, within the text type, MIME also supports the plain subtype, which is usually standard 7-bit ASCII. In addition, MIME supports the rich text subtype, which allows for some simple formatting features, such as page breaks.

The image type supports image files, and its subtypes include GIF (Graphics Interchange Format) and JPG (a compressed image format developed by the Joint Photographic Experts Group). In the words of RFC 2045, the video type supports "time-varying picture images," and for now, MPEG (a compressed video format developed by the Motion Picture Experts Group) is its only subtype. The audio type supports audio data, and its only subtype is basic.

According to the RFC, there is no one ideal audio format in use today. So the developers of MIME tried to define a subtype that would be the lowest common denominator. The basic subtype for audio signifies "single-channel audio encoded using 8-bit ISDN mu-law" at a sampling rate of 8 KHz.

The application type supports two types of data—data that's meant to be processed by an application and data that doesn't fall into any of the other categories. For now, the application type supports the octet-stream subtype, which means the message can carry arbitrary binary data. Also, it supports the PostScript subtype, meaning the message can be sent to print as a PostScript file.

NOTE: If a mail agent receives a message whose content subtype it doesn't recognize, by default it will attempt to pass the message on as an application-type message with a subtype of octet stream (or application/octet stream).

The remaining two content types allow for special handling of an email message. For instance, the message type allows an email to contain an encapsulated message. The external-body subtype allows an email to indicate an external location where the intended body of the message resides. That way, the user can choose whether or not to retrieve the message body. The message type also allows MIME to send a large email message as several small ones (the subtype for this is partial). The receiving MIME-enabled mail agent can then open the smaller email messages and reassemble them into the original long version.

The multipart type allows an email message to contain more than one body of data. The mixed subtype allows users to mix different data formats into one email message. The alternative subtype allows a message to contain different versions of the same data, each version in a different format. MIME mail agents can then select the version that works best with the local computing environment. The digest subtype allows users to send a collection of messages in one email, such as the kind used with Internet mailing lists sent in digest form. Finally, the parallel subtype allows mixed body parts, but the ordering of the body parts is not important.

To package the different data formats into the 7-bit ASCII format, MIME uses five different encoding schemes: 7 bit, 8 bit, binary, quoted-printable, and base 64. But as you'll soon see, only the quoted-printable and base 64 schemes actually encode data.

The 7-bit scheme tells mail agents that the message contents are in plain ASCII. For this reason, no encoding is necessary, as all mail systems should support ASCII. The 8-bit scheme indicates that the contents contain 8-bit characters. It's up to the mail agents to encode this information using their preferred means, if they have any. Because not all mail agents use the same encoding method for 8-bit characters, there's a good chance the 8-bit characters won't appear correctly when the email is opened. For this reason, the 8-bit scheme currently isn't a reliable encoding scheme. The binary scheme, because it is similar to the 8-bit scheme, shares the same problem.

The quoted-printable scheme is used for text that contains a mixture of 7-bit and 8-bit characters. Essentially, it allows 7-bit characters to go unencoded and converts each 8-bit character into a set of three 7-bit characters. As a result, mail servers and mail agents see an email containing only 7-bit characters.

The base 64 scheme is used for data that isn't text, such as data that constitutes an executable file. It works by breaking the data down into sets of 3 octets, each set containing 24 bits. Then, it converts each set of 24 bits into a 4-character sequence. (In other words, every 6 bits of data is represented by a character.) The characters used in the sequencing come from a set of 65 characters, all of which can be found in any version of ASCII. Because it's in ASCII, data encoded by base 64 should be readable by any mail server or mail agent.

Using MIME is simple. If your mail system supports MIME, it automatically chooses the data type and encoding scheme. It then adds MIME headers to the body of the message. These headers tell receiving mail agents that they've received a MIME message and indicate how the mail agents should handle the message.

The main headers are MIME-Version, Content-Type, and Content-Transfer-Encoding. The last two headers refer to the data type found in the message and the scheme used to encode the data, respectively. Some other headers are Content-Description, which lets you type in a description of the message (much like SMTP's Subject header), and Content-ID, which is similar to SMTP's Message-ID.

8.7.4   Retrieving Email

What happens when the email arrives at the recipient's Internet mail server? The user agent can employ different methods to access, or retrieve, its email from the MTA. For instance, the majority of companies rely on proprietary email packages, such as Microsoft Exchange or cc:Mail, to handle email operations on the local network. There are a few different models for this operation and a few different protocols that can handle the task, including proprietary protocols found in commercial email packages and two Internet standards-based protocols: Post Office Protocol 3 (POP3) and the Internet Message Access Protocol 4 (IMAP4).

When an email reaches the designated recipient's mail server, the message is placed in a message store, which is also called a post office. In its most basic form, the message store, located on a server, is usually some type of file system that holds delivered mail for access by users. You can think of the message store as one large directory with subdirectories dedicated to each user. These subdirectories are also known as mailboxes. A more advanced form of message store would allow users to create personal folders to store read and unread messages, and it would allow users to create archives for groups of messages. Other advanced features include the ability to perform keyword searches and to create a hierarchy of folders.

The message store component of an email system is usually located on a different machine than the message transport component, which handles the delivery of incoming messages and the transmission of outgoing ones. The reason for their separation is that the transport component usually handles a high volume of operations. So if both systems were on the same machine, the traffic of inbound and outbound mail would experience slower performance, as would users when they try to retrieve and manipulate mail in their mailboxes.

Once an email message is delivered to the message store, it's ready for retrieval by the recipient. There are three basic models for message access: offline, online, and disconnected.

The offline model is the most basic of the three. In the offline model, a client connects to the mail server and downloads all messages designated for that particular recipient. Once downloaded, the messages are erased from the server. Then the user processes, manipulates, and stores the messages locally at the client machine. The advantages of the offline model are obvious. It requires a minimal amount of server connect time since the client accesses the server only periodically for mail downloads. Also, because the mail processing is performed at the client, the offline model won't eat up a lot of server resources.

And it requires less server storage space because mail is deleted from the server once it has been downloaded by a client.

Of course, the offline model does have drawbacks. For instance, because mail is downloaded to a particular machine, you must use that machine to access processed mail. So if you downloaded email to your desktop computer and then went on the road with your laptop, you wouldn't be able to read those same messages off the mail server from your laptop. And because all processing and storage is performed at the client machine, the client must have enough resources to be up to the task.

With the online model, all email processing and manipulation is performed at the server. In fact, all the messages remain on the server even after users have read them. In some implementations, users can save email to the local client.

As you might expect, the online model requires a constant connection to the server whenever users need to access and work with their email. Consequently, this model's high connect time could mean that more users are sharing bandwidth at any given time. Because the server is burdened with all message processing and storage, you'll need a more power-ful server to handle an online email system than you would with an offline system. In addi-tion, users can't work with their email unless they are online, meaning they must be at a client that can access the mail server even to do the simplest task, such as reread an old message.

However, because all mail is stored and processed at the server, you can access your email from any client that can connect to the server. For the same reason, this model re-quires fewer resources at the client machine. In addition, online-type systems usually of-fer enhanced features such as the ability to create a multitude of personal folders to organize email and the ability to create archives for storing messages.

The disconnected model combines elements of both the offline model and the online model. Using this system, a user connects to a mail server to retrieve and download email. Then, the users can process the mail on the local client. Once the user is finished with the messages, the user once again connects to the mail server and uploads any changes. With this model, the mail server acts as the main repository of the user's email.

The disconnected model's strengths are that users can access email from a variety of clients that have access to the mail server. Users can also process mail offline. This also translates to a shorter server connect time. However, the model does require a sufficient amount of resources on both the server and the client.

The majority of email packages, Microsoft Mail, Microsoft Exchange, and Lotus cc:Mail, follow the online model of email access. These products use their own protocols to retrieve messages from Internet mail servers and store them on local servers. Users can then use the products' client programs to access and work with mail from message stores on the proprietary email servers. The attractiveness of these proprietary packages lies in the added features they offer. For instance, they usually offer advanced mail manipulation such as folders, hierarchical folders, and archives. Some offer the ability to create bulletin board systems for general corporate communications. In addition, some offer message no-tification when a new email arrives for a particular user. Also, they usually offer search-ing abilities, so users can find emails with particular headings, or look up the addresses of other users on the system.

One of the first email access protocols developed was the **Post Office Protocol (POP),** the latest version of which is POP3. In essence, **POP3** follows the offline model of message access, and it revolves around send-and-receive types of operation. However, there have been recent attempts to remodel POP3 so that it has some online capabilities, such as saved-message folders and status flags that display message states. But using POP3 in online mode usually requires the additional presence of some type of remote file system protocol.

NOTE: While earlier versions of POP require SMTP, POP3 can be used with or without SMTP.

IMAP4 is the more advanced of the standards-based message access protocols. It follows the online model of message access, although it does support offline and disconnected modes. IMAP4 offers an array of up-to-date features, including support for the creation and management of remote folders and folder hierarchies, message status flags, new-mail notification, retrieval of individual MIME body parts, and server-based searches to minimize the amount of data that must be transferred over the connection.

8.7.5   Messaging Application Programming Interface

Messaging application programming interface (MAPI) is a standardized set of C functions placed into a code library known as a dynamic link library (DLL). The MAPI library is also available to Visual Basic application writers through a Basic-to-C translation layer. The functions were originally designed by Microsoft, but they have received the support of many third-party vendors. The standard library of messaging functions allows Windows applications developers to take advantage of the Windows messaging subsystem, supported by default with Microsoft Mail or Microsoft Exchange. By writing to the generic MAPI, any Windows application can become "mail-enabled," where an email can be sent from within a Windows application with the document you are working on as an attachment. Since MAPI standardizes the way messages are handled by mail-enabled applications, each such application does not have to include vendor-specific code for each target messaging system.

8.7.6   X.400

X.400 is a set of standards relating to the exchange of electronic messages (messages can be email, fax, voice mail, telex, etc.). It was designed to let you exchange email and files with the confidence that no one besides the sender and the recipient will ever see the message, that delivery is assured, and that proof of delivery is available if desired.

One goal of X.400 is to enable the creation of a global electronic messaging network. Just as you can make a telephone call from almost anywhere in the world to almost anywhere in the world, X.400 hopes to make that a reality for electronic messaging. X.400 only defines application-level protocols. It relies on other standards for the physical transportation of data such as X.25, Frame Relay, or ATM.

The address scheme that X.400 uses is called the originator/recipient address (O/R address). It is similar to a postal address in that it uses a hierarchical format. While a postal address hierarchy is country, zip code, state, city, street, and recipient's name, the O/R address hierarchy consists of countries, communication providers (administrative domain name), companies (private domain name), organizational units, surname, and given name. For example, while you might use Pat.Regan@acme.com for an Internet email address, X.400 would use:

G=Pat; S=Regan; OU=acme; PRMD=acme.com; ADMD=mci; C=us

8.8   FILE AND PRINT SHARING

As mentioned in chapter 1, two common services available on networks are file and print sharing. **File sharing** allows you to access files on another computer without using a floppy disk or other forms of removable media. To ensure that the files are secure, most networks can limit the access to a directory or file and can limit the kind of access (permissions or rights) that a person or a group of people have. For example, if you have full access to your home directory (personal directory on the network to store files), you can list, read, execute, create, change, and delete files in your home directory.

Depending on the contents of a directory or file, you could specify who has access to the directory or file, and you can specify what permissions or rights those people have over the directory or file. For example, you could specify that a certain group of people will not be able to see or execute the files, while giving a second group of people the ability to see or execute the file but not make changes to the files and not delete the files. Lastly, you could give rights to a third group of people so that they can see, execute, and change the files.

Print sharing allows several people to send documents to a centrally located printer in the office. Therefore, not everyone requires his or her own personal laser printer. Much like for files, networks can limit who has access to the printer. For example, if you have two laser printers—a standard laser printer and an expensive high-resolution color laser printer—you can assign everyone access to the standard laser printer while only assigning a handful of people access to the expensive printer.

When a file or printer is shared so that it can be accessed from the network by any user, a mapping or logical connection must be made to the device. This way, the file or printer will appear as a local resource even though it is elsewhere on the network. The mappings or logical connections are often established with login scripts or through profiles. The advantage of login scripts and profiles is that when a user logs off or reboots his or her computer, the mappings and logical connections are reestablished.

If a user cannot access a network file share or printer share, check to see if the problem exists with other people—including you if you can access the shared resource. If no one else is having a problem, either the network resource is not mapped on the user's computer or the user does not have the necessary rights and permissions to access the resource.

If a person is having problems opening an application or document and the person's network drives are not mapped on the workstation, you should have the user log off and log back on to see if the problem remains. If the problem remains, check the login script or profile to make sure it is set up with the necessary commands to establish the mapping or logical connection.

SUMMARY

1. If a computer knows the destination's NetBIOS name, it can send the NetBIOS name to a WINS and the WINS will send back the corresponding IP address of the destination.

2. If a computer knows the Fully Qualified Domain Name (FQDN), it can send that to a DNS service and the DNS service will return the corresponding IP address.

3. A Domain Name System (DNS) is a hierarchical client/server-based distributed database management system that translates Internet domain names such as MICROSOFT.COM to an IP address.

4. DNS clients are called resolvers, and DNS servers are called name servers.

5. The first TCP/IP networks used HOSTS files to translate from domain names (such as MICROSOFT.COM, ACME.COM, or MIT.EDU) to IP addresses.

6. The most popular implementation of the DNS Protocol is the Berkeley Internet Name Domain (BIND), which was developed for UC Berkeley's BSD UNIX operating system.

7. A host name is a name assigned to a specific computer within a domain or subdomain by an administrator to identify the TCP/IP host.

8. A Fully Qualified Domain Name describes the exact position of a host (computer) within the domain hierarchy, and it is considered to be complete.

9. A DNS zone is a portion of the DNS name space whose database records exist and are managed in a particular DNS database file.

10. The primary name server is a name server that stores and maintains the zone file locally. Changes to a zone, such as adding domains or hosts, are done by changing files at the primary name server.

11. A secondary name server gets the data for its zone from another name server, either a primary name server or another secondary name server.

12. The Active Directory integrated zone (used in Windows 2000 and .NET networks) has the zone defined using the Active Directory, not the zone files.

13. DNS servers use a mechanism called round-robin or load sharing to share and distribute loads for network resources.

14. Reverse query is when a resolver knows an IP address and wants to know the host name.

15. A WINS server contains a database of IP addresses and NetBIOS (computer) names that update dynamically.

16. A WINS client can be configured to use one of four NetBIOS name resolution methods. They include B (broadcast) node, P (point-to-point) node, M (mixed) node, and H (hybrid) node.

17. A WINS proxy agent is a WINS-enabled computer configured to act on behalf of other host computers that cannot directly use WINS.

18. A DHCP server maintains a list of IP addresses called a pool. When a user needs an IP address, the server removes the address from the pool and issues it to the user for a limited amount of time.

19. Created for diskless workstations, the Bootstrap Protocol (BOOTP) enables a booting host to configure itself dynamically.

20. DHCP, which stands for Dynamic Host Configuration Protocol, is an extension of BOOTP. It is used to automatically configure a host during bootup on a TCP/IP network and to change settings while the host is attached.

21. A DHCP relay agent is a computer that relays DHCP and BOOTP messages between clients and servers on different subnets. This way, you can have a single DHCP server handle several subnets without the DHCP server being connected directly to those subnets.

22. The Simple Network Management Protocol (SNMP) has become the de facto standard for internetwork management. It allows configuring remote devices, monitoring network performance, detecting network faults, detecting inappropriate access, and auditing network usage.

23. A trap is an unsolicited message sent by an SNMP agent to an SNMP management system when the agent detects that a certain type of event has occurred locally on the managed host.

24. The World Wide Web (WWW) refers to the means of organizing, presenting, and accessing information over the Internet.

25. Web pages are written using the HyperText Markup Language (HTML).

26. HTML is a simple language. It is implemented as special ASCII tags or codes that you embed within your document to give the browser a general idea of how the information should be displayed.

27. A web server is a computer equipped with the server software that uses Internet protocols such as HyperText Transfer Protocol (HTTP) and File Transfer Protocol (FTP) to respond to web client requests on a TCP/IP network via web browsers.

28. The web browser is the client program/software that you run on your local machine to gain access to a web server. It receives the HTML commands, interprets the HTML, and displays the results.

29. HTTP is the protocol that web browsers and web servers use to communicate with each other over the Internet.

30. File Transfer Protocol (FTP) is used on the Internet to send files.

31. Typically when accessing a web page, you will specify the web page location by using a Uniform Resource Locator, or URL.

32. A gopher is a system that predates the World Wide Web for organizing and displaying files on Internet servers.

33. A search engine is a program that searches documents for specified keywords and returns a list of the documents where the keywords were found.

34. For its web server, Microsoft offers the Internet Information Server (IIS), while Linux uses Apache for its web server.

35. Electronic mail (email) is the transmission of messages over communications networks.

36. An email message is treated as having two parts: an envelope and its contents. The envelope contains the email address of the sender, the email address of the receiver, and a delivery mode. The contents are the message that the sender wants delivered to the recipient.

37. The transmission of an email message through the Internet relies on the Simple Mail Transfer Protocol, which specifies the way a UA establishes a connection with an MTA and the way it transmits its email message.

38. MIME (Multipurpose Internet Mail Extensions) makes the data appear as standard email messages to the Internet's SMTP servers regardless of the data it contains.

39. One of the first email access protocols developed was the Post Office Protocol, the latest version of which is POP3.

40. IMAP4 is the more advanced of the standards-based message access protocols. It follows the on-line model of message access, although it does support offline and disconnected modes.

41. X.400 is a set of standards relating to the exchange of electronic messages (messages can be email, fax, voice mail, telex, etc.).

QUESTIONS

1. Pat installs and configures the Internet Information Server (IIS) web server on a Windows 2000 Server. Pat configures IIS for both WWW and FTP. He wants users to access the same IIS server as either WWW.DATA.ACME.COM or FTP.DATA.ACME.COM. Which resource record type must Pat add?
   - a. AFSDB
   - b. CNAME
   - c. MB
   - d. MG
   - e. MINFO

2. Pat's network contains four Windows 2000 Server computers that set up as web servers. Pat wants to enable Windows 98 and Apple Macintosh computers running Internet Explorer to access each of these web servers by using a host name. What should Pat use?
   - a. DHCP
   - b. DNS
   - c. FTP
   - d. WINS

3. Pat has just installed the DNS service on a Linux Server computer. Pat needs to add a resource record for her domain's mail server. Which resource record must Pat add?
   - a. CNAME
   - b. MX
   - c. PTR
   - d. WKS

4. A _____ is a portion of the DNS name space whose database records exist and are managed in a particular DNS database file.
   a. FQDN   b. zone
   c. subdomain   d. host name

5. The top of a DNS name space is referred to as the _____.
   a. root directory
   b. root container
   c. root domain
   d. master domain

6. Which of the following are NOT a second-level domain name? (Choose all that apply.)
   a. COM   b. " "
   c. EDU   d. MIL
   e. WWW   f. FTP

7. _____ names are additional names that an organization can create that are derived from the registered second-level domain name so that they can divide their domain.
   a. Second-level domain
   b. Zone
   c. Subzone
   d. Subdomain

8. What is the maximum length of an FQDN?
   a. 11   b. 63
   c. 128   d. 255

9. In Windows 2000, there are three types of zones. Which is NOT one of these?
   a. Self-replicating
   b. Standard primary
   c. Active Directory integrated
   d. Standard secondary

10. When a client does not know the address of a host, what kind of query will it send to its preferred DNS server to find out the address?
   a. Interactive   b. Iterative
   c. Recursive   d. Broadcast

11. When trying to resolve a host name from its IP address, DNS uses a special domain called _____.
   a. in-addr.arpa
   b. caching-only
   c. PTR
   d. conical

12. From his Windows 98 computer, Pat tries to connect to a Windows 2000 Server using http://support.acme.com. He receives the message: "Internet Explorer cannot open the Internet site http://support.acme.com. A connection with the server could not be established." Pat has no problems when connecting to other remote servers using host names. Pat's computer does not use a HOSTS file for name resolution. Which of the following are the most likely causes of the problem? (Choose two answers.)
   a. The ARP cache on the Windows 98 computer does not have an entry for support.acme.com.
   b. The DNS server has no entry for support.acme.com.
   c. The IP address on the Windows 2000 computer for the default gateway is incorrect.
   d. The Windows 98 computer was provided with an incorrect IP address for support.acme.com from the DNS server.

13. Pat wants to see a list of all NetBIOS names currently cached on his Windows 2000 Server. What command should he use?
   a. ARP
   b. nbtstat
   c. netstat
   d. Network monitor
   e. ping
   f. tracert

14. Pat moves a Windows 2000 Professional computer originally configured at a DHCP client from one subnet to another on a DHCP network. He discovers that the workstation cannot connect to the local Windows 2000 Server. What is the most likely cause of the problem?
   a. The Windows 2000 Professional computer's default gateway address has been manually set.
   b. The Windows 2000 Professional computer's default gateway address is set as a local-level option on the DHCP server.
   c. The Windows 2000 Professional computer's IP address has been manually set.
   d. The workstation's IP address is set as a scope-level option on the DHCP server.

15. You want DHCP to assign IP addresses to all the Windows-based computers on your network. Each Windows 2000 Server is to be assigned that same unique IP address each time that server is booted up. How should you proceed?

   a. Exclude a range of IP addresses that will be assigned to the servers.

   b. For each server, create a separate scope that contains that server's IP address.

   c. For each server, implement a client reservation.

   d. For each server, specify an unlimited lease period.

16. Pat administers several hundred computers on a TCP/IP network with six subnets. Many users have notebook computers that run Windows 98, Windows 2000, and Linux. Pat wants to automatically assign IP addresses to these laptop computers each time they connect. Which service must Pat configure?

   a. DHCP   b. DNS

   c. FTP   d. SNMP

   e. WINS

17. What kind of server would you use to resolve NetBIOS names to IP addresses?

   a. DHCP   b. DNS

   c. FTP   d. SNMP

   e. WINS

18. Which service must be started on a Windows 2000 Professional computer so that trap messages can be forwarded to various host names on a network?

   a. The Alerter service

   b. The EventLog service

   c. The SAP service

   d. The SNMP service

19. Email can include which of the following? (Choose all that apply.)

   a. Text

   b. Pictures

   c. Video clips

   d. Sound clips

   e. Executable files

20. Which protocol is responsible for transmitting the email message through the intranet?

   a. MTA

   b. SNMP

   c. SMTP

   d. DNS

21. When you send an email message to pregan@ acme.com, how does the MTA know which server to send the message to?

   a. It looks up the address using the WINS server.

   b. It looks up the address using the DNS.

   c. It looks in the internal database to find the address.

   d. It sends it to the router, and the router will figure out what to do with it.

22. Which protocol allows the embedding of pictures and video clips into an email message?

   a. SNMP   b. DNS

   c. IP   d. MIME

23. Which protocol is responsible for receiving an email message and forwarding it to the appropriate mailbox?

   a. SNMP   b. MIME

   c. DNS   d. POP

24. What are the two main parts of an email message? (Choose two answers.)

   a. Envelope

   b. Contents

   c. Header

   d. Sending address

   e. Destination address

25. Which service does TFTP use to send and receive data?

   a. User Datagram Protocol (UDP)

   b. Internetwork Packet Exchange (IPX)

   c. Sequenced Packet Exchange (SPX)

   d. Transmission Control Protocol (TCP)

26. Which service does FTP use to send and receive data?

   a. User Datagram Protocol (UDP)

   b. Internetwork Packet Exchange (IPX)

   c. Sequenced Packet Exchange (SPX)

   d. Transmission Control Protocol (TCP)

27. Which well-known port is used by HTTP?

   a. 21   b. 25

   c. 80   d. 110

28. Which port is commonly used by Simple Mail Transfer Protocol (SMTP)?

   a. 21   b. 80

   c. 25   d. 110

29. Which well-known port is used by Post Office Protocol-3 (POP3)?

   a. 21   b. 80

   c. 25   d. 110

30. Which layer of the Open Systems Interconnection (OSI) model relays email messages?

   a. The application layer

   b. The network layer

   c. The transport layer

   d. The data link layer

31. On a UNIX computer, which file stores Internet Protocol (IP) addresses and the Internet names that these IP addresses represent?

a. Hosts   b. Login

c. Alias   d. Group

32. Pat is using a Windows NT Server 4.0 computer to connect to the Internet. Zachary is a webmaster with his own Internet domain named Acme.com. He wants to view material on his friend's web server at www.jr.org. After the DNS server has checked the current domain and the current top-level domain, in which order will the DNS search for the Internet Protocol address for www.jr.org?

a. root, jr.org, .org

b. root, .org, jr.org

c. .org, jr.org, root

d. .org, root, jr.org

33. The local area network at Rosalyn's company uses TCP/IP. Rosalyn is using the Simple Network Management Protocol to collect data about a router that she thinks may have failed. Which TCP/IP data service provides SNMP with connectionless communications?

a. ICMP   b. POP3

c. IMAP   d. UDP

34. Which protocol can configure the IP address, the subnet mask, and the default gateway for a host on a network?

a. ARP   b. DHCP

c. TCP   d. UDP

35. What services can the Dynamic Host Configuration Protocol provide to TCP/IP hosts? (Choose all correct answers.)

a. It can resolve IP addresses to media access control (MAC) addresses.

b. It can resolve TCP/IP host names to IP addresses.

c. It can provide an IP address to a client on a TCP/IP network.

d. It can provide a default gateway address to a host on a TCP/IP network.

36. You are using Microsoft Outlook to send and receive email messages. You set up an email account using the Internet Connection Wizard. Which service does Microsoft Outlook use for outgoing email messages?

a. POP3   b. SMTP

c. IMAP   d. SNMP

37. Which of the following characterizes the application protocol SMTP? (Choose two answers.)

a. Clients use it to upload email messages to email servers.

b. Clients use it to download email messages.

c. It sends email messages from one email server to another.

d. It provides Kerberos authentication.

38. What is TFTP?

a. Telephony Free Transport Protocol

b. Telsys File Transfer Protocol

c. Test File Transfer Protocol

d. Trivial File Transfer Protocol

39. What protocol would a POP-3 client use to send mail?

a. TCP   b. MAPI

c. SNMP   d. SMTP

40. Which TCP port does FTP use?

a. 25   b. 23

c. 110   d. 21

41. What TCP/IP is used to publish web documents?

a. HTML   b. THTP

c. TPHP   d. HTTP

42. What TCP/IP is used to receive mail from a message store?

a. SMTP   b. RPC

c. FTP   d. POP3

43. Which TCP port does POP-3 use?

a. 25   b. 53

c. 110   d. 23

44. Bidirectional file transfer between hosts is dependent on what TCP/IP?

a. IP   b. UDP

c. TCP   d. FTP

45. You have a small local area network. For security reasons you have the network divided into two subnets, each with two servers and ten workstations. A dedicated router connects the two subnets. On one of your servers you install the DHCP service, which will dynamically assign IP addresses and subnet information along with DNS and WINS address information. You find that the workstations on the subnet without the DHCP server are not getting IP addresses. What are two possible solutions to this problem? (Choose two answers.)

a. Install a DHCP relay agent on the subnet with the DHCP server.

b. Disable ICMP on the router.

c. Install a DHCP relay agent on the subnet without the DHCP server.

d. Enable BOOTP broadcasts on your router.

46. You have a large network divided by switches, bridges, routers, and brouters that covers multiple physical locations. Your company has purchased an expensive application that will give you the ability to remotely probe and reconfigure network devices, as well as inventory hardware. What protocol will this application use?

a. HTTP
b. SNMP
c. HTML
d. SMTP

47. You have a multisegmented LAN and wish to provide the ability for all computers to see the entire browse list of all computer names. You are not allowing any broadcasts through your routers. What service do you need on your network?

a. DHCP
b. DNS
c. NAT
d. WINS

48. Over the weekend an ISP has given your LAN access to the Internet. It provided no other services. On Monday morning no one can access the web. Later you find that a few people are getting onto the web because they know the IP address of the site they want to get to. What service did your ISP fail to provide?

a. DHCP
b. WINS
c. HTTP
d. DNS

49. Your company has an intranet spanning three servers. Because the company has never needed to connect to the Internet, no DNS service is running on the network. Due to an IP network ID change on one of the subnets, none of your users are able to access any of the company's intranet links to the IIS on that particular subnet. What is the remedy?

a. Add a new subnet to the network.
b. Remap the intranet links.
c. Reconfigure the HOSTS file on all the IISs.
d. Reconfigure the HOSTS file on all the workstations.

50. Servers on the Internet that receive queries for IP data and then return that data to the requestor are known as?

a. Intranet servers
b. NetBIOS servers
c. IISs
d. DNS servers

51. Which type of proxy handles Internet email?

a. SMTP
b. IP
c. NAT
d. HTTP

52. A user complains about not being able to open an application. You notice that one of the usual network drives is not mapped on the workstation. What is your next step in troubleshooting the problem?

a. Manually remap the network drive.
b. Have the user log off and then log back on.
c. Disconnect and reconnect the network cable from the workstation's NIC.
d. Check the user rights on the server to that application.

53. One of your users complains that he cannot save a particular file to a network file share. What is the first thing you would do, assuming you can see the file share from the user's computer?

a. Try saving the same file to the share.
b. Check the file size.
c. Check the user's rights to the share.
d. Try saving a different file to the share.

54. One of your local users cannot access a file resource that's on a server in one of your company's offices in another state. What should you check first?

a. Whether the user's network cable is bad
b. Whether the user has rights to the resource
c. Whether your router's WAN port is down
d. Whether the user has access to any resources on the LAN

55. You work for a small, but rapidly growing, corporation. You are currently using a 10Base2 Ethernet LAN, which uses the NetBIOS Enhanced User Interface (NetBEUI) network communications protocol. The owner of your company wants you to improve the speed of the LAN, connect the LAN to the Internet, optimize Internet access speed, and protect the LAN from Internet hackers. You need to implement a strategy for upgrading the LAN that meets the following goals:

- Increase the overall speed of the LAN.
- Connect the LAN to the Internet.
- Optimize Internet access speed.
- Use port filtering to protect the LAN from infiltration by hackers from the Internet.

You plan to take the following actions:

- Place a router on the network and configure it as the default gateway.
- Replace the coaxial network cable with Category 5 unshielded twisted-pair (CAT 5 UTP) cable and install 100BaseTX network interface cards in the network nodes.
- Configure the network to use the TCP/IP suite.
- Place IP proxy software on one of the network servers.

Which of the objectives does your plan meet? (Choose all that apply.)

a. The overall speed of the LAN is increased.
b. The LAN is successfully connected to the Internet.
c. Internet access speed is optimized.
d. The LAN uses port filtering for protection from Internet hackers.

Protocol Analyzers

Topics Covered in this Chapter

9.1 PROTOCOL ANALYZERS

Introduction

One of the most valuable tools available to a person trying to troubleshoot a network problem is the protocol analyzer. The protocol analyzer gives you the opportunity to look at the individual packets that are being sent on your network. Of course, it takes an experienced person to make sense of all the networks.

Objectives

- Use a protocol analyzer to capture a packet.

- Explain how to use a protocol analyzer to troubleshoot a network problem.

9.1   PROTOCOL ANALYZERS

Sometimes when you have to troubleshoot a network problem, you need to take a good look at what is being sent through your network to determine the cause. This is where **protocol analyzers** come in. A protocol analyzer, also known as a network analyzer, is software or a hardware/software device that allows you to capture or receive all the packets on your media, store them in a trace buffer, and then show a breakdown of each of the packets by protocol in the order they appeared. Therefore, it can help you analyze all levels of the OSI model to determine the cause of the problem. Network analysis is the art of listening in on a network's communications to examine how devices communicate and determine the health of that network.

The operation of a protocol analyzer is actually quite simple:

- It receives a copy of every packet on a piece of wire by operating in a promiscuous capture mode (a mode that captures all packets on the wire, not just broadcast packets and packets addressed to the analyzer's adapter).
- It time-stamps the packets.
- It filters out the stuff you're not interested in.
- It shows a breakdown of the various layers of protocols, bit by bit.

These packet traces can be saved and retrieved for further analysis. Once a packet is captured from the wire, the analyzer breaks down the headers and describes each bit of every header in detail.

While it is easy to capture the packets and time-stamp them, often people have a tendency to capture every packet on a segment when trying to troubleshoot a specific problem. The only problem with doing that is the large amount of traffic being sent through a wire. Therefore, when you analyze the packets, you can easily become overwhelmed. To get around this, set up filters and do some basic statistical analysis. This lets you isolate the problem with more ease and less time.

For example, if the problem only affects users communicating with a certain server, put an analyzer on that segment and put a filter on traffic to and from the server. If the problem only affects users who go through a router, start by putting an analyzer on one of the routed segments and putting a filter on traffic to and from the router's MAC addresses. If you are troubleshooting a slow login process, you should start by analyzing from the client's segment and putting a filter on the client's MAC address.

A protocol analyzer can be used for more than just troubleshooting. You can use it to monitor your network performance and head off problems before they occur, and you can use it to determine when your network is becoming too congested. This way, you can update your network hardware or subdivide your network before your network performance becomes a problem. In addition, you can find out what is using the bandwidth of your network so that you can optimize or tweak your network.

As well, you can baseline the throughput of a particular application. Therefore, you can determine how much traffic the application is causing. This is particularly important if you can actually test the application before purchasing it to see if your current network can handle the additional generated traffic. This would include looking at how much traffic is generated when you log onto the new application or server, querying and updating a database, and handling the transfer of files.

Of course, if you analyze your network when everything is working properly, it will be easier to identify packet anomalies, and you can compare the difference in performance when you update or replace drivers, network interfaces, platform upgrades, and so forth. This will even give you concrete numbers that you can use in reports to management so that you can justify equipment costs and upgrades.

A good analyzer should have some alerts/alarms that notify you of unusual or faulty traffic patterns. Some useful alarms should include:

- **Utilization percentage**—Shown as a percentage of the bandwidth that is used up. On Ethernet networks especially, the performance degrades significantly when the utilization gets up above 40 percent. Watch the collision/fragment error count in relationship to utilization.
- **Packets per second**—Shown in number of packets seen per second on the network. This number can give you an idea of how many packets an interconnecting device (such as a router or switch) will need to process per second.
- **Broadcasts per second**—This number tells you how much broadcast (processed by all devices regardless of their operating system or protocol) is on your network. Excessive broadcast will slow your network dramatically.
- **Server/router down**—When the server and/or router goes down.
- **MAC-layer errors**—Shown in per-second increments. These MAC-layer errors are defined as layers 1 and 2 errors that corrupt packet formats or make access to the network impossible.

The analyzer should also be able to build trend graphs to illustrate the current and long-term traffic patterns (such as utilization and packets per second). In order to make the communications information useful to you, the analyzer decodes, or interprets, the actual packet information received.

9.1.1   **Placement of Protocol Analyzers**

In order for these analyzers to capture all the traffic, you must have a network interface card and driver that supports promiscuous-mode operations. A promiscuous-mode card is able to capture error packets and packets that are not addressed to the local card. Of course, broadcast and multicast traffic should also be visible to the analyzer.

Most current analyzers require NDIS version 3.0+ drivers to support promiscuous-mode operations. In some cases, the analyzer manufacturer offers a specialized driver to enhance the card and provide error reporting. This does not mean that all cards can run in promiscuous mode. Some cards use special functionality to enhance performance at the expense of promiscuous-mode operations. For example, some cards support promiscuous-mode operation, but they will not report bad packets so that it will increase network performance. Since the card does not process the bad packets, it can perform other functions so that it will increase performance.

To analyze multiple LAN and/or WAN segments simultaneously, you have to figure where you want to place your network analyzers. Certainly, you don't necessarily need an analyzer on every network segment. Then again, having multiple analyzers has its advantages. The devices that connect your network and the layout of the network will affect where you need to place the routers:

- Since a hub is a multiport repeater, the traffic generated or received by one computer can be seen by all computers connected to the hub. Therefore, if you have a hub, you can connect your protocol analyzer to any port on the hub.
- Since a bridge isolates and localizes traffic, you should consider placing protocol analyzers on both sides of the bridge.
- If you have a switch (a switch is a multiport bridge that isolates and localizes traffic), you will have to use a hub to establish a connection between the switch port and the PC, or the host that you want to analyze and the protocol analyzer. This lets you see all the traffic being sent or received by the PC or host. You can also accomplish this by using analyzer agents or configuring your switch for port spanning or mirroring (if your switch supports it). Port spanning or mirroring configures the switch to send a copy of any port's traffic to another port.
- Since routers isolate and direct traffic based on the network address, if you place an analyzer on one side of the router, you would only see traffic that is destined to that network. Therefore, you should consider placing an analyzer on each side of the router or loading analyzer agents on the router.

   NOTE: The analyzer agent should be a multisegment agent that can capture packets from both connected networks.

- To test WAN links, you can connect analyzers or agents to both sides of the WAN links, and you can also connect analyzers or agents to the IP router and/or CSU. Of course, you may need assistance from your WAN link provider on how to make the physical connection to your WAN link.

Analyzer agents are typically configuration options or software programs that are loaded on switches or PCs to enable them to capture traffic from the wire and send the data to a management console. This way, you can set up the agents and tap into them as needed without actually being there, and it allows you to analyze all the traffic from a central location.

9.1.2   Using Microsoft's Network Monitor

In this chapter, the protocol analyzer we will discuss is Microsoft's Network Monitor, which is included with Windows NT, 2000, and Windows Server 2003 Servers. Also included is a Windows Network Monitor driver, which is needed to pass the packets to the Network Monitor program and can act as an analyzer agent that you can load on other machines to direct the packets to your computer (as long as it is running Windows Network Monitor).

NOTE: The version that comes with Microsoft Systems Management Server has additional features than the one that comes with Microsoft servers.

While Microsoft Network Monitor is not the best protocol analyzer available, it is widely available for no extra cost. Other protocol analyzers include Novell analyzer (http://www.novell.com) and Wild Packet Etherpeek (http://wildpackets.com).

To install Network Monitor, you start the Add/Remove applet in the control panel and select Add/Remove Windows Components. In the Windows Components Wizard, highlight Management and Monitoring Tools and then click Details. In the Management and Monitoring Tools windows, select the Network Monitor Tools check box and click the OK button. Click the Next button in the Windows Components Wizard to continue. You may be prompted for additional files, in which case you will have to insert your Windows 2000 Server disk or type a path to the location of the files on the network. Lastly, click the Finish button.

To install the Network Monitor on a Windows 2000, Windows Server 2003, or Windows XP computer, double-click the Network and Dial-Up Connections (or Network Connections) Control Panel applet. Right-click the local area connection that you want to monitor and then click Properties. In the Local Area Connection Properties dialog box, click Install. In the Select Network Component Type dialog box, click Protocol and then click Add. In the Select Network Protocol dialog box, click Network Monitor Driver and then click OK. Again, you may be prompted for files.

To capture network frames, click the Start button, select Programs, select Administrative Tools, and then choose Network Monitor. If you are prompted for a default network on which to capture frames, select the local network from which you want to capture data by default. On the Capture menu, click Start.

After you have captured data, you can view it in the Network Monitor user interface. Network Monitor performs some data analysis automatically because it translates raw capture data and organizes it into the structure of a logical frame. Network Monitor also displays overall network segment statistics, including:

- Broadcast frames
- Multicast frames
- Network utilization
- Total bytes received per second
- Total frames received per second

See figure 9.1.

Figure 9.1   Network Monitor Showing the Network Statistics Taken While Capturing Packets

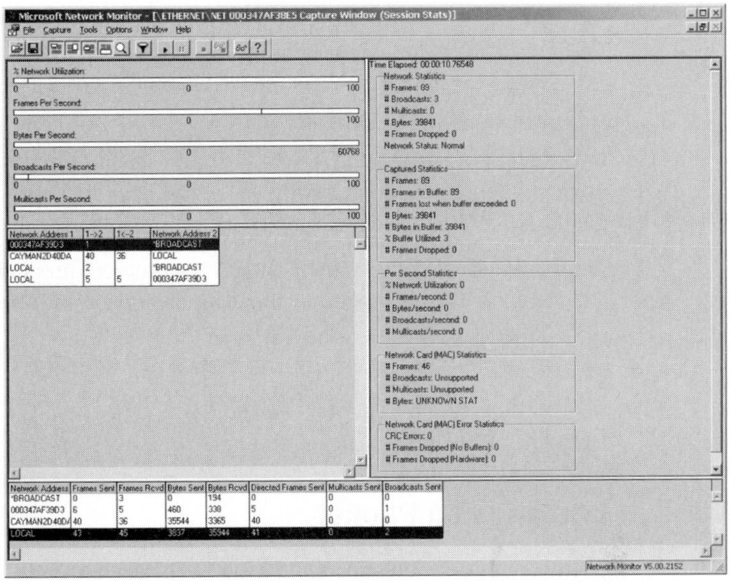

Network Monitor acts as a Network Driver Interface Specification (NDIS)-compliant driver to copy frames to the capture buffer, a resizable storage area in memory. The default size is 1 MB. Because Network Monitor uses the local-only mode of NDIS instead of promiscuous mode, you can use Network Monitor even if your network adapter does not support promiscuous mode. For security reasons, Windows 2000 Network Monitor captures only those frames, including broadcast and multicast frames, sent to or from the local computer that has the Network Monitor driver.

To capture only those frames that originate with specific computers, determine the address of the computer on your network and associate the address with its Domain Name System (DNS) or NetBIOS name. After the associations are made, you can save the name to an address database (.adr) file that can be used to design capture filters and display filters. A capture filter allows you to specify criteria for inclusion in or exclusion from the capture. To show the Capture Filter dialog box, open the Capture menu and select Capture Filter or press F8 in the Capture window. When you design a capture filter, you can limit a capture to frames consisting of specific data types, capture frames sent using a specific protocol, or use a capture trigger to initiate actions following the capture.

After capturing data, you might want to save it. For example, it is useful to save captures before starting another capture (to prevent loss of the captured data) if you think you might need to analyze the data later or if you need to document network use or problems. When you save captured data, the data in the capture buffer is written to a capture (.cap) file.

Similar to a capture filter, you can use a display filter like a database query to specify which frames to display. Because a display filter operates on data that has already been captured, it does not affect the contents of the Network Monitor capture buffer. A frame can be filtered based on the following data:

■ The frame's data-link-layer or network-layer source or destination address
■ The protocols used to send the frame or packet
■ The properties and values the frame contains

To design a display filter, specify decision statements in the Display Filter dialog box. Information in the Display Filter dialog box is in the form of a decision tree, which is a graphical representation of a filter's logic. When you modify display filter specifications, the decision tree reflects these modifications. You can specify:

■ **Protocol**—Specifies the protocols or protocol properties
■ **Address filter**—Specifies the computer addresses on which you want to capture data (default is ANY < – –> ANY)
■ **Property**—Specifies property instances that match your display criteria

With display filters, you can use AND, OR, and NOT logic, and unlike a capture filter, you can use more than four address filter expressions.

9.1.3 Looking at a Packet

If you recall from chapter 2, the data is encapsulated at the transport layer, which is encapsulated at the network layer, which is encapsulated at the data link layer and then converted to bits and sent over some form of media. Of course, at the destination device, the packet is received and processed by the network interface card, and the data is stripped out as you go through the data link layer, the network layer, and the transport layer. The protocol analyzer will typically show the data and the headers that were added by each layer.

Every frame that is captured by a protocol analyzer will have up to four parts, as discussed in chapter 2. Figure 9.2 shows a capture for a computer trying to ping another computer. As you can see by looking at the packet, the fifth frame is 74 bytes in size. It is an IP packet that is sent on an Ethernet network.

If you recall back in chapter 4, the Ethernet (data link layer) header includes the destination and source MAC addresses. The destination MAC address is 000347AF39D3, and the source MAC address is 000347AF38E5. The Ethernet header also indicates that it is carrying an IP packet.

The IP header shows the packet as an IPv4 packet, which contains an ICMP packet. Its source IP address is 192.168.1.100, and its destination IP address is 192.168.1.101. Of course, ARP was used earlier to determine the MAC addresses for the two IP addresses.

The ICMP packet is an echo packet from 192.168.1.100 to 192.168.1.101. When the packet is received and processed by 192.168.1.101, the host assigned to the 192.168.1.101 address will, it is hoped, return an echo reply showing that the two computers have a connection.

9.1.4 Troubleshooting Ethernet Problems

On an Ethernet network, the following types of problems can occur:

■ Collisions
■ CRC errors

Figure 9.2   Looking at a Capture Packet

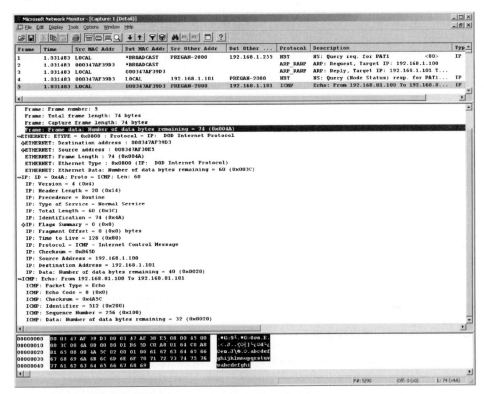

- Short frames
- Long frames
- Jabber

Collisions

The preamble of an Ethernet packet is an alternating series of 1s or 0s ending with two 1s, so it looks like AAAAAAAAAAAA in hex, because hex A is 1010 in binary. Normally an analyzer doesn't capture the preamble unless two preambles collide and the analyzer starts receiving a packet prematurely.

As you learned earlier, collisions occur when two or more packets run into each other. The result of a collision is a fragment. It should also be noted that Ethernet collisions often show up as CRC error packets whenever the collision extends beyond the preamble of the Ethernet frame. If your protocol analyzer is capable of capturing and displaying packets with CRC errors, you can do some basic collision analysis. By capturing all the packets on a segment with a high rate of collisions, you can see if any of these preamble collisions makes it into the capture buffer. A small percentage usually do, which is adequate to perform an analysis.

Because two preambles collided, the fragment in the analyzer's capture buffer may not be a perfect AAAA pattern. It may also be a string of hex 5555. . . because hex 5 is 0101 in binary, or it may be a combination of As and 5s. Regardless, these fragments are very easy to spot. Some analyzers also have a predefined display filter to show only the error packets. Of course, you should not set your capture filter to capture only error packets, because valuable information immediately follows a collision.

In addition, when a collision occurs, a station will attempt to resend the frame very quickly, usually within 100 microseconds of a collision. This means that two or more packets are guaranteed to appear immediately following the fragment in the buffer.

By looking at the source address of the next two or three frames following the collision, you can be fairly sure that at least two of the addresses identify stations that were involved in the collision. By analyzing several samples, a pattern may emerge where one station's address appears much more frequently than others. If not, the collisions are probably truly random.

If one address sticks out more than others, you may want to increase the traffic load on that device. If it is a router, do a file transfer using FTP across that router and monitor the collision rates at the same time. If it's a workstation, drag and drop a file from the workstation to a server. If the collision rates go way up, there may be something wrong with that workstation's NIC, or there may be something wrong with how it's wired into the network, including whether it exceeds maximum cable lengths or violates the 5-4-3 rule.

CRC Errors

Before the Ethernet packet is placed out on the wire, the network interface card runs an algorithm on the contents of the frame. The result of this equation is placed in the Frame Check Sequence filed at the end of the frame. Upon receipt of a frame, a recipient network interface card performs a CRC calculation on the packet. It compares its result with the value at the end of the incoming packet. If the two numbers match, the packet is considered valid. If not, the packet is considered corrupt. If CRC errors are recorded in a protocol analysis session above the 2 to 3 percent level of overall traffic on a network, they are considered excessive.

Lots of CRC errors from a single source address indicate a possible problem with the NIC or the cable to the hub/switch port. Lots of CRC errors from various sources indicate cabling problems.

Short and Long Frames

The IEEE 802.3 specification dictates a minimum packet size of 64 bytes and a maximum packet size of 1518 bytes. A short frame (also known as a runt) is less than 64 bytes with a valid CRC, while a long frame (also known as a giant) is greater than 1518 bytes. These frames are typically formed by faulty or out-of-specification LAN drivers.

You should identify the addresses attempting to communicate in the trace that are right before and after the long or short packets. Next, one by one, remove the suspect nodes and reanalyze the network until the problem subsides. After the node area is located, troubleshoot the NIC, transceiver, and NIC driver.

Jabber

At times, Ethernet network interface cards and external transceivers can generate a problem called jabbering. This is when garbled bits of data are emitted within the frame sequence in a continuous transmission fashion. The packet length is usually more than 1518 bytes. Jabbering is often identified by a protocol analyzer as a CRC error. When nodes detect collisions, they emit a normal JAM signal on the network segment to clear transmission. Sometimes certain nodes attempt to keep jamming the network due to excessive high collision rates; this problem also can be captured as high CRC or late collision error rates. The cause can be overloaded traffic levels. If the bandwidth-utilization levels are normal or low for the particular Ethernet segment, it is possible that the collision detection pair of a jamming node's NIC or transceiver cannot hear the network signal and may not know a collision has stopped. If this occurs, the node continues to jam the network. If a certain node on an Ethernet segment emits a lot of jabber, the node's NIC and transceiver should be troubleshot through consecutive replacement and reanalysis.

9.1.5   Broadcasts

Broadcasts are used to reach all devices on a network or subnetwork. There are two types of broadcasts: an all-nets broadcast and a subnet broadcast.

All-nets broadcast packets are addressed to 255.255.255.255 in the IP header. Literally, the packets are addressed to all networks. Subnet broadcasts contain the subnet address in the broadcast packet. For example, the subnet broadcast designation on network 10.0.0.0/8 would be 10.255.255.255. On a network 200.23.34.0/24, the subnet broadcast would be 200.23.34.255.

Of course, while broadcasts are not forwarded by most routers, broadcasts can take up valuable bandwidth and processing power in the receiving devices. To view the number of broadcasts, you should list the packets by the most popular destination addresses and look for the broadcast addresses. Unsolicited broadcasts (announcements) should not account for more than 10 percent of all traffic during hours that most data exchange is occurring. During hours of low or no traffic, the number of broadcasts is probably higher because the network is mostly idle except for periodic broadcasts and multicasts. If you think you have a high number of broadcasts, you then need to figure out what type of broadcasts they are, including if they are routing broadcasts or unanswered ARP packets.

9.1.6   Troubleshooting TCP/IP Problems

By understanding how a workstation creates and transmits a packet, you can identify the step a workstation is performing when an error occurs. For example, the workstation may need to send a DNS query or an ARP broadcast. Knowing when the workstation may need to send a packet and using a network analyzer to track what the workstation actually does is an extremely effective troubleshooting method: You can determine what steps a workstation has completed successfully and identify a relationship between the information contained in the packets sent and the processes that occurred at the workstation.

For example, suppose a workstation sent ARP broadcasts requesting a local device's hardware address, and suppose the requests were unanswered. In this case, you could assume that the workstation had resolved the application port number, obtained the destination IP address, and determined the route (local) to the destination. If you were troubleshooting this problem, you could investigate several possible causes:

■ The workstation might not have performed the route resolution process properly, and the destination device might actually be on another network.
■ The workstation might have received incorrect DNS information.
■ The destination device is not available.

Therefore, to analyze TCP/IP, you need to know what to expect when everything is working properly. Then you can analyze the packet properly. Of course, many of these transactions will have some things in common.

As a review from chapter 4, TCP/IP uses the following resolution process when it needs to communicate with a computer:

1. Resolve the application to the port number.
2. Resolve the host name to the IP address (host file or DNS server). If there is no answer, go to the secondary DNS server. If there is no response from the secondary DNS server, you cannot finish communication.
3. Determine if the destination address indicates that the destination device is local or remote. Upon receipt of the IP address, use the network mask to determine the network portion of your address. In addition, the network mask determines the network portion of your address to the desired target.
   a. If the destination address is local (the network addresses are the same), resolve the MAC address of the local target by first checking the ARP cache for the information. If it does not exist, send an ARP broadcast looking for the target's hardware address.
   b. If the destination address is remote, perform route resolution to identify the appropriate "next-hop" router. Look in the local routing tables to determine if you have a host or network entry for the target. If neither entry is available, check for a default gateway entry. Then resolve the MAC address of the next-hop router. Check your ARP cache first. If the information does not exist in the cache, send an ARP broadcast out to get the information.

If a protocol utilizes TCP to establish a logical connection between devices, the procedure uses a three-way handshake. The device that initiates the handshake process is called device 1, and the destination device, or the target of the connection, is called device 2.

1. Device 1 sends its TCP sequence number and maximum segment size to device 2 (device 1 SYN to device 2).
2. Device 2 responds by sending its synchronize sequence number and maximum segment size to device 1 (device 1 SYN ACK to device 2).
3. Device 1 acknowledges receipt of the sequence number and segment size information (ACK).

By looking inside the handshake packets, you can see what a TCP header actually contains during the handshake process and understand how to troubleshoot this process.

For example, let's look at a DHCP client trying to get an address from a DHCP server. DHCP clients use the following basic start-up routine:

1. DHCP discover
2. DHCP offer
3. DHCP request
4. DHCP ACK

Look at your DHCP communications to identify that pattern. What if the discovery process fails? What if more than one address is offered (more than one DHCP server)? Did the client NACK (negative acknowledgment to not accept the address) the second address? What did the client ask for in the request? Did the server reply with all required information? What if the DHCP release gets literally smashed to bits along the way? There is no ACK required for those packets, so the DHCP client just assumes that if it sent the packet, it got there. The DHCP server won't release this address until the lease time has expired.

9.1.7   Using ICMP Packets to Troubleshoot

The TCP/IP suite includes the Internet Control Message Protocol (ICMP), a message protocol that can help you identify network problems such as incorrect gateway settings, unavailable applications or processes, and fragmentation problems. If your job responsibilities include troubleshooting or testing TCP/IP networks, you should know ICMP.

ICMP messages provide feedback on communication problems such as the following:

■ A client has been configured with the wrong IP address for its DNS server. The destination device sends an ICMP message, indicating that this device does not support the DNS port (port 53).

■ An application does not permit fragmentation of its communications, but fragmentation is required to communicate with the destination device. The router that would normally fragment the packet sends the source device an ICMP message, indicating that the packet could not be forwarded because the packet's "don't fragment" bit was set.

■ A client sends all communications to a default router although another router offers the best route. The default router sends an ICMP message that includes the IP address of the router that provides the best route.

■ A packet arrives at a router with a time to live (TTL) value of 1. The IP TTL value decrements as the IP packet is forwarded through each router. If an IP packet has a TTL value of 1, the router cannot decrement the TTL value by 1 and then forward the packet. Instead, the router discards the packet and sends an ICMP message, indicating that the packet's TTL expired in transit.

In addition, certain utilities use ICMP for testing and diagnostics. Chapter 6 has already discussed the ping and tracert commands.

ICMP packets are a little different from other packets. For example, ICMP packets do not rely on User Datagram Protocol (UDP) or Transmission Control Protocol (TCP). Instead, ICMP packets sit directly on the IP header. The structure of the ICMP packet depends on the type of information being exchanged and the function of the packet. (See table 9.1.)

Before you can use ICMP to troubleshoot your company's network, you must capture the ICMP traffic on that network. You can set up a network analyzer to capture all TCP/IP

Table 9.1   Troubleshooting with ICMP Packet Structure

ICMP Type	Message	Description
3 (destination unreachable codes)	Network unreachable	The sending device knows about the network but believes it is not available at this time. Perhaps the network is too far away through the known route.
	Host unreachable	The sending device knows about host but doesn't get an ARP reply, indicating the host is not available at this time.
	Protocol unreachable	The protocol defined in the IP header cannot be forwarded.
	Port unreachable	The sending device does not support the port number you are trying to reach.
	Fragmentation needed but Don't fragment bit was set	The router needs to fragment the packet to forward it across a link that supports a smaller maximum transmission unit (MTU) size. However, application set the don't fragment bit.
	Source route failed	The ICMP sender can't use the strict or loose source routing path specified in the original packet.
	Destination network unknown	The ICMP sender does not have a route entry for the destination network, indicating this network may never have been available.
	Destination host unknown	The ICMP sender does not have a host entry, indicating the host may never have been available on the connected network.
	Source host isolated	The ICMP sender (router) has been configured to not forward packets from the source (the old electronic pink slip).
	Communication with destination network is administratively prohibited	The ICMP sender (router) has been configured to block access to the desired destination network.
	Communication with destination host is administratively prohibited	The ICMP sender (router) has been configured to block access to the desired destination host.
	Destination network unreachable for type of service	The sender is using a type of service (TOS) that is not available through this router for that specific network.
	Destination host unreachable for type of service	The sender is using a TOS that is not available through this router for that specific host.

Continued

Table 9.1 Continued

ICMP Type	Message	Description
3 (destination unreachable codes)	Communication administratively prohibited	The ICMP sender is not available for communications at this time.
	Host precedence violation	The precedence value defined in the sender's original IP header is not allowed (for example, using flash override precedence).
5 (redirect codes)	Redirect datagram for the network (or subnet)	The ICMP sender (router) is not the best way to get to the desired network. The reply contains the IP address of the best router to the destination. Dynamically adds a network entry in the original sender's route tables.
	Redirect datagram for the host	The ICMP sender (router) is not the best way to get to the desired host. The reply contains the IP address of the best router to the destination. Dynamically adds a host entry in the original sender's route tables.
	Redirect datagram for the type of service and network	The ICMP sender (router) does not offer a path to the destination network using the TOS requested. Dynamically adds a network entry in the original sender's route tables.
	Redirect datagram for the type of service and host	The ICMP sender (router) does not offer a path to the destination host using the TOS requested. Dynamically adds a host entry in the original sender's route tables.
6 (alternate host address codes)	Alternate address for host	The reply indicates that another host address should be used for the desired service. Should redirect the application to another host.
11 (time exceeded codes)	Time to live exceeded in transit	The ICMP sender (router) indicates that originator's packet arrived with a TTL of 1. Routers cannot decrement the TTL value to 0 and forward the packet.
	Fragment reassembly time exceeded	The ICMP sender (destination host) did not receive all fragment parts before the expiration (in seconds of holding time) of the TTL value of the first fragment received.
12 (parameter problem codes)	Pointer indicates the error	The error is defined in greater detail within the ICMP packet.
	Missing a required option	The ICMP sender expected some additional information in the Option field of the original packet.
	Bad length	The original packet structure had an invalid length.

traffic and filter just the ICMP traffic (postfiltering), or you can set up a prefilter to capture just ICMP traffic (if the network analyzer you are using provides prefiltering capabilities).

After setting up the network analyzer to filter ICMP traffic, you should take a good look at the ICMP traffic that crosses the network. How many ICMP redirect messages do you see? It is typical to have some redirect messages (especially during start-up hours in the morning), but if one device is constantly being redirected before communicating with other devices on the network, you may need to assign that device a different default gateway.

"Network unreachable" and "host unreachable" messages may indicate a route or routing failure. For example, if a router cannot forward a packet addressed to a certain device or network because that device or network is considered down, the router will send a "network unreachable" or "host unreachable" message to the source device. This problem could be caused by a faulty IP stack on the destination device or by routing failures that have made a network unreachable.

"Port unreachable" messages, on the other hand, may indicate that a device is configured incorrectly. For example, if a device continually sends DNS queries to a specific IP address and receives "port unreachable" messages, the IP address for the DNS server may not be valid.

Although ICMP messages are invaluable for troubleshooting TCP/IP networks, you should be aware that hackers find ICMP messages equally useful. For example, excessive "port unreachable" messages may be the first sign that a hacker is trying to discover what network services are running on a network. Port scanning utilities often use the simplistic approach of sending packets to a device and incrementing the destination port number by 1 in each packet. "Port unreachable" messages help determine which ports are not active, thereby identifying the ports or processes that are available on a system. Because hackers sometimes use "port unreachable" messages in this way, you should carefully examine these types of messages on your company's network. You should also examine the "echo request" and "echo reply" messages being transmitted on the network. Hackers sometimes use echo requests to "discover" IP addresses of live devices on the network. If echo requests are being used in this way, the destination IP address is typically incremented by 1 in each message. For example, you will see an echo request sent to 10.0.0.1, an echo request sent to 10.0.0.2, an echo request sent to 10.0.0.3, and so on.

These types of requests may also be sent by a management product that is building a map of your company's network (and, therefore, has a legitimate reason for discovering devices). However, if an unknown or suspect device is performing this type of discovery, it can be the first sign that a hacker is attempting to get information about your company's network.

In addition, hackers use ICMP messages to cripple network devices. For example, if you find an excessive number of ICMP echo packets on a network, you may have cause for concern. An excessive number of ICMP echo packets may indicate a denial-of-service attack. A denial-of-service attack focuses on overloading or crippling a device to the point that the device cannot provide services to other devices.

Because hackers can use ICMP messages to gain information about a network or to actually harm a network, many companies restrict devices from transmitting specific ICMP messages across their connection to the Internet. If your company's security policy doesn't cover ICMP messages, you may want to revise it to include such a restriction.

9.1.8   Analyzing Security Issues

Protocol analyers can also be used to help detect security breaches. Denial-of-service (DoS) attacks are far too common these days. A DoS attack is based on the theory that if you hog all the CPU of a device, that device must deny services to anyone else.

There are two primary types of DoS attacks:

■ Brute force attacks
■ Stealthy attacks

Brute force attacks are typically easy to spot with an analyzer. We'd think of a brute force attack if a device were sending out 10,000 pings in rapid succession to another device.

Stealthy attacks typically only require one or two packets to kill the target. For example, a TCP SYN packet that uses the target's IP address in both the source and destination address field of the IP header can kill some implementations of IP.

Lastly, protocol analyzers can be used to capture packets that include passwords. This is particularly dangerous when passwords are sent in clear text in which the password can be read without being cracked.

SUMMARY

1. A protocol analyzer, also known as a network analyzer, is software or a hardware/software device that allows you to capture or receive all the packets on your media, store them in a trace buffer, and then show a breakdown of each of the packets by protocol in the order they appeared.

2. By setting up filters and doing some basic statistical analysis, you can isolate a specific problem with more ease and less time.

3. You can use a protocol analyzer to monitor your network performance and head off problems before they occur, and you can use it to determine when your network is becoming too congested.

4. You can also use a protocol analyzer to baseline the throughput of a particular application. Therefore, you can determine how much traffic the application is causing. This is particularly important if you can actually test the application before purchasing it to see if your current network can handle the additional generated traffic.

5. If you analyze your network when everything is working properly, it will be easier to identify packet anomalies, and you can compare the difference in performance when you update or replace drivers, network interfaces, platform upgrades, and so forth. This will even give you concrete numbers that you can use in reports to management so that you can justify equipment costs and upgrades.

6. In order for these analyzers to capture all the traffic, you must have a network interface card and driver that supports promiscuous-mode operations.

7. To analyze multiple LAN and/or WAN segments simultaneously, you have to figure where you want to place your network analyzers.

8. Analyzer agents are typically configuration options or software programs that are loaded on switches or PCs to enable them to capture traffic from the wire and send the data to a management console. This way, you can set up the agents and tap into them as needed without actually being there, and it allows you to analyze all the traffic from a central location.

9. Collisions occur when two or more packets run into each other. The result of a collision is a fragment.

10. It should be noted that Ethernet collisions often show up as CRC error packets whenever the collision extends beyond the preamble of the Ethernet frame.

11. If CRC errors are recorded in a protocol analysis session above the 2 to 3 percent level of overall traffic on a network, they are considered excessive.

12. A short frame (also known as a runt) is less than 64 bytes with a valid CRC, while a long frame (also known as a giant) is greater than 1518 bytes. These frames are typically formed by faulty or out-of-specification LAN drivers.

13. At times, Ethernet network interface cards and external transceivers can generate a problem called jabbering. This is when garbled bits of data are emitted within the frame sequence in a continuous transmission fashion.

14. Broadcasts are used to reach all devices on a network or subnetwork.

15. Unsolicited broadcasts (announcements) should not account for more than 10 percent of all traffic during hours that data exchange is occurring.

16. The TCP/IP suite includes the Internet Control Message Protocol (ICMP), a message protocol that can help you identify network problems such as incorrect gateway settings, unavailable applications or processes, and fragmentation problems.

17. Although ICMP messages are invaluable for troubleshooting TCP/IP networks, you should be aware that hackers find ICMP messages equally useful.

18. Protocol analyzers can also be used to help detect security breaches.

QUESTIONS

1. You administer a LAN that uses TCP/IP as its network communications protocol. You want to view the number of UDP packets that have been sent to SERVER_1 from CLIENT_A. Which tool should you use to view this information?

  a. A hardware loopback

  b. A Performance Monitor

  c. Monitor.nlm

  d. A protocol analyzer

2. You need to look inside your network packets to determine which of your routers is bad. Which of the tools below can do the job?

  a. Advanced cable tester

  b. Voltmeter

  c. TDR

  d. Protocol analyzer

3. Your network has been slow lately, and after looking into various possible causes and installing both client and server patches and upgrades, you decide to look at packet traffic. You want to be able to capture a packet stream and decode it to determine each packet's origination and destination, as well as its contents. What device will allow you to do this?

  a. Protocol analyzer

  b. Tone locator

  c. Tone generator

  d. TDR

4. In planning for the expansion of your company's network, you want to capture and decode TCP/IP packets on your Windows 2000 Server computer. What must you use?

  a. ICMP

  b. Performance Monitor

  c. Network Monitor

  d. UDP

HANDS-ON EXERCISE

Exercise 1: Using Network Monitoring

1. On the Windows 2000 Server, click the Start button, select Settings, select Control Panel, and then select Add/Remove Programs.

2. In Add/Remove Programs, click Add/Remove Windows Components.

3. In the Windows Component Wizard, highlight Management and Monitoring Tools and then click the Details button.

4. In the Management and Monitoring Tools window, select the Network Monitor Tools check box and then click OK.

5. Click the Next button in the Windows Components Wizard to continue. If you are prompted for additional files, insert your Windows 2000 Server disk or type a path to the location of the files on the network.

6. Click Finish to complete the installation.

7. Reboot the computer to clear the cache.

8. Click the Start button, select Programs, select Administrative Tools, and choose Network Monitor. If you are prompted for a default network on which to capture frames, select the local network from which you want to capture data by default.

9. On the Capture menu, click Start.

10. Open a DOS Window and ping your partner's computer by computer name.

11. Change back to Network Monitor and click on the Stop button (forth from the right on the tool bar).

12. Look at the statistics—specifically, how many frames, how many broadcasts, how many frames dropped, and the %network utilization.

13. Click on the Display Captured Data button (second from the right on the tool bar).

14. Double-click a packet and analyze its parts. Look for Ethernet, IP, TCP, and upper-layer protocols.

    NOTE: Not all packets will have all these.

15. Look for the packets that resolved the Host address to the IP address and the packets that resolved the IP address to the MAC address. What type of packets were they?

16. Look for the ping packets from your computer to your partner's computer. What type of packets were they?

17. Close Network Monitor.

PART III

NETWORK OPERATING SYSTEMS

CHAPTER **10**

Networking with Windows

Topics Covered in this Chapter

Introduction

Most computer users today are familiar with Windows and the Windows environment. For those people who are familiar with the network, they already know that Windows NT Server, Windows 2000 Server, and Windows Server 2003 are currently the most popular networking operating systems. While Windows was designed as a network file, print, and application server, it offers many other features and services. This chapter is meant as an introduction to Windows as a network operating system.

Objectives

- Identify the basic capabilities (client support, interoperability, authentication, file and print services, application support, and security) for Windows servers.
- Identify the basic capabilities (client connectivity, local security mechanism, and authentication) of client Windows workstations.
- Given specific parameters, configure a Windows client to connect to a Windows server.
- Configure the TCP/IP, IPX, and NetBEUI protocols on a Windows computer.

- Share a drive or directory on a Windows machine.
- Share a printer on a Windows machine.
- Use Network Neighborhood in Windows to navigate to a network resource.
- Given a troubleshooting scenario involving a client connectivity problem (e.g., incorrect protocol/client software/authentication configuration, or insufficient rights/permission), identify the cause of the problem.
- Describe the shutdown procedure for the Windows operating systems.

10.1   WHAT IS WINDOWS?

Windows 9X (Windows 95, Windows 98, and Windows Me) were innovative operating systems for the PC. The first of these, Windows 95, was built using part 16-bit/part 32-bit code designed to replace DOS and Windows 3.XX and yet provide compatibility with most 16-bit applications (Windows 3.XX applications) and also provide the ability to run the newer 32-bit applications. Behind the intuitive interfaces used by Windows 9X is a preemptive, multithreading, multitasking environment with dynamically loaded device drivers and memory paging. In addition, Windows 9X includes built-in network support for common protocols including TCP/IP and IPX/SPX. It also has the capability to load additional protocols, and it can run a wide range of network client software packages so that it can talk to a wide range of servers including Windows NT, Windows 2000, and Windows Server 2003, Novell NetWare, and Linux running Samba. See table 10.1 for the 9X system requirements.

Table 10.1 Windows 9X Requirements

	Windows 95	Windows 98	Windows Me
Processor	• Minimum—386DX processor (or higher) • Recommended—Pentium (or higher) • Supports one processor	• Minimum—486DX/66 MHz or higher • Recommended—Pentium processor • Supports one processor	• Pentium 150-MHz processor or better • Supports one processor
RAM	• Minimum—4 MB • Recommended minimum—16–32 MB	• Minimum—16 MB of memory (RAM) • Recommended minimum—32–64 MB	• 32 MB of RAM or better
Drives	• A high-density floppy disk drive • CD-ROM drive • Hard drive with a minimum of 55 MB of free disk space	• 120 MB of free hard disk space. (Typical installation requires 195 MB. Depending on your system configuration and the options you choose to install, you may need between 120 MB and 355 MB.) • CD-ROM or DVD-ROM drive	• Hard drive with a minimum of 480 to 645 MB of free disk space • A high-density floppy disk drive or CD-ROM drive (recommended)
Video system	• Minimum—VGA or better • Recommended—Super VGA (16- or 24-bit color)	• Minimum—VGA or better • Recommended—Super VGA (16- or 24-bit color)	• Minimum—VGA or better • Recommended—Super VGA (16- or 24-bit color)
Pointing devices	• Not required, but highly recommended	• Microsoft mouse or compatible pointing device	• Microsoft mouse or compatible pointing device

Windows 9X is considered a desktop operating system. While Windows 9X can request services from a server, it can also provide some services as well. For example, you can share a drive, directory, or printer, which can then be accessed by other machines running Microsoft Windows or Linux with Samba. In addition, some Windows 9X versions provide a personal web server.

Windows NT Server, Windows 2000 Server, and Windows Server 2003 are operating systems that can function as a desktop operating system but were designed as a network file, print, or application server. Earlier versions of Windows NT used the Windows 3.XX interface, and Windows NT 4.0 uses the Windows 95 interface. Windows 2000 and Windows Server 2003 use Windows 98's Active Desktop interface (merging of the desktop interface and browser interface) that lets you put active content from web pages on your desktop. Different from Windows 9X, Windows NT, Windows 2000, and Windows

Table 10.2   System Requirements for Windows NT and Windows 2000 Servers

	Windows NT 4.0 Server	Windows 2000 Server	Windows 2000 Advanced Server
Processor	• Minimum—Intel 486DX/33 or higher • Recommended—Pentium • Supports up to four processors	• Pentium 133 MHz (or higher) • Supports up to four processors	• Pentium 133 MHz (or higher) • Supports up to eight processors
RAM	• 16 MB of RAM • 32 MB recommended • 4 GB maximum	• A minimum of 128 MB • 256 MB or higher recommended for most network environments • 4 GB maximum	• A minimum of 128 MB • 256 MB or higher recommended for most network environments • 8 GB maximum
Disk space	• 125 MB of free disk space	• 2-GB hard disk with a minimum of 1.0 GB free space. (Additional free hard disk space is required if you are installing over a network.)	• 2-GB hard disk with a minimum of 1.0 GB free space. (Additional free hard disk space is required if you are installing over a network.)
Display	• VGA or higher	• VGA or higher	• VGA or higher
Input devices	• Keyboard and pointing device (such as mouse, trackball, or glide pad)	• Keyboard and pointing device (such as mouse, trackball, or glide pad)	• Keyboard and pointing device (such as mouse, trackball, or glide pad)
Networking	• Network card if you need network connectivity	• Network card if you need network connectivity	• Network card if you need network connectivity
Other	• CD-ROM (or network card for installation over the network)	• CD-ROM (or network card for installation over the network)	• CD-ROM (or network card for installation over the network)

Note: Another version called Windows 2000 Data Center Server will support up to 64 GB of memory and 32 processors. In addition, Microsoft also offers 64-bit versions of Windows.

Server 2003 use a 32-bit architecture that was designed with power, reliability, and security in mind. See table 10.2 and table 10.3 for the system requirements for Windows NT, Windows 2000, and Windows Server 2003.

NOTE: Before installing Windows NT, Windows 2000, Windows Server 2003, and Windows XP, you should make sure that the hardware is on Microsoft's hardware compatibility list (http://www.microsoft.com/hwtest/hcl).

Windows NT Workstation, Windows 2000 Professional, and Windows XP (Home and Professional) are workstation versions of Windows NT, Windows 2000, and Windows

Table 10.3 Windows Server 2003

	Windows Server 2003 Web Server	Windows Server 2003 Standard Server	Windows Server 2003 Enterprise Server
Processor	• Pentium 133 MHz (or higher) • 550 MHz or better • Supports up to two processors	• Pentium 133 MHz (or higher) • 550 MHz or better • Supports up to two processors	• Pentium 133 MHz (or higher) • 733 MHz or better • Supports up to eight processors
RAM	• 128 MB of RAM • 256 MB recommended • 2 GB maximum	• 128 MB of RAM • 256 MB recommended • 4 GB maximum	• 128 MB of RAM • 256 MB recommended • 32 GB maximum for x86-based computers • 64 GB maximum for Itanium-based computers
Disk space	• 1.5 GB of free disk space	• 1.5 GB of free disk space	• 1.5 GB of free disk space for x86-based computers • 2.0 GB of free disk space for Itanium-based computers
Display	• VGA or higher	• VGA or higher	• VGA or higher
Input devices	• Keyboard and pointing device (such as mouse, trackball, or glide pad)	• Keyboard and pointing device (such as mouse, trackball, or glide pad)	• Keyboard and pointing device (such as mouse, trackball, or glide pad)
Networking	• Network card if you need network connectivity	• Network card if you need network connectivity	• Network card if you need network connectivity
Other	• CD-ROM (or network card for installation over the network)	• CD-ROM (or network card for installation over the network)	• CD-ROM (or network card for installation over the network)

Note: Another version called Windows Server 2003 Data Center Server will support up to 64 GB (x86-based computers) of memory or 128 GB (Itanium-based computers). It requires a minimum of 8 processors (400 MHz) and supports up to 32 processors. In addition, Microsoft also offers 64-bit versions of Windows.

Server 2003. They are designed as desktop operating systems while still having the same architecture as their server counterparts. (See table 10.4 for system requirements.) All of these can connect to a Microsoft domain except Windows XP Home. Although Windows XP Home cannot be part of a domain, it can still connect to the Internet, and it can be part of a peer-to-peer network called a workgroup. Domains and workgroups will be explained later in this chapter.

Table 10.4 System Requirements for Windows NT, Windows 2000, and Windows XP

	Windows NT 4.0 Workstation	Windows 2000 Professional	Windows XP Home Edition	Windows XP Professional
Processor	• Minimum—Intel 486DX/33 or higher • Recommended—Pentium • Supports up to two processors	• Minimum—Pentium 133 MHz (or higher) • Recommended—Pentium II, 350 MHz or better • Supports up to two processors	• 233 MHz or higher • Recommended—300 MHz or higher • Supports one processor	• 233 MHz or higher • Recommended—300 MHz or higher • Supports up to two processors
RAM	• 12 MB of RAM • 16 MB recommended • 4 GB maximum	• A minimum of 64 MB • 128 MB or higher recommended • 4 GB maximum	• A minimum of 64 MB of RAM • 128 MB of RAM or higher recommended • 4 GB maximum	• A minimum of 64 MB of RAM • 128 MB of RAM or higher recommended • 4 GB maximum
Disk space	• 110 MB of free disk space	• 2-GB hard disk with a minimum of 650 MB free space. (Additional free hard disk space is required if you are installing over a network.)	• Minimum of 1.5 GB of available hard disk space	• Minimum of 1.5 GB of available hard disk space
Display	• VGA or higher	• VGA or higher	• Minimum of VGA resolution • SVGA (800 × 600) or higher video adapter and monitor recommended	• Minimum of VGA resolution • SVGA (800 × 600) or higher video adapter and monitor recommended
Input devices	• Keyboard and pointing device (such as mouse, trackball, or glide pad)	• Keyboard and pointing device (such as mouse, trackball, or glide pad)	• Keyboard and pointing device (such as mouse, trackball, or glide pad)	• Keyboard and pointing device (such as mouse, trackball, or glide pad)
Networking	• Network card if you need network connectivity	• Network card if you need network connectivity	• Network card if you need network connectivity	• Network card if you need network connectivity
Other	• CD-ROM (or network card for installation over the network)	• CD-ROM (or network card for installation over the network)	• CD-ROM (or network card for installation over the network) • Sound card and speakers or headphones recommended	• CD-ROM (or network card for installation over the network) • Sound card and speakers or headphones recommended

Windows XP is the long-awaited merger of Microsoft's consumer and business operating systems. Different from all versions of Windows 2000 and all versions of Windows Server 2003, Windows XP uses a new interface that has a cleaner default desktop and a more intuitive interface. Yet it contains the 32-bit kernel and driver set from Windows NT and Windows 2000. In addition, it is the first consumer Windows that includes:

- A personal firewall
- Remote assistance that lets you invite someone to connect to your computer and help you with problems
- Remote desktop connection that allows you to connect to a computer's desktop from a remote location and lets you run applications as if you were sitting at the console

10.2   WINDOWS NT FAMILY ARCHITECTURE

The Windows NT family OSs (Windows NT, Windows 2000, Windows XP, and Windows Server 2003) are modular operating systems that have many small self-contained software components that work together to perform the various operating system tasks. Each component provides a set of functions that acts as an interface to the rest of the system. (See figure 10.1.)

Figure 10.1   Windows 2000/2003 Architecture

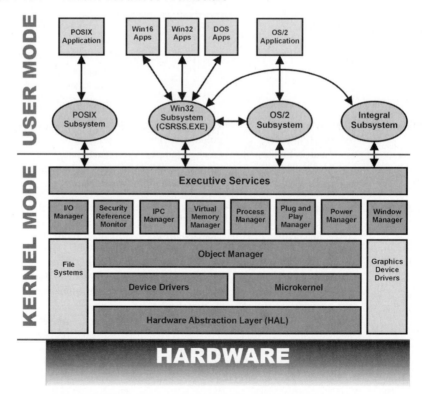

10.2.1   User Mode

The Windows NT family architecture can be divided into two major layers: user mode and kernel mode (also referred to as privileged mode). The **user-mode** programs run in Ring 3 of the Intel **386 microprocessor protection model** and are protected by the operating system. The user mode is a less privileged processor mode that has no direct access to hardware and can only access its own address space. Since programs running in Ring 3 have very little privilege to programs running in Ring 0 (kernel mode), the programs in user mode should not be able to cause problems with components in kernel mode. Since most of the user's applications run in Ring 3, an application can be terminated without causing problems with the other applications running on the computer. In addition, programs executed in user mode do not have direct access to hardware. If a program tries to access the hardware directly, the program will be terminated so the system can maintain reliability and security.

The user-mode layer contains the environment subsystems and the integral subsystems. An environment subsystem provides the application programming interfaces (APIs) for the programs and convert them to the proper calls for Windows. APIs are a set of routines, protocols, and tools for building software applications. In the Windows environment, APIs provide building blocks that a program can then use as a common interface. For example, when a Win32-based application such as Microsoft Word 2000 asks the operating system to do something, the Win32 subsystem captures that request and converts it to commands that Windows 2000 understands. The environment subsystems include:

- A 32-bit Windows-based system (Client Server Runtime Subsystem, or CSRSS. EXE)—Handles windows and graphic functions for all subsystems. It provides a working environment for Win32; Win16-, and DOS-based applications.
- An OS/2 subsystem—Provides APIs for 16-bit character-mode (DOS mode or bounded) OS/2 applications.
- A POSIX subsystem—Provides API for POSIX- compatible UNIX applications.

NOTE: The 32-bit Windows-based subsystem controls the input/output (I/O) between subsystems so that all applications have a consistent user interface.

The integral subsystem performs essential operating system functions such as the security subsystem, the workstation service, and the server service. The security subsystem tracks rights and permissions associated with user accounts, tracks which system resources are audited, and performs login authentication. The workstation services allow access to the network, while the server service provides API to access the computer as a server.

10.2.2   Kernel Mode

The **kernel-mode** components run in Ring 0 of the Intel 386 microprocessor protection model. While user-mode components are protected by the OS, the kernel-mode components are protected by the processor. The kernel mode has direct access to all hardware and all memory including the address space of all user-mode processes. It includes the Windows NT family Executive Services, the hardware abstraction layer, and the Microkernel.

Table 10.5 Windows Executive

Component	Function
I/O Manager	Manages input from and delivers output to the file systems, device drivers, and software cache.
Security Reference Monitor	Enforces security policies on the local computer.
Interprocess Communication (IPC) Manager	Manages communications between clients and servers. It includes: Local procedure call (LPC)—Manages communications when clients and servers exist on the same computer Remote procedure call (RPC)—Manages communications when clients and servers exist on separate computers
Virtual Memory Manager (VMM)	Implements and controls memory (physical and virtual).
Process Manager	Creates and terminates processes and threads.
Plug and Play Manager	Maintains central control of the plug and play process. It communicates with device drivers so that drivers can be directed to add and start devices.
Power Manager	Controls power management APIs, coordinates power events, and generates power management requests.
Window Manager and Graphical Device Interface (GDI)	A device driver (WIN32K.SYS) that manages the display system, and it contains the Window Manager and GDI. The Window Manager will control window displays and manages screen output and receives input from devices such as the keyboard and mouse. The GDI contains the functions that are required for drawing and manipulating graphics.
Object Manager	Creates, manages, and deletes objects that represent operating system resources such as processes, threads, and data structures.

The Windows NT family Executive Services consist of managers and device drivers. They include the I/O Manager, Object Manager, Security Reference Manager, Local Procedure Call facility, Virtual Memory Manager (VMM), Win32K Window Manager and GDI, hardware device drivers, and graphics device drivers. (See table 10.5.)

The **hardware abstraction layer (HAL)** is a library of hardware manipulating routines that hides the hardware interface details. It contains the hardware-specific code that handles I/O interfaces, interrupt controllers, and multiprocessor operations so that it can act as the translator between specific hardware architectures and the rest of the Windows software. As a result, programs written for Windows can work on other architectures, making those programs portable. Therefore, when you load Windows, you need to load the appropriate HAL.

A **kernel** is the central module of an operating system. It is the part of the operating system that loads first, and it remains in RAM. Because it stays in memory, it is important for the kernel to be as small as possible while still providing all the essential services required by other parts of the operating system and applications. Typically, the kernel is responsible for memory management, process and task management, and disk management.

The Microkernel is the central part of Windows, which coordinates all I/O functions and synchronizes the activities of the Executive Services. Much like the other kernels in Windows 9X, the Microkernel determines what is to be performed and when it is to be performed, while handling interrupts and exceptions. Lastly, it is designed to keep the processor(s) busy at all times.

Device drivers are programs that control a device. A device driver acts like a translator between the device and the programs that use the device. Each device has its own set of specialized commands that only its driver knows. While most programs access devices by using generic commands, the driver accepts the generic commands from the program and translates them into specialized commands for the device. Device drivers are installed into the Executive Services section of Windows. A device driver is a portion of kernel-mode code that implements the Windows NT Driver Model Specification. Many drivers used by Windows use the **Windows Driver Model (WDM),** which is also compatible with Windows 98 and Windows Me.

NOTE: To manage your devices and device drivers, you will use the **device manager.**

A **service** is a program, routine, or process that performs a specific system function to support other programs. Different from a device driver, a service is a user-mode process that implements the Service Controller Specification. The service controller is the component of the system that controls starting, pausing, stopping, and continuing services in the system. It also starts and stops (loads and unloads) device drivers. In other words, the service controller provides a common user interface and application programming interface for services and device drivers.

10.3   WINDOWS FILE AND DISK SYSTEMS

Throughout the life of the PC, there have been a few different file systems. Today, the most common include FAT, FAT32, and NTFS. (See table 10.6.)

10.3.1   FAT

The **File Allocation Table (FAT)** is a simple and reliable file system that uses minimal memory. It supports file names of 11 characters, which include the 8 characters for the file name and 3 characters for the file extension.

VFAT, an enhanced version of the FAT structure, allows Windows to support long file names (LFNs) up to 255 characters. When people refer to FAT, they probably mean VFAT. Since it is built on ordinary FAT, each file has to have an 8-character name and a 3-character extension to be backward-compatible for DOS and Windows 3.XX applications. Therefore, programs running in DOS and Windows 3.XX will not see the longer file names. When running a Win32 program (programs made for Windows 9X and the Windows NT family), they can see and make use of the longer names.

Table 10.6   Common Microsoft File Systems

	FAT	FAT32	NTFS
Operating systems	Used by DOS and all Microsoft Windows versions	Used by Windows versions since Windows 95B/OSR2	Used by the Windows NT family
Volume size	Floppy disk to 4 GB	512 MB to 32 GB	10 MB to 2 TB
Maximum file size	2 GB	4 GB	Limited by size of volume
Limit on entries in root directory	Yes	No	No
File and directory security	No	No	Yes
Directory and file compression	No	No	Yes
Cluster remapping	No	No	Yes

After the file name and file extension, 1 byte is used for attributes. The **file attribute** field stores a number of characteristics about each file. Attributes can be either on or off. The most common attributes include read-only, hidden, system, and archive.

NOTE: One of these attributes indicates if the file is a real file or a directory.

Since DOS reserves 1 byte for attributes, it can keep track of up to 8 attributes. Remember, 1 byte equals 8 bits (on/off switch). To change the attributes, you would use the ATTRIB command (right-click the file and select the Properties option). (See table 10.7.)

10.3.2   FAT32

FAT32, which uses 32-bit FAT entries, was introduced in the second major release of Windows 95 (OSR2/Windows 95B) and is an enhancement of the FAT/VFAT file system. It supports hard drives up to 2 TB. It uses space more efficiently, such as 4-KB clusters for drives up to 8 GB, which has resulted in 15 percent more efficient use of disk space relative to large FAT drives.

The root directory is an ordinary cluster chain. Therefore, it can be located anywhere in the drive. In addition, it allows dynamic resizing of FAT32 partitions (without losing data) and allows the FAT mirroring to be disabled, which allows a copy of the FAT other than the first to be active. Consequently, FAT32 drives are less prone to failure of critical data areas such as FAT.

Table 10.7   DOS File Attributes

Attributes	Abbreviations	Description
Read-only	R or RO	When a file is marked as read-only, it cannot be deleted or modified. *Note:* The opposite of read-only is read-write.
Hidden	H	When a file is marked as hidden, it cannot be seen during normal directory listings.
System	S or Sy	When a file is marked as system, it should not be moved. In addition, it usually can't be seen during normal directory listings.
Volume label		The name of the volume.
Subdirectory		A table that contains information about files and subdirectories.
Archive	A	When a file is marked as archive, the file has not been backed up. Anytime a file is new or has been changed, the archive attribute comes on automatically. When the archive attribute is off, the file has been backed up.

10.3.3   NTFS

NTFS is a file system for the Windows NT family operating systems (OSs) designed for both the server and workstation. It provides a combination of performance, reliability, security, and compatibility. It supports long file names, yet maintains an 8.3 name for DOS and Windows 3.XX programs. Since NTFS is a 64-bit architecture, NTFS is designed to support up to 2^{64} bytes, which equals 18,446,744,073,709,551,616 bytes, which equals 16 exabytes.

Since Windows NT family OSs include enhanced security, NTFS supports a variety of multiuser security models and allows computers running other operating systems to save files to the NTFS volume on a server. These include DOS, Microsoft Windows, UNIX, Linux, and even Macintosh computers. It does not allow DOS to access an NTFS volume directly, only through the network (assuming you have the proper permissions or rights to access the volume).

To make an NTFS volume more resistant to failure, NTFS write updates to a log area (making NTFS a journaled file system) and supports remapped clusters. If a system crash occurs, the log area can be used to quickly clean up problems. If a cluster is found to be bad, NTFS can move the data to another cluster and mark the cluster as bad so that the operating system doesn't use it.

FAT is simpler and smaller than NTFS and uses an unsorted directory structure. NTFS is generally faster because NTFS uses a B-tree directory structure, which minimizes the number of disk accesses required to find a file; this makes access to the file faster, especially if it is a larger folder.

NTFS supports volume set and directory/file compression. A volume set combines several hard drives (or parts of hard drives) into a single volume. If the volume needs to be expanded again, you just add another hard drive and expand the volume. NTFS allows an individual file or directory to be compressed without compressing the entire drive.

To convert FAT or FAT32 to NTFS, you would use the following command at the command prompt.

```
convert x: /fs:ntfs
```

where `x:` is the drive that you are trying to convert.

NOTE: You cannot convert NTFS volume to FAT or FAT32 using Microsoft utilities.

10.3.4 CDFS

The **CDFS (CD-ROM file system)** is the read-only file system used to access resources on a CD-ROM disk. Windows supports CDFS so as to allow CD-ROM file sharing. Because a CD-ROM is read-only, you cannot assign specific permissions to files through CDFS.

10.3.5 Volumes

Windows NT 4.0 supports basic storage, and Windows 2000, Windows XP, and Windows Server 2003 support basic storage and dynamic storage. Basic storage is the division of a hard disk into partitions. A partition is a port of the disk that functions as a physically separate unit of storage. A disk that is initialized for basic storage is called a basic disk. It contains primary partitions, extended partitions, and logical drives.

Dynamic storage creates a single partition that includes the entire disk. You then divide the dynamic disk into **volumes,** which can consist of a portion, or portions, of one or more physical disks. On a dynamic disk, you can create simple volumes, spanned volumes, and stripped volumes, which are forms of the software RAID. (RAID is explained in chapter 14.) Dynamic storage doesn't have the restrictions of basic storage; for example, you can size and resize a dynamic disk without restarting Windows 2000.

10.3.6 Disk Management Tools

To manage disks in Windows NT, you use the Disk Administrator program, which is located under Administrative Tools. For Windows 2000, Windows XP, and Windows Server 2003, you use the Disk Management console, which can be found in the Computer Management console. (See figure 10.2.) Both of these utilities display your storage system in either graphical view or list view.

10.4 WINDOWS NT FAMILY NETWORK ARCHITECTURE

In chapter 2, the OSI model was discussed as a modular approach to networking. The OSI model was made of layers that have specific tasks to perform. While the OSI model is a common learning tool in the network, it often does not correspond to any common network architecture. Windows NT family network architecture is arranged in a similar fash-

Figure 10.2   The Disk Management Console

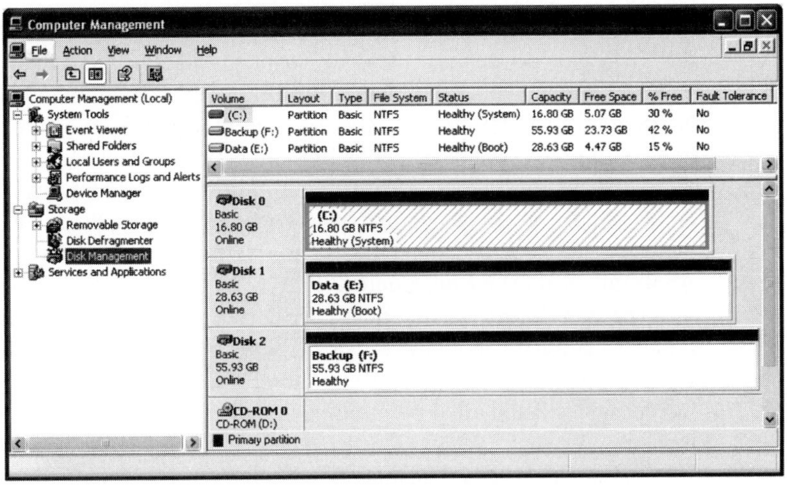

Figure 10.3   Windows 2000 Network Architecture

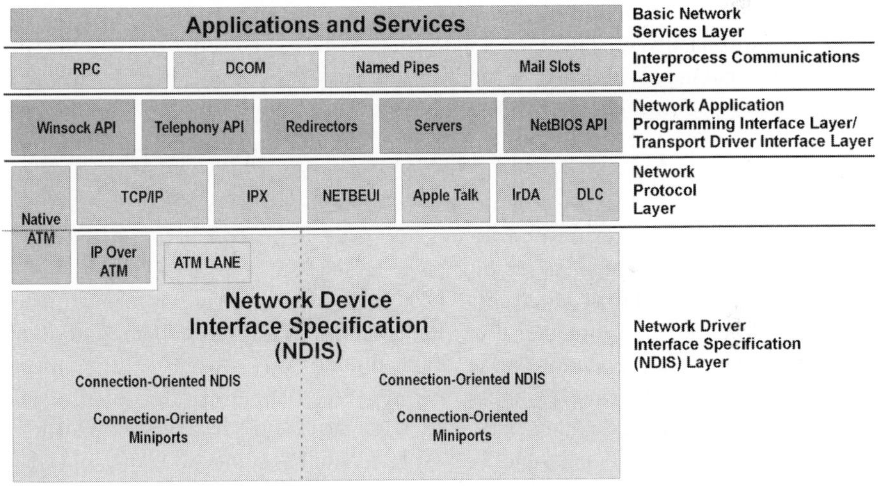

ion, where the Windows network components are arranged in layers and each layer has specific tasks to perform within its assigned layer. (See figure 10.3.)

The Network Driver Interface Specification (NDIS) layer is the layer that provides a communication path between two network devices. It would be similar to the network layer in the OSI model since it acts as the link between the network adapters and the network protocols above it and manages the binding of network interface connections to the protocol. When a packet gets to the card, the NDIS will route it to the appropriate protocol such as TCP/IP or IPX. This layer contains the network media, including ISDN lines, Ethernet, Token Ring, FDDI, and other fiber-optic pathways. It also includes the miniport drivers that control the network interface adapter, and it connects the hardware device to the protocol stack.

Windows NDIS is implemented by a file called NDIS.SYS. It is also referred to as the NDIS wrapper because the code surrounds all NDIS device drivers. When a protocol is bound to a network component (**binding**), the protocol is linked to a network device. NDIS allows an unlimited number of network adapters in a computer and an unlimited number of protocols bound to one or more network adapters.

The network protocol layer, similar to the data and network layer of the OSI model, includes the protocols that allow clients and applications to send data over the network. It includes TCP/IP, NWLink (IPX/SPX), NetBEUI, Infrared Data Association (IrDA), AppleTalk, and Data Link Control (DLC).

The transport driver interface layer provides a standard interface between network protocols and the network services (network applications, network redirectors, and network application programming interfaces. The network application programming interface provides standard programming interfaces for network applications and services including Winsock, NetBIOS, telephony API (TAPI), messaging API (MAPI), and others. The interprocess communications layer is the protocol that manages communications between clients and servers, including RPC, DCOM, name pipes, and mailslots.

Applications that split processing between networked computers are called distributed applications. A client/server application that uses distributed processing has its processes divided between a workstation and a more powerful server (application server). Typically, the client portion formats requests and sends them to the server for processing, and the server runs the requests and passes the results back to the client.

Interprocess communication (IPC) allows bidirectional communication between clients and servers using distributed applications. IPC is a mechanism used by programs and multiprocesses. IPCs allow concurrently running tasks to communicate among themselves on a local computer or between the local computer and a remote computer. Windows provides many different IPC mechanisms, including local procedure calls, remote procedure calls, named pipes, and mailslots.

Local procedure calls (LPCs) are used to transfer information between applications on the same computer. **Remote procedure calls (RPCs)** are used to transfer information between applications on separate computers. The Microsoft RPC mechanism is unique in that it uses other IPC mechanisms, such as named pipes, NetBIOS, or Winsock, to establish communications between the client and the server. Since the RPC is already defined, a programmer can send a message to another server with the appropriate arguments, and the server returns a message containing the results of the program executed.

Named pipes and mailslots are high-level IPC mechanisms used by network computers. Different from the other mechanism, both are written as file system drivers. A **named pipe** is connection-oriented messaging—pipes set up a virtual circuit between two points to maintain reliable and sequential data transfer. A pipe connects two processes so that the output of one can be used as input to the other. A **mailslot** is connectionless messaging, which means that messaging is not guaranteed. Connectionless messaging is useful for identifying other computers or services on a network, such as the browser service offered in Windows 2000. Connecting to a named pipe and mailslot is done through the common Internet file system (CIFS) redirector. A redirector intercepts file input/output requests and directs them to a drive or resource on another computer. The redirector allows a CIFS client to locate, open, read, write, and delete files on another network computer running CIFS.

A **component object model (COM)** is an object-based programming model designed to promote interoperability by allowing two or more applications or components to easily cooperate with one another, even if they were written by different vendors, at different times, in different programming languages, or if they are running on different computers running different operating systems. COM is the foundation technology upon which broader technologies can be built, such as object linking and embedding (OLE) technology and ActiveX.

COM+ is an extension of the component object model. COM+ is both an object-oriented programming architecture and a set of operating system services. It adds to COM a new set of system services for application components while they are running, such as notifying them of significant events or ensuring they are authorized to run. COM+ is intended to provide a model that makes it relatively easy to create business applications that work well with the Microsoft Transaction Server (MTS) in a Windows NT system.

A **distributed component object model (DCOM)** is a set of Microsoft concepts and program interfaces in which client program objects can request services from server program objects in other computers in a network. Since DCOM uses TCP/IP and HTTP, it can work with a TCP/IP network or the Internet. For example, you can create a page for a website that contains a script or program that can be processed before being sent to a requesting user by a more specialized server in the network. Using DCOM interfaces, the web server site program, which is now acting as a client, forwards an RPC to the specialized server. The specialized server provides the necessary processing and returns the result to the web server. It passes the result on to the web page viewer.

The top layer is the basic network services layer. It supports network user applications by providing needed services such as file services, print services, network address management, and name services, just to name a few.

10.5   WORKGROUPS

Even before the release of Windows 2000, Windows NT had the ability to work with workgroups and domains. With the release of Windows 2000, domains were combined to form the Active Directory, Microsoft's directory service.

Computers and devices on a peer-to-peer network are usually organized into logical subgroups called **workgroups.** In a workgroup, each computer has a local security database, so that it can track its own user and group account information. The user information is not shared with other workgroup computers. Since the user information is not shared with the other computers, you must create users on each workgroup computer if you want to use the other computers' network resources. A workgroup is used more as a basic grouping of the computers and is only intended to help users find objects such as printers and shared folders within that group.

Workgroup Advantages

- A workgroup does not require a computer running Windows server to hold centralized security information.
- A workgroup is simple to design and implement.

- A workgroup is convenient for a limited number of computers in close physical proximity—it should not consist of more than 10 computers.
- Each user must manage his or her computer resources.

Workgroup Disadvantages

- A user must have a user account of each computer to which he or she wants to gain access.
- Any changes to user accounts, such as changing a user's password or adding a new user account, must be made on each computer.

10.6   WINDOWS NT DOMAIN

The Windows **domain** is a logical unit of computers and network resources that define a security boundary. It is typically found on medium- or large-size networks or networks that require a secure environment. Different from a workgroup, a domain uses one database to share its common security and user account information for all computers within the domain. Therefore, it allows centralized network administration of all users, groups, and resources on the network.

Domain Benefits

- A domain provides centralized administration because all user information is stored centrally.
- A domain provides a single logon process for users to gain access to network resources, and users only need to log on once to gain those resources.
- A domain provides scalability, and so it can create very large security databases.

In Windows NT, the Security Accounts Manager (SAM) database (also called the domain database) contains information about all the users and groups within a domain. The system of domains and trusts for a Windows NT Server network is known as **Windows NT Directory Service (NTDS).** To manage the users and groups of a domain and to set up trust with other domains (trusts allow users to access resources in other domains), you use the User Manager for Domains program.

In an NTDS network, Windows NT servers can become loaded and configured to have a copy of the domain database (domain controller). However, only one copy of the database can be considered the master copy. The master copy is on the **primary domain controller (PDC),** and the backup copies are located on **backup domain controllers (BDCs).**

NOTE: You can promote a BDC to a primary PDC without reinstalling the server. But a member server (a server that is not a domain controller and does not have a copy of the domain database) has to be completely reinstalled if you want to change it to a domain controller.

A trust relationship (or simply trust) is a relationship between domains that makes it possible for users in one domain to access resources in another domain. The domain that grants access to its resources is known as the trusting domain. The domain that accesses the resources is known as the trusted domain.

A one-way domain trust relationship is established when domain A allows the users in domain B to access its resources (assuming that the users have the appropriate permissions). In a one-way trust relationship, the domain with the resources to share trusts the domain with the users who want to access those resources—not the other way around.

If users from domain B should be able to access resources on domain A, and users from domain A should be able to access resources on domain B, you need to establish a two-way domain trust relationship. Two-way domain trust relationships are common in WAN situations where two or more locations manage their own domains but need to share information.

Even though trust relationships exist between domains, users and groups within those domains must still be assigned permissions on trusted servers. You assign user and group permissions only after you create trust relationships between servers.

When you establish trust relationships in Windows NT domains, the trust relationships are nontransitive. That means that even if domain A trusts domain B and domain B trusts domain C, domain A will not trust domain C unless you establish a trust between the two.

To configure trust relationships, you need at least two domains. In addition, you must have administrator rights to the domains. To alter the trust relationships for a particular domain, click Start, point to Programs, point to Administrative Tools, and click User Manager for Domains. In the User Manager for Domains dialog box, click Policies on the menu bar and then click Trust Relationships. The Trust Relationship dialog box opens, allowing you to add or remove trust relationships between domains.

With Windows NT domains, Microsoft recommends using one of four domain models:

- **Single**—The single-domain model is the simplest Windows NT domain model. As its name implies, a single-domain model consists of one domain that services every user and resource in an organization. This model works well for small networks.
- **Master**—The master domain model uses a single domain to contain and control user account information. In addition, it uses separate resource domains to manage resources such as networked printers and shared files. Each resource domain trusts the master domain. This model suits an environment where each department in an organization controls its files and print sharing and a central department manages the user IDs and groups.
- **Multiple master**—The multiple master domain model uses two or more master domains (domains that contain the users) that are joined in two-way trusts to manage many resource domains. A multiple master domain provides the option of centrally managing all user IDs, groups, and account information, or it allows decentralized administration. Because this model lets you manage complex relationships, it's useful for large corporations or when companies merge.
- **Complete trust**—In the complete trust domain model, all domains trust all other domains. To enable domains to communicate with each other, two-way trusts must be established between all domains. Of course, this model is the most difficult to establish. Each domain still administers its resources.

The domain model you select will depend on the division of your organization (either by function or geographically), the existing infrastructure, the number of users you support, the site of network administrators, or decisions made by management.

10.7 ACTIVE DIRECTORY

Active Directory was introduced with Windows 2000 and is used in Windows 2000 and Windows Server 2003 domains. **Active Directory** is a directory service that combines domains, X.500 naming services, DNS, X.509 digital certificates, and Kerberos authentication. It stores all information about the network resources and services such as user data, printers, servers, databases, groups, computers, and security policies. In addition, it identifies all resources on a network and makes them accessible to users and applications. Active Directory supports several name formats so that it can be compatible with several standards. They are shown in table 10.8.

Active Directory uses **Lightweight Directory Access Protocol (LDAP)** to communicate. LDAP performs basic read, write, and modify operations; it can query an object in the Active Directory; and it allows Windows 2000 and Windows Server 2003 to communicate with Windows 3.51 and Windows 4.0 Servers.

The Active Directory database uses a data store to hold all the information about the objects. The data store is sometimes referred to as the directory. It is stored on domain controllers and is accessed by network applications and services. The Active Directory uses the Extensible Storage Engine (ESE), which allows the Active Directory object database to grow to 17 TB, giving it the ability to hold up to 10 million objects. It is best, though, to have no more than a million objects.

Active Directory, similar to a file system found on your hard drive, is a structured hierarchy consisting of domains and organizational units connected together to form a tree of objects (computers, users, and network resources). (See figure 10.4.) The top of the Active Directory is called the root domain. Domains that are under the root domain are child domains to the root domain, while the root domain is the parent domain to its child domains. In addition, a child domain could have another domain under it. Therefore, the child domain of the root domain is the parent domain to its child domain.

Table 10.8 Name Format Supported by Windows 2000 Active Directory

Format	Description
RFC 822	Internet email address that follows a user_name@domain.xxx format
HTTP Uniform Resource Locator (URL)	Hypertext Transfer Protocol web page address that uses DNS and follows the http://domain.xxx/path_to_page
Universal Naming Convention (UNC)	A NetBIOS address that follows the \\domainname\sharename
LDAP URL	An address that specifies the server or resource on the Active Directory services tree that follows the LDAP://Domain.xxx/CN= Commonname_of_Resource,OU= Organization_Name, DC= DomainComponentName.

Figure 10.4   Active Directory Structure

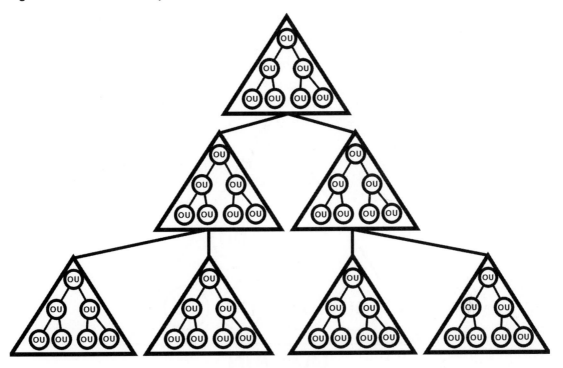

When the Active Directory tree is partitioned into several domains, the Active Directory database is broken up in smaller parts and stored on different domain controllers. Since the parts are smaller and more manageable, there is a lesser load on the individual domain controllers.

One advantage to the administrator and the user stems from the fact that since all the computers and network resources are tracked in a single central database (stored in a domain controller), the administrator only has to create one user account and password for each user. Therefore, the user only has to log on once to the domain and is allowed to use any available resource located within the domain, assuming that the user has the necessary permissions to use the resource. When the user logs on, the user is logging onto the domain, not a specific server within the domain. This means that any server that does authentication within the domain can do the authentication for the user.

The domain is not limited to a single network or a specific type of network configuration. For example, the domain can be made up of the computers located within a single LAN, or computers that are part of a LAN, or computers that are spread out over several LANs connected with any number of physical connections, including Ethernet, Token Ring, dial-up lines, ISDN, fiber optics, and wireless technology.

Figure 10.5   Active Directory Organizational Units and Objects

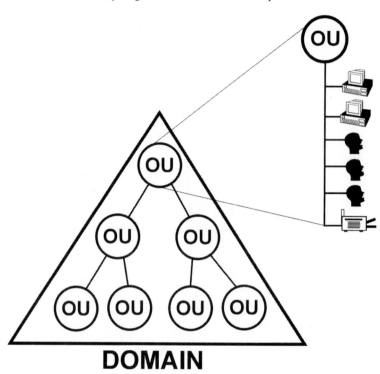

10.7.1   Organizational Units

To help organize the objects within the Active Directory domain and minimize the number of domains created to organize objects, the domain can contain organizational units. **Organizational units (OUs)** are used to hold users, groups, computers, and other organizational units. (See figure 10.5.)

NOTE: An organizational unit can only contain objects that are located in the domain that the organizational unit belongs to.

There are no restrictions on the depth (layers) of the OU hierarchy. However, a shallow hierarchy performs better than a deep one. Therefore, you should not create an OU hierarchy any deeper than necessary. In addition, you should create organizational units that mirror your organization's function or business structure.

Since domains and organizational units are used to contain objects (users, network resources, and other domains), they are sometimes referred to as containers. Containers can be assigned group policy settings or delegate administrative authority. Group policies are a set of configuration settings that control the work environments for users in the domain or organizational unit. Delegating administrative authority allows assigned users to manage objects within the organizational unit.

Table 10.9 Object Classes

Object Type	Description
User account	The information that allows a user to log onto Windows 2000, including the user logon name.
Computer account	The information about a computer that is a member of the domain.
Domain controllers	The information about a domain controller including its DNS name, the version of the OS loaded on the controller, the location, and the person responsible for managing the domain controller.
Groups	A collection of user accounts, groups, or computers that you can create and use to simplify administration.
Organizational unit	An OU can contain other objects, including other OUs.
Shared folders	A pointer to a shared folder on a computer. A pointer contains the address of certain data, not the data itself.
Printers	A pointer to a printer on a computer.

10.7.2   Objects

An **object** is a distinct, named set of attributes or characteristics that represent a network resource, including computers, people, groups, and printers. Attributes have values that define specific objects. For example, the attributes of a user account might include the user's first and last names, department, and email address, while the attributes of printers might include the printer name and location. The most common object types include user account, computer, domain controllers, groups, shared folders, and printers. (See table 10.9.)

As new objects are created in Active Directory, they are assigned a 128-bit unique number called a globally unique identifier, or GUID (sometimes referred to as a security identifier, or SID). While objects have several names including a common name, a relative name, or some other identifier and can be moved to another container whenever you want, the GUID stays the same. Therefore, the GUID can be used to locate an object no matter what name it is using or where it has been moved.

The **schema** of the Active Directory contains a formal definition and set of rules for all objects and attributes of those objects. The default schema contains definitions of commonly used objects, such as user accounts, computers, printers, and groups.

A **name space** is a set of unique names for resources or items used in a shared computing environment. A distinguished name (DN) for an object is a name that uniquely identifies an object by using the actual name of the object plus the names of container objects and domains that contain the object. The distinguished name identifies the object as well as its location in a tree.

A DNS distinguished name would have a format that would look like:

CN=<Object Name>,OU=<Organizational Unit(s)>,O=<Organization>,
C=<CountryCode>

For example:

CN=pregan,OU=users,OU=support,O=Acme.com,C=US

An LDAP distinguished name would have a format that would look like:

/O=<Organization>/DC=<DomainComponent(s)>/CN=<Object Name(s)>

For example:

/O=Internet/DC=Com/DC=Acme/DC=Support/CN=Users/CN=pregan

10.7.3   Forests

A **forest** consists of one or more trees that are connected by two-way, transitive trust relationships, which allow users to access resources in the other domain/tree. In a forest, each tree still has its own unique name space. Therefore, a forest is useful in organizations that need to maintain organization structures, such as a company that needs distinct public identities for its subsidiaries. By default, the name of the root tree, or the first tree that is created in the forest, is used to refer to a given forest. A comparison between Active Directory tree and an Active Directory forest is:

Active Directory Tree	Active Directory Forest
■ A hierarchy of domains	■ One or more sets of trees
■ A contiguous name space	■ Disjointed name spaces between these trees
■ Transitive Kerberos trust relations between domains	■ Transitive Kerberos trust relationships between the trees
■ A common schema	■ A common schema
■ A common global catalog	■ A common global catalog

A domain controller has a replica (copy) of every object in its own domain. Therefore, when you search for an object or retrieve information about an object within a domain, you can retrieve all the information from a domain controller. If you need to search for an object or retrieve information about an object in another domain that is part of the forest, you would access a replica of the global catalog within the current domain to find the object.

The **global catalog** holds a replica of every object in the Active Directory. But instead of storing the entire object, it stores those attributes most frequently used in search operations (such as a user's first and last names). It can even be configured to store additional properties as needed. In addition, the global catalog has sufficient information about each object to locate a full replica of the object at the object's respective domain controller. Of course, since all objects in the forest are listed in the global catalog, all domains within a forest share a single global catalog.

The domain controller that contains the global catalog is known as the global catalog server. By default, a global catalog is created automatically on the initial domain controller in the forest.

Clients must have access to a global catalog to log on. In addition, a global catalog is necessary to determine group memberships during the logon process.

NOTE: A user who is a member of the Domain Admins group is able to log onto the network even when a global catalog is not available.

If your network has any slow or unreliable links, you should enable at least one global catalog on each side of the link for maximum availability and fault tolerance.

10.7.4   Planning Your Domain Structure

When deciding between a single domain with organizational units or a forest, keep in mind that a single domain is the easiest to administer because all users and network resources are located within this domain. If you have a small network or a simple medium-size network, you should probably use the single-domain model.

If you determine that the single-domain model no longer meets your needs, you should then consider using the tree structure based on your organization structure. And as noted earlier, the organization structure could be based on geography or function.

As a general rule, when a network is a far-reaching WAN and you decide to use multiple domains to organize your network resources, you should organize your domain by geography. This is because your domain should not span the WAN. The exception to this rule would be if you have fast links (256 KB or faster) that connect the different sites. Then you can organize your domain by function. This is because of the added traffic that must go back and forth to keep the domain controllers in sync.

NOTE: Windows Active Directory does define sites so that you optimize the synchronization of the domain controllers. Of course, it is always a good idea to have a local domain controller and global catalog for all the domains at each site so that the users can access the Active Directory database quickly.

Another consideration when designing your domain structure is basing your design on organizational units that allow you to delegate administrative control over smaller groups of users, groups, and resources (decentralized administration). If you are using domains for the various WAN sites, you can assign site administrators to each site that would have full authority over the site, or you can limit the authority. In any case, those site administrators would not have any control over the other sites. If you are using a domain design based on function, you can then assign the administrator by function. For example, you can have an administrator for all salespeople and an administrator for all research people. Lastly, when choosing your directory structure, try to look for future growth. This way, you can plan for future organization units if needed.

10.7.5   Domain Controllers and Sites

Forests, domains, and organizational units are considered logical structures because they don't follow any subnet or network boundary. The physical structure of the Active Directory, which uses subnet/network boundaries, consists of domain controllers and sites.

The computer that stores a replica (copy) of the account and security information of the domain and defines the domain is known as the **domain controller.** A Windows

2000/Windows Server 2003 domain controller is a Windows server with an NTFS partition running Active Directory services. The directory data (account and security information) is stored in the NTDS.DIT file on an NTFS partition on the domain controller. Access to domain objects is controlled by access control lists (ACLs). ACLs contain the permissions associated with objects that control which users can gain access to an object and what type of access users can gain to the objects. Lastly, the domain controller manages user-domain interactions including user logon processes, authentication, and directory searches.

Active Directory uses multimaster replication. This means that there is no master domain controller/primary domain controller, as there is in Windows NT. Instead, all domain controllers store writable copies of the directory. When a change is made to one of the domain controllers, it is the job of the domain controller to replicate those changes to other domain controllers within the same domain within a short period of time. By adding a domain controller to a domain, the server is automatically configured for replication.

Another type of server worth mentioning on the domain is a **member server.** A member server does not store copies of the directory database, and therefore it does not authenticate accounts or receive synchronized copies of the directory database. These servers are used to run applications dedicated to specific tasks, such as managing print servers, managing file servers, running database applications, or running a web server.

For each domain, it is recommended to have more than one domain controller. If you only have one domain controller and it goes down for any reason, the users cannot use any of the network resources because there is no server to authenticate them and to give them the permissions needed to use the network resources. An extra domain controller does not provide fault tolerance for the different network services. It only provides fault tolerance for the Active Directory that contains the account and security information.

NOTE: While Windows NT has primary and secondary domain controllers, domain controllers within a Windows 2000 or Windows Server 2003 domain are all equal to one another.

In addition, another domain controller would divide the workload of authentication between the different controllers, allowing for faster network response time.

NOTE: Because domain controllers store all user account information for a domain, each domain controller should be locked in a secure room. Also, only administrators should be allowed to log on interactively to the console of a domain controller. (Logging on interactively will be explained later.)

In Active Directory, a site is defined as one or more IP subnets that have LAN-speed network connectivity. By default, Windows 2000/Windows Server 2003 places domain controllers into a single default site object named DEFAULT-FIRST-SITE-NAME. Intrasite replication occurs on an as-needed basis because it is assumed that domain controllers within a site can transfer data between themselves at great speed and with minimal cost. However, when areas of a network are separated by low-bandwidth WAN connections, it is advantageous to partition these areas into separate sites. Intersite replication occurs based on the administrator-defined schedules rather than on an as-needed basis because connections between domain controllers are assumed to be slow and expensive. Creating multiple sites also provides for a high degree of control over domain replication traffic, workstation logon traffic, and global catalog replication traffic.

Figure 10.6   Patrick E. Regan User Account and WS1 Computer Account Shown in the Organizational Unit Corporate

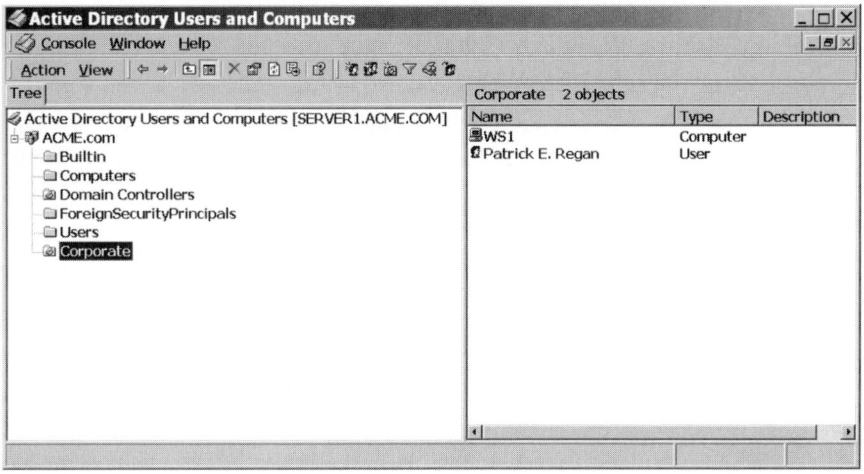

10.8   USER, GROUP, AND COMPUTER ACCOUNTS

As with any modern network operating system (NOS), Windows uses the concept of users and groups to grant access to the various network resources. Unlike other NOSs, Windows also creates computer accounts. Windows NT uses the User Manager for Domain, whereas Windows 2000 Server and Windows Server 2003 use the Active Directory Users and Computers console. (See figure 10.6.)

10.8.1   User Accounts

A **user account** enables a user to log onto a computer and domain with an identity that can be authenticated and authorized for access to domain resources. Each user who logs onto the network should have his or her own unique user account and password.

In Windows domain, there are two different types of user accounts. The **domain user account** can log onto a domain to gain access to the network resources. Of course, a domain controller must authenticate the domain user account.

A **local user account** allows users to log on at and gain resources on only the computer where you create the local user account. The local user account is stored in the local security database. When using a local user account, you will not be able to access any of the network resources. In addition, for security reasons, you cannot log on as a local user account on a domain/domain controller.

10.8.2   Built-in User Accounts

Windows NT, Windows 2000, and Windows Server 2003 start with two built-in accounts: the administrator and guest accounts. The administrator account is used to manage the overall computer and domain configuration; create and maintain computer, user accounts, and groups; manage security policies; assign rights and permissions; and create and maintain printers.

Since the administrator is the most important account within the domain, there should be certain precautions and procedures that you should follow.

■ Since every Window NT, Windows 2000, and Windows Server 2003 computer begins with having an administrator account, it is a good idea to rename the administrator account. This way, if hackers are trying to get in, they will have to guess both the name of the administrator and the password.

NOTE: You can only rename the administrator account; you cannot delete it.

■ You should not use the administrator as a general user. Instead you should use a second account and use it as a general user, and use the administrator account for administrative tasks only.

The guest account is supposed to be used for the occasional users that need to log on and gain access to resources. The guest account, as well as any other account that performs the same function as the guest account, should only be used in low-security networks. Unlike the administrator account, the guest account is disabled by default. Of course, if you enable it, you should assign it a password. Similar to the administrator account, you cannot delete the guest account, but you can rename it.

10.8.3   Home Directories

The **home directory** is a folder used to hold or store a user's personal documents. It can be accessed from any Microsoft operating system including DOS or Windows operating system computers. While a home directory is typically assigned to an individual user, it can be made available to many users if needed.

The home directory can be stored on the user's computer or in a shared folder on the file system. By having home directories on a file server, a user can gain access to his or her home directory no matter which server that user is logging in from. In addition, if all the home directories are located in one location, it is easy to back up all the user's data files. Lastly, if the home files are on an NTFS partition, you can make access to the user's personal files more secure.

10.8.4   User Profiles

A **user profile** is a collection of folders and data that stores the user's current desktop environment and application settings. (See table 10.10.) A user profile also contains all the network connections that are established when a user logs onto a computer, such as mapped drives to shared folders on a network server. This way, when the user logs onto a system, the user will get the same desktop environment that he or she had the last time on the computer. On computers running Windows, user profiles are automatically created and maintained on the local computer when the user logs onto a computer for the first time.

Table 10.10   User Profile Folders and Data

Source	Parameters Saved
Accessories	All user-specific program settings affecting the user's Windows environment, including Calculator, Clock, Notepad, and Paint.
Application data and registry hive	Application data and user-defined configuration settings such as a custom dictionary. The program manufacturer decides what data to store in the user profile folder.
Cookies	Cookies store user information and preferences. Cookies are mostly generated when using a browser such as Microsoft Internet Explorer.
Control panel	All user-defined settings made in the control panel.
Desktop contents	Items stored on the desktop including files, shortcuts, and folders.
Favorites	Shortcuts to favorite locations on the Internet.
Mapped network drive	Any user-created mapped network drive.
My Documents	User-stored documents. The My Documents shortcut can be redirected to the home directory.
My Network Places	The NetHood folder stores shortcuts to other computers on the network.
My Pictures	User-stored picture items.
Online user education bookmarks	Any bookmarks placed in the Windows 2000 Help system.
Printer settings	Printhood folder stores shortcuts to defined network printer connections.
Screen colors and fonts	All user-definable computer screen colors and display text settings.
Start menu	Shortcuts to program items stored in the Start menu. *Note:* There are several Start menus (User Start menu, All User Start menu, and Default Start menu) that are merged into the Start menu that is shown to the user.
Templates	User template items.
Windows Explorer	All user-definable settings for Windows Explorer.
Windows 2000–based programs	Any program written specifically for Windows 2000 can be designed so that it tracks program settings on a per-user basis. If this information exists, it is saved in the user profile.

There are three types of user profiles:

- **Local user profile**—A local user profile is created the first time you log onto a computer. The profile is stored on a computer's local hard disk (C:\Documents and Settings\user_logon_name folder). Any changes made to your local user profile are specific to the computer on which you made the changes.
- **Roaming user profile**—A roaming user profile is created and stored on a server. Since the profile is on a server, the profile can be accessed from any computer. Therefore, this profile is available every time you log onto any computer on the network.
- **Mandatory user profile**—A mandatory user profile is a roaming profile that can be used to specify particular settings for individuals or an entire group of users. When the user logs off, Windows 2000/Windows Server 2003 does not save any changes that the user made during the session. A mandatory user profile can only be changed by the administrator.

10.8.5   Groups

A **group** is a collection of user accounts. By using groups, you can simplify administration by assigning rights and permissions to the group, and everyone listed in the group will be assigned these rights and permissions. Users can be members of multiple groups, and groups can be members of other groups.

NOTE: Groups are not containers. They list members, but they do not contain the members.

In Windows 2000 and Windows Server 2003 domains, there are two group types: security and distribution. The security group is used to assign permissions and to gain access to resources. Security groups can also be used for nonsecurity purposes such as those that you would find in a distribution group. The distribution group is used only for nonsecurity functions such as those used to distribute email messages to many people. Typically distribution groups are only used for special applications that have been designed to use them—for example, the Microsoft Exchange Mail Server.

The security group can be divided into group scopes, which define how the permissions and rights are assigned to the group. In Windows 2000 and Windows Server 2003, there are three scopes: the global group, domain local group, and universal group. If you have a single domain, it is best to use global and domain local groups to assign permissions to network resources. In a domain tree, it is best to use global and universal groups if you only are not using any Windows NT domain controllers.

10.8.6   Global and Domain Local Group Scopes

The **global group** is usually used to group people within its own domain. Therefore, it can list user accounts and global groups from the same domain. The global group can be assigned access to resources in any domain.

The **domain local group** is usually used to assign rights and permissions to network resources that are in the domain of the local group. Different from a domain local group, it can list user accounts, universal groups, and global groups from any domain and local groups from the same domain.

Example:

Let's say that we have a printer and some files that need to be accessed in the SALES domain. We have users in the FINANCE and SALES domains that need to access these resources. Here's what to do:

- First create two global groups called ANALYSTS and MANAGERS in the FINANCE domain, and create a global group called SALESPEOPLE in the SALES domain. Then assign the respective users from the FINANCE domain to the ANALYSTS group and MANAGERS group and the users from the SALES domain to the SALESPEOPLE group.
- Next, create a domain local group called SALES_RESOURCES in the SALES domain. Assign rights and permissions to the SALES_RESOURCES group to the printer and directories in the SALES domain.
- Then assign users from either domain and assign the ANALYSTS group, MANAGERS group, and SALESPEOPLE group. Since groups consist of many people, it is best to assign the groups first. If, after that, you still have certain individuals who need different or special access, you can assign them as needed.

As you can see, the global groups are used to group people that have common purposes or that have the need for the same rights and permissions. The advantage of using this scheme is that when you add members to the two global groups, they will automatically get access to the resources.

10.8.7   Universal Groups

The universal group was introduced with Windows 2000. The **universal security group** is only available in native mode (not mixed mode, whereas a domain contains both Windows NT and Windows 2000/Windows Server 2003 domain controllers). Note that in a mixed-mode network, the universal group, option will be grayed out. It can contain users, universal groups, and global groups from any domain, and it can be assigned rights and permissions to any network resource in any domain in the domain tree or forest. As you can easily see, the universal group is much easier to use and more flexible.

NOTE: You can change a global group into a universal group if the global group is not a member of another global group, and you can change a domain local group to a universal group if it doesn't contain any other domain local group.

When choosing a group name, use intuitive names. To help identify groups, groups are assigned unique security identifiers (SIDs), much like the SIDs assigned to domain controllers. If you delete a group and re-create a group with the same name, the second group would have a different SID. Since these are different groups, the users, rights, and permissions will have to be reassigned.

10.8.8   Computer Accounts

Computers that are running Windows NT, Windows 2000, or Windows Server 2003 are, by design, much more secure than Windows 9X. For example, to use a Windows NT, Windows 2000, or Windows Server 2003 computer, you must log onto it with a user name and password. If you don't log on, you cannot bypass the logon screen by pressing the Escape key and access the computer's files and programs, and you will not able to access any network resources.

NOTE: Windows 2000/Windows Server 2003 computers are more secure than Windows NT computers.

To use a Windows NT, Windows 2000, or Windows Server 2003 computer that is not a domain controller, you must create a computer account for the computer. The computer account is used to uniquely identify the computer on the domain. By having the computer account, you can then audit the computer; and to make sure only authorized people access the computer, you can assign permissions to gain access to network resources.

A computer account, which matches the name of the computer, is an account created by an administrator to uniquely identify the computer on the domain. By being able to identify the computer, a secure communications channel can be created between the client computer and the domain controller. Lastly, the computer account allows the administrator to remotely manage the computer user environment and manage the computer user and group accounts.

10.9   POLICIES

Policies are a tool used by administrators to define and control how programs, network resources, and the operating system behave for users and computers in a domain or Active Directory structure. It also defines user rights.

Windows NT only has one level of policies. To access the policies in Windows NT, you access the Users Manager for Domains program. The account policies define what the minimum password length will be, how many times an invalid password can be entered before the account is locked and for how long the account will be locked, and how often a person has to change his or her password. The other policies are located in the policy editor, in which you can define policies for specific users or for specific computers.

For Windows 2000 and Windows Server 2003 domains, since the Active Directory is a structured hierarchy, there are different levels of policies so that you can have a customized configuration. The different levels of policies are applied in this order:

1. Windows NT 4.0-style policies
2. Unique local group policy object
3. Site group policy objects, in administratively specified order
4. Domain group policy objects, in administratively specified order
5. Organizational unit group policy objects, from the highest to lowest organizational unit and in administratively specified order

Figure 10.7   Windows 2000 Group Policies Shown in a Customized MMC

These policies are configured with a Microsoft Management console (MMC), which can be customized to contain any policy level you desire. (See figure 10.7.)

NOTE: To refresh the group policies so that they will go in effect immediately, do the following:

WINDOWS 2000:

1. Click on the Start button and select the Run option.
2. To refresh the user configuration settings, in the Run dialog box, input `secedit/ refreshpolicy user_policy`. To refresh the computer user configuration settings, in the Run dialog box, `input secedit/refreshpolicy machine_ policy`.

WINDOWS SERVER 2003

1. Click on the Start button and select the Run option.
2. To refresh the user configuration settings, in the Run dialog box, input `gpupdate/ target:User/force`. To refresh the computer configuration settings in the Run dialog box, input `gpupdate/target:computer/force`. To refresh the user and computer configuration settings, in the Run dialog box, input `gpupdate/force`.

10.10   RIGHTS AND PERMISSIONS

A **right** authorizes a user to perform certain actions on a computer, such as logging onto a system interactively, logging on locally to the computer, backing up files and directories, performing a system shutdown, or adding or removing a device driver. Administrators can assign specific rights to individual user accounts or group accounts. Rights are managed with the user rights policy. For Windows NT 4.0, you open the User Manager

Table 10.11 Common Object Types

Object Type	Object Manager	Management Tool
Files and folders	NTFS	Windows Explorer
Shares	Server service	Windows Explorer
Active Directory objects	Active Directory	Active Directory Users and Computers console or snap-in
Registry keys	The registry	Registry editor (REGEDT32.EXE)
Services	Service controllers	Security templates, Security configuration, and analysis
Printer	Print spooler	Printer folder

for Domains, open the Policies menu, and select the User Rights Policy. For Windows 2000 and Windows Server 2003, user rights can be found by opening the Group Policy, opening Computer Configuration, opening Windows Settings, opening Security Settings, opening Local Policies, and opening User Rights Assignment.

One right that should be pointed out is the logon locally right. This right allows a user to log on at the computer's keyboard. Since most protection can be bypassed by being able to log on directly to a machine without going through the network, this right should only being given to a few people.

Permission defines the type of access granted to an object or object attribute. The permissions available for an object depend on the type of object. For example, a user has different permissions than a printer. When a user or service tries to access an object, its access will be granted or denied by an object manager. Common object managers are shown in table 10.11.

When a computer, user, or group is assigned rights or permissions, the computer, user, or group is assigned a security identifier (SID). Similar to the SID that is assigned to a domain, a SID assigned to a computer, user, or group is a unique alphanumeric structure. The first part of the SID identifies the domain in which the SID was issued, and the second part identifies an account object within the domain. Therefore, when a computer, user, or group accesses an object, the computer, user, or group is identified by its SID and not its user name.

Each object uses an access control list (ACL) to list users and groups. The ACL is divided into discretionary access control lists (DACLs) and system access control lists (SACLs). The DACL contains the access control permissions for an object and its attributes and also contains the SIDs, which can use the object. The permissions and rights that a user has are referred to as access control entries (ACEs). The SACL contains a list of events that can be audited for an object. The access control entries include Deny Access and Grant Access.

Every object in Active Directory has an owner, which controls how permissions are set on an object and to whom permissions are assigned. The person who creates the object automatically becomes the owner and, by default, has full control over the object,

Table 10.12   Standard Permissions

Object Permission	Description
Full control	Contains all permissions for the object, including take ownership
Read	Views objects and object attributes, including the object owner and the Active Directory permissions
Write	Able to change all object attributes
Create all child objects	Able to add any child object to an OU
Delete all child objects	Able to delete any child object from an OU

Figure 10.8   Permissions for the User Account

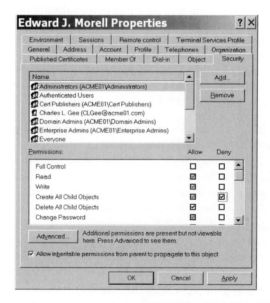

even if the access control list does not grant the person access. If the member of the administrator group creates an object or takes ownership of an object, the administrator group becomes the object group. A member of the domain administrator group has the ability to take ownership of any object in the domain and then change permissions.

Standard permissions are the most commonly used, frequently assigned permissions that apply to the entire object. Standard permissions are sufficient for most administrative tasks. The standard permissions are made of special permissions, which provide a finer degree of control. (See table 10.12.)

Explicit permissions are those that are specifically given to the object when the object is created or assigned by another user. (See figure 10.8.) Inherited permissions are those rights that are given to a container and apply to all of the child objects under the container. By using inherited permissions, you can manage permissions more easily, and you can ensure consistency of permission among all objects within a given container.

The permissions can be allowed or denied for each user or group. To explicitly allow or deny the permission, click the appropriate check box. If a check box is shaded, the permission was granted to the user or group for a container that the object is in and the permission was inherited. If the Allow or the Deny box is not checked for a permission, the permission may be still granted from a group. You would then have to check the groups of which the user or group is a member to determine if the rights are granted or denied.

Every object, whether it is in Active Directory or it is an NTFS volume, has an owner. The owner controls how permissions are set on the object and to whom permissions are granted. When an object is created, the person creating the object automatically becomes the owner. Administrators will create and own most objects in Active Directory and on network servers (when installing programs on the server). Users will create and own data files in their home directories and will create and own some data files on network servers.

Ownership can be transferred in the following ways:

- The current owner can grant the take ownership permission to other users, allowing those users to take ownership at any time.
- An administrator can take ownership of any object under his or her administrative control. For example, if an employee leaves the company suddenly, the administrator can take control of the employee's files.

Although an administrator can take ownership, the administrator cannot transfer ownership to others. This restriction keeps administrators accountable for their actions.

The best way to give sufficient permissions to an organizational unit is to delegate administrative control to the container (decentralized administration), so that the user or group will have administrative control for the organizational unit and the objects in the organizational unit. To delegate control to an organizational unit, run the Delegation of Control Wizard. To start the Delegation of Control Wizard, right-click the organizational unit and select Delegate Control. You can then select the user or group to which you want to delegate control, the organizational units and objects you want to grant those users the right to control, and the permissions to access and modify objects. For example, a user can be given the right to modify the Owner of Accounts property without being granted the right to delete accounts in that organizational unit. (See figure 10.9.)

10.11 NTFS PERMISSIONS

A primary advantage of NTFS over FAT and FAT32 is that NTFS volumes have the ability to apply NTFS permissions to secure folders and files. By setting the permissions, you specify the level of access for groups and users for accessing files or directories. For example, to one user or group of users, you can specify that they can only read the file; another user or group of users can read and write to the file, while others have no access. No matter if you are logged on locally at the computer or accessing a computer through the network, NTFS permissions always apply.

Figure 10.9 Delegation of Control Wizard

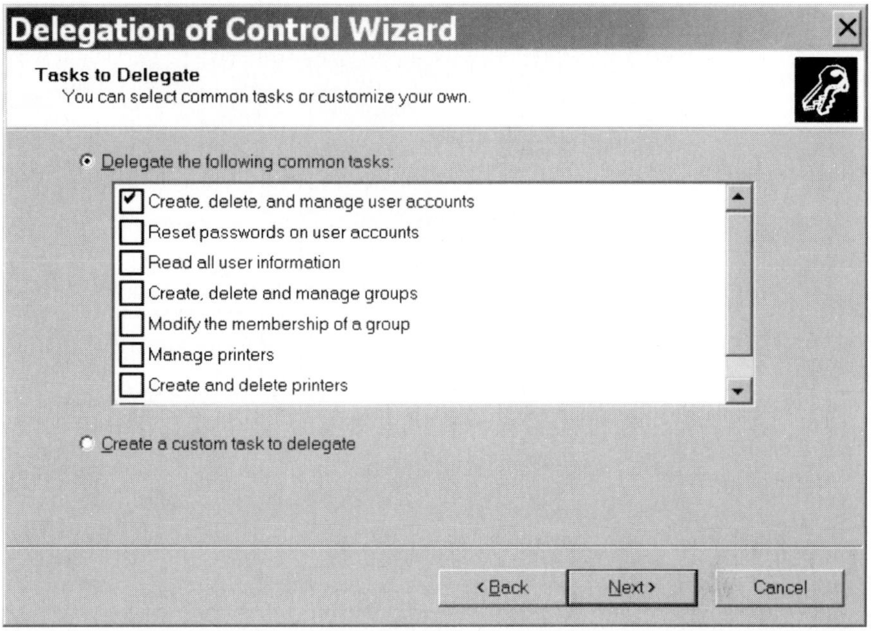

All the folder and file NTFS permissions (also known as special permissions) are listed in tables 10.14 and 10.16. To simplify the task of administration, the permissions have been logically grouped into the standard folder and file NTFS permissions as shown in tables 10.13 and 10.15.The standard folder permissions include read, write, list folder contents, read & execute, modify, and full control. The standard file permissions include read, write, read & execute, modify, and full control.

The NTFS permissions that are granted are stored in an **access control list (ACL)** with every file and folder on an NTFS volume. The ACL contains an access control entry (ACE) for each user account and group that has been granted access for the file or folder as well as the permissions granted to each user and group.

To assign NTFS permissions you would right-click a drive, folder, and file (using My Computer or Windows Explorer), select the Properties option, and click on the Security tab. To assign the special permissions, click on the Advanced button within the Security tab and click on the View/Edit button. (See figure 10.10.)

Permissions are given to a folder or file as explicit permissions and inherited permissions. Explicit permissions are those that are granted directly to the folder or file. Some of these permissions are granted automatically, such as when a file or folder is created, while others have to be assigned manually.

To explicitly grant a permission to a folder or file, you would select the permission by putting a check in the respective check box. To remove an explicit permission, deselect the permission to remove the check in the respective check box. To remove a user or group from being assigned explicit permissions, click to highlight the user and then click on the Remove button.

Figure 10.10   NTFS Permissions

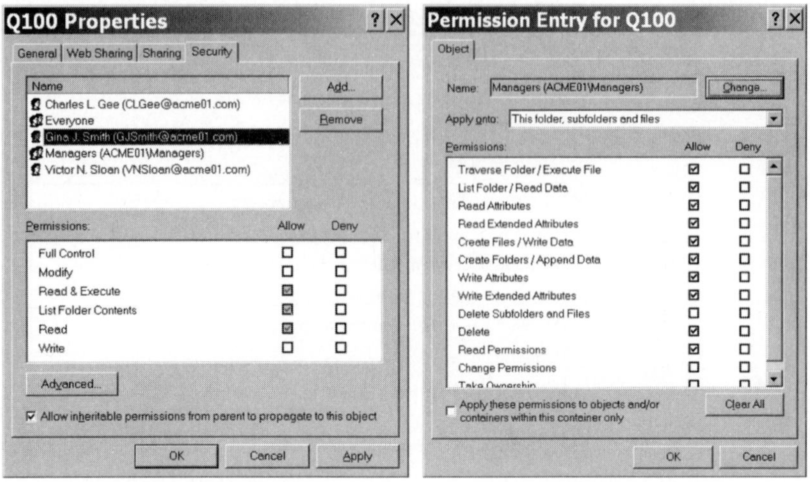

Table 10.13   Standard NTFS Folder Permissions

NTFS Folder Permissions	Allows the User to:
Read	Display the file's data, attributes, owner, and permissions.
Write	Create new files and subfolders within the folder, write to the file, append to the file, read and change folder attributes, and view ownership.
List folder contents	See the names of folders and subfolders in the folder.
Read & execute	Display the folder's contents and display the data, attributes, owner, and permissions for files within the folder; it also allows the user to execute files within the folder. In addition, it allows the user to navigate a folder to reach other files and folders, even if the user does not have permissions for these folders.
Modify	Read files, execute files, write and modify files, create folders and subfolders, delete subfolders and files, and change attributes of subfolders and files.
Full control	Read files, execute files, write and modify files, create folders and subfolders, delete subfolders and files, change attributes of subfolders and files, change permissions, and take ownership of files.

Note: Groups or users granted full control for a folder can delete files and subfolders within that folder regardless of the permissions protecting the files and subfolders.

Table 10.14   NTFS Folder Permissions

Special Permissions	Full Control	Modify	Read & Execute	List Folder Contents	Read	Write
Traverse folder/execute file	X	X	X	X		
List folder/read data	X	X	X	X	X	
Read attributes	X	X	X	X	X	
Read extended attributes	X	X	X	X	X	
Create files/write data	X	X				X
Create folders/append data	X	X				X
Write attributes	X	X				X
Write extended attributes	X	X				X
Delete subfolders and files	X					
Delete	X	X				
Read permissions	X	X	X	X	X	X
Change permissions	X					
Take ownership	X					
Synchronize	X	X	X	X	X	X

Note: Although list folder contents and read & execute appear to have the same permissions, these permissions are inherited differently. List folder contents is inherited by folders but not files, and it should only appear when you view folder permissions. Read & execute is inherited by both files and folders and is always present when you view file or folder permissions.

Since a user can be a member of several groups, it is possible for the user to have several sets of explicit permissions to a folder or file. When this happens, the permissions are added together to form the effective permissions. The effective permissions are the actual permissions when logging in and accessing a file or folder. They consist of explicit permissions plus any inherited permissions. Inherited permissions will be discussed a little bit later.

NTFS file permissions override folder permissions. Therefore, if a user has access to a file, the user will still be able to gain access to that file even if he or she does not have access to the folder containing the file. Of course, since the user doesn't have access to the folder, the user cannot navigate or browse through the folder to get to the file. Therefore, a user would have to use the universal naming convention (UNC) or local path to open the file.

Table 10.15 Standard NTFS File Permissions

NTFS File Permissions	Allows the User to:
Read	Display the file's data, attributes, owner, and permissions
Write	Write to the file, append to the file, overwrite the file, change file attributes, and view file ownership and permissions
Read & execute	Display the data, attributes, owner, and permissions for files and execute files
Modify	Read files; execute files; write, modify, and delete files; and change attributes of files
Full control	Read files; execute files; write, modify, and delete files; and change attributes, change permissions, and take ownership of files

Note: Groups or users granted full control for a folder can delete files and subfolders within that folder regardless of the permissions protecting the files and subfolders.

Table 10.16 NTFS File Permissions

Special Permissions	Full Control	Modify	Read & Execute	Read	Write
Traverse folder/execute file	x	x	x		
List folder/read data	x	x	x	x	
Read attributes	x	x	x	x	
Read extended attributes	x	x	x	x	
Create files/write data	x	x			x
Create folders/append data	x	x			x
Write attributes	x	x			x
Write extended attributes	x	x			x
Delete subfolders and files	x				
Delete	x	x			
Read permissions	x	x	x	x	x
Change permissions	x				
Take ownership	x				
Synchronize	x	x	x	x	x

When you set permissions to a folder (explicit permissions), the files and subfolders created in the folder inherit these permissions (inherited permissions). In other words, the permissions flow down from the folder into the subfolders and files, indirectly giving permissions to a user or group. Inherited permissions ease the task of managing permissions and ensure consistency of permissions among the subfolders and files within the folder.

When viewing the permissions, the permissions will be checked, cleared (unchecked), or shaded. If a permission is checked, the permission was explicitly assigned to the folder or file. If the permission is clear, the user or group does not have that permission explicitly granted to the folder or file.

NOTE: A user may still obtain permission, through a group permission, or a group may still obtain permission through another group.

If the check box is shaded, the permission was granted through inheritance from a parent folder.

When you copy and move files and folders from one location to another, you need to understand how the NTFS folder and file permissions are affected. If you copy a file or folder, the new folder or file will automatically acquire the permissions of the drive or folder that the folder or file is being copied to.

If the folder or file is moved within the same volume, the folder or file will retain the same permissions that were already assigned. When the folder or file is moved from one volume to another volume, the folder or file will automatically acquire the permissions of the drive or folder that the folder or file is being copied to. An easy way to remember the difference is to remember that when you move a folder or file from within the same volume, the folder or file is not physically moved, but the Master File Table is adjusted to indicate a different folder. When you move a folder or file from one volume to another, Windows copies the folder or file to the new location and then deletes the old location. Therefore, the moved folder or file is new to the volume and acquires the new permissions.

10.12   NETWORK SERVICES

To get full access to the network, you would need the following network software, installed by right-clicking the local area connection for which you want to install the software, and then clicking the Properties button (see figure 10.11):

1. A client software choice such as Client for Microsoft Networks or Client for NetWare Networks
2. A network adapter driver or dial-adapter driver (such as a modem)
3. A protocol stack such as TCP/IP, IPX/NWLINK, or NetBEUI

10.12.1   Client Software

Client software often uses redirectors and designators. A **redirector** is a small section of code in the network operating system that intercepts requests in the computer and determines if they should be left alone to continue in the local computer's bus or be redirected

Figure 10.11 Local Area Connection Properties Showing the Driver and Networking Components

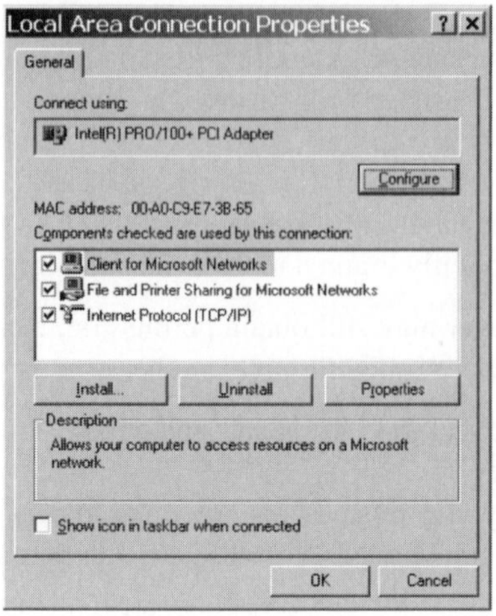

out to the network to another server. For example, when a user requests a file from the server or sends a print job to a network printer, the request is intercepted by the redirector and forwarded out onto the network.

A **designator** keeps track of which drive designations are assigned to network resources. For example, when a user attaches or maps to a network drive, the user and the software applications may see a drive letter such as drive M or drive P just like any other local drive. When these drives are accessed, a redirector will route the request to the network.

The **Client for Microsoft Networks** provides the redirector (VREDIR.VXD) to support all Microsoft networking products that use the Server Message Block (SMB) Protocol. This includes support for connecting computers to and accessing files and printers on Windows for Workgroups, Windows 9X, Windows NT, Windows XP, Windows 2000, and Windows Server 2003 computers.

Under the properties of the Client for Microsoft Network, you can specify a domain name to log onto, and you can specify the network logon options. When a user logs onto a workstation/client computer on the network, the password is validated by one of the domain controllers, and access is granted to the assigned network resources.

Client for NetWare Networks, run with the IPX Protocol, is used to process login scripts, support all NetWare 3.XX command-line utilities and some 4.X/5.X/6.X command-line utilities, connect and browse to NetWare servers, access printers on the NetWare server, and process login scripts on the NetWare server.

NOTE: It is recommended to use the Novell Client for Windows provided by Novell instead of the Client for NetWare Networks provided through Microsoft. The Novell Client for Windows will allow access to Novell servers using TCP/IP and will give you full access to the server and its utilities.

10.12.2   **Network Card Drivers**

While the network card or adapter is the hardware device that allows you to connect to the network, it requires a software driver before it can be used. If the driver that you need is not listed, you will have to use the Have Disk button, and you will have to indicate where the installation files are for the driver that you want to install.

If the card is a legacy card, it will go into the properties of the network card and specify the network card's resources such as I/O address and IRQ. While most protocols are automatically bonded to the network card, this can be configured in the network card's Properties dialog box under the Bindings tab. In addition, if a card has advanced features, it will usually be configured using the Adapter Properties dialog box.

10.12.3   **TCP/IP**

The default protocol is the Internet Protocol (TCP/IP). If you access the properties of TCP/IP, you can select Obtain an IP address automatically, which is handed to the computer using a DHCP server, or you can manually set the IP address, the subnet mask, and the default gateway.

The IP address identifies the computer on the IP network, while the subnet mask specifies which bits of the IP address are used as the network address and which bits are used for the host address. Of course, to communicate on a TCP/IP network, you must use an IP address and a subnet mask. The default gateway is the gateway or router address that connects your local area network to another network. Therefore, if you need to connect to another subnet or network, you must use the default gateway. At the bottom of the Internet Protocol (TCP/IP) Properties dialog box, you can specify the DNS server address. The DNS server is for name resolution to resolve DNS names such as MICROSOFT.COM or MIT.EDU to an IP address.

If the computer is set to accept a DHCP server and one does not respond, the computer will use Automatic Private IP Addressing (also available in Windows 98 and Windows Me), which generates an IP address in the form of 169.254.x.y (where x.y is the unique host address) and the subnet mask as 255.255.0.0. After the computer generates the address, it broadcasts this address until it can find a DHCP server.

NOTE: When you have an Automatic Private IP address, you can only communicate with computers on a network/subnet that also has an Automatic Private IP address.

If you click on the Advanced. . . button, you can add additional gateways, additional DNS servers, and additional WINS servers.

NOTE: The additional gateway addresses, DNS servers, and WINS servers are only accessed when the first address is inaccessible. (See figure 10.12.)

Several utilities can be used to test and troubleshoot the TCP/IP network. To verify the TCP/IP configuration, use either the IP configuration program (WINIPCFG.EXE) command (available in Windows 95 and 98) or the IPCONFIG.EXE command (available in Windows 98, Windows Me, Windows NT, Windows 2000, Windows XP, and Windows Server 2003). The ping utility can be used to verify a TCP/IP connection, while the trac-

Figure 10.12   TCP/IP Properties

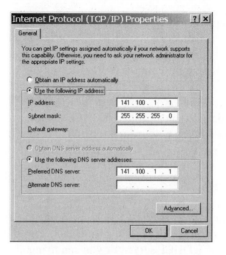

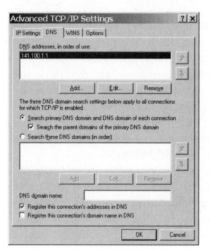

ert will show all router addresses as it connects to another computer. Lastly, the Address Resolution Protocol (ARP) can be used to display the MAC address.

10.12.4   Windows IPCONFIG and WINIPCFG Programs

The IPCONFIG.EXE and WINIPCFG.EXE will display the addressing and settings of TCP/IP. To execute the Windows IP configuration utility, you would type WINIPCFG.EXE from the Windows command prompt or from the Run. . . option from the Start button. (See figure 10.13.) If you have more than one network interface, you can

Figure 10.13   WINIPCFG Utility

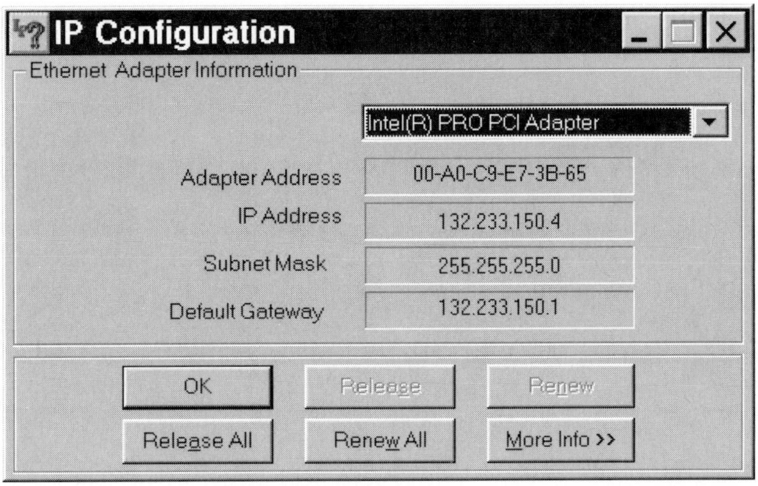

Note: If there is more than one network adapter, the various cards can be selected by the pull-down box.

select the appropriate adapter from the pull-down menu. You can view the IP address, the subnet mask, and the default gateway/router for that interface. If you click on the More Info>> button, you also view other information such as addresses and configurations of the WINS, DNS, and DHCP server.

The IPCONFIG command can only be executed at the command prompt. (See figure 10.14.) Much like the IP configuration utility, the IPCONFIG command also shows the IP address, the subnet mask, and the default gateway/router address. To see additional information, you can type in IPCONFIG/ALL.

To make assigning the addresses easier, you can automatically assign the IP addresses to the computers using a DHCP server. To release the address assigned by the DHCP server, you could click on the IP Configurations Renew or Renew All buttons, or you could type in IPCONFIG/RELEASE. To request new addresses from the DHCP server, you can click on the Renew or Renew All button.

NOTE: The Release All and Renew All buttons do all settings including the WINS and DNS server addresses.

Any time that you use the IP configuration utility or the IPCONFIG command and you are having trouble connecting to the network, you should verify the IP address (to make sure it is on the correct subnet), the subnet mask, and the router. If the address is not in the correct subnet, you would not be able to communicate with any other computer on the network. If the subnet mask is wrong, you may not be able to communicate with other computers. If the gateway address is wrong, you can communicate with computers on your subnet, but not on other subnets.

Figure 10.14   IPCONFIG Command

```
C:\>ipconfig

Windows 98 IP Configuration

0 Ethernet adapter :

    IP Address. . . . . . . . . : 132.233.150.4
    Subnet Mask . . . . . . . . : 255.255.255.0
    Default Gateway . . . . . . : 132.233.150.1
```

```
C:\>IPCONFIG /ALL

  Windows 98 IP Configuration

    Host Name . . . . . . . . . : HOSTPC.DOMAIIN.COM
    DNS Servers . . . . . . . . : 132.233.150.10
    Node Type . . . . . . . . . : Hybrid
    NetBIOS Scope ID. . . . . . :
    IP Routing Enabled. . . . . : No
    WINS Proxy Enabled. . . . . : No
    NetBIOS Resolution Uses DNS : Yes

  0 Ethernet adapter :
    Description . . . . . . . . : Intel(R) PRO PCI Adapter
    Physical Address. . . . . . : 00-A0-C9-E7-3B-65
    DHCP Enabled. . . . . . . . : No
    IP Address. . . . . . . . . : 132.233.150.4
    Subnet Mask . . . . . . . . : 255.255.255.0
    Default Gateway . . . . . . : 132.233.150.1
    Primary WINS Server . . . . : 132.133.150.10
    Secondary WINS Server . . . :
    Lease Obtained. . . . . . . :
    Lease Expires . . . . . . . :
```

Figure 10.15 NWLink IPX/SPX NetBIOS Dialog Box

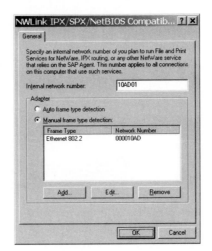

10.12.5 Other Network Protocols

Two common network protocols are NWLink (IPX/SPX) and NetBEUI, both of which are much simpler to configure than TCP/IP. For the NetBEUI Protocol, nothing has to be done. Remember that the NetBEUI Protocol is nonroutable.

For the NWLink, you need to specify the internal network number (8-digit hexadecimal number) that identifies the server. In addition, you need to specify the frame type and external number that identifies the subnet. If you select the auto frame type detection, it will automatically detect the frame type used by the network adapter to which it is bound. If NWLink detects no network traffic or if multiple frame types are detected in addition to the 802.2 frame type, NWLink sets the frame type to 802.2. (See figure 10.15.) To help troubleshoot the IPX Protocol in Windows 2000, Windows Server 2003, and Windows XP, you can use the IPXROUTE command to determine computer settings and perform diagnostic tests to resolve communication problems.

10.13 SHARING DRIVES, DIRECTORIES, AND PRINTERS

As mentioned earlier in this chapter, File and Print Sharing for Microsoft networks gives you the ability to share your files or printers with Windows computers using the SMB Protocol.

NOTE: Even though you enable file and print sharing for an individual computer, you must still share a drive, directory, or printer before it can be accessed through the network.

Figure 10.16   Sharing Drive/Directory

10.13.1   Sharing Drives and Directories

If you enable file sharing, you can share any drive or directory by right-clicking the drive or directory and selecting the Sharing option from the File menu or by selecting the Sharing option from the shortcut menu of the drive or directory. You would then provide a share name (the name seen by other clients) and the type of access that users can have. A shared drive and directory will be indicated with a hand under the drive or directory icon. (See figure 10.16.)

NOTE: If you are using Windows XP, you need to open a folder, select the Tools menu, and select the Folder Options option. Then click the View tab and deselect the simple file

Figure 10.17   Access a Shared Folder Using Windows Explorer

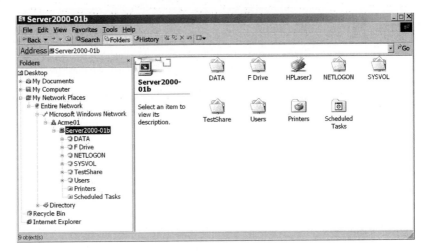

sharing. This will allow you to access the same file sharing and NTFS permissions interface as available in the other Windows version.

The shared drive or directory can then be accessed in one of three ways:

1. Using Network Neighborhood and accessing the resources under the server. (See figure 10.17.)
2. Specifying the Universal Naming Convention (UNC) using the Run. . . option under the Start button. The UNC format is specified as \\computername\sharename.
3. Selecting the Map Drive button on the tool bar of Network Neighborhood or Microsoft Explorer on machines using Active Desktop.

10.13.2   User and Share Security Models

Depending on the version of Windows you are using and the options you choose, Windows can provide share-level security or user-level security. Peer-to-peer networks normally incorporate share-level security, in which a network administrator assigns passwords to network resources. In a network that uses share-level security, users must remember many passwords. When users have to remember many passwords, network security can be violated because some users will write down their passwords and post the passwords near their computers. Share-level security is a decentralized network security method, which is only practical in small networks.

In the user-level security model, a user has a user account that includes a user name and password. The user account is provided with an access control list each time the user logs onto the network. A user can only access a network resource if that resource is on the ACL. The user-level security model allows a network administrator to manage network security from a central network location. An administrator would normally

Table 10.17   Share Permissions

Share Permissions	Description
Read	The read permission allows the user to view folder names and file names, open and view subfolder files and their attributes, and navigate the tree structures.
Change	The change permission allows all permissions granted by the read permission, and it also allows the user to create folders, add files to folders, change data in files, append data to files, change file attributes, and delete folders and files.
Full control	The full control allows all permissions granted by the change permission, and it also allows the user to change file permissions and take ownership of files.

assign users to group accounts and then grant the group accounts permissions to resources on the network. Groups ease the administrative burden because the administrator must only change the permissions granted to a group account to change the permissions for the users in that group. Of course, servers would typically use the user-level security mode.

10.13.3   Share Permissions

To control how users gain access to a shared folder, you assign shared folder permissions.

NOTE: These permissions are not needed or used if the user is accessing the directory locally (logged onto the computer that has the shared drive or directory). The shared folder/drive permissions are shown in table 10.17.

To grant or change share permissions to a shared folder, you would right-click the shared folder or drive, select the Sharing option, and click on the Permissions button. To add a user or group, click on the Add button. To remove a user or group, click to highlight the user or group and click on the Remove button. To specify which permissions to grant, click to highlight the user or group and select or deselect the Allow and Deny options.

Different from NTFS permissions, there is only one level of permissions assigned to a shared folder. For example, on the SERVER1 server, suppose you share a folder called FOLDER1 and assign the full control permission. In the FOLDER1 folder, suppose you share a folder called FOLDER2 and assign the modify permission. Therefore, if you access the \\SERVER1\FOLDER1 folder, you are using the full control share permission, even when accessing the \\SERVER1\FOLDER1\FOLDER2 folder. If you access the \\SERVER folder, you are using the change share permission. The permissions assigned to FOLDER1 are not an issue.

Since a user can be a member of several groups, it is possible for the user to have several sets of explicit permissions to a shared drive or folder. The effective permissions are the combination of all of the user and group permissions. For example, if the user has

Table 10.18   Special Shares

Special Share	Description
Drive letter$	A shared folder that allows administrative personnel to connect to the root directory of a drive, also known as an administrative share. It is shown as A$, B$, C$, D$, and so on. For example, C$ is a shared folder name by which drive C might be accessed by an administrator over the network.
ADMIN$	A resource used by the system during remote administration of a computer. The path of this resource is always the path to the Windows system root (the directory in which Windows is installed: for example, C:\Winnt).
IPC$	A resource sharing the named pipes that are essential for communication between programs. It is used during remote administration of a computer and when viewing a computer's shared resources.
PRINT$	A resource used during remote administration of printers.
NETLOGON	A resource used by the Net Logon service of a Windows Server computer while processing domain logon requests. This resource is provided only for Windows 2000 Server and Windows Server 2003 computers. It is not provided for Windows 2000 Professional and Windows XP. computers.
FAX$	A shared folder on a server used by fax clients in the process of sending a fax. The shared folder is used to temporarily cache files and access cover pages stored on the server.

a write permission to the user and the read permission to the group that a user is a member of, the effective permission would be the write permission.

NOTE: Like NTFS permissions, deny permissions override the granted permission.

When accessing a shared folder on an NTFS volume, the effective permissions that a person can do in the shared folder are calculated by combining the shared folder permissions and the NTFS permissions. When combining the two, you first determine the cumulative NTFS permissions and the cumulative shared permissions and apply the more restrictive permissions or the permissions that give the person the lesser permissions.

In Windows 2000 and Windows Server 2003 several special shared folders are automatically created by Windows for administrative and system use. (See table 10.18.) Different from regular shares, these shares do not show when a user browses the computer resources using Network Neighborhood, My Network Place, or similar software. In most cases, special shared folders should not be deleted or modified. For Windows 2000 Professional and Windows XP Professional computers, only members of the administrators or backup operators group can connect to these shares. For Windows 2000 and Windows Server 2003, members of administrators, backup operators, and server operators groups are the only members that can connect to these shares.

An **administrative share** is a shared folder typically used for administrative purposes. To make a shared folder or drive into an administrative share, the share name must have a $ at the end of it. Since the shared folder or drive cannot be seen during browsing, you would have to use a UNC name, which includes the share name (including the $). And the folder would have to be accessed using the Start button, selecting the Run option and typing the UNC name, and clicking the OK button. By default, all volumes with drive letters automatically have administrative shares (C$, D$, E$, and so on). Other administrative shares can be created as needed for individual folders.

10.13.4   Home Directories Permissions

To create a home folder on a network file server, the first step is to create and share a folder such as a home or user's folder in which to store all home folders on a network server. The home folder for each user will reside below the shared folder. The second step is to remove the default permission for the default full control from the everyone group and assign full control to the users group. This ensures that only users with domain user accounts can gain access to the shared folder.

To provide the home folder path, you open the properties of the user and input the path in the Home folder section. Since the home folder is on a network server, click Connect and specify a drive letter to use to connect. As a result, when the user logs onto the network, the drive letter you assign will appear in the My Computer. In the text box, a UNC name appears in the form of \\servername\sharename\userlogonname. If you use the %username% variable as the user's logon name, it will automatically name and create the user's home folder as the same name as the user's logon name. In addition, the user and the built-in administrator group are assigned the NTFS full control permission. All other permissions are removed from the folder, including those for the everyone group. This way, the folder is secure for the individual users. Lastly, you can further enhance the home folder feature by redirecting the user's My Documents pointer to the location of his or her home directory.

10.13.5   Sharing Printers

To make a printer connected to a Windows machine available to other people, you would "share" the printer. To share a printer is very similar to sharing files. First you would install the print driver using the Add Printer icon in the Printer folder. Next you would select the printer, and then you would choose the Sharing option. After assigning the printer share name and option password, you would click on the OK button. A hand will appear under the Printer icon to indicate that the printer is shared.

You would then go to the client computers and run the Add Printer Wizard. Instead of choosing the Local Printer option, you will be choosing the Network Printer option. You would then be prompted for the UNC name (\\computername\sharename) of the printer. When you print to this printer, the print job will be automatically redirected to the Windows 9X machine and sent to the printer.

Figure 10.18   Network Neighborhood Showing a Computer With Network Services

10.13.6   Browsing the Network

Windows uses a computer browser service to easily find network computers and their services. An example of using the computer browser service is when you use Network Neighborhood, My Network Places, the NET VIEW command, and Windows Explorer to show the available network resources such as shared drives, directories, and printers in a graphical environment. The computer browser service allows easy access to these resources. (See figure 10.18.)

After installing the appropriate network software (client software, network card driver, and protocol), the Network Neighborhood is automatically placed on the desktop. When you double-click on the icon, a window will open that displays computers that have network resources available. By double-clicking on one of the computers listed in the Network Neighborhood, a new window will open and display the network resources (mostly shared drives, directories, and printers) provided by that computer. (See figure 10.19.)

Windows uses a browser service to discover all shared devices on the network and compile a database of those resources. A browser is a network computer that tracks the location, availability, and identity of shared devices. Several kinds of browsers exist including a domain master browser, master browser, and backup browser. A domain master browser tracks resources for a group of domains. A master browser maintains a database (called a browse list) of shared resources for its domain. Every time a computer on the network starts, it registers itself with the master browser. In an environment containing two or more domains, each master browser will pass along its browse list to the domain master browser. A backup browser keeps a copy of the master browser browse list in case the master browser goes down. Although every server on a Windows NT network has the potential to act as a browser, by default the PDC is the master browser for its domain. In Windows 2000 and Windows Server 2003 networks, Active Directory can provide computer browser service by using global catalogs.

Figure 10.19   Network Resources Shown in Network Neighborhood

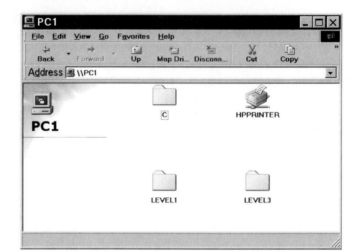

When Windows clients need to perform NetBIOS name resolution, the client will first look in its local cache to see if it knows its IP address by talking to it recently. If not, it will then check the LMHOSTS file. If the entry is not in the file, the client will, by default, do a broadcast looking for that computer (if the client is a B node) or by checking with a WINS server (if the client is a P or H node).

B-node broadcasts don't go through routers. So typically NT4 domains with B nodes that don't have any WINS servers to access will have a truncated browse list featuring only those workstations on the same side of the router as they were located. Additionally, master browser election wars would frequently occur when more than one domain was on the same subnet, causing more confusion. So with Windows NT4, the Network Neighborhood list was created via a highly unreliable network of master browsers on each subnet.

The Windows 2000/Windows Server 2003 domain master browser (the DC with the PDC emulator role) actively queries its configured WINS server for a list of all registered domains. It uses this list to contact the other domain master browsers in order to coordinate browse lists with them. This means that the browse list presented to the workstation should be more complete than before. All workstations should be represented, assuming that the domain master browser of every domain (or workgroup) is configured to use a central WINS server and also assuming that at least one workstation from every domain (or workgroup) on each router segment is able to contact its domain master browser.

10.13.7   Internet Information Services

Internet Information Services (IIS) is a built-in web and FTP server found in Windows 2000, Windows Server 2003, and Windows XP. IIS makes it easy to share documents and information across a computer in an intranet or the Internet. To manage IIS, you use the Internet Information Service console or the browser-based Internet Services Manager (HTML).

10.14   EVENT VIEWER

The **event viewer** is a very useful utility for viewing and managing logs of system, program, and security events on your computer. The event viewer gathers information about hardware and software problems, and it monitors Windows security events. The event viewer can be executed by clicking on the Start button, clicking on Programs, clicking on Administrative Tools, and clicking on Event Viewer or by adding it to the Microsoft Management console.

The Windows event viewer starts with three kinds of logs:

- **Application log**—The application log contains events logged by programs. For example, a database program might record a file error in the programs log. Program developers decide which events to monitor. The application log can be viewed by all users.
- **Security log**—The security log contains valid and invalid logon attempts as well as events related to resource use, such as creating, opening, or deleting files or other objects. For example, if you have enabled logon and logoff auditing, attempts to log onto the system are recorded in the security log. By default, security logging is turned off. To enable security logging, you can use Group Policy to set the audit policy, or you can change the registry. To audit files and folders, you must be logged on as a member of the administrators group or have been granted the Manage auditing and security log right in Group Policy.
- **System log**—The system log contains events logged by the Windows system components. For example, the failure of a driver or other system component to load during start-up is recorded in the system log. The event types logged by system components are predetermined by Windows. The system application log can be viewed by all users.

All BackOffice family applications (Microsoft applications designed to work on the Windows NT family) can post security events to the Windows 2000/Windows Server 2003 event log. In addition, several servers also make their own logs, which can also be viewed by the Windows 2000/Windows Server 2003 event viewer.

SUMMARY

1. Windows NT Server, Windows 2000 Server, and Windows Server 2003 are operating systems that can function as a desktop operating system but were designed as a network file, print, or application server.

2. The Windows NT family OSs (Windows NT, Windows 2000, Windows XP, and Windows Server 2003) are modular operating systems that have many small self-contained software components that work together to perform the various operating system tasks.

3. The Windows NT family architecture can be divided into two major layers: user mode and kernel mode (also referred to as privileged mode).

4. A kernel is the central module of an operating system. Typically, the kernel is responsible for memory management, process and task management, and disk management.

5. A service is a program, routine, or process that performs a specific system function to support other programs.

6. The Microsoft Management console (MMC) is one of the primary administrative tools used to manage Windows 2000 and Windows Server 2003 services.

7. Windows 2000 and Windows Server 2003 support FAT, FAT32, and NTFS file systems.

8. NTFS is a file system for the Windows NT, family operating systems (OSs) designed for both the server and workstation. It provides a combination of performance, reliability, security, and compatibility.

9. In a workgroup, each computer has a local security database, so that it can track its own user and group account information.

10. The Windows domain is a logical unit of computers and network resources that define a security boundary. A domain uses one database to share its common security and user account information for all computers within the domain.

11. The system of domains and trusts for a Windows NT Server network is known as Windows NT Directory Service (NTDS).

12. Active Directory is a directory service that combines domains, X.500 naming services, DNS, X.509 digital certificates, and Kerberos authentication.

13. Active Directory, similar to a file system found on your hard drive, is a structured hierarchy consisting of domains and organizational units connected together to form a tree of objects (computers, users, and network resources).

14. An object in the Active Directory tree is a distinct, named set of attributes or characteristics that represent a network resource, including computers, people, groups, and printers.

15. A forest consists of one or more trees that are connected by two-way, transitive trust relationships, which allow users to access resources in the other domain/tree.

16. The computer that stores a replica (copy) of the account and security information of the domain and defines the domain is known as the domain controller.

17. A member server does not store copies of the directory database and therefore does not authenticate accounts or receive synchronized copies of the directory database.

18. A user account enables a user to log onto a computer and domain with an identity that can be authenticated and authorized for access to domain resources.

19. The home directory is a folder used to hold or store a user's personal documents.

20. A user profile is a collection of folders and data that stores the user's current desktop environment and application settings.

21. A group is a collection of user accounts.

22. By using groups, you can simplify administration by assigning rights and permissions to the group, and everyone listed in the group will be assigned these rights and permissions.

23. The computer account is used to uniquely identify the computer on the domain.

24. Policies are a tool used by administrators to define and control how programs, network resources, and the operating system behave for users and computers in the Active Directory structure.

25. Group policy settings are inherited, are cumulative, and affect all computers and user accounts in the Active Directory container with which the group policy is associated.

26. A right authorizes a user to perform certain actions on a computer, such as logging onto a system interactively or backing up files and directories.

27. The best way to give sufficient permissions to an organizational unit is to delegate administrative control to the container (decentralized administration), so that the user or group will have administrative control for the organizational unit and the objects in the organizational unit.

28. A primary advantage of NTFS over FAT and FAT32 is that NTFS volumes have the ability to apply NTFS permissions to secure folders and files.

29. To get full access to the network, you would need a client software, a network adapter driver or dial-adapter driver (such as a modem), and a protocol stack.

30. The Client for Microsoft Networks provides the redirector (VREDIR.VXD) to support all Microsoft networking products that use the Server Message Block (SMB) Protocol.

31. The IPCONFIG.EXE and WINIPCFG.EXE will display the addressing and settings of TCP/IP.

32. If you enable file sharing, you can share any drive or directory by right-clicking the drive or directory and selecting the Sharing option from the File menu or by selecting the Sharing option from the shortcut menu of the drive or directory.

33. To control how users gain access to a shared folder, you assign shared folder permissions.

34. An administrative share is a shared folder typically used for administrative purposes.

35. The Network Neighborhood (and My Network Places) is a browser service that shows the available network resources such as shared drives, directories, and printers in a graphical environment and allows easy access to these resources.

QUESTIONS

1. You just installed Windows 2000 on a computer on which you had Windows NT. Unfortunately some of the components are not being detected by Windows 2000. What should you do?
   a. Try to reinstall it again.
   b. Make sure that the hardware devices are on the compatibility list.
   c. Reformat the system and reinstall again.
   d. Make sure that you have the newest BIOS.

2. What acts as a translator between your hardware and the Windows 2000 OS specific to the computer architecture?
   a. Device drivers   b. TSRs
   c. ACPI   d. HAL
   e. Kernel   f. Microkernel

3. _____ is a less privileged mode that has no direct access to hardware and can only access its own address space.
   a. User mode
   b. Protected mode
   c. Isolated mode
   d. Kernel mode

4. Windows 2000 primarily uses _____ multitasking.
   a. cooperating
   b. preemptive
   c. task switching
   d. context switching

5. What type of multitasking does Windows 2000 use?
   a. Symmetric   b. 32 bit
   c. Asymmetric   d. 64 bit

6. Which service allows you to access network resources?
   a. PnP   b. Workstation
   c. Server   d. Print spooler

7. To add or configure the network components, you would use _____.
   a. Network and Dial-up Connections
   b. Add/Remove Hardware
   c. Network Neighborhood
   d. Device Manager
   e. System

8. Which is the default client for Windows 2000?
   a. Client Services for Microsoft
   b. Client Services for NetWare
   c. File and Print Sharing for Microsoft Networks
   d. File and Print Sharing for NetWare

9. What is the default protocol or protocols loaded in Windows 2000?
   a. TCP/IP
   b. NetBEUI
   c. AppleTalk
   d. IPX/NWlink
   e. DLC

10. Automatic Private IP Addressing can provide an IP address in which range?
    a. 169.254.0.0 to 169.254.255.255
    b. 130.0.0.0 to 130.255.255.255
    c. 150.101.0.0 to 150.101.255.255
    d. 127.0.0.0 to 127.255.255.255

11. Which utility can you use to find the MAC and TCP/IP address of your Windows 95/98 workstation?
    a. Ping   b. IPCONFIG
    c. WINIPCFG   d. Tracert

12. Which Windows NT utility will display the current TCP/IP configuration of that host?
    a. ARP
    b. WINIPCFG
    c. IPCONFIG
    d. WINIPCONFIG

13. Which IPCONFIG switch will display the most complete list of IP configuration information for that station for a Windows XP computer?
    a. /All   b. /Release
    c. /Renew   d. /?

14. Which utility on a Windows 95 computer can be used to view IP address configuration information?
    a. WINIPCFG   b. IPCONFIG
    c. WINIPCONFIG   d. netstat

15. By default, on which directory service does Windows NT Server 4 rely?
    a. NDS   b. AD
    c. NIS   d. NTDS

16. What is the name of the domain user and group administration program for a Windows NT Server?
    a. NTADMIN
    b. User Manager for Domains
    c. Domain Administrator
    d. NT Domain Manager (NTDM)

17. In Windows 2000, to see all domains, you would use _____.
    a. Add Network Place
    b. Computers Near Me

c. Entire Network

d. Network Neighborhood

18. You want to examine the IP Address–to–Media Access Control resolution of outgoing packets on your Windows 2000 Server computer. Which of the following should you use?

a. Tracert   b. Ping

c. ARP   d. nbtstat

19. If you combine domain trees, you form a domain _____.

a. forest   b. collective

c. jungle   d. group

20. Which of the following is used in a peer-to-peer network that has each computer keep its own security database?

a. Forest

b. Domain

c. Workgroup

d. Replication partner

21. Which of the following uses a centralized security database to authenticate users on the network?

a. Client

b. Workgroup

c. Domain

d. Replication partner

22. Which of the following is true of a global catalog in Windows 2000?

a. A global catalog contains only the commonly queried objects and ALL attributes for a single domain only.

b. A global catalog contains only the commonly queried objects and ALL attributes for a forest.

c. A global catalog contains ALL objects and ALL attributes for a forest.

d. A global catalog contains ALL objects and common attributes for a forest.

e. A global catalog contains ALL objects and common attributes for a single domain only.

23. Which of the following are containers? (Choose all that apply.)

a. Organization units

b. Groups

c. Domains

d. Domain controllers

24. The computer that stores a replica (copy) of the account and security information of the domain and defines the domain is known as the _____.

a. member server

b. ACL

c. domain controller

d. Windows 2000 Advanced Server

25. Which of the following are true of organizational units (OUs) in a Windows 2000 domain? (Choose all that apply.)

a. Administrative control of OUs can be delegated to other users.

b. An OU can contain other OUs.

c. Administrators have permission to create OUs by default.

d. Users and computers are two of the built-in OUs in a domain.

26. Which feature of Windows 2000 allows an administrator to enforce desktop settings for users?

a. System policy

b. User policy

c. Group policy

d. Universal policy

27. You want to control the permissions of files and directories on an NTFS drive on the network. Which application must you use?

a. Windows Explorer

b. Active Directory Users and Computers console

c. Computer Management console

d. Disk Administrator console

28. A Windows 2000 Server contains a shared folder on an NTFS partition. Which of the following statements concerning access to the folder is correct?

a. Compared with accessing the folder locally, a user accessing the folder remotely has the same or more restrictive access permissions.

b. Compared with accessing the folder locally, a user accessing the folder remotely has less restrictive access permissions.

c. Compared with accessing the folder locally, a user accessing the folder remotely has the same access permissions.

d. Compared with accessing the folder locally, a user accessing the folder remotely has more restrictive access permissions.

29. What feature of Windows NT 4.0 stores user and group information for a domain?

a. The bindery service   b. NDS

c. NTDS   d. SAM

30. Which network client will allow your Windows computer to participate in a Microsoft workgroup?

a. Client for NetWare Networks

b. Banyan DOS/Windows 3.1 Client

c. Client for Microsoft Networks

d. NetWare Client32

31. Your company uses a Windows NT 4.0 domain. You attempt to connect to a shared folder, but your computer displays an error message indicating that you cannot access the folder. What is the most likely cause of this problem?
   a. You do not have the password for the shared folder.
   b. You are not the owner of the shared folder.
   c. You are a member of a group that has been denied access to the shared folder.
   d. You have been assigned a user policy that has been denied access to the shared folder.

32. You have recently transferred from the marketing department to the documentation department. Your company uses a Windows NT 4.0 network. You attempt to print the documentation guidelines to a printer named DOC_PRINTER1 in the documentation department. Your computer displays an error message indicating that you cannot print to DOC_PRINTER1. Paul, one of your coworkers, indicates that he recently printed a document using DOC_PRINTER1 and that the printer seemed to be in working order. What is the most likely reason you cannot print to DOC_PRINTER1?
   a. Your user account has not been assigned to the documentation group policy.
   b. Your user account is not a member of the documentation group.
   c. You have not been given the share-level password to the printer.
   d. Your user account has not been assigned a computer policy that allows you to print to DOC_PRINTER1.

33. What is IIS?
   a. International Internet Service
   b. Internetwork Internet Server
   c. Intelligent Intranet System
   d. Internet Information Services

34. Which of the following Microsoft file systems supports file-level security?
   a. FAT32   b. NTFS
   c. FAT12   d. FAT16

35. What is the term for a Windows NT 4.0 Server that contains the accounts database?
   a. BDC   b. TCD
   c. RDC   d. PDC

36. How many BDCs (backup domain controllers) are required in an NT 4.0 network?
   a. As many as there are PDCs.
   b. One

   c. Two
   d. None

37. You'll be setting up a network where your client machines will be running various operating systems such as Windows 95, Macintosh, and Linux. Which server operating systems could you use? (Choose two answers.)
   a. NetWare
   b. Macintosh
   c. Windows .NET Server
   d. Windows 2000 Server

38. Your LAN's only NOS is NT 4.0 Server. Your main concerns are network security and ease of policy distribution. Which client operating system do you choose?
   a. DOS 6.22
   b. Windows NT 4.0 Workstation
   c. Windows for Workgroups 3.11
   d. Windows 95

39. You manage a Windows 2000 network that consists of 25 Windows 2000 servers, 2 NetWare 4 servers, and 1 NetWare 5 server. Your client machines are all running Windows 98. You want your client computers to have the best possible access to services on the network. What client software would you install on your workstations to allow this? (Pick two answers.)
   a. Microsoft Client for NetWare Networks
   b. Netware Client for Microsoft Networks
   c. Microsoft Client for Microsoft Networks
   d. Novell Client for Windows 95/98

40. What is a Windows NT domain?
   a. A physical boundary
   b. A logical grouping of resources
   c. A group of servers only
   d. A logical security boundary

41. Which Windows NT Server utility would you use to view the account properties of a user within an NT domain?
   a. User Manager for Domains
   b. User Manager
   c. NDS
   d. NWADMIN

42. You are running NetWare on all your servers. You are concerned about securing local data on your workstations. What operating system do you install on your workstations?
   a. Windows NT (or Windows 2000)
   b. Windows 98
   c. NetWare
   d. Windows 95

43. What is the term for a logical drive resource that can be accessed through a drive letter or name and that spans multiple physical drives or parts of multiple physical drives?
   a. Share
   b. HDD
   c. Volume
   d. None of the choices listed

44. On a Windows 95 or 98 computer you suspect a resource conflict is causing the NIC to malfunction. To find the problem, what would be a good place to look first?
   a. Device Manager
   b. System Management
   c. Device Editor
   d. System Resource Editor

45. Which of the following file systems will support long file names?
   a. HPFS
   b. FAT16
   c. NTFS
   d. All of the above

46. You install a 16-bit ISA sound card in a client's computer, and now the PCI network card won't work. What should you do to correct the situation?
   a. Reserve an IRQ and I/O port addresses for your sound card in Device Manager.
   b. Run the Windows Conflict Manager.
   c. Set the PCI network card to use different system resources.
   d. Remove the sound card since ISA and PCI devices cannot coexist in the same PC.

47. There are a group of Windows 98 computers on your network that were recently installed by a contractor. They are unable to browse any servers. All servers are running the Windows 2000 Server operating system. You check the network cards and cabling on your Windows 98 workstations and they are fine. Which network component did the contractor probably forget to install?
   a. Microsoft Client for NetWare Networks
   b. Client for Microsoft Networks
   c. Client/Server Connection Client
   d. Microsoft Client for Networks

48. What is ACL?
   a. Actuarial calibration length
   b. Active cable length
   c. Access centralized list
   d. Access control list

49. On a Windows NT or 2000 server, where would you find the system, security, and application logs?
   a. Server.log
   b. Sys.log
   c. c:\winnt\system32\logfiles
   d. Event viewer

50. On a Windows 98 computer, where do you enter the NetBIOS or computer name for the system? (Choose two answers.)
   a. In the TCP/IP Properties\Identification tab
   b. In the System Properties\Identification Properties tab
   c. In the Network Neighborhood Properties\Network Identification tab
   d. In the Control Panel\Network applet\Network Identification tab

HANDS-ON EXERCISES

Exercise 1: Computer Management

1. Click on the Start button, select the Programs option, select the Administrative Tools option, and select the Computer Management option.
2. Open the Event Viewer. Click on the System option under Event Viewer. Look for any errors. If you find one, double-click it to view it.
3. Go back to the Event Viewer and click on the Application option under the Event Viewer. Look for errors. If you find one, double-click to view it.
4. Close the Event Viewer branch.
5. Click on the System Summary under System information. Record the amount of virtual memory currently being used.

6. Double-click on the Hardware resources and click on the IRQs to show which devices are using which other devices.
7. Click on the Services option located under Services and Applications. This will display all the services loaded on your system.
8. To get a better view of the services, open the View menu and select the Detail option. Record the status of the service, workstation, and print spooler service.
9. Right-click the Server option and select the Properties option.
10. Record the Startup type.
11. Close the Computer Management console.

Exercise 2: Navigating the Network

1. Right-click the desktop and create a folder called SHARED.
2. Right-click the SHARED directory and select the Sharing option.
3. Select the Share this folder option and click the OK button. A hand should appear underneath the SHARED directory icon.
4. Open the Shared directory and create a text file with the your name. To create a text file, right-click the Open folder, select the New option, and select the Text Document option.
5. Close all windows.
6. Double-click My Network Places to open it.
7. Double-click Computers Near Me.
8. Double-click your computer to see all your shared resources.
9. Double-click the SHARED share to see the text file that you created.
10. Click the Up button twice on the tool bar to go back to the Computers Near Me window.
11. Double-click your partner's computer. Double-click the SHARED share to see your partner's text file.
12. Click the Up button three times on the tool bar to go back to My Network Places.
13. Double-click the Entire Network icon and then click on the entire contents hyperlink on the left side of the window.
14. Double-click the Microsoft Windows Network icon. You should see your workgroup and any domains that might be out there.
15. Double-click your workgroup to see all the computers assigned to the workgroup.
16. Use the Up arrow button on the tool bar to go back to the My Network Places window.
17. Double-click the Add Network Place icon.
18. In the Type the location of the Network Place text box, type the following:

    partnercomputername\SHARED

19. Click on the Next button and then click on the Finish button.
20. Close the SHARED window.
21. Notice the new icon in the My Network Places. Double-click on it.
22. Close all windows.

Exercise 3: Install and Configure a WINS

Installing the WINS

Computer A should do the following:
1. Open the Add/Remove Programs applet in the Control Panel.
2. Click on the Add/Remove Windows Components.

3. Click on the Next button.
4. Click to highlight the Networking Services option.
5. Click on the Details button.
6. In the Subcomponents of Networking Services, make sure that there is a check mark in the check box next to the Windows Internet Name Service (WINS).
7. Click on the OK button.
8. Click on the Next button.
9. If the Insert Disk dialog box appears, insert the Windows 2000 installation CD-ROM, ensure that the path to the source files is correct, and click on the OK button.
10. Click on the Finish button.

Configuring the WINS Client

Computer A should do the following:
1. To make the WINS server register itself in its WINS database, you need to make the WINS server a client. To make a WINS server a client, right-click My Network places and select the Properties option. Right-click the Local Area Connection and select Properties.
2. Click Internet Protocol (TCP/IP) and click on the Properties button.
3. In the Internet Protocol (TCP/IP) Properties dialog box, click the Advanced button.
4. In the WINS tab, click the Add button.
5. In the TCP/IP WINS Server dialog box, type the IP address of the WINS server and click the Add button.
6. Click the OK button to close the Advanced TCP/IP Settings dialog box and click OK to close the Internet Protocol (TCP/IP) Properties dialog box.
7. Click OK to close the Local Area Connection Properties dialog box.

Computer B should do the following:
1. To configure the client to use the WINS server, right-click My Network places and select the Properties option. Right-click the Local Area Connection and select Properties.
2. Click Internet Protocol (TCP/IP) and click on the Properties button.
3. In the Internet Protocol (TCP/IP) Properties dialog box, click the Advanced button.
4. In the WINS tab, click the Add button.
5. In the TCP/IP WINS Server dialog box, type the IP address of the WINS server and click the Add button.
6. Click the OK button to close the Advanced TCP/IP Settings dialog box and click OK to close the Internet Protocol (TCP/IP) Properties dialog box.
7. Click OK to close the Local Area Connection Properties dialog box.

Managing the WINS Server

Computer A should do the following:
1. From the Administrative Tools, start the WINS console.
2. Right-click Active Registration and select the Find by Name option.
3. In the Find names beginning with text box, type in an asterisk (*) and click on the Find Now button. Notice all your entries.
4. To add a static WINS record, right-click Active Registration and select the New Static option.
5. In the Computer name text box, type in TESTPC. In the IP address, type in 100.100.100.100. Click on the OK button.
6. Use the Find names option to show all entries in the WINS table. Notice the entries in the WINS table, especially the Static column.

Computer B should do the following:

1. From the command prompt, ping TESTPC. Since the TESTPC does not exist, the ping will respond to Destination host unreachable. Of course, the important thing at this time is to look at the address resolution.

Computer A should do the following:

1. Right-click the three entries for TESTPC and select the Delete option.

Exercise 4: Install and Configure a DHCP Server

Installing a DHCP Server

Computer A should do the following:

1. Open the Add/Remove Programs applet in the Control Panel.
2. Click on the Add/Remove Windows Components.
3. Click on the Next button.
4. Click to highlight the Networking Services option.
5. Click on the Details button.
6. In the Subcomponents of Networking Services, make sure that there is a check mark in the check box next to the Windows Internet Name Service (WINS).
7. Click on the OK button.
8. Click on the Next button.
9. If the Insert Disk dialog box appears, insert the Windows 2000 installation CD-ROM, ensure that the path to the source files is correct, and click on the OK button.
10. Click on the Finish button.

Creating a Scope

Computer A should do the following:

1. Open the DHCP console from the Administrative Tools.
2. Right-click the server and click the New Scope option. Click on the Next button.
3. On the Scope Name page, type the name of your server in the Name text box and click on the Next button.
4. Use 192.168.XXX.YY for the Start address and 192.168.XXX.YY+1 for the End address, where XXX is the room number and YY is your computer number. Change the subnet mask to 255.255.255.0. Click on the Next button.
5. Since we have no exclusions, click on the Next button.
6. On the Lease Duration, click the Next button to keep the default of 8 days.
7. On the Configure DHCP Options page, select the Yes, I want to configure these options now and click the Next button.
8. On the Router (Default Gateway) page, type in the address of your gateway or local router. If you don't have one, for now type in 192.168.1.254. Click on the Add button and then click the Next button.
9. On the Domain Name and DNS Servers page, type in ACME.COM for the parent domain and type in the address of your DNS server. Click the Add button and then the Next button.
10. On the WINS Servers page, type in the address of your primary WINS Server address. Click the Add button and then the Next button.
11. Select the Yes, I want to activate this scope now and click the Next button. Click the Finish button.
12. Click on the Scope option and look at the various options that were configured with the wizard.

Making a Client Reservation

Computer A should do the following:

1. From the DHCP console, right-click Reservations and select the New Reservation option.
2. Put in the MAC address of your partner's network card. You can find this out by executing the IPCONFIG /ALL command at your partner's computer.

Testing the DHCP Server

Computer B should do the following:

1. At the command prompt, execute the IPCONFIG /ALL command.
2. Right-click My Network Places and select the Properties option.
3. Right-click Local Area Connection and select the Properties option.
4. Click the Internet Protocol (TCP/IP) and click on the Properties button.
5. In the Internet Protocol (TCP/IP) dialog box, click the Obtain an IP address automatically option.
6. Click the Obtain DNS server address automatically option.
7. Click on the Advanced button.
8. In the WINS tab, click the WINS address and click the Remove button.
9. Click the OK button to close the Advanced TCP/IP Settings, click OK to close the Internet Protocol (TCP/IP) Properties dialog box, and click OK to close the Local Area Connection Properties dialog box. Remember, any settings set manually will override settings given by a DHCP server.
10. At the command prompt, execute the IPCONFIG /ALL command and study the current setting.

Exercise 5: Installing Active Directory

Computer A should do the following:

1. To start the Active Directory Installation Wizard and to make the stand-alone server into a domain controller, you will need to execute the DCPROMO.EXE file. Click on the Start button and select the Run option. Enter DCPROMO in the open text box and click on the OK button. Click on the Next button.
2. For the Domain Controller Type page, select the Domain controller for a new domain option and click on the Next button.
3. For the Create a Tree or Child Domain page, select the Create a new domain tree option and click on the Next button.
4. For the Create or Join Forest page, select the Create a new forest of domain trees option and click on the Next button.
5. For the New Domain Name page, enter ACMEYY.COM in the Full DNS name for the new domain text box. YY is your computer number. Click on the Next button.
6. On the NetBIOS Domain Name page, leave the default Domain NetBIOS name and click on the Next button.
7. On the Database and Log Locations page, leave the default file locations and click on the Next button.
8. On the Shared System Volume page, leave the default location for the SYSVOL folder and click on the Next button.
9. On the Configure DNS page, click on the Yes, install and configure DNS on this computer (recommended) option and click on the Next button.
10. On the Permissions page, select the Permissions compatible only with Windows 2000 servers option and click on the Next button.
11. On the Directory Services Restore Mode Administrator Password page, enter the administrator password and click on the Next button.
12. On the Summary page, click on the Next button.

13. Click on the Finish button. If the screen asks you to reboot the computer, do so.

NOTE: It may take a few minutes to reboot.

14. Log in as administrator.
15. Open your TCP/IP settings and change your DNS address to the address of your computer.

Exercise 6: Adding a Second Domain Controller

Computer B should do the following:
1. Open your TCP/IP settings and change your DNS address to the address of your partner's computer.
2. To start the Active Directory Installation Wizard and to make the stand-alone server into a domain controller, you will need to execute the DCPROMO.EXE file. Click on the Start button and select the Run option. Enter DCPROMO in the open text box and click on the OK button. Click on the Next button.
3. For the Domain Controller Type page, select the Additional domain controller for an existing domain option and click on the Next button.
4. For the Create a Tree or Child Domain page, select the Create a new domain tree option and click on the Next button.
5. For the Create or Join Forest page, select the Create a new forest of domain trees option and click on the Next button.
6. For the New Domain Name page, enter ACMExx.COM in the Full DNS name for new domain text box. Click on the Next button.
7. On the NetBIOS Domain Name page, leave the default Domain NetBIOS name and click on the Next button.
8. On the Database and Log Locations page, leave the default file locations and click on the Next button.
9. On the Shared System Volume page, leave the default location for the SYSVOL folder and click on the Next button.
10. On the Configure DNS page, click on the Yes, install and configure DNS on this computer (recommended) option and click on the Next button.
11. On the Permissions page, select the Permissions compatible only with Windows 2000 servers option and click on the Next button.
12. On the Directory Services Restore Mode Administrator Password page, enter the administrator password and click on the Next button.
13. On the Summary page, click on the Next button.
14. Click on the Finish button.

NOTE: If you have a DHCP server still running, it will need to be authorized for the domain. To authorize a domain, do the following:

15. From the DHCP console, right-click DHCP and select the Browse authorized servers . . . option.
16. In the Authorized Servers in the Directory dialog box, click the Add button.
17. In the Authorized DHCP Server dialog box, enter the name or IP address of the DHCP server to authorize, and then click OK.
18. In the DHCP dialog box, click Yes to confirm the authorization.

Exercise 7: Creating Organizational Units

1. Start the Active Directory User and Computers console.
2. To create an organizational unit called Sales, right-click the ACMExx.COM domain, select New option, and select Organizational Unit. In the Name text box, input SALES and click on the OK button.

3. Using the same method, create the following organizational units under ACMExx.COM:

RESEARCHX
EDUCATIONX
MANUFACTURINGX

Exercise 8: Creating Users

1. To create a user in the ACMExx.COM domain, right-click the ACMExx.COM domain, select New option, and select User. In the New Object–User dialog box input your first name, middle initial, and last name. For your User Logon Name, use your first initial, middle initial, and last name without spaces. Therefore, if your name is Paul G. Rogers, your login name would be PGRogers. Click on the Next button.

2. Enter the password of PW (uppercase) and enable the Password must change password on next logon option. Click on the Finish button.

3. After your account has been created, right-click on your account and select the Properties option. Input your description as domain administrator and your office as the server room. Input your telephone number, email address, and web page URL if you have one. Click on the Address tab and type in your address. Click on the Telephones tab and input your phone numbers. Click on the Organization tab and input administrator for the Title, IT for the Department, and Acme Corporation for the Company. Click on the OK button.

4. Create the following users in the appropriate organizational unit.

First Name	Middle Initial	Last Name	User Logon Name	Title	Department	Organizational Unit
Charles	L	Gee	CLGeeX	Sales Manager	Sales	Sales
Frank	J	Biggs	FJBiggsX	Sales Rep.	Sales	Sales
Herold	W	Jones	HWJonesX	Sales Rep.	Sales	Sales
Paul	L	Ray	PLRayX	Sales Rep.	Sales	Sales
Juan	O	Hermes	JOHermesX	Sales Rep.	Sales	Sales
Jill	K	Knight	JKKnightX	Sales Admin. Asst.	Sales	Sales
Jean	A	Mao	JAMaoX	Training Manager	Education	Education
Edward	J	Morell	EJMorellX	Trainer	Education	Education
Donna	L	Starr	DLStarrX	Manufact. Manager	Manufacturing	Manufacturing
Eric	O	Skow	EOSkowX	Manufact. Technician	Manufacturing	Manufacturing
Victor	N	Sloan	VNSloanX	Manufact. Technician	Manufacturing	Manufacturing
Sonny	K	Wong	SKWongX	Research Engineer	Manufacturing	Research
Gina	J	Smith	GJSmithX	Research Engineer	Manufacturing	Research

5. Right-click on the SALES organizational unit, select the New option, and select the Computer option. Enter computer1 for the name. Click on the OK button.
6. Right-click each of these users and input their appropriate departments.

Exercise 9: Home Directories

1. Open My Computer and open the C drive. In the C drive, right-click on empty space within the C drive window, select the New option, and select the Folder option. Name the folder USERS and press the Enter key.

   NOTE: In a real working environment, you should not store data files on the system and/or boot partition.

2. Right-click the USERS folder and select the Sharing . . . option. Select the Share this folder. Click on the Permissions button. Since Everyone is already highlighted, click on the Remove button.
3. Click on the Add button. Select the USERS group and click on the OK button. In the Permissions section, make sure there is a check mark for Allow Full Control. Click on the OK button to close the Sharing window and click OK to close the Properties window.
4. Within the Active Directory Users and Computers console, open the properties window of Charles L. Gee. Click on the Profile tab. Select the Connect to option under Login scripts, select the Z drive, and in the To: text box, enter:

   \\SERVER2000-xxy.acmeXX.com\Users\%USERNAME%

5. Open the USERS folder, and notice the CLGeeX folder.
6. Do the same for Frank J. Biggs, Herold W. Jones, and Paul L. Ray.

Exercise 10: Finding an Object

1. Right-click the ACMExx.COM domain and select the Find option.
2. Under the Users, Contacts and Groups tab, enter Jill in the Name: text box and click on the Find Now button. Notice that the computer had no problem finding Jill Knight.
3. Double-click on Jill K. Knight at the bottom of the window. Click on the OK button.
4. Click on the Advanced tab. Click on the Field button, select the User option, and select the Department field. Click on the Condition pull-down menu and look at the various conditions available. For the value, input Sales. Click on the Add button and click on the Find Now. . . button.

Exercise 11: Creating Groups

1. Right-click the ACMExx.COM domain, select the New option, and select the Group option. Click on the Global Group scope and Security Group type. Enter the Managers in the Group name text box and click on the OK button.
2. Right-click the Managers group and select the Properties option. Click on the Members tab, click on the Add . . . button, click on Charles L. Gee, and click on the Add button. Click on and add Jean A. Mao and Donna L. Starr. Click on the OK button to close the Select Users, Contacts, or Computers dialog box. Click on the OK button to close the Managers property window.
3. Right-click the ACMExx.COM domain, select the New option, and select the Group option. Click on the Global Group scope and Security Group type. Enter the ManagersX in the Group name text box and click on the OK button.

4. Right-click the ManagersX group and select the Properties option. Click on the Members tab, click on the Add . . . button, click on Charles L. Gee, and click on the Add button. Click on and add Jean A. Mao and Donna L. Starr. Click on the OK button to close the Select Users, Contacts, or Computers dialog box. Click on the OK button to close the Managers property window.

5. Right-click the ACMExx.COM domain, select the New option, and select the Group option. Click on the Global Group scope and Security Group type. Enter the ProductStaffX in the Group name text box and click on the OK button.

6. Right-click the ProductStaffX group and select the Properties option. Add all staff members from the manufacturing department.

7. Right-click the ACMExx.COM domain, select the New option, and select the Group option. Click on the Domain Local group scope and Security Group Type. Enter the ProductResourcesX in the Group name text box and click on the OK button.

8. Right-click the ProductResourcesX group and click on the Properties option. Add the Managers group, ProductStaffX group, and Jill K. Knight to the group.

9. Create a Universal group called SalesX.

   NOTE: The only thing left to do is to assign rights and permissions to the Product Resources group. When new employees are added, they only have to be added to the appropriate group and they will automatically inherit these rights.

10. Open My Computer and open the C drive. In the C drive, right-click on empty space within the C drive window, select the New option, and select the Folder option. Name the folder TestShare and press the Enter key.

11. Right-click the TestShare folder and select the Sharing. . . option. Select the Share this folder. Click on the Permissions button. Since Everyone is already highlighted, click on the Remove button.

12. Click on the Add button. Select the ProductResourcesX and click on the OK button. Click on the OK button to close the Sharing window.

   NOTE: In this case, we didn't need to use the local and global group because both the users and resources are in the same domain. Remember that global groups are meant to group people within a specific domain. The domain local group is intended to give rights and permissions to objects.

Exercise 12: Disable, Rename, and Delete an Account

1. Paul Ray just got fired. To disable the account, right-click on the Paul L. Ray account and select the Disable Account option. Notice the small X that appears.

2. Right-click the Paul L. Ray account and select the Rename option. While the entire name is highlighted, press the Delete key and press the Enter key.

3. Change the name to Tom J. Landers.

   Full name: Tom J. Landers
   First name: Tom
   Last name: Landers
   Display name: Tom J. Landers
   User logon name: TJLanders
   User logon name: (pre-Windows2000): TJLanders

4. Click on the OK button.

5. Right-click on Tom J. Landers account and select the Reset Password option. Change the password to TEST. Click on the OK button to close the windows and click on the OK button to close the confirmation dialog box.

6. Right-click on the Tom J. Landers account and select the Enable account option.
7. Right-click on the Tom J. Landers account and select the Properties option. Select the Profile tab and notice the location of the home folder. While the account has been changed, Tom Landers was able to get access to everything that Paul L. Ray had including his home directory, even though the home directory is still called Paul L. Ray. Eventually, I would rename the directory and make sure that Tom's profile tab indicates the new name of the home folder.
8. Right-click on Tom Landers's folder and select the Delete option. Say Yes to the Are you sure dialog box.
9. Paul Ray got hired back into his original position. Unfortunately, since his account was deleted, a new one will have to be created. Therefore, in the Sales organizational unit, create a new Paul L. Ray account.
10. Using the Profile tab located within Paul Ray's properties, re-create his home folder.
11. When the Home directory was not created warning appears, click on the OK button. Paul doesn't automatically get the same rights that he had before because his account is being re-created. All of Paul's previous rights, permissions, and group memberships will have to be given again.

Example 13: Using Templates

1. Right-click the Sales organizational unit and select New User. Specify the following:

First name: Sales
Last name: Template
User logon name: _SALES_TEMPLATE

Click on the Next button.

NOTE: The reason you would use the first underscore (_) is because it will show first on the list if the items are alphabetized by user name.

2. Specify the password of password and enable the User must change password at next logon options. Click on the Next button.
3. Click on the Finish button.
4. Right-click the Sales Template and select the Properties option.
5. Put an address and phone number in the Address and Telephones tab.
6. Using the Member of tab, assign the template to the Sales group.
7. Right-click the Sales template again and select the Copy command to create a new user called Charlie Brown. Use the password of password for Charlie Brown.
8. Right-click the Charlie Brown account and select the Properties option. View the various tabs that you modified when creating the _SALES_TEMPLATE.

Exercise 14: User Rights

1. Log out as the administrator.
2. Try to log in as Charlie Brown. It shouldn't work because Charlie Brown has not been given the right to log on locally to a domain controller.
3. Log in as the administrator.
4. In the Administrative Tools, open the Domain Controller Security Policy console.
5. If you open Computer configuration, Window settings Security Settings, Local Policies, User Rights Assignment, you will find Allow log on locally. Double-click Allow log on locally.
6. At the command prompt, execute the following command: secedit /refreshpolicy machine_policy.
7. Click the Add button, click on the Browse button, select the Charlie Brown account, click on the Add button, and click on the OK button twice.

8. Log out as the administrator and log in as Charlie Brown. When you log in, change the password to PW.

9. As Charlie Brown, try to disable Frank Biggs's account.

10. Log out as Charlie Brown and log in as administrator.

11. Right-click the Sales Organizational Unit and select the Delegate Control option. Click the Next button.

12. Click the Add button, select Charlie Brown, and click the Add button. Click the OK button. Click on the Next button.

13. Select the Create, Delete, and Manage User Accounts and Modify the membership of a group options. Click on the Next button. Click the Finish button.

14. Log out as the administrator and log in as Charlie Brown.

15. Disable Frank Biggs's account.

16. Log out as Charlie Brown.

Example 15: Group Policies

1. Log in as the administrator.

2. Using the Run option, start a Microsoft Management console.

3. In the MMC, open the Console menu and select the Add/Remove Snap-in option.

4. In the Add/Remove Snap-in dialog box, click on the Add button.

5. Click on the Group policy snap-in and then click on the Add button. In the Select Group Policy Object dialog box, click on the Browse button.

6. In the Browse for a Group Policy Object dialog box, click the Domain Controllers.AcmeXX. com. Click the OK button.

7. Click the Default Domain Controllers Policy and click the OK button. Click the Finish button.

8. Click on the Group policy snap-in again and click on the Add button. In the Select Group Policy Object dialog box, click on the Browse button.

9. In the Browse for a Group Policy Object dialog box, click the Sales.AcmeXX.com. Click the OK button.

10. Click the New Group Policy Object button (next to the Up folder button). Since the New Group Object is highlighted, rename it to the Sales Group Policy and click the OK button. Click the Finish button.

11. Click on the Group policy snap-in and then click on the Add button. With the Local Computer selected for the Group Policy Object, click on the Finish button.

12. Click on the Close button. Click on the OK button.

13. Under the Default Domain Controllers Policy, open Computer Configuration, Windows Settings, Security Policy, Account Policies. Click on Password Policy.

14. In the Detail pane, double-click the Minimum password length option.

15. In the Template Security Policy Setting dialog box, enable Define this policy setting in the template. Specify at least 8 characters. Click the OK button.

16. Under Password Policy, click on Account Policy option.

17. In the Detail pane, double-click the Account Lockout Duration.

18. Enable the Define this policy setting in the template. Keep the default of 30 minutes and click on the OK button. Click on the OK button.

19. At the command prompt, execute the following command: secedit /refreshpolicy machine_policy.

20. Log out as the administrator.

21. Log in as Charlie Brown.

22. Log in six times with the password of test. This should lock out the account.

23. Log in as the administrator.

24. Using the Active Directory Users and Computers console, right-click Charlie Brown's user account and select the Properties option.
25. In the Account tab, remove the X in the Accounts locked out box.
26. Try to change Charlie Brown's password to LETMEIN. Then change the password to PASSWORD.
27. In the Active Directory Uses and Computers console, right-click the Sales department and select the Properties option.
28. In the Group Policy tab, click the Add button. Click the Up folder button and double-click the Domain Controllers.acmexx.com folder. Select the Default Domain Controllers Policy. Click the OK button.
29. Click the Default Domain Controllers Policy and click on the Up button to make the Default Domain Controller Policy have a higher priority than the Sales Policy. Although we are not going to use the option, notice the Block Policy Inheritance. Click the OK button.

Exercise 16: Understanding NTFS Rights

1. Delete the M drive and create a new FAT32 volume. Assign the M drive.
2. On your C drive (which should be NTFS volume) create a folder called DIRN1.
3. In the DIRN1 folder, create a DIRN2 folder.
4. In the DIRN2 folder, create a DIRN3 folder.
5. In the DIRN3 folder, create a text file called FILE1.TXT. To create a text file, open the DIRN3 folder, right-click the empty space, and select New Text file. In the FILE1.TXT file, put in your first name.
6. Right-click the DIRN1 folder and select the Properties option. Click on the Security tab.
7. Click on the Add button. Pick Charlie Brown and click on the Add button. Click on the OK button.
8. Log out as the administrator and log in as Charlie Brown.
9. Try to open the FILE1.TXT file.
10. Try to add your last name to the FILE1.TXT file, save the changes, and exit the program.
11. Try to delete the FILE1.TXT file and try to delete the DIRN2 folder.
12. Try to create a new text file called FILE2.TXT in the DIRN1 folder.
13. Log out as Charlie Brown and log in as the administrator.
14. Right-click the DIRN1 folder and click the Properties option.
15. From the Security tab, click to highlight Charlie Brown. Give Charlie Brown the Write right.
16. Log out as administrator and log in as Charlie Brown
17. Try to open the FILE1.TXT file.
18. Try to add your last name to the FILE1.TXT file, save the changes, and exit the program.
19. Try to delete the FILE1.TXT file and try to delete the DIRN2 folder.
20. Try to create a new text file called FILE2.TXT in the DIRN1 folder.
21. Log out as administrator and log in as Charlie Brown.
22. For the DIRN1 folder, assign the Modify right also to Charlie Brown.
23. Log out as Charlie Brown and log in as the administrator.
24. Try to open the FILE1.TXT file.
25. Try to add your last name to the FILE1.TXT file, save the changes, and exit the program.
26. Try to delete the FILE1.TXT file and try to delete the DIRN2 folder.
27. Try to create a new text file called FILE2.TXT in the DIRN1 folder.
28. Log out as Charlie Brown and log in as the administrator.
29. In the DIRN1 folder, re-create the DIRN2 folder. In the DIRN2 folder, re-create the DIRN3 folder. In the DIRN3 folder, re-create the FILE1.TXT file.

30. Right-click the DIRN2 folder and select the Properties option. From the Security tab, specify Charlie Brown to have the Read permission.
31. Log out as the administrator and log in as Charlie Brown.
32. Open the FILE1.TXT file and try to add your last name to it. Again try to save it.

Exercise 17: Share Rights

1. Log in as the administrator.
2. In the M drive, create a folder called DIRS1.
3. In the DIRS1 folder, create a folder called DIRS2.
4. In the DIRS2 folder, create a text file called FILE1.TXT, with your first name listed in the file.
5. Right-click the DIRS1 folder and select the Sharing option.
6. Click the Share this folder option. Keep the default share name and click on the Permissions button to set the Share rights.
7. With the Everyone account highlighted, click on the Remove button.
8. Click on the Add button and add Charlie Brown.
9. For Charlie Brown share rights, assign only the Read permission. Click on the OK button.
10. From your partner's computer, log in as Charlie Brown and access the DIRS1 share through the My Network Places.
11. Close the DIRS1.
12. Click on the Start button, select the Run option, and execute \\SERVER2000-XXx\DIRS1.
13. Try to open the FILE1.TXT file.
14. Try to add your last name to the FILE1.TXT file, save the changes, and exit the program.
15. Try to delete the FILE1.TXT file.
16. Try to create a new text file called FILE2.TXT in the DIRS1 folder.
17. As the administrator back at your own computer, change the share rights to include Read and Change permissions for Charlie Brown.
18. From your partner's computer, try to open the FILE1.TXT file.
19. Try to add your last name to the FILE1.TXT file, save the changes, and exit the program.
20. Try to delete the FILE1.TXT file.
21. Try to create a new text file called FILE2.TXT in the DIRS1 folder.
22. As the administrator back at your own computer, change the rights to Deny all rights for Charlie Brown.
23. From your partner's computer, try to access the shared folder.
24. Back at your own computer, log out as the administrator and log in as Charlie Brown.
25. Try to delete the FILE2.TXT file. Try to figure out why Charlie Brown was able to delete the folder.

Exercise 18: Groups and Permissions

1. Log in as the administrator. Make sure that Charlie Brown is a member of the Sales group.
2. In the C drive, create a folder called GROUP1.
3. In the GROUP1 folder, create a GROUP2 folder.
4. In the GROUP2 folder, create a new text file with your name in it.
5. Assign the NTFS Read and Execute and List Folder Contents to the Sales group for the GROUP1 folder.
6. Assign the NTFS Modify to Charlie Brown for the GROUP1 folder.
7. Assign the NTFS List Folder Contents to Everyone for the GROUP1 folder.

8. Right-click the text file and select the Properties option. Click on the Permissions tab. Notice the rights assigned to Charlie Brown. Of those rights, you should notice which ones are grayed out.
9. Assign Full Control to Charlie Brown for the GROUP1 folder.
10. Assign Full Control to Sales for the GROUP1 folder.
11. Deny all rights to Everyone for the GROUP1 folder.
12. Share the GROUP1 folder and assign Full Control share permissions to Charlie Brown, Sales group, and Everyone.
13. From your partner's computer, log in as Charlie Brown and try to access the text file.
14. Back on your own computer, log out as administrator and log in as Charlie Brown. See if you can access the text file in the GROUP2 folder.
15. Log out as Charlie Brown and log in as the administrator.
16. Remove Everyone from the NTFS permissions for the GROUP1 folder.
17. Log out as the administrator and log in as Charlie Brown. Try to access the text file.
18. Back on your partner's computer, logged in as Charlie Brown, try to access the text file.

Exercise 19: Taking Ownership of a Directory

1. Log in as Charlie Brown.
2. Create a folder called OWNER in the C drive.
3. Inside the OWNER folder, create a text file.
4. Right-click the OWNER folder and assign Charlie Brown all NTFS permissions.
5. Notice that the check marks for the NTFS rights for the Everyone group are grayed out. Click on the Everyone group and click on the Remove button.
6. The error appeared because the rights are inherited from above the folder. Click on the OK button.
7. Disable the Allow inheritable permissions from parent to propagate to this object option. Select the Remove button so that the inherited rights are not allowed to flow into the OWNER folder.
8. Click on the OK button.
9. Log out as Charlie Brown and log in as the administrator.
10. Try to access the text file.
11. Right-click the OWNER folder and select the Properties option.
12. Click the Security tab. Read the message and click on the OK button.
13. Click on the Advanced button.
14. Click on the Owner tab. Click on the Administrators and enable the Replace owner on sub-containers and objects. Click on the OK button. Click the Yes button to replace all permissions.
15. Click on the OK button.
16. Right-click the OWNER folder and select the Properties option. Click on the Security tab to view the NTFS permissions.
17. Click the OK button to close the OWNER properties dialog box.
18. Open the OWNER folder.
19. Try to access the text document.
20. Right-click the text document and select the Properties option. Try to figure out why the administrator cannot access the text document.
21. In the Security tab, assign all NTFS permissions to the Administrators group. Click the OK button.
22. Try to access the text document.

CHAPTER **11**

Networking with Novell NetWare

Topics Covered in this Chapter

Introduction

NetWare, developed by Novell, was the first network operating system to gain wide acceptance in the PC market. Unfortunately, as a result of a combination of circumstances—the success of Windows NT, with its ease of use, and a couple of missed opportunities by Novell—NetWare has lost most of its network operating system market share. Yet NetWare is one of the more powerful network operating systems on the market today.

Objectives

- Identify the basic capabilities (client support, interoperability, authentication, file and print services, application support, and security) of NetWare network operating systems.
- Given specific parameters, configure a Windows client to connect to a NetWare server.

- Given a troubleshooting scenario involving a client connectivity problem (e.g., incorrect protocol/client software/authentication configuration, or insufficient rights/permission), identify the cause of the problem.
- Describe the shutdown procedure for the NetWare network operating systems.

11.1   WHAT IS NETWARE?

Different from Windows and Linux, NetWare consists of servers only and is not available as a desktop operating system. As a server though, it allows DOS, Windows, Macintosh, UNIX, and Linux computers to communicate with it. Also different from the other network operating systems (NOSs), most of the administrative features are typically done on a workstation that is connected to the server. Four major versions of NetWare are in use today. (See tables 11.1 and 11.2.)

NetWare 3.X, which includes NetWare 3.11 and 3.12, was based on the NetWare 386 (a version of NetWare that was made to be used by the Intel 386-based machine). At the time of its release, NetWare 3.X made NetWare the de facto standard for business networks. NetWare 3.X supports multiple, cross-platform clients and has minimal hardware requirements (4 MB of RAM, 75 MB of hard disk space). It uses a flat database called the **bindery** to keep track of users and groups on that server, and it is administered with a menu-based DOS utility known as SYSCON. To configure and set up NetWare as a print server, you would use PCONSOLE. To manage your files, you would use FILER. The default protocol for NetWare 3.X was IPX; NetWare 3.11 and earlier versions defaulted to the 802.3 frame type, and NetWare 3.12 and later versions defaulted to the 802.2 frame type.

Table 11.1   Various Versions of NetWare

	Default Protocol	Type of Server-Based Access	Interface	Comments
NetWare 3.11	IPX (Ethernet uses 802.3 frame)	Bindery (single-server database)	Command interface	
NetWare 3.12/3.2	IPX (Ethernet uses 802.2 frame)	Bindery (single-server database)	Command interface	
NetWare 4.XX	IPX (Ethernet uses 802.2 frame)	NDS (distributed replicated database)	Command interface	Introduced directory services
NetWare 5.XX	TCP/IP	NDS (distributed replicated database)	Command interface/ GUI	Added many Internet and NT-like utilities and allowed management of network resources at the server
NetWare 6.0	TCP/IP	NDS (distributed replicated database)	Command interface/ GUI	Added CIFS, AFP, and NFS

Table 11.2   System Requirements for Windows NT and Windows 2000 Servers

	NetWare 4.X	NetWare 5.0	NetWare 6.0
Processor	• 386SX or better • Supports up to 32 processors	• Pentium processor or better • Supports up to 32 processors	• Pentium II or better • Supports up to 32 processors
RAM	• 20 MB • 64 MB (recommended)	• 64 MB • 128 MB (recommended)	• 256 MB • 512 MB (recommended)
Disk space	• 15-MB DOS • 100-MB NetWare partition	• 30-MB DOS partition • 550-MB NetWare partition	• 200-MB DOS partition • 2-GB NetWare partition

NetWare 4 was introduced in 1993. It featured the first version of the Novell Directory Services (NDS), which replaced the bindery used in NetWare 3.X. As an administrator, it allowed simplified administration of multiple servers such as creating one account for all the servers that are part of the NDS tree. Later, Novell introduced 4.02, 4.1, 4.11, and 4.2.

NOTE: Netware 4.11 was also known as IntraNetWare, a name meant to capitalize on the Internet craze.

NetWare 5 continued on with NDS (of course, with an improved version of NDS). Different from previous versions of NetWare, it made TCP/IP the default protocol. One of the most obvious differences to the server interface is the addition of a graphic user interface (GUI) to the NetWare server based on Java, using X Window. This allows you to do common administrative tasks at the server such as creating users and assigning permissions.

NOTE: The Java interface is given a low priority on the server so that the server can use most of its processing power to act as a server and not a desktop interface. In addition, the NetWare 5 operating system kernel provides multiprocessor support, memory protection, and virtual memory. In addition, ZENWorks lets you configure, update, and maintain network workstations without leaving your desk.

Novell NetWare 6 leverages the powerful Novell eDirectory, giving you a way to easily manage your network from virtually any web-enabled, wireless device or traditional desktop computer. NetWare 6 also supports open, Internet standards and includes innovative, browser-based net services.

NetWare 6 incorporates Novell iFolder, a unique net services solution that enables you to access personal files from anywhere, at anytime, through virtually any web-enabled device. When you use Novell iFolder, you don't have to worry about emailing critical presentation files from your corporate desktop to your laptop. Anything you save in iFolder on one machine will be seamlessly synchronized in iFolder on all your other machines. You can work online, offline, at home, in the office, or on the road, and iFolder will ensure that you always have the most recent version of your file. Moreover, iFolder provides powerful security for your data, ensuring that your files are always protected from prying eyes.

To further extend your network's boundaries, NetWare 6 supports various file protocols, including AppleTalk Filing Protocol (AFP), for Macintosh, Network File System (NFS), Common Internet File System (CIFS), and systems application architecture (SAA) for mainframes and AS/400. CIFS is the same protocol that allows you to create, view, and access shared drives, directories, and printers with Microsoft Windows. NetWare's support for these diverse protocols eliminates the need for client software: Communication is accomplished over TCP/IP, and native applications are run over the network. This protocol support also enables you to seamlessly leverage the power of eDirectory through virtually any third-party system including Windows, Macintosh, and UNIX for central administration and file access.

11.2   STARTING AND STOPPING NETWARE

To start the server, change to the start-up directory (usually C:\nwserver) and enter the following at the DOS prompt:

```
SERVER
```

You may also place the command to execute SERVER.EXE in the DOS's AUTOEXEC. BAT file. Then, whenever the server computer is powered up, the server starts automatically.

If the splash screen is active and you want to see a list of modules as they load, you can still display the system console screen by pressing Alt+Esc and tapping Esc until the system console is displayed. After the server is running, you can display a list of all loaded modules by entering the following at the server console prompt:

```
MODULES
```

To ensure data integrity, use the DOWN command at the server console before turning off power to the server. DOWN writes all cache buffers to disk, closes all files, updates the appropriate Directory Entry Tables and File Allocation Tables, and exits the network. If you want to bring the server down and then restart it immediately, without exiting to DOS, use the RESTART SERVER command instead of DOWN.

11.3   NETWARE ARCHITECTURE

As with any other operating system, the core of NetWare is the kernel. NetWare's 32-bit kernel is responsible for overseeing all critical server processes. The program SERVER. EXE runs the kernel from a server's DOS partition.

NetWare, like most other modern network operating systems, is modular. It consists of a core component and other pieces that are loaded into memory as necessary. The core component is called the core OS or kernel, and the other modules are called **NetWare loadable modules (NLMs).** With this design, NetWare can run more efficiently by loading only what it needs. During boot, the SERVER.EXE loads the critical NLMs that the kernel needs to run the NetWare operating system.

NLMs fall into four categories:

- Disk drivers
- LAN drivers
- Name space modules
- Utilities

The first two NLMs (disk drivers and LAN drivers) are just ordinary drivers for NetWare. The older disk drivers have a .DSK extension, whereas newer disk drivers have the extension .HAM or .CDM. For NetWare 4, 5, and 6, a combination of IDEATA.HAM and IDEHD.CDM provides access to IDE hard drives and CD drives. The LAN drives have a .LAN extension. For example, the 3C5X9.LAN is the driver that allows NetWare to access 3Com EtherLink III network cards.

The name space module controls how files look and how they are stored on a disk. For example, automatically during installing, NetWare by default loads LONG.NAM so that the system can store long file names such as those generated by Microsoft Windows. If not, they would be stored using the DOS naming convention (8.3 format). Different operating systems use different naming conventions for files, and the name space modules make it possible to store those files on volumes on a NetWare server. The extension of these name space modules is .NAM. For example, the NFS.NAM name space module enables NetWare to store files on a disk using the UNIX NFS (Network File System) naming convention.

In the early versions of NetWare, including 3.X, you had to manually type in these commands in the appropriate parameters, which by today's standard is not very user-friendly. But with NetWare 4.XX and later, the disk driver, LAN drivers, and name space modules can be loaded using other programs, including the Install program and the NWCONFIG.NLM.

A server batch file is an executable file containing server console commands. You execute the file by entering its name at the System Console prompt. NetWare then executes all the commands in the file in order. By default the NetWare installation process creates two batch files that are always executed whenever the server is booted:

- **STARTUP.NCF**—This file is executed first. It loads the server's disk drivers and name spaces, and can execute certain server parameters that cannot be executed anywhere else. This file resides in the server's start-up directory, C:\NWSERVER.
- **AUTOEXEC.NCF**—This file is executed after STARTUP.NCF and after volume SYS: is mounted. It sets time synchronization parameters and the bindery context, stores the server name, loads the communication protocol, loads the LAN drivers and settings for the network boards, binds the protocol to the installed drivers, loads other NLM programs, and executes server parameters. This file resides in the SYS:SYSTEM directory.

The bulk of the NLMs don't fit in the driver category. Instead, they are utilities or programs (which have an .NLM extension) that you can run on the NetWare server. The most popular NLMs are:

- **MONITOR.NLM**—Monitors the server
- **NWCONFIG.NLM**—NetWare configuration NLM
- **NDPSMGR.NLM**—Novell Distributed Print Services that loads the NetWare printing manager

NOTE: With NetWare 5 and 6, in most cases, you can load an NLM by entering just the name of the NLM at the System Console prompt. The exception is if there is an .NCF file of the same name as the NLM. In this case, you must use the load command. Using the load command directs the server to load the NLM instead of executing the file of the same name. In addition, you must use the load command to load NLMs for the older versions of NetWare.

Assuming that you have sufficient memory, today's NetWare manages physical memory very efficiently without making any configuration changes. Any memory that is not used by the kernel, NLM, or other programs running on NetWare is used for caching. Caching is the process of saving frequently used data to an area of RAM where it will be readily available for future requests. Of course, if the information is found in the RAM, the access is much faster than if it had to pull the data from the disks. The more RAM your system has, the more additional RAM will be used for caching.

While early versions of NetWare servers only had a command interface, the newer NetWare versions can use either a command interface or a GUI. One of the reasons that the command interface is used is so that the server can use its processing power to act as a server instead of providing a GUI. If you do run the GUI on the NetWare server, it tries to run the GUI so that it does not sacrifice the performance of the server services.

11.4   NETWARE DISK STRUCTURE

NetWare offers its own high-performance file system that supports the long file names of DOS, Macintosh, UNIX, OS/2, and Windows. While NetWare supports DOS file names by default, you have to load the proper NLM to support other file names including the long ones.

The NetWare file system is organized much like the DOS file system. In NetWare, volumes are the major division of the NetWare file system. A **volume** is the physical amount of storage on a hard disk or other storage device, such as a CD-ROM. A volume may be fully contained on a single hard drive or may be spread across multiple hard drives.

The physical storage space that makes up a volume is allocated and named during the installation of the NetWare operating system on the server. These are some of the rules for naming a volume:

- The name must be two to fifteen characters long.
- The physical volume name must be followed by a colon (:).
- Spaces, commas, back slashes, and periods are not permitted within the volume name.
- You can use only the letters A–Z, the numbers 0 through 9, and the following symbols: ~, !, #, $, %, ^, &, (), −, _, and { }.
- Two physical volumes on the same NetWare server cannot have the same name.
- The back slash (\) or the forward slash (/) must be used to separate the NetWare server name from the physical volume name.

The main volume for NetWare is the **SYS volume.** During installation, it is automatically created, and several directories are placed in it:

- **LOGIN**—Contains the programs necessary for logging into the network.
- **PUBLIC**—Holds the NetWare commands and utilities available to network users.
- **SYSTEM**—Stores files used by the NetWare server operating system or by the SUPERVISOR. The SYSTEM directory holds NLMs and files specific to the NetWare server.
- **MAIL**—Used by the NetWare operating system. It contains subdirectories for each user. They are named after the user's hexadecimal ID number and contain the user's login script.
- **ETC**—Contains sample files to help configure the server for TCP/IPs.
- **DELETED.SAV**—Contains deleted files from directories that have been removed.

You need to add more volumes to hold the directories and files for your users. Novell breaks down these directories into the following categories:

- **Home directories**—Users should have a "home" directory in which to store individual user-created files. Usually, individual users are given full access to their own directories so that they can manage (create, delete, modify, move, copy, and so on) their own files.
- **Application directories**—These contain application files and establish security over the application files. A user must be given access to an application directory in order to use the application. User-created data files generated using the applications are usually not kept here.

- **Configuration file directories**—Many applications use configuration files to allow a user to set up or customize how an application functions for him or her without affecting how the application functions for the other users. Configuration files allow users to make changes to an application without affecting the application for other users.
- **Shared data directories**—Shared data directories facilitate the distribution of information needed by groups of users. These directories are also used to ensure data security.

Accessing the NetWare file system starts with accessing the volume containing the needed directories and files. Applications that use NetWare volume-naming procedures can access file systems through the server\volume or volume object name. For example, NetWare utilities can access the NetWare file system in this manner. Many legacy applications cannot access NetWare volumes by their volume names. These applications require the use of a drive mapping. Newer Windows-based programs do not require drive mappings. Instead, they locate workstation and network resources by object name as displayed in My Computer or Network Neighborhood.

The full path to a directory can be used when identifying a directory. The path consists of the names of the levels in the file system structure, beginning with the server name. Next is the volume name or the name of the volume object in the NDS tree, followed by the names of all directories leading to the specific directory or file. The directory name with a full path would look like this:

SERVER\VOLUME:DIRECTORY\SUBDIRECTORY

or

VOLUME_OBJECT:DIRECTORY\SUBDIRECTORY

To mount a CD on a Novell server, you need to load the cdrom.nlm. This would be done by executing the following command at the server console:

```
load cdrom
```

11.5   NOVELL DIRECTORY SERVICES

With the **Novell Directory Services (NDS),** which are based primarily on the X.500 Internet directory standard, all the network resources are represented as objects and placed into a hierarchical structure called an NDS tree. This tree is a logical representation of a network and encompasses all network resources, including users, groups, printers, and servers. Since the tree can have multiple servers, you only have to create one user on the tree, and you can give access to the server and its resources to that one user without creating multiple users for the various servers. Therefore, NDS provides central management of network information, resources, and servers. In addition, it gives a standard method of managing, viewing, and accessing network information, resources, and servers. Lastly, like the Active Directory, the NDS directory follows the X.500 standard and the Domain Name System. To change the directory database, a NetWare network administrator uses a program called NetWare Administrator (32-bit Windows version—NWADMIN32) and

Figure 11.1   Novell ConsoleOne is to Administer Network Users and Resources

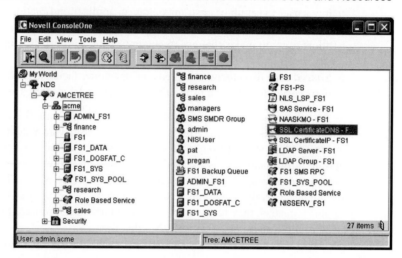

Novell ConsoleOne. NetWare Administrator is considered the legacy software, and ConsoleOne is the utility that is available on the NetWare server. (See figures 11.1 and 11.2.)

The directory can be partitioned and replicated on multiple servers. A server could contain the entire directory, pieces of it, or none at all. This way, if the directory becomes too large, you can divide it into smaller pieces and put them on several servers. When planning the tree structure, you should make sure that the part of the tree that has the resources needed by a group of users has a server with the required information near to them. This way, when a user needs to access a network resource, he or she does not have to connect to a computer that is far away to get authenticated.

NOTE: NDS for Solaris 2.0 implements NDS on Sparc Solaris servers and workstations so that the Solaris server or workstation acts as a directory server. The product allows you to place replicas of the NDS database on a Solaris server. These replicas utilize the same directory services provided on the NetWare server. The ability to add replicas provides access to the NDS database on local Solaris servers, enabling accessibility on remote networks.

11.5.1   The NDS Tree Structure

The directory consists of objects. An object is a unit of information about a resource, comparable to a record in a conventional database. NDS represents each network resource as an object in the directory. Different types or categories of objects exist.

Much like the root directory found on your C drive in DOS, the top of the NDS tree is the [Root] object. The objects that are your network resources are called leaf objects. To organize your leafs, you use containers, which are analogous to the directories. As a network administrator, you can make trustee assignments and grant rights to the entire NDS tree from the [Root] object. Country, organization, and alias objects can be created directly under the [Root] object.

Figure 11.2   Administering an Organizational Unit Using ConsoleOne

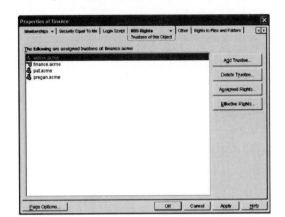

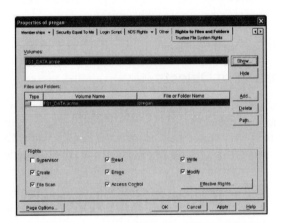

Container objects contain leaf or other container objects. They are used to logically group and organize the objects of your directory. They can represent countries, locations within countries, companies, departments, responsibility centers, workgroups, and shared resources.

- The country container organizes the NDS tree by countries. It is identified by a predetermined two-character country abbreviation.
- An organization container organizes objects by organization groups, such as company, university, or department.
- The organizational unit organizes objects by subunit groups such as division, business unit, project team, or department. It can exist in organizational and organizational unit objects only.

NDS version 8 also includes locality and domain containers. The domain container can be contained by all the other containers in the operational schema including the tree root, and it can contain all the other operational container objects, except the tree root.

The leaf objects represent network resources, such as users, servers, and NetWare volumes. Common leaf objects include:

- **NetWare Server**—Represents a NetWare server, which is used by many objects, such as volume objects, to identify the server that provides a service.
- **User**—Represents a person who uses the network. It contains the password and other login restriction properties.
- **Alias**—Points to another object at a different location in the NDS tree so that it allows a user to access an object outside of the user's normal working context.
- **Volume**—Represents a physical volume.

NOTE: A volume is not a container although the NetWare Administrator gives you this impression because when you open it, it shows the files on the volume. When you are managing the files, the files are not part of the NDS tree.

- **Groups**—A leaf object that allows you to group users together. By having people grouped together, you can give permissions for an object to the group and all users in the group get access to the object.

NOTE: A group is not a container; it is just a list of users.

11.5.2   Naming of Objects

A leaf object's common name (CN) is the name shown to the leaf object in the NDS tree. When users need to access a resource such as a server in the NDS tree, the common name for the server object must be included in the request. To use a resource outside the user's parent container, the user must change his or her location, or context, in the NDS tree.

Context is represented as either of the following:

- An object's position in the NDS tree
- A position you navigate to in the NDS tree after logging in (similar to your current directory when using DOS)

Contexts may be expressed in two ways: typeful and typeless. The typeful notation is a relatively lengthy way of expressing context that includes identifiers for the organization (O=) and organizational unit (OU=). Typeless notation eliminates the "O" and "OU" designations.

When used to define an object's position in the NDS tree, context is a list of container objects leading from the object to the [Root]. Locating an object through context is similar to locating a file using the directory path.

For example, if you have the ACME organization located under the [Root] container, an NY under the ACME organization, and an OU under CORP, the context or location on the NDS tree of the CORP object would be represented by

Typeful:

OU=CORP.OU=NY.O=ACME

Typeless:

CORP.NY.ACME

An objects' distinguished name is a combination of its common name and its context. To make it easily seen as a distinguished name, it always starts with a leading period (.). For example, if you have a user called PRegan located in the CORP organizational unit, the distinguished name of the PRegan object would be:

Typeful:
.CN=PRegan.OU=CORP.OU=NY.O=ACME
Typeless:
.PRegan.CORP.NY.ACME

Since a distinguished name must be unique, you cannot have two leaf objects with the same name in the same container. However, an NDS tree can have two leaf objects with the same name in different containers.

A relative distinguished name lists the path of objects leading from the object being named to the container representing the object's current context, or current location, in the NDS tree. It does not start with a leading period. When you use a relative distinguished name, NDS builds a distinguished name from:

current context + relative distinguished name = distinguished name

11.5.3   The Admin

The admin object is the only default user account. It is created automatically when the NetWare is installed. It has the authority to manage all aspects of the network and is the primary user object in the initial network setup. The admin account/object is a user object; it can be deleted or modified, and it can have its security access revoked. Never delete or modify the admin account/object unless you know that additional user objects have the same rights. If you delete the admin object and no other user object has the same rights, you won't be able to change security access to any object in the tree. You are not limited to one object with supervisory authority; you can create additional user objects with the same rights as the admin object.

11.5.4   Trustees

Every object in the NDS and every directory and file has an **access control list (ACL)** that shows who has what rights to the object, directory, and file. The list of users that have been given access is known as a trustee list. A trustee can be a user, a group (a list of users), or a container. A container actually acts as a natural group, which contains groups and users. When a container is made a trustee, any user or group also acquires those rights to the object, directory, or file.

Any of the following NDS objects can be made a trustee of an object:

■ User object
■ Group object
■ Organizational role object

Table 11.3   Object Rights

Right	Description
Supervisor (S)	Grants all access privileges. A trustee with the supervisor right also has complete access to all the object's property rights.
Browse (B)	Enables an object trustee to see the object in the NDS tree.
Create (C)	Enables an object trustee to create objects below this object in the NDS tree. This right is available only on container objects.
Delete (D)	Enables an object trustee to delete the object from the NDS tree.
Rename (R)	Enables an object trustee to change the name of the object.
Inheritable (I)	Enables an object trustee of a container to inherit the assigned object rights to objects and subcontainers within a container. This right is granted by default to facilitate inheritance of rights to container objects and subcontainers. Revoking this right limits trustees to the rights assigned for the trustee to the container object only. No rights to the container's objects and subcontainers are inherited. This right is available only for container objects.

- Containers and parent containers
- [Root]
- [Public] trustee

Assigning any of these objects, except the user object, as object trustees of other objects allows access to be granted to multiple users at once. With the exception of the user object, all users who are members of the objects listed above receive rights simultaneously when assigned as a trustee. NDS security is easier to implement and manage when you grant rights to objects, such as group objects and container objects that pass their rights to multiple users. In addition, a user object can receive rights by being granted security equivalence to another user object.

NDS security has two distinct sets of rights: object (or entry) and property (or attribute). Rights, in all but one case, do not flow from NDS into the file system. One important note is that both the supervisor object right and the supervisor property right can be blocked by an inherited rights filter (IRF).

Because NDS security and file system security are separate, file system administration and administration of NDS objects can be handled by one network administrator or divided among various network administrators.

Object rights are rights granted to a trustee of an object. They control what a trustee can do with the object, such as browsing, renaming, or deleting the object, and they control access to the object as a whole but not to the object's property values. (See table 11.3.)

An **object** contains information about its associated network resource. This information makes up the object's attributes and is stored in the properties of the object. For example, the user object contains information such as the person's address and phone number. Property rights are rights that grant access to the attributes of an object. (See table 11.4.)

Table 11.4   Properties Rights

Right	Description
Supervisor (S)	Grants all rights to the object's properties.
Compare (C)	Enables the comparison of any value to a value of the property, returning a true or false response. The value of the property cannot be viewed, only the true or false status of the value submitted for comparison. If the read right is granted, the compare right is automatically granted.
Read (R)	Enables the display of the property value. Granting the read right automatically grants the compare right.
Write (W)	Enables the object trustee to modify, add, change, and delete a property value. Granting the write right automatically grants the add/remove self right.
Add/Remove Self (A)	Enables an object trustee to add or remove itself as a value of the object property. If the write right is granted, the add/remove self right is automatically granted.
Inheritable (I)	Enables an object trustee of a container to inherit the assigned property rights to objects within the container. Without this right, the remaining property rights assigned to the trustee apply only to the container object's properties and not to the properties of the objects in the container. This right is granted by default when the All Properties option is selected, and removed by default when the Selected Properties option is selected. This right is available only on container objects.

They control access to the attribute values stored within an NDS object's properties, allowing users to see, search for, or change the attribute values; and they also control a user's ability to use a network resource represented by an NDS object.

NDS, like the file system, uses rights inheritance. Inheritance minimizes the individual rights assignments needed to administer the network. Because NDS is a hierarchical structure, rights flow downward from a container to subcontainers. To keep objects from having some or all of the NDS rights they inherit, you can stop rights from inheriting either by making a new object trustee assignment at a lower level in the NDS tree and assigning new rights (which overwrite the inherited rights) or by blocking rights with an inherited rights filter. The IRF of an NDS object does not block rights that are granted to an object trustee of the same object. Only those rights inherited from object trustee assignments made higher in the tree are affected.

An IRF filters rights inheritance in the NDS tree. An IRF can be placed to block the inheritance of either object rights or property rights. An IRF placed on an object can block property rights granted through either the All Properties option or the Selected Properties option. Property rights granted through the Selected Properties option will only be filtered if the inheritable property rights are granted to a selected property.

If a trustee is not granted the inheritable property right for a selected property, that right will not be inherited by other objects lower in the NDS tree regardless of IRFs placed on objects lower in the NDS tree. The IRF of an NDS object does not block rights that are granted to an object trustee of the same object. Only those rights inherited from object trustee assignments made higher in the tree are affected.

The default trustees and rights established during installation are:

- Admin (the first NDS server in the tree) supervisor object right to [Root].
- [Public] (the first NDS server in the tree) browse object right to [Root].
- NetWare Server admin has the supervisor object right to the NetWare server object, which means that admin also has the supervisor right to the root directory of the file system of any NetWare volumes on the server.
- Volumes - [Root] has the read property right to the Host Server Name and Host Resource properties on all volume objects. This gives all objects access to the physical volume.
- Root directory of the volume - Admin has the supervisor right to the root directory of the file systems on the volume. For volume SYS, the container object has read and file scan rights to the volume's\PUBLIC directory. This allows user objects under the container to access NetWare utilities in\PUBLIC.
- Home directories - If home directories are automatically created for users, they have the supervisor right to those directories.

[Public] is not a normal object in NDS. It refers to all objects in the tree and any user that is not authenticated as well. That means if rights are granted to [Public], users who are not logged in have those rights. Because of this, you must be very careful when granting rights to [Public]. By default, [Public] includes the read right and the file scan right. By making [Public] a trustee of an object, directory, or file, you effectively grant all objects in NDS the rights to that object, directory, or file. [Public] is only used in trustee assignments and must always be entered within square brackets. [Public] can be added or deleted like any other trustee.

11.5.5 NDS Design

When designing the NDS structure, you need to consider using the tree structure based on your organizational structure. The organizational structure could be based on geography or on function. As a general rule, when a network is a far-reaching WAN, you should organize your tree structure by geography. This is because your domain should not span the WAN. The exception to this rule would be if you have fast links (256 KB or faster) that connect the different sites. You can then organize your domain by function. This is because of the added traffic that must go back and forth to keep the NetWare servers in sync. Of course, it is always a good idea to have a server with a replica of the NDS tree (or at least the part of the tree that contains the site of the users and local resources) so that the users can access the NDS database quickly.

Another consideration when designing your tree structure is to base your design on containers that allow you to delegate administrative control over smaller groups of users, groups, and resources (decentralized administration). If you are using containers for location for the various WAN sites, you can assign site administrators to each site that would have full authority over the site, or you can limit the authority. In any case, those site administrators would not have any control over the other sites. If you are using a tree design by function, you can then assign the administrator by function. For example, you can have an administrator for all salespeople and an administrator for all research people. Lastly, when choosing your directory structure, try to look for future growth. This way, you can plan for future organizational units if needed.

An overall concept for a good NDS tree design is to have the tree resemble the shape of a pyramid, with most of the containers and objects at the bottom of the structure and fewer containers at the top. At the top of the tree, you should create one organization. This gives you one place where rights can be granted to everyone for select network resources. You could choose the [Root] object as the place to grant rights to everyone, but it is not recommended. If you give rights to the root, everyone receives those rights including those people who are not logged in.

11.5.6   Interacting between Windows and NetWare

Both Novell and Microsoft have made great efforts to enable their operating systems to interact with each other. Microsoft developed a Gateway Services for NetWare, and Novell has created NDS for NT.

When Windows NT Server was released in 1993, Microsoft released two additional programs to facilitate the integration of Windows NT and NetWare. The Gateway Services for NetWare (GSNW) installs as a service on a Windows NT Server and translates requests for Windows NT resources into NetWare requests. At a lower level, GSNW is translating SMB/CIFS protocol requests into NCP requests. GSNW allows multiple Windows NT clients to connect through a Windows NT Server to NetWare servers using Windows NT client software and protocols; no Novell NetWare client software is needed.

NOTE: Gateway Services for NetWare depends on and works with NWLink.

File and Print Services for NetWare (FPNW) is really a method of providing files and printers hosted by a Windows NT Server to Novell clients. When installed and configured on a Windows NT Server, this service makes a Windows NT Server look like a NetWare server to Novell clients. This service is good when you have a small number of NT servers and a larger number of NetWare servers.

The NDS for NT program enables the Windows NT domains to appear as container objects in NWAdmin. Windows NT Servers appear as server objects and groups, and users from Windows NT domains appear as NDS group and user objects, respectively. With NDS for NT enabled, users who require service from both Windows NT and NetWare servers can simply log into the DNS tree, rather than logging into both types of servers.

11.6   DIRECTORY AND FILE RIGHTS

Rights determine the type of access a user has to network directories and files. A user cannot do anything to a directory or a file without the assignment of rights. (See table 11.5.) If rights are granted to a directory for a trustee, the rights will flow down to, or are inherited by, all files and directories within and below the directory. Because directory and file rights are the same, all rights to a directory can be inherited by the files in the directory.

Table 11.5   NetWare Directory/File Rights

Right	Abbreviation	Description
Supervisor	S	Grants users all rights to the directory and its files and subdirectories and their files. Users grant any right to other users. This right cannot be blocked by an inherited rights filter (IRF).
Read	R	Grants users the right to open files in the directory and read their contents or run the programs.
Write	W	Grants users the right to open and change the contents of files.
Create	C	Grants users the right to create new files and subdirectories.
Erase	E	Grants users the right to delete the directory and its files and subdirectories.
Modify	M	Grants users the right to change the attributes or name of a file or directory.
File scan	F	Grants users the right to see files and directories.
Access control	A	Grants users the right to change trustee assignments and the IRF so that all rights, except the supervisor right, are given.

There are two ways to block the inheritance of rights. The first way is to make a trustee assignment lower in the directory structure. The rights inherited by an object are overwritten when the same object is given a trustee assignment lower in the directory structure, unless the assignment above the new assignment is the supervisor right.

The second way to block the inheritance of rights is to set up an IRF. An IRF controls the rights that a trustee can inherit from parent directories. The IRF of a directory or file does not block rights that are granted to a trustee of the same directory or file. Only those rights inherited from trustee assignments made higher in the file structure are affected.

NOTE: In file system security, the supervisor right cannot be blocked by an IRF.

11.7   NETWARE CLIENT SOFTWARE

To connect to a Novell network, it is best to use the Novell NetWare Client software provided by Novell. You should not use the client software (Client for NetWare Networks) distributed with Microsoft Windows because you will not get full support when connecting to the Novell network. You can find the Novell Client for NetWare (such as NetWare Client32 which would be used on Windows 9X, Windows NT, and Windows 2000 machines) from the following locations:

■ Novell's website at http://www.novell.com
■ NetWare Client CD as part of the NetWare installation CD set

- The ZENworks CD
- The SYS volume of a NetWare server

It is always recommended to download the latest copy from the Novell site.

NOTE: You also need the client software loaded to use NetWare Administrator and ConsoleOne.

11.8   TROUBLESHOOTING WITH NETWARE

If you are using NetWare 4.X, you would type in the following command to load and bind LAN drivers at the server console command prompt:

```
load install
```

For NetWare 5.X and 6.0, you would use the following command:

```
load nwconfig
```

See figure 11.3.

To check your network configuration including TCP/IP addresses, subnet masks and default gateway, and IPX Internet and external network numbers and frame types, use the `config` command. (See figure 11.4.) To run a ping program from the NetWare server, type in the following command at the server console command prompt:

```
load ping
```

Then specify the IP address or name that you want to ping. (See figure 11.5.)

Figure 11.3   NWCONFIG is Used to Configure the Server

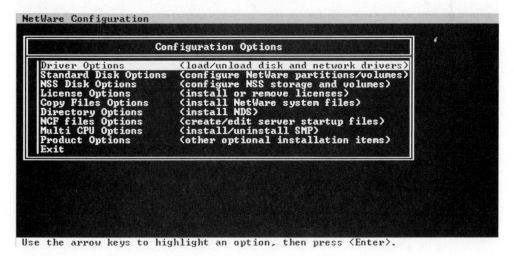

Figure 11.4   The Config Program Shows the Network Configuration

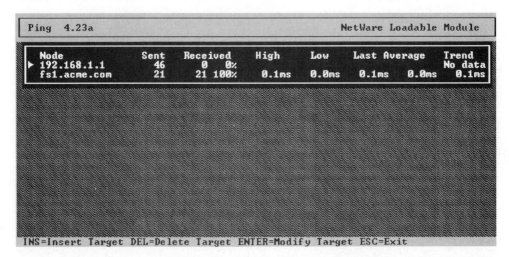

```
File server name: FS1
IPX internal network number: 03C8D1D1
Server Up Time:  7 Minutes 24 Seconds

Intel(R) PRO PCI Adapter for CHSM spec 1.11
       Version 1.62    November 24, 1999
       Hardware setting: Slot 3, I/O ports B800h to B81Fh,
                         Memory E3000000h to E3000FFFh,
                         Interrupt Ah
       Node address: 00A0C9E73B65
       Frame type: ETHERNET_802.2
       Board name: CE100B_1_E82
       LAN protocol: IPX network A050F389
Intel(R) PRO PCI Adapter for CHSM spec 1.11
       Version 1.62    November 24, 1999
       Hardware setting: Slot 3, I/O ports B800h to B81Fh,
                         Memory E3000000h to E3000FFFh,
                         Interrupt Ah
       Node address: 00A0C9E73B65
       Frame type: ETHERNET_II
       Board name: CE100B_1_EII
       LAN protocol: ARP
       LAN protocol: IP Address 192.168.1.2 Mask FF.FF.FF.0(255.25

Tree Name: ACMETREE
Bindery Context(s):
    .acme
```

Figure 11.5   Ping Command

```
 Ping  4.23a                                    NetWare Loadable Module

  Node              Sent    Received    High     Low    Last Average      Trend
▶ 192.168.1.1        46      0   0%                                      No data
  fs1.acme.com       21     21 100%     0.1ms   0.0ms   0.1ms   0.0ms    0.1ms
```

```
INS=Insert Target  DEL=Delete Target  ENTER=Modify Target  ESC=Exit
```

NetWare uses three log files that can help you diagnose problems on a NetWare server:

- CONSOLE.LOG
- ABEND.LOG
- SYS$LOG.ERR

The Console Log file (CONSOLE.LOG) keeps a history of all errors and information that have been displayed on the server's console. It is located in the SYS:\ETC directory on the server and is created and maintained by the utility CONLOG.NLM that comes with

NetWare 3.12 and later versions. You must load this utility manually (or place the load command in the AUTOEXEC.NCF file so that it starts automatically upon server start-up) by typing the following at the console prompt:

```
LOAD CONLOG
```

Once the utility is loaded, it erases whatever CONSOLE.LOG file currently exists and starts logging to the new file.

The ABEND.LOG file registers all Abends on a NetWare server. An Abend, short for "abnormal end," is an error condition that can halt the proper operating of the NetWare server. You can think of the Abend as being similar to a Window's stop error, also referred to as the Blue Screen of Death. Since an Abend error causes the server to reboot, the ABEND.LOG contains all the information that is output to the console screen during an Abend, such as the flags and registers of the processors at the time of the Abend and the NLMs that were in memory, including the versions, descriptions, memory settings, and exact time and date. The ABEND.LOG is located in the SYS:SYSTEM directory on the server.

The general Server Log file (SYS$LOG.ERR), found in the SYS:SYSTEM directory, lists any errors that occur on the server, including Abends and NDS errors and the time and date of their occurrence.

SUMMARY

1. Different from Windows and Linux, NetWare consists of servers only and is not available as a desktop operating system.

2. NetWare 3.X uses a flat database called the bindery to keep track of users and groups on that server, and it is administered with a menu-based DOS utility known as SYSCON.

3. NetWare 4 introduced the first version of the Novell Directory Services (NDS), which replaced the bindery used in NetWare 3.X. As an administrator, it allowed simplified administration of multiple servers such as creating one account for all the servers that are part of the NDS tree.

4. The program SERVER.EXE runs the kernel from a server's DOS partition.

5. NetWare, like most other modern network operating systems, is modular. It consists of a core component and other pieces that are loaded into memory as necessary. The core component is called the core OS or kernel, and the other modules are called NetWare loadable modules (NLMs).

6. A server batch file is an executable file containing server console commands. You execute the file by entering its name at the System Console prompt.

7. NetWare offers its own high-performance file system that supports the long file names of DOS, Macintosh, UNIX, OS/2, and Windows.

8. While NetWare supports DOS file names by default, you have to load the proper NLM to support other file names including the long ones.

9. In NetWare, volumes are the major division of the NetWare file system. A volume is the physical amount of storage on a hard disk or other storage device, such as a CD-ROM. A volume may be fully contained on a single hard drive or may be spread across multiple hard drives.

10. The main volume for NetWare is the SYS volume.

11. With the Novell Directory Services (NDS), which are based primarily on the X.500 Internet directory standard, all the network resources are represented as objects and placed into a hierarchical structure, called an NDS tree.

12. The NDS tree is a logical representation of a network and encompasses all network resources, including users, groups, printers, and servers.

13. To change the directory database, a NetWare network administrator uses a program called NetWare Administrator (32-bit Windows version—NWADMIN32) and Novell ConsoleOne.

14. The directory can be partitioned and replicated on multiple servers. A server could contain the entire directory, pieces of it, or none at all. This way, if the directory becomes too large, you can divide it into smaller pieces and put them on several servers.

15. To organize your leafs, you use containers, which are analogous to the directories.

16. The leaf objects represent network resources, such as users, servers, and NetWare volumes.

17. Container objects (country container, organization container, and organizational unit) contain leaf or other container objects. They are used to logically group and organize the objects of your directory. They can represent countries, locations within countries, companies, departments, responsibility centers, workgroups, and shared resources.

18. A leaf object's common name (CN) is the name shown to the leaf object in the NDS tree.

19. Context is represented as either an object's position in the NDS tree or a position you navigate to in the NDS tree after logging in (similar to your current directory when using DOS).

20. The admin object is the only default user account. It is created automatically when the NetWare is installed. It has the authority to manage all aspects of the network and is the primary user object in the initial network setup.

21. Every object in the NDS and every directory and file has an access control list (ACL) that shows who has what rights to the object, directory, and file.

22. The list of users that have been given access to an object is known as a trustee list.

23. A trustee can be a user, a group (a list of users), or a container.

24. An object contains information about its associated network resource. This information makes up the object's attributes and is stored in the properties of the object.

25. NDS, like the file system, uses rights inheritance. Inheritance minimizes the individual rights assignments needed to administer the network. Because NDS is a hierarchical structure, rights flow downward from a container to subcontainers.

26. An inherited rights filter (IRF) filters rights inheritance in the NDS tree. An IRF can be placed to block the inheritance of either object rights or property rights.

27. The Gateway Services for NetWare (GSNW) installs as a service on a Windows NT server and translates requests for Windows NT resources into NetWare requests.

28. File and Print Services for NetWare (FPNW) is really a method of providing files and printers hosted by a Windows NT Server to Novell clients.

29. The NDS for NT program enables the Windows NT domains to appear as container objects in NWAdmin.

30. Rights determine the type of access a user has to network directories and files.

31. To connect to a Novell network, it is best to use the Novell NetWare Client software provided by Novell. You should not use the client software (Client for NetWare Networks) distributed with Microsoft Windows because you will not get full support when connecting to the Novell network.

32. NetWare uses three log files (CONSOLE.LOG, ABEND.LOG, and SYS$LOG.ERR) that can help you diagnose problems on a NetWare server.

QUESTIONS

1. You use a Windows 98 computer to browse a Novell NetWare version 5.0 LAN. Which network client will allow you to use the full functionality of NDS?
   a. Client for NetWare Networks
   b. NetWare Client32
   c. Client for Microsoft Networks
   d. Banyan DOS/Windows 3.1 Client

2. Which clients allow a Windows computer to connect to a Novell NetWare 5.0 network? (Select all correct answers.)
   a. Client for Microsoft Networks
   b. NetWare Client32
   c. Client for NetWare Networks
   d. Banyan DOS/Windows 3.1 Client

3. Which clients allow a Windows computer to connect to a Novell NetWare 6.0 network? (Select all correct answers.)
   a. Client for Microsoft Networks
   b. NetWare Client32
   c. Client for NetWare Networks
   d. Banyan DOS/Windows 3.1 Client

4. You are a network engineer for the Acme Corporation. The company's LAN uses Novell NetWare version 4.11. You have divided the company's LAN into four subnets and configured the LAN to use the communication protocol that is native to Novell NetWare version 4.11 networks. Which protocol provides routing services among the subnets in this LAN?
   a. SPX   b. IPX
   c. IP   d. NetBEUI

5. Which protocol is native to Novell NetWare version 4.11?
   a. TCP   b. IP
   c. IPX   d. NetBEUI

6. Which protocol is native to Novell NetWare version 6.0?
   a. TCP   b. IP
   c. IPX   d. NetBEUI

7. What is the default IPX frame type for Novell NetWare 3.11?
   a. 802.2   b. SNAP
   c. 802.3   d. II

8. What is the default IPX frame type for Novell NetWare 4.1?
   a. 802.2   b. SNAP
   c. 802.3   d. II

9. What feature of Novell NetWare version 3.12 allows you to administer users and groups?
   a. Novell bindery   b. NTDS
   c. NDS   d. SAM

10. What feature of Novell NetWare 4.11 allows you to administer a WAN from one Novell NetWare 4.11 computer?
    a. The bindery service   b. NTDS
    c. NDS   d. SAM

11. You are using a Windows NT 4.0 Workstation computer on a Novell NetWare 5.0 network. Which two protocols can you use on the Windows NT computer to connect to a NetWare 5.0 server on the network? (Choose two answers.)
    a. NetBEUI   b. IP
    c. NetBIOS   d. IPX

12. Which tool should you use to view software and operating system performance in a Novell NetWare 4.11 network?
    a. Device Manager
    b. Performance Monitor
    c. Monitor.nlm
    d. Task Manager

13. To manage NDS, what utilities would you use? (Select two answers.)
    a. NetWare Administrator
    b. Monitor.nlm
    c. Novell ConsoleOne
    d. Config.nlm

14. Which directory service is based mainly on the Internet directory standard X.500?
    a. NTDS   b. NDS
    c. X.25   d. AppleTalk

15. Which of the following network operating systems have a graphical interface? (Choose all that apply.)
    a. Linux   b. NetWare 4
    c. NetWare 5   d. Windows NT

16. Which component of the NetWare server architecture can be loaded and unloaded as required, thus conserving memory?
    a. NLM   b. VLM
    c. OSI   d. ISO
    e. SCO

17. Which category of NetWare loadable modules (NLMs) is used to interface between the NetWare core operating system and the disk subsystem?
    a. LAN drivers
    b. Disk drivers
    c. Utility NLMs
    d. Maintenance NLMs

18. Which category of NLMs is used to interface between the NetWare core operating system and the network interface card?
    a. LAN drivers
    b. Disk drivers
    c. Name space modules
    d. Maintenance NLMs

19. Which category of NLMs is used to make NetWare capable of storing files with different naming conventions?
    a. LAN drivers
    b. Disk drivers
    c. Name space modules
    d. Maintenance NLMs

20. Which of these operating systems does NOT provide a graphical interface?
   a. NetWare 4
   b. UNIX
   c. NetWare 5
   d. Windows 2000 Server

21. _____ are NetWare NLMs that provide communication between the NetWare core OS and any installed NICs.
   a. PCI drivers
   b. Disk drivers
   c. Microprocessor modules
   d. LAN drivers

22. Your Windows 2000 Workstation users are complaining that when they save file with long file names to their NetWare home directory, the file names are truncated each time. What NLM do you need to load on your NetWare server to keep this from happening?
   a. NLM nontruncating module
   b. LAN driver NLM
   c. Name space module for long file name support
   d. All of the choices listed

23. Other than NetWare version 4.X and greater, which network operating systems will support the native running of NDS?
   a. NetWare 3.X
   b. Solaris UNIX
   c. Windows NT
   d. All of the choices listed

24. Which log file keeps a history of all information and errors that have been displayed on a NetWare server's console?
   a. Console.log
   b. Netware.log
   c. Display.log
   d. Server.log

25. Which file holds information on all abnormal ends on a Netware server?
   a. Abend.log   b. Crash.log
   c. Stop.log   d. Quit.log

26. Your network runs a combination of NetWare 4 and 5 servers. All your client workstations are Windows 98 and Windows 2000 computers. You plan on installing the Novell Client for NetWare to enable your workstations to communicate with your servers. Which of the following does NOT specify when configuring the NetWare client software?
   a. Tree
   b. Server to log into
   c. Context
   d. Domain

HANDS-ON EXERCISES

Exercise 1: Installing Novell NetWare 6.0

The following exercises will be done on computer A.

1. If your computer supports bootable CD, insert the NetWare CD into the drive. If not, boot the computer, access the CD-ROM, and execute the install.bat file.

2. If the screen asks what type of CD drive you have, specify the appropriate drive by pressing the appropriate key (I for "IDE" and S for "SCSI").

3. Select the appropriate language by using the arrow key and pressing the Enter key.

4. When the screen asks about the license agreement, select the Accept License Agreement option and press the Enter key.

5. If the screen states that there are no bootable partitions found on the computer's hard disk, select the Create a new boot partition option and press the Enter key. Since Novell recommends a 200-MB boot partition, 200 MB is selected by default. Since Continue is already selected, press the Enter key. When the screen asks if you are sure, select the Continue option and press the Enter key. When the new boot partition has been created, press any key to reboot.

6. Press I for "install."
7. If the screen asks what type of CD drive you have, specify the appropriate drive by pressing the appropriate key (I for "IDE" and S for "SCSI").
8. After formatting the partition, the screen will show a license agreement for JReport Runtime. Press the F10 key.
9. Since the Express Install and new server are already selected, select the Continue option by using the arrow keys and press the Enter key.
10. A 4-GB SYS volume will be created. Press the Enter key.
11. After several minutes of copying files and starting a GUI, enter NWXXX-YY for the server name, where XXX is your room number and YY is your computer number. Press the Enter key. Then select the Next key.
12. When the screen asks for the NetWare 6 License/Cryptographic diskette, click on the Browse button next to the Location text box and browse to NETWARE6:\License\Demo directory. Click on the OK button. Click on the Next button.
13. Select the IP and IPX box. Enter the following TCP/IP address:

    IP address: 192.168.XXX.1YY
    Subnet mask: 255.255.255.0
    Gateway: 192.168.XXX.254

Click on the Next button.
14. When the screen asks to specify the parameters for the Domain Name Service, enter the NWXXX-YY.ACMEXX.COM. If you have a DNS server, enter the DNS address. Click on the Next button. If you don't have a DNS server, you will get a warning telling you that you will have to click the OK button to continue.
15. Select your time zone and click on the Next button.
16. If the screen warns you that no bindery or NDS information is available on this server and that NDS will be installed, click on the OK button.
17. Select the New NDS tree option and click on the Next button.
18. Type NWTreeYY in the Tree Name text box.
19. Click the Browse button next to the Context for the server object text box. Click the Add button. Organization is already selected. Type in Corp in the Container Name. Click on the OK button. The O=Corp is in the Target context text box. Click the OK button.
20. Enter the password for the password and click on the Next button.
21. After NDS is installed, click on the Next button.
22. If the screen asks you to install the license disk, click the Next box.
23. When the screen lists the products to be installed, click the Finish button.
24. After a few minutes, the screen will state that installation is complete. Click the Yes button to reboot the computer.

    NOTE: Don't forget to remove the CD and floppies.

Exercise 2: Commands at the Server Console

The following exercise will be done on computer A.
1. To exit the GUI, click on the Novell button and select the Close GUI option. If the screen asks are you sure, click on the Yes button.
2. From the command prompt, type in nwconfig.
3. Select Driver options and press the Enter key.

4. Select the Configure disk and storage device drivers and press the Enter key. Notice the drivers loaded.
5. Use the arrow keys to highlight the Return to previous menu option and press the Enter key.
6. Select Configure network drivers.
7. Select the Select an additional driver option and press the Enter key.
8. Select the Novell Ethernet NE2000 driver and press the Enter key.
9. Use the arrow keys to highlight the TCP/IP option and press the space bar to select. For the IP address, enter the following information:

   IP address: 192.168.XXX.253
   Subnet: 255.255.255.0

   where XXX is the classroom number. Select the Novell NE2000 card that is configured for interrupt 3 and port number 300. Press F10 to save the protocol settings.
10. Highlight the Save parameters and load drivers option and press the Enter key.
11. When an error occurs, press the Enter key. Of course, it failed to find the card.
12. Press the Enter key. What key would you use to load an unlisted driver?
13. Press F10 to continue without selecting.
14. Select the Return to previous menu and select Enter.
15. Select the Return to previous menu and select Enter.
16. Select the NCF files option. Select the Edit Autoexec.ncf file. Notice the NIC card drivers and bindings loaded.

    NOTE: You may need to scroll down.

17. What is the last command executed in the autoexec.ncf?
18. Press the Esc key.
19. Select the Edit startup.ncf option and press the Enter key.
20. When the screen asks the server boot path, press the Enter key.
21. Notice the drivers loaded in the startup.ncf.
22. Press the Esc key.
23. Press Esc again and exit NetWare configuration.
24. Execute the monitor command at the prompt. What is the processor utilization? How many open files are there?
25. Press the Tab key for the next window. Press the Tab key.
26. Select the Connections option and see if anyone is logged in under Active connections.
27. Press the Esc key.
28. Select the Volumes option and press the Enter key.
29. How much space was used in the SYS volume?
30. Press the Esc key.
31. Select the LAN/WAN drivers and press Enter.
32. Press the Esc key.
33. Select the loaded modules and press the Enter key. How many bytes is the SERVER.NLM using?
34. Press the Esc key.
35. View the System resources, Virtual parameters, and Kernel options.
36. Press Alt+F10 and exit the monitor.
37. Execute the volume command.
38. Execute the modules command.
39. Press the space bar a couple of times.
40. Press the Esc key.

41. Execute the load conlog command.
42. Exit the console log program.
43. Execute the startx command to start the GUI.

Exercise 3: Mounting a CD

The following exercise will be done on computer A.
1. Select the Utilities option and then the Server console. Click on the Server console window.
2. Open the Novell menu.
3. Insert the Novell NetWare 6 CD in the CD drive.
4. Execute the volume command.
5. Execute the Load CD-ROM command.
6. Execute the volume command.
7. Execute the nwconfig command.
8. Select NCF files options and then select the Edit AUTOEXEC.NCF file option.
9. Add the `load cdrom` command before the startx command. This should make the cdrom.nlm load automatically during boot.
10. Press F10 to save the changes.
11. Exit NWConfig.
12. Exit the GUI.
13. Execute the down command.
14. At the command prompt, execute the server command.
15. Open a server console and execute the volume command to make sure the CD was mounted.

Exercise 4: Using ConsoleOne at the Server

The following exercise will be done on computer A.
1. On the NetWare server, click the Novell button at the bottom left corner of the screen. Select the Settings option and select the GUI environment option. Choose the appropriate video driver, and select 800 × 600 resolution and 256 colors. Click the Test button. When the warning appears, click OK and then click the Test button. If a grid appears on your screen and then your screen goes back to the video board screen, click the OK button and click on the Yes button to save the current configuration. Click on the Yes button to restart the GUI.
2. Click on the Novell button and select the ConsoleOne option.
3. Select the NDS tree and click on the NDS Authenticate button (last button on the tool bar).
4. Type in the user name of admin with the password of password. Click the Browse button next to the Tree text box and select your tree. If it does not show up, you will have to enter it manually. Type in O= corp in the context text box.
5. Right-click the corp organization and select New followed by Organizational Unit. Enter Sales for the name and click the OK button.
6. Using ConsoleOne, select the Logical Volume Disk Management button.

   NOTE: If you move the mouse over the buttons and pause without clicking or moving the mouse, a text message will appear identifying the box.

7. Select corp for the NDS context and your server. Click the OK button.
8. Click the New button. Enter Data in the Name text box. Click the Next button.
9. Select the Unpartitioned space and enter 1024 MB for the Volume Quota. Click the Next button.

10. Enter NWPool for the name and 1024 for the pool. Click the OK button. Click the Finish button. Click the Yes button. Close the Properties box by clicking on the Cancel button.
11. In the corp organization, click the Data volume (name will be *servername*_data). On the right pane, right-click the empty space and select the New button. Select the Object option. Select the Directory option and click the OK button.
12. Enter Home in the Name text box. Click the OK button.
13. Right-click the sales organization unit and select the New option and then User. Enter jsmith in the Name text box and Smith in the Surname text box. Select the Create Home directory and click the Path browse button. Navigate to the Data volume. Double-click the Data volume and select the Home folder. Click the OK button.
14. Type password for the new password and click the Set password button.
15. Under the Sales OU, double-click jsmith. Click on the General tab and enter an address and phone number.
16. Select the Restriction tab and select Require a password option. Set the minimum password length to six characters.
17. Click the small down arrow on the Restrictions tab and select Intruder Lock. Change back to password restrictions.
18. Click the Change password option and change the password to letmein.
19. Click the Security equal to me tab. Click the Add button. Use the navigation buttons to find and select the Admin the corp organization. Click the Apply button.
20. Click the Rights to Files and Folders tab. Click the Show button. Select the Data volume in the corp organization. Click the OK button. Notice the permissions that jsmith has.
21. Enable the supervisor right and click the Apply button and the Effective rights button. Click the Close button.
22. Close ConsoleOne.

Exercise 5: Install the NetWare Client Software

The following exercise will be done on computer B, which should be a Windows computer such as Windows 2000. If you are using the computer from the Windows chapter, it would probably be best to run DCPromo to remove Active Directory.

1. Follow the instructions from your instructor or download the NetWare client software from Novell.
2. Run the executable file and unzip the files into the C:\Client directory.
3. In the c:\client\winnt\i386 directory, double-click the setupnw.exe.
4. When the software license agreement appears, click on the Yes button.
5. Typical installation is already selected. Click on the Install button.
6. When installation is complete, click on the Reboot button to reboot the computer.
7. Press Ctrl+Alt+Del to log in.
8. Enter admin and password for the user name and password. Click on the Advanced button. Click on the Trees button. Select your tree and click the OK button. Click on the OK button. Click the Context button. Select the Corp button.

   NOTE: If you want to specify the Sales organizational unit, you have to double-click the Corp container.

9. Open My computer. Notice the driver mappings to system and public in the sys volume.
10. Open My Network Places and double-click Novell Connections.
11. Double-click your server and open the data volume.
12. Double-click the home directory. You should see the home directory for jsmith.
13. Close the window.

Exercise 6: Loading and Using ConsoleOne from the Client Computer

The following exercise will be done on computer B.

1. Get instructions from your instructor on the location of the Console installation program or download it from the Internet.
2. Double-click the exe file. Click on the Setup button. Click on the Next button.
3. When the license agreement appears, click the Accept button.
4. When the installation path appears, click the Next button.
5. When the components list appears, click on the Next button.
6. Select your language and click on the Next button.

   NOTE: English is already selected.

7. When the JReport Runtime license appears, click the I Do Accept option and click on the Next button.
8. Click the Finish button.
9. When the installation is complete, click the Close button.
10. On the desktop, double-click the ConsoleOne icon to start ConsoleOne.
11. Open NDS and your tree.
12. Open the Corp organizational unit.
13. Double-click the Data volume. In the left pane, right-click the white empty space and select the New option. Then select the Object option. Select the Directory option and select the OK button. Type projects in the Name text box. Click the OK button.
14. Right-click the projects and select Properties.
15. Click the Trustees tab.
16. Click the Add Trustee button and find and select jsmith. Remember, jsmith is in the sales organizational unit.
17. Automatically, John Smith gets the read and file scan. Assign write, create, erase, and modify permissions. Click the Apply button.
18. With jsmith still highlighted, click the Effective Rights button.
19. Click the Close button to close the Effective Rights dialog box.
20. Click the Close button to close the Properties of projects dialog box.
21. Create an Engineer organization organizational unit in the Corp organization.

    NOTE: You may need to open the View menu and select the Refresh option.

22. In the Engineer organization, create the following users:

    Daffy Duck (DDaffy)
    Bugs Bunny (BBunny)
    Tweety Bird (TBird)

23. Assign the Engineer organization the read, file scan, write, create, erase, and modify permissions to the Project folder by right-clicking the Project option and selecting the Trustees tab.
24. Right-click BBunny and select Properties. Show the Effective Rights for BBunny for the Project folder.
25. Create a Managers group in the Sales group. Assign Tweety Bird and JSmith to the Managers group by opening the Properties, selecting the Members tab, and clicking on the Add button.
26. Assign Tweety Bird to the Engineers organization. Assign him full control for the organization. Open the Engineer Properties, select the NDS Rights tab, and select the Supervisor permission. Click the OK button.

CHAPTER **12**

Networking with Linux

Topics Covered in this Chapter

Introduction

Linux (pronounced LIH-nuhks with a short "i") is a UNIX-like operating system that was designed to provide personal computer users a free or very low-cost operating system comparable to traditional and usually more expensive UNIX systems. Linux comes in versions for all the major microprocessor platforms including the Intel, PowerPC, Sparc, and Alpha platforms. Because Linux conforms to the POSIX standard user and programming interfaces, developers can write programs that can be ported to other platforms running Linux (or UNIX, which also conforms to the POSIX standard).

Objectives

- Identify the basic capabilities (client support, interoperability, authentication, file and print services, application support, and security) for the UNIX/Linux operating systems.
- Given specific parameters, configure a Windows or a Linux client to connect to a Windows or a Linux server.
- Given a troubleshooting scenario involving a client connectivity problem (e.g., incorrect protocol/client software/authentication configuration, or insufficient rights/permission), identify the cause of the problem.
- Describe the shutdown procedure for the Linux (command prompt and X Window interfaces) operating systems.

12.1   WHAT IS LINUX?

Different from the other network operating systems discussed in this book, Linux is licensed through GNU (pronounced with a hard "g"—kind of rhymes with "canoe"). **GNU,** which is self-referential, is short for "GNU's not UNIX." It was started in 1983 by Richard Stallman at the Massachusetts Institute of Technology. GNU refers to a UNIX-compatible software system developed by the Free Software Foundation (FSF). As GNU was started, hundreds of programmers created new, open-source versions of all major UNIX utility programs with Linux providing the kernel. Many of the GNU utilities were so powerful that they have become the virtual standard on all UNIX systems. For example, gcc became the dominant C compiler, and GNU emacs became the dominant programmer's text editor.

The philosophy behind GNU is to produce software that is nonproprietary. Anyone can download, modify, and redistribute GNU software. The only restriction is that the person cannot limit further redistribution. In other words, you can download GNU from the Internet at no charge, pass on copies to friends, and even modify its internals, as long as you make the source code available. In addition, the **General Public License (GPL)** is a form of copyright used to protect the software from being taken over.

When people refer to GNU software as free software, they are referring to the freedom to run, copy, distribute, study, change, and improve the software—but not necessarily free of cost. Linux is distributed commercially by a number of companies. Most distributions are available to download for free, without support, purchased as an inexpensive CD or available as a boxed set with manuals and different levels of support. Several distributions worth noting are:

- **Red Hat Linux**—This is the most popular Linux distribution. It includes a slew of personal productivity tools, as well as server essentials like Apache, Samba, and Sendmail. It ships with both KDE and GNOME interfaces, while advanced features like LDAP integration make Red Hat Linux an attractive tool for the enterprise. Red Hat Linux is the most portable version of Linux, with code that runs natively on the Intel, Alpha, and Sparc processors.
- **Slackware Linux**—This is designed specifically for the Intel platform and was the first commercial Linux distribution. Slackware Linux includes kernel, XFree86, KDE, and October GNOME, and it supports many PC hardware devices, including Ethernet and multiple (up to 16) processors.
- **SuSE**—SuSE works closely with the XFree86 Project, and its distribution has the most recent modifications to, and device drivers for, the X11R6 X Window system. Also, the SuSE CD-ROM distribution has the most bundled applications. It includes the YaST2 installation and configuration tool, KDE 1.1.2, October GNOME, VMWare (limited version), Hummingbird PC X-server for Windows, SuSE Proxy Suite, StarOffice, and more. SuSE Linux is particularly popular in Europe, although the U.S. release is growing in popularity.
- **Debian Linux**—This is available on Intel, Sparc, Alpha, and Motorola 680$\times$0 platforms. It's based on the Debian package format (.deb), which uses the dpkg package manager, dselect, and apt for package management interfaces.
- **Mandrake Linux**—Mandrake Linux is built on top of Red Hat Linux. It stresses ease of use for both personal users and server installation, offering customer support with each purchase. It includes a host of add-on packages.

12.1.1 Linux Minimum Requirements

While Linux works with a lot of hardware, it does not work with every piece of hardware. Therefore, hardware compatibility is particularly important if you have an older system or a system that you may have to build yourself. To determine if your computer is compatible for Linux, you can use the hardware compatibility list (HCL).

Of course, before installing Linux, you should look at what are the minimum requirements to run Linux. Then, when choosing the system and its components, you will need to check to see if the system and its components are compatible with Linux.

Linux Hardware Compatibility HOWTO

http://users.bart.nl/~patrickr/hardware-howto/Hardware-HOWTO.html
http://www.linuxdoc.org/HOWTO/Hardware-HOWTO.html

Table 12.1   Memory Requirements

Application	Minimum	Typical	Swap Space
Network gateway	4 MB	8 MB	None required
Network server	8 MB	16 MB	Usually none required
Network multiuser	16 MB	32 MB	64 MB
X workstation	32 MB	64 MB	100 MB
X with image processing	64 MB	128 MB	100 MB
X with StarOffice Suite	128 MB	256 MB	100 MB

Red Hat Linux Hardware Compatibility List

http://hardware.redhat.com/hcl/

Red Hat Hardware Compatibility Lists (Older Linux–4.2, 5.2, 6.0, 6.1, and 6.2)

http://www.redhat.com/support/hardware/hcl-old.html

Since Linux requires task-switching and memory management facilities found on the 80386 and later processors, it will require a 386 or greater. Therefore, you can use an Intel 386, 486, Pentium, Pentium with MMX, Pentium Pro, Pentium II, Pentium II Xeon, Pentium III, Pentium III Xeon, Pentium 4, Intel Xeon, or Celeron processor. Most non-Intel clones will work, including the AMD Athlon, Duron, and K6 processors. Since Linux can emulate a math processor, you don't necessarily need a math processor. But you will get better performance if you have one.

Linux supports ISA, EISA, VESA local bus, and PCI. The MCA bus architecture found on most IBM PS/2 machines has had minimal support since the 2.1.x kernel, but it would not be recommended.

Linux will run on 2 MB, but you will need to use a special installation procedure until the disk swap space is installed. Therefore, you should run with at least 4 MB of memory. If you are planning on using X Window, 32 MB should be minimum, but it would be recommended to have at least 64 MB. X11 components use approximately 30 MB. Netscape uses approximately 10 MB of RAM. Image processors that require large bitmaps use tens of MB. The StarOffice Suite version 5.0 uses a minimum of 50 MB of RAM. As a general rule, if you have more memory, Linux will run faster. (See table 12.1.)

Another consideration involves the use of hardware emulators such as vmware to run MS-Windows or X11 Windows. In this case, 128 MB may be required for each virtual machine to achieve higher performance levels in both Linux and MS-Windows applications.

Just about any standard graphics card, including MDA, Hercules, CGA, EGA, VGA, or Super VGA video card, and monitor will do. Linux will work with all video cards in text mode. However, if you wish to run the X Window system, you will need VGA or better. You should note that X Window does not run on all video card chipsets. Therefore, you

Table 12.2   System Requirements for Red Hat Linux 7.2

	Minimum Requirements	Recommended Workstation Requirements[*]	Recommended Server Requirements[*]
RAM	32 MB or more	64 MB or more	64 MB or more
Disk space	500 MB or more	1.2 GB or more	600 MB or more
Display	VGA or higher	VGA or higher	VGA or higher
Disk drives	Bootable CD-ROM drive or 3 1/2" floppy drive	Bootable CD-ROM drive or 3 1/2" floppy drive	Bootable CD-ROM drive or 3 1/2" floppy drive

[*]Recommended requirements allow for swap space and data storage.

should either run SuperProbe (a program included in the XFree86 distribution) or choose one by looking on the hardware compatibility list.

You will need a 3.5-inch floppy drive. A stripped-down Linux can actually run on a single floppy, but that's only useful for installation and certain troubleshooting tasks. If you want to use a CD-ROM drive, make sure it is either IDE ATAPI or SCSI.

To install a minimal copy of Linux, you will need at least 10 MB. Of course, you can try Linux and not much else. You can also fit an installation that includes X Window in 80 MB. Most Linux distributions will take between 500 MB and 1.7 GB depending on which options you choose, including the kernel source code, some space for user files, and a spool area. To install a commercial distribution that has a desktop GUI environment, a commercial word processor, and a front-office productivity suite will take 2 GB of disk space or more.

Table 12.2 shows the system requirements for Red Hat Linux 7.2. Of course, for best performance, I would recommend a Pentium II processor with a minimum of 128 MB of RAM and a large hard drive. If it is going to be a busy server or workstation, I would consider multiprocessors with even more RAM and a SCSI HD or SCSI RAID system.

12.1.2   Linux Kernel

The Linux kernel acts as a mediator for your programs and your hardware. Like the UNIX kernel, the Linux kernel is designed to do one thing well. It handles low-level functions like managing memory, files, programs that are running, networking, and various hardware devices. (See figure 12.1.) For example, it arranges for the memory management of all the running programs (processes) and makes sure that they all get a fair share of the processor clock cycles. In addition, it provides a nice, fairly portable interface for programs to talk to your hardware. Therefore, you can often think of the kernel as a cop directing traffic. Unlike Windows, it does not include a windowing system or GUI. Instead, Linux users can choose among a number of X servers and window managers.

Figure 12.1   The Linux Kernel

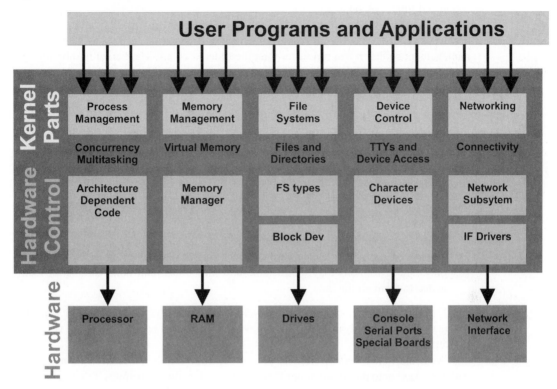

On March 14, 1994, three years after Torvalds started with his project, Linux version 1.0 was released. By this time, several Linux distributors had begun packaging the Linux kernel with basic support programs, the GNU utilities and compilers, the X Window system, and other useful programs. In 1995, the Linux 1.2 kernel was released, which supported kernel modules, the PCI bus, kernel-level firewalls, and non-TCP/IP networking protocols. By this time, Linux had become as stable as any commercial version of UNIX on the Intel x86 platforms. It was even ported to the DEC Alpha, Sun Sparc, and SGI MIPS platforms.

In 1996, the Linux 2.0 kernel was released. Version 2 offered preliminary support for symmetric multiprocessor (SMP) machines. It also was fully functional on multiple platforms, including the X86, Alpha, Sparc, Motorola 68K, PowerPC, and MIPS platforms. In 1999, version 2.2 was released. The 2.2 kernel fully supports the SMP machines, and it more fully supports multiple platforms, including the 64-bit platforms. In addition, version 2.2 is significantly faster than previous versions.

In 2001, the Linux 2.4 kernel was released. To make it more suitable for enterprise-level applications, the new kernel allows for better symmetric multiprocessing scalability up to 32 processors on large X86 Intel servers. In addition, it supports up to 64 GB of physical memory. Other enhancements include new drivers for hardware including USB, IEEE

1394 (FireWire), and 3D-accelerated graphics cards and support for IBM's S/390 mainframe and Intel's IA-64 architecture.

As each new version of the kernel is released, bugs are found and fixed, more device drivers are added so that it can talk to more types of hardware, and it has better process management so that it can run faster than the older versions. The kernel numbering system is relatively simple to follow. The versions are expressed in the following format:

```
major.minor.patch
```

such as 2.4.2 or 2.4.7.

Because Linux follows the "open development model," all new versions are released to the public, whether or not they are considered "production quality." To help people tell whether they are getting a stable version or not, the following scheme has been implemented:

- Versions n.x.y, where x (the minor version) is an even number, are stable versions, and only bug fixes will be applied as y is incremented. So from version 1.2.2 to 1.2.3, there were only bug fixes, and no new features.
- Versions n.x.y, where x is an odd number, are beta-quality releases for developers only. These versions may be unstable and may crash, and new features are being added to them all the time.

From time to time, as the correct development kernel stabilizes, it will be frozen as the new "stable" kernel, and development will continue on a new development version of the kernel.

After the minor version number comes the build number, which is the number of patches added onto the major.minor kernel version thus far. Sometimes if there are minor changes or distribution-specific alterations to the kernel, there will be a dash on the end with another number listing how many changes were made there.

To find the latest stable version of the Linux kernel, the latest beta version of the Linux kernel, and the latest prepatch (alpha) version, go to the Linux Kernel Archives website (http://www.kernel.org). Of course, for reliability, especially in servers, I would only suggest using stable versions of the Linux kernel. Since a Linux user has access to the source code, you can compile a custom kernel designed especially for your system and needs, cut your boot time and system RAM usage significantly, and improve performance in other ways and make a system more secure.

12.1.3   Linux Interfaces

Virtually every utility you would expect to find on standard implementations of UNIX has been ported to Linux. This includes basic commands such as *ls, awk, tr, sed, bc, more,* and so on. Therefore, you can expect your familiar working environment on other UNIX systems to be duplicated on Linux.

The shell is a program that reads and executes commands from the user. You can think of a shell as a command interpreter, functioning much like the COMMAND.COM does in DOS. A shell uses a command-driven interface. Since the shell serves as the primary interface between the operating system and the user, many users identify the shell with Linux. But, in reality, the shell is not part of the kernel; instead it is the interface. Many shells provide features such as job control (allowing the user to manage several running processes at once), input and output redirection, and a command language for writing shell scripts. A shell script is a file containing a program in the shell command language, analogous to a batch file under MS-DOS. The default shell for Linux is the GNU Bourne again shell (bash).

As you type in a command at the keyboard, you must know the correct spelling, syntax, parameters, and punctuation for the command to work. If not, you will receive an error message or an incorrect response. Therefore, to be successful at using the shell, it will require a little memorization. Different from DOS, all commands in Linux are case-sensitive. This means that you have to type in the correct lowercase and uppercase letters. In addition, Linux uses the forward slash (/) rather than the backslash (\). See table 12.3 for common Linux commands.

The **X Window** system was the result of research efforts in the early 1980s at MIT to develop a platform-independent graphics protocol. The X Window system is an open standard that is managed by the X.ORG consortium. Although Microsoft has its own platform-dependent windowing system (an integral part of the Windows 95/98/NT operating systems), there are vendor-supplied X Window products that can be installed to run on these systems.

Since X Window is a standard platform-independent graphics interface that is freely distributable, it has been a particularly attractive system for UNIX vendors. As a result, almost all UNIX graphical interfaces including Motif and OpenLook are based on it. However, many commercial vendors have distributed proprietary enhancements to the original X Window software. The version of X Window available for Linux is known as XFree86.

As for an intuitive graphical interface, Linux has at least a dozen different, highly configurable graphical interfaces (known as window managers), which run on top of XFree86. The most popular window managers at the moment are KDE (K Desktop Environment) and GNOME (GNU Network Object Model Environment; the "G" in GNOME is pronounced as a hard "g"). These offer the point-and-click, drag-and-drop functionality associated with other user-friendly environments such as Microsoft Windows and Apple Macintosh, but are extremely flexible and can take on a number of different looks and feels. Today, complex tasks like system administration, package installation, upgrading, and network configuration can be done easily through graphical programs. Programs that work with one window manager nearly always work with all the others.

GNOME is a powerful graphics-driven environment, which includes a panel (for starting applications and displaying status), a desktop (where data and applications can be placed), multiple window managers (which control the look and feel of your desktop), and a standard set of desktop tools and applications. GNOME's session manager remembers settings and currently running programs, so once you've set things the way you like, they'll stay that way. (See figure 12.2.) GNOME is the default desktop for Red Hat Linux.

KDE provides a complete desktop environment, including a file manager, a window manager, an integrated help system, a configuration system, numerous tools and utilities, and an ever-increasing number of applications. It uses a contemporary desktop, a searchable help system with convenient access to help on the use of the KDE desktop and its applications, standardized menu and tool bars, color schemes, and more. (See figure 12.3.) KDE is the most common desktop used in Linux.

Table 12.3 Commonly Used Linux Commands

Command	Function
date	Display the current date and time
ls –la	Display all files in the current directory, with details
ps –ax	Display details of the current running programs
find *dir* filename–print	Search for file name in the directory *dir* and display the path to the name after finding the file
cat *file*	Display the contents of *file*
cd */d1/d2/d3*	Change the current directory to *d3*, located in */d1/d2*
cp *file1 file2*	Make a copy of *file1* named *file2*
rm *file*	Remove (delete) *file*
mv *file1 file2*	Move (or rename) *file1* to *file2*
mkdir *dir*	Make a new directory named *dir*
rmdir *dir*	Remove the directory named *dir*
who	Display a list of users who are currently logged in
vi *file*	Use the visual editor named vi to edit *file*
grep "*string*" *file*	Search for the string of characters given in *string* in the file named *file*
sort *filename*	Sort alphabetically the contents of *filename*
man *command*	Display the manual page entry for *command*
chmod *rights file*	Change the access rights of *file* to *rights*
telnet *host*	Start a virtual terminal connection to *host* (where *host* may be an IP address or a host name)
ftp *host*	Start an interactive file transfer to (or from) *host* using FTP (where *host* may be an IP address or a host name)
startx	Start the X Window system
kill *process*	Attempt to stop a running program with the process ID *process*
tail *file*	Display the last 10 lines of *file*
exit	Stop the current running command interpreter, and log off the system if it is the initial command interpreter started upon logging in

Figure 12.2   GNOME Desktop

Besides the X Window system being a graphical interface used to run applications, it is also a powerful client/server system that allows applications to be run and shared across a network. In addition, since Linux is a multitasking and a multiuser operating system, it allows you to have more than one logon session running at the same time so that you can run several applications from different computers.

12.1.4   Network Programs

Since Linux is often used as a server, it has a complete implementation of TCP/IP networking software including drivers for the popular Ethernet cards and the ability to use serial line protocols such as SLIP and PPP to provide access to a TCP/IP network via a modem. With Linux, TCP/IP, and a connection to the network, you can communicate with users and machines across the Internet via electronic mail, USENET news, file transfers, and more.

Figure 12.3   KDE Desktop

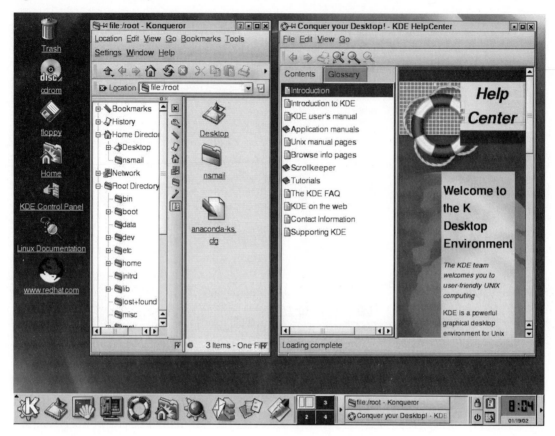

In fact, Linux owes its versatility to the wide availability of software that runs on it. Some of the more important network software includes:

- **Web server**—Apache (http://www.apache.org) is the most popular web server.
- **Web proxy**—To better control web usage and to allow for caching of frequently accessed pages, Squid (http://www.squid-cache.org) is used.
- **File sharing**—Linux can be made to look like an NT server with respect to file and print sharing. Samba (http://www.samba.org) is the software that does this.
- **Email**—Linux excels at handling email. Sendmail (http://www.sendmail.org) is the most widely used mail transfer agent (MTA). Qmail (http://www.qmail.org) and PostFix (http://www.postfix.org) are alternatives. Regardless of your choice, a basic knowledge of Sendmail is required both for the exam and for everyday work.
- **DNS**—The Domain Name Service provides mappings between names and IP addresses, along with distributing network information (i.e., mail servers). BIND (http://www.isc.org/products/BIND/) is the most widely used name server.

12.1.5   Linux Documentation Project

While many Linux distributors provide extensive documentation for their distribution of Linux, the Linux operating system is not owned by any one person or company. To provide free, high-quality documentation for the GNU/Linux operating system, the **Linux Documentation Project (LDP)** was created. The LDP is a loose team of writers, proofreaders, and editors who are working on a set of definitive Linux manuals.

> **Linux Documentation Project**
>
> http://www.linuxdoc.org/
>
> http://www.ibiblio.org/mdw/index.html
>
> http://www.linux.org/docs/index.html

Within the LDP, you will find:

- **Guides**—Longer, more in-depth books on various Linux topics
- **HOWTOs and mini-HOWTOs**—Subject-specific help on various Linux topics
- **FAQs**—Frequently asked questions
- **Man pages**—Help on individual commands
- *Linux Gazette*—Online magazine

12.2   STARTING AND SHUTTING DOWN LINUX

When you boot Linux, you will typically start with a LILO boot prompt or a LILO GUI prompt. At the LILO boot prompt, you can do one of the following:

- Press the Enter key—this will cause LILO to boot its default boot entry.
- Type in the name of the boot label to specify which image to boot.
- Wait until the LILO's timeout period expires (default is 5 seconds), at which time, LILO will automatically boot the default boot entry.

However, if you forget the boot labels defined on your system, you can always press the Tab key at the LILO's boot prompt to display a list of defined boot labels.

If your system boots with a LILO GUI prompt, you can do any of the following:

- Press Enter—this causes LILO's default boot entry to be booted.
- From the list of boot labels, select one with the arrow keys and press the Enter key. This causes LILO to boot the operating system corresponding to the boot label.
- Wait until the LILO's timeout period expires (default is 5 seconds), at which time LILO will automatically boot the default boot entry.

Before you can start using Linux, you will need to log in. Like Windows NT or Windows 2000, Linux uses accounts to manage privileges, maintain security, and more. Not all accounts are created equal; some have fewer rights to access files or services than others. Earlier in this chapter, I mentioned the root account. The root account is the all-powerful account to the Linux system. It is always recommended that you create a second account for yourself and use this account when you need to access the computer as a normal user. When you need to modify the configuration of Linux, you would then use the root account.

NOTE: Linux is case-sensitive. Therefore, Linux account names and passwords are case-sensitive. This means that root refers to a different account than Root does. Of course, the lowercase root is the name of the root login, or system administrator.

The login command prompt for Linux will include the name of the computer. The name of the computer will probably be called localhost unless you gave the machine a name in the network setting. You type in the account name and press the Enter key, followed by typing the account password and pressing the Enter key. (See figure 12.4.) The prompt will then change to something similar to [accountname@localhost/root].

If you're using the graphical login screen, type the account name in the Login: text box and press the Enter key. (See figure 12.5.)

Figure 12.4 Logging Into Linux Using the Text-Based Interface

```
Red Hat Linux release 7.0

Kernel 2.2.16 on an i686


localhost login:accountname↵

Password:accountpassword↵


[accountname@localhost/accountname]
```

Figure 12.5 The Graphical Login Screen

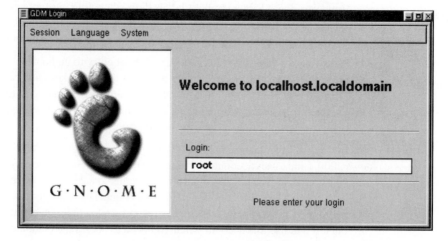

NOTE: If you still see a console screen instead of a graphical desktop, you can start the X Window system by typing startx and pressing the Enter key after logging in at the command prompt.

Logging out of the system depends on how the shell has been configured. By default, the shell will accept the end-of-file (EOF) character Ctrl+D as the termination command and return the user to the Login: prompt. Alternatively, some shells may be configured to prompt the user to use the built-in shell command Logout or Exit.

To log out from GNOME, click once on the Main Menu button on the panel and drag your mouse cursor to the first item, labeled "Logout." When the confirmation dialog appears, select the Logout option and click the Yes button. If you want to save the configuration of your panel, as well as any programs that are running, check the Save current setup option as well. To log out in KDE, you can log out from the Menu button on the panel. It is located near the task bar, at the center of the panel. You'll return to either the graphical login screen or the shell prompt, depending on how you chose to log in. (See figure 12.6.)

The fastest way to stop X Window and go back to the command prompt is to press Ctrl+Alt+Backspace. If you press the Ctrl+Alt+Del keys or select Log out from the panel's Main Menu button, you will get a Logout dialog box asking if you want to really log out. You then have three options to select:

- **Logout**—Logs you out of your account and returns you to the login screen, leaving the system running
- **Halt**—Logs you out of your account and shuts down the system
- **Reboot**—Logs you out of your account and restarts the system

Select any of the options and press the Yes button to continue. (See figure 12.7.) Also, select the Save current setup option if you want to save your session. Saving your session will preserve your current configuration of the panel and save the programs you might have open. If you don't wish to proceed, choose the No button to continue with your session.

Like other modern operating systems, Linux uses a data cache to reduce long disk I/O delays by saving previously accessed disk blocks in main memory so that they are quickly accessible when needed. The blocks are flushed to the hard disk when they are no longer needed. In Linux, the data cache grows and shrinks in size depending on the amount of memory not used by other programs. Approximately, 1.5 MB of memory is left unallocated for new programs, and the remainder is given over to the data cache. As a result, it not uncommon to see the data cache occupy one-third to two-thirds of physical memory.

While this system makes Linux one of the fastest operating systems used today, a problem may occur in the event that there is a power failure or the computer has been improperly shut down. The disk blocks in main memory are lost, and the file system can be corrupted. Linux will minimize this problem by periodically saving any outstanding modified disk blocks. Therefore, to turn off Linux, you must first terminate the system programs that reset open files and flush the data cache.

There are three ways to shut down the Linux system properly. Shutting down using the Halt or Reboot option using GNOME or KDE desktop was already discussed. On

Figure 12.6   GNOME and KDE Main Menus

many Linux versions, this Ctrl+Alt+Del keyboard combination issues a shutdown command that closes all the processes properly and then reboots the machine. Not all versions of Linux support this sequence, though, so check your documentation carefully.

NOTE: If there are Windows NT or Windows 2000 users around, this option may be removed from the /etc/inittab file to prevent them from shutting down the Linux box inadvertently when using the NT login technique.

Another method of shutting down Linux is with the UNIX shutdown command. When you issue the shutdown command, Linux starts to terminate all processes and

Figure 12.7   Configuration Logout Box

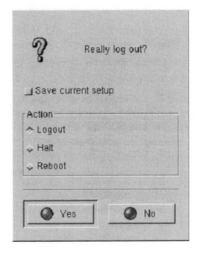

then shuts down the kernel. The shutdown command displays several different messages, depending on the version of Linux, but all inform you of the process or check that you really want to shut down the system. The shutdown command allows you to specify a time until shutdown, as well as an optional warning message to be displayed to all users logged in.

`shutdown -r now` `shutdown -r 5`	The -r means reboot, and will restart your machine. It reboots the system after 5 minutes.
`shutdown -h now` `shutdown 15 'Backup Time!'`	The -h means halt, and will shut down the system. It shuts down the system after 15 minutes and displays the message "Backup Time!" to all users on the system, prompting them to log off. This command is handy when you enforce a policy of shutting down at specific intervals, either for maintenance or for backups.
`shutdown -k now`	This does not perform a shutdown but sends the shutdown announcement to all users.
`shutdown -c now`	This cancels a running shutdown.

In most cases, using Ctrl+Alt+Del or the shutdown command results in the display of a number of status messages on the main console. When Linux has finished shutting down the system, you see the message "The system is halted." When this message appears on-screen, it is safe to shut off the system power or reboot the machine.

NOTE: You can also use the `init 0` to perform a shutdown, issue the `halt` command, which is the same as issuing the `shutdown -h now` command or the `reboot` command, which is the same as issuing the `shutdown -r now` command.

12.3   FILE SYSTEMS

Linux supports various file systems for storing data including ext2 file systems (developed specifically for Linux), journaled file systems including ext3 and reiserfs, Microsoft's VFAT file systems (used in Windows 95/98), and ISO 9660 CD-ROM file systems. In addition, since Linux computers can act as a server, various computer systems can store files on the server including DOS, Windows, Apple Macintosh, UNIX, and Linux.

12.3.1   Disk Names in Linux

Linux treats all devices as files and has actual files that represent each device. In Linux, these device files are located in the /dev directory. Linux file names are similar to MS-DOS file names except that the Linux file names do not use drive letters such as A and C and they substitute the slash (/) for the MS-DOS backslash as the separator between directory names.

Because Linux treats a device as a file in the /dev directory, the hard disk names start with /dev. Table 12.4 lists the hard disk and floppy drive names that you may have to use. Of course, when you use the Red Hat Disk Druid or Linux fdisk program to prepare the Linux partitions, you have to identify the disk drive by its name such as /dev/hda.

Table 12.4   Hard Disk and Floppy Drive Names

Device Type	Name	Description
IDE hard drives	/dev/hda	First IDE hard drive (typically the C drive)
	/dev/hdb	Second IDE hard drive
SCSI hard drives	/dev/sda	First SCSI hard drive
	/dev/sdb	Second SCSI hard drive
Floppy disk	/dev/fd0	First floppy drive (A drive)
	/dev/fd1	Second floppy drive (B drive)

When Disk Druid or fdisk displays the list of partitions, the partition names are of the form /dev/hda1, /dev/hda2, etc. Linux constructs each partition name by appending the partition number (1 through 4 for the four primary partitions on a hard disk) to the disk name. Therefore, if your PC's single IDE hard drive has two partitions, the installation program uses /dev/hda1 and /dev/hda2 as the names of these partitions.

12.3.2   Linux File System

Under most operating systems (including Linux), there is the concept of a file, which is just a bundle of information given a file name. Examples of files might be your paper or report, an email message, or an actual program that can be executed. Essentially, anything saved on disk is saved as an individual file.

Files are identified by their file names. These names usually identify the file and its contents in some form that is meaningful to you. There is no standard format for file names as there is under MS-DOS and some other operating systems; in general, a file name can contain any character (except the / character) and is limited to 256 characters in length.

With the concept of files comes the concept of directories. A directory is a collection of files. It can be thought of as a folder that contains many different files. Directories are given names with which you can identify them. Furthermore, directories are maintained in a treelike structure, in which directories may contain other directories.

Most Linux systems use a standard layout for files so that system resources and programs can be easily located. This layout forms a directory tree, which starts at the root directory (designated as the / directory).

NOTE: This is different from DOS and the command prompt under Windows, which use the backslash (\) as the root directory and as a separator for directories.

Directly underneath / are important subdirectories: /bin, /etc, /dev, and /usr, among others. (See figure 12.8.)

Red Hat and other Linux distributors are committed to the Filesystem Hierarchy Standard (FHS), a collaborative document that defines the names and locations of many files and directories. The current FHS document is the authoritative reference for any FHS-compliant file system. The complete standard can be viewed at http://www.pathname.com/fhs/. Unfortunately, the standard leaves many areas undefined or extensible. See table 12.5 for a list of well-known directories used in Linux.

12.3.3   Mount Points

DOS and Windows use drive letters to identify logical drives or partitions that may be on one hard drive or several hard drives. Each drive has a root directory and its hierarchical file system.

In Linux, all the drives and their partitions make up a single directory file system. If you have more than one physical disk partition, it is associated with a specific part of the file system. All you have to do is decide which part of the Linux directory tree should be located on each partition. This process is known in Linux as mounting a file system on a device (the disk partition). The term *mount point* refers to the directory you associate with

Figure 12.8   Sample File System

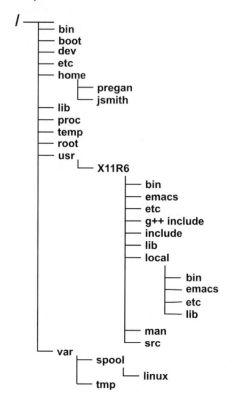

a disk partition or any other device. In other words, each mount point is a disk partition, and that disk partition is mounted on the directory of the limb above it.

Although you can get by with a single large partition for the entire Linux file system and another for the swap space, you can better manage the disk space if you create separate partitions for the key parts of the Linux file system. This is probably more important if Linux is running as a server where you need to do a better job of managing disk space. (See figure 12.9.) Some of the recommended partitions are:

- /bin
- /boot
- /home
- /lib
- /root
- /tmp
- /var

At the very minimum, you will need two partitions for your system: root (/) and swap. For performance reasons, Linux likes to have swap on its own partition. For most other installations, I would recommend three partitions: swap, root, and home. If you require

Table 12.5   Key Directories Used in Red Hat Linux File System Hierarchy

Directory Name	Description
root (/)	In this setup, all files (except those stored in /boot) reside on the root partition. For Red Hat Linux, a 900-MB root partition will permit the equivalent of a workstation-class installation with very little free space. A 1.7-GB root partition will let you install every package in Red Hat Linux.
/bin	Contains the executable programs that are part of the Linux operating system and that are available to all users. Many commands, such as cat, cp, ls, more, and tar, are located in /bin.
/boot	The /boot directory contains the operating system kernel (which allows your system to boot Linux), along with files used during the bootstrap process. Due to the limitations of older PC BIOSs, creating a small partition to hold these files is a good idea. This partition should be no larger than 16 MB.
/dev	Contains all device files. The /dev directory contains files divided into subdirectories that contain files for a specific device. These directories are named based on the type of device files they contain. Some examples of these are hda1 for the first IDE hard drive, fd0 for the floppy disk, and cdrom, which is the symbolic link created to the CD-ROM device that Linux was installed from.
/etc	The location for configuration files and for the system boot scripts. These boot scripts are located under the /etc/rc.d/ directory. The services that run at bootup are located under /etc/rc.d/init.d/.
/home	The /home directory is where users store their own files. When planning out the size of the /home partition, you must consider how many users you have and how much space you want to allocate for each user.
/lib	Contains library files for C and other programming languages.
/lost+found	The system's directory for placing lost files that are found or created by disk errors or improper system shutdowns. At boot, programs such as fsck find the inodes, which have no directory entries, and reattach them as files in this directory. Every disk partition has a lost+found directory.
/mnt	An empty directory, typically used to mount devices temporarily such as floppy disks and disk partitions. Also contains the /mnt/floppy directory for mounting floppy disks and the /mnt/cdrom directory for mounting the CD-ROM drive. *NOTE:* You also can mount the CD-ROM drive on another directory.
/opt	The location for optional packages.
/proc	A special directory that contains information about various aspects of the Linux system configuration information (interrupt usage, I/O port use, and CPU type) and the running processes (programs).
/root	The home directory for the root user.

Continued

Table 12.5 Continued

Directory Name	Description
/sbin	The location of some system configuration and system administrator executable files. Most of the general administration commands such as LILO, fdisk, fsck, and route are located here and are typically available only to root.
/swap	The /swap directory is where the swap files (virtual memory) are. If you have more swap space, it will allow more programs to run simultaneously or larger programs to run with more data. If you have 16 MB or less, you must have a swap partition. Even if you have more memory, a swap partition is still recommended. The minimum size of your swap partition should be equal to the amount of RAM in your system or 16 MB, whichever is larger. Linux allows up to 16 swap partitions.
/tmp	The /tmp directory is used to hold temporary files, typically needed in larger multiuser systems and network file servers.
/usr	The /usr directory is where most of the software on the Linux system resides. It should be about 150 to 600 MB, depending on which packages are installed. */usr/bin*–Contains executable files for many Linux commands including utility programs commonly available in Linux but not part of the core Linux operating system */usr/doc*–Contains the documentation files for the Linux operating system, as well as many utility programs such as bash, mtools, the xfm file manager, and the xv image viewer */usr/games*–Contains some old Linux games such as fortune, banner, and trojka */usr/include*–Contains the header files (files with names ending in .h) for the C and C++ programming languages */usr/lib*–Contains the libraries for C and C++ programming languages */usr/local*–Used for storing programs you want to separate from the rest of the Red Hat Linux software */usr/man*–Contains the online manual, which can be read by using the man command */usr/sbin*–Contains many administrative commands, such as commands for electronic mail and networking */usr/share*–Contains shared data such as default configuration files and images for many applications */usr/src*–Contains much of the Linux source code */usr/X11R6*–Contains the XFree86 (X Window system) software
/var	Short for "various." It is used to store files that change often, such as for spooling, logging, and other data.

Figure 12.9   Sample Linux File System Using Mounted Points

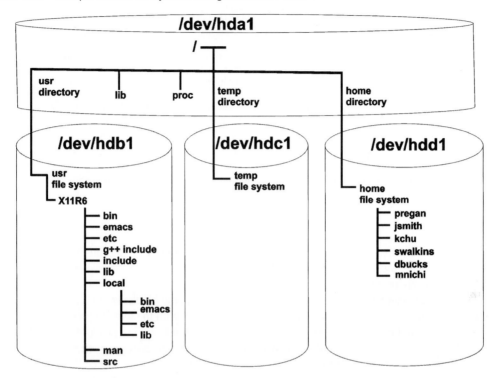

complicated system requirements and decide to use multiple partitions, you should refer to table 12.6.

The size of the swap partition usually varies between the amount of physical memory (RAM) and twice the amount of physical memory. There are situations, such as certain databases, where an increased swap partition is desired. But if the swap file is too big, you will most likely see a degradation in performance since disk is much slower than RAM.

Partitions like /usr and /home tend to fill up quickly, so if you have extra space, it should be put there. Depending on the function of the computer, extra space could go to other partitions. For example, if you are using Linux as a mail server, you would have little need for the /home partition, but you would want lots of space in /var to store all the mail. If you have a file server, you would not have many binary files in /bin, but its /home will most likely need to be quite large.

12.3.4   Second Extended File System

The second extended file system, referred to as ext2, is considered the native Linux file system. It is very similar to other modern UNIX file systems, but it most closely resembles the Berkeley Fast Filesystem used by BSD systems. The maximum size of an ext2 file system is 4 TB, while the maximum file size is currently limited to 2 GB by the Linux kernel.

Table 12.6   Directory Guidelines for a Complicated Disk Structure for Linux

Name	Min Size	Usage
Swap	128 MB	Virtual memory, which is used to store inactive memory to disk until it is later used
/	250 MB	Root file system including basic libraries, programs, and configuration
/var	250 MB	For files that change frequently including logs, spool files, and lock files
/usr	500 MB	Used by most applications
/boot	16 MB	Used to store the kernel
/home	500 MB	Home directory of users including user-specific configuration and data

The ext2 file system, like a lot of other file systems, is built on the premise that the data held in files is kept in data blocks. You can think of data blocks as being similar to clusters or allocation units. These data blocks are all the same length, and although that length can vary among different ext2 file systems, the block size of a particular ext2 file system is set when it is created (during formatting with the mke2fs program).

So far as each file system is concerned, block devices are just a series of blocks that can be read and written. A file system does not need to concern itself with where on the physical medium a block should be put; that is the job of the device driver. Whenever a file system needs to read information or data from the block device containing it, it requests that its supporting device driver read an integral number of blocks.

The ext2 defines the file system topology by describing each file in the system with an inode data structure. An inode specifies which blocks of data belong to which files and describes the access rights of the file, the file's modification times, and the type of file.

12.3.5   Journaling File Systems

The ext2 is not perfect. It is a static file system, which can not guarantee that all updates to your hard drive are performed safely. If the computer is shut down improperly, for example because of a power outage or a system crash, it can take several minutes for Linux to verify the integrity of the partition when the computer reboots. In addition, to make things worse, the ext2 is not fault-tolerant. To overcome such situations, the journaling file system was created, which becomes more important with mission-critical servers and the necessity for maintaining large data sets.

Journaling file systems are superior to static file systems when it comes to guaranteeing data integrity and increasing overall file system performance. Journaling and logging file systems can keep track of the changes either to a file's "metadata" (information such as ownership, creation dates, and so on) or to the data blocks associated with a file, or to both, rather than maintaining a single static snapshot of the state of a file.

When modifying the blocks in the middle of a file and then adding new blocks to the end, a journaling file system would first store the pending changes (modified and new blocks) in a special section of the disk known as the "log." The file system would then update the actual file and directory inodes using the data from the log, and would then mark that log operation as having been completed ("committed," in logging terms).

Whenever a file is accessed, the last snapshot of the file is retrieved from the disk, and the log is consulted to see if any uncommitted changes have been made to the file since the snapshot was taken. Every so often, the file system will update file snapshots and record the changes in the log, thereby "trimming" the log and reducing access time. The process of committing operations from the log and synchronizing the log and its associated file system is called a checkpoint.

Journaling and logging file systems get around the problem of inconsistencies introduced during a system crash by using the log. Before any on-disk structures are changed, an "intent-to-commit" record is written to the log. The directory structure is then updated, and the log entry is marked as committed. Since every change to the file system structure is recorded in the log, file system consistency can be checked by looking in the log without the need for verifying the entire file system. When disks are mounted, if an intent-to-commit entry is found but not marked as committed, then the file structure for that block is checked and fixed if necessary.

After a crash, file systems can come online almost immediately because only the log entries after the last checkpoint need to be examined. Any changes in the log can be quickly "replayed," and the corrupted part of the disk will always correspond to the last change added to the log. The log can then be truncated since it will be invalid, and no data is lost except for any changes that were being logged when the system went down. Mounting a heavily populated directory that requires subsequent validation for database partitions might take 10 to 20 minutes to perform a file system check with fsck with a standard static file system. A journaled file system can reduce that time to a few seconds.

The disadvantage of logging is it generally requires more disk writes because you have to first append log records to the log and then replay them against the file system. However, in practice, the system can operate more efficiently by using its "free time" to commit entries from the log and checkpoint the file system records. Also, because logs are stored separately on the disk from file system data and are only appended to, logging changes happen much faster than actually making those changes.

There are multiple offerings from commercial software vendors that have released journaled file systems for Linux, and there are a few contenders from the open-source community. The most popular two are ext3 and reiserfs.

The ext3 is designed to make the migration from static ext2fs file systems to ext3fs as easy as possible. The initial proposal of ext3 had simply included the addition of logging capabilities to ext2fs through a log file in that file system. Although it is slow, it is extremely reliable.

Reiserfs is a new, general-purpose file system for Linux that is designed for flexibility and efficiency. In some cases it still extracts a slight performance penalty in the interests of increased reliability and faster restart times. Reiserfs is more space efficient than most files, as it can pack many small files into a single block. In addition, reiserfs supports file system plug-ins that make it easy to create your own types of directories and files. This guarantees reiserfs a place in the Linux file systems of the future by making it easy to extend reiserfs to support the requirements for protocols that are still being finalized, such as streaming audio and video. For example, a system administrator can create a special file system object for streaming audio or video files and then create her own special item and search handlers for the new object types. The content of such files can already be stored in TCP/IP packet format, reducing processing latency during subsequent transmission of the actual file.

12.3.6   Virtual File System

The virtual file system (VFS) is a kernel software layer that handles all system calls related to the Linux file system. It must manage all the different file systems that are mounted at any given time. It provides a common interface to several kinds of file systems including disk-based file systems, network-based file systems, and special file systems such as the /proc file system and /dev file system. To do this it maintains data structures that describe the whole (virtual) file system and the real mounted file systems. VFS describes the system's files in terms of superblocks and inodes in much the same way as the ext2 file system uses superblocks and inodes.

As each file system is initialized, it registers itself with the VFS. This happens as the operating system initializes itself at system boot time. The file system's drivers are either built into the kernel itself or built as loadable modules. File system modules are loaded as the system needs them. So, for example, if the VFAT file system is implemented as a kernel module, then it is only loaded when a VFAT file system is mounted. When a block device–based file system is mounted, and this includes the root file system, the VFS must read its superblock. Each file system type's superblock read routine must work out the file system's topology and map that information onto a VFS superblock data structure. The VFS keeps a list of the mounted file systems in the system together with their VFS superblocks. Each VFS superblock contains information and pointers to routines that perform particular functions.

For example, the superblock representing a mounted ext2 file system contains a pointer to the ext2 specific inode read routine. This ext2 inode read routine, like all the other file system specific inode read routines, fills out the fields in a VFS inode. Each VFS superblock contains a pointer to the first VFS inode on the file system. For the root file system, this is the inode that represents the / directory. This mapping of information is very efficient for the ext2 file system but moderately less so for other file systems.

12.3.7   The /proc and /dev File Systems

The flexibility of the virtual file system is shown in the /proc and /dev file systems. The file systems and their files and directories do not actually exist, and yet they act like real file systems for quick and easy access.

The /proc file system, like a real file system, registers itself with the virtual file system. The /proc file system does not store data; rather, its contents are computed on demand according to user file I/O requests. When the VFS makes calls to it requesting inodes as its files and directories are opened, the /proc file system collects the appropriate information, formats it into text form, and places it into the requesting process's read buffer.

NOTE: If you need to analyze your system and its performance, much of the information can be found in the /proc file system.

Linux represents its hardware devices as special files. A device file does not use any data space in the file system; it is only an access point to the device driver. The ext2 file system and the Linux VFS both implement device files as special types of inode. There are two types of device file: character and block special files. Within the kernel itself, the device drivers implement file semantics; you can open them, close them, and so on. Character devices allow I/O operations in character mode, and block devices require that all I/O is via the buffer cache. When an I/O request is made to a device file, it is forwarded to the appropriate device driver within the system. Often this is not a real device driver but a psuedo-device driver for some subsystem such as the SCSI device driver layer. Device files are referenced by a major number and a minor number. The major number identifies the device type, and the minor number identifies the unit, or instance, of that major type. For example, the IDE disks on the first IDE controller in the system have a major number of 3, and the first partition of an IDE disk would have a minor number of 1.

12.4 RUN LEVELS

A run level is an identifier that specifies the current system state such as single-user mode, multiuser mode, multiuser mode with network modes, and others. By specifying a run level, you are specifying a software configuration that will permit only a select group of processes (programs) to run, and you are specifying what daemons (service) to start or stop.

While the exact run-level definitions vary between distributions, they all serve the same purpose, which is to bring the system up to run in the desired state by using the /etc/inittab file. However, thanks to the Linux Standard Base Specification the various distributions will standardize on run levels.

Levels 0, 1, and 6 are reserved. Run level 0 is used to halt the system, run level 1 is used to place the system in single-user mode (similar to a Windows safe command prompt mode), and run level 6 is used to reboot the system.

NOTE: The default run level should never be 0 or 6.

(See table 12.7.) Generally Linux operates in full multiuser mode with network support.

12.4.1 Daemons

In Linux, a **daemon** is a background process or service that monitors and performs many critical system functions and services. Typically, a daemon is started when the system boots, and the daemon processes run as long as the system is running. Most daemons have the capability to restart copies of themselves to handle specific tasks. Daemons usually have names that end with d. Some of the common daemons are:

- **anacron**—Checks cron jobs that were left out due to downtime and executes them. Useful if you have cron jobs scheduled but don't run your machine all the time— anacron will detect that during bootup.
- **amd**—Automount daemon (automatically mounts removable media).
- **apmd**—Advanced Power Management BIOS daemon. For use on machines, especially laptops, that support apm.

Table 12.7 Red Hat Linux Run Levels

Run Level	Red Hat Linux Run Levels	Linux Standard Base Specification Run Levels
0	Halt	Halt
1	Single-user mode	Single-user mode
2	Multiuser mode, without NFS	Multiuser with no network services exported
3	Full multiuser mode	Normal/full multiuser
4	Not used	Reserved for local use; default is normal/full multiuser
5	Full multiuser mode (with an X-based login screen)	Multiuser with xdm or equivalent
6	Reboot	Reboot

- **arpwatch**—Keeps watch for Ethernet/IP address pairings.
- **autofs**—Control the operation of automount daemons (competition to amd).
- **crond**—Automatic task scheduler. Manages the execution of tasks that are executed at regular but infrequent intervals, such as rotating log files and cleaning up /tmp directories.
- **cupsd**—The Common UNIX Printing System (CUPS) daemon. CUPS is an advanced printer spooling system that allows setting of printer options and automatic availability of a printer configured on one server in the whole network. The default printing system of Linux Mandrake.
- **dhcpd**—Implements the Dynamic Host Configuration Protocol (DHCP) and the Internet Bootstrap Protocol (BOOTP).
- **httpd**—Daemon for the Apache web server.
- **inetd**—Listens for service requests on network connections, particularly dial-in services. This daemon can automatically load and unload other daemons (ftpd, telnetd, etc.), thereby economizing on system resources.
- **isdn**—Controls ISDN services.
- **kerneld**—Automatically loads and unloads kernel modules.
- **klogd**—The daemon that intercepts and displays/logs the kernel messages depending on the priority level of the messages. The messages typically go to the appropriately named files in the directory /var/log/kernel.
- **Kudzu**—Detects and configures new or changed hardware during boot.
- **Linuxconf**—The Linuxconf configuration tool. The automated part is run if you want Linuxconf to perform various tasks at boot time to maintain the system configuration.
- **lpd**—Printing daemon.
- **named**—The Internet Domain Name Server (DNS) daemon.
- **netfs**—Network File System mounter. Used for mounting NFS, smb, and NCP shares on boot.

- **network**—Activates all network interfaces at boot time by calling scripts in /etc/sysconfig/network-scripts.
- **nfsd**—Used for exporting NFS shares when requested by remote systems.
- **nfslock**—Starts and stops NFS file locking service.
- **pcmcia**—Generic services for pcmcia cards in laptops.
- **Portmap**—Needed for remote procedure calls.
- **routed**—Manages routing tables.
- **Sendmail**—Mail transfer agent. This is the agent that comes with Red Hat.
- **smbd**—The Samba (or smb) daemon, a network connectivity service to MS Windows computers on your network (hard drive sharing, printers, etc.).
- **smtpd**—Simple Mail Transfer Protocol, designed for the exchange of electronic mail messages. Several daemons that support SMTP are available, including Sendmail, smtpd, rsmtpd, Qmail, and Zmail.
- **Squid**—An HTTP proxy with caching. Proxies relay requests from clients to the outside world and return the results. You would use this particular proxy if you wanted to use your Linux computer as a gateway to the Internet for other computers on your network. Another (and probably safer at home) way to do it is to set up masquerading.
- **syslogd**—Manages system activity logging. The configuration file is /etc/syslog.conf.
- **usb**—Daemon for devices on Universal Serial Bus.
- **xfs**—X font server.
- **xntpd**—Finds the server for an NIS domain and stores the information about it in a binding file.
- **ypbind**—NIS binder. Needed if computer is part of a Network Information Service domain.

A typical Linux configuration employs a minimum of 15 separate daemons running in the background and may run 50 or more daemons.

12.4.2   Starting, Stopping, and Restarting Daemons

To shut down a daemon under Linux, type the name of the daemon, followed by "stop." For example, to stop the smb daemon, type in the following command:

```
smb stop
```

To start the daemon, you would use "start" instead of "stop." To start the smb daemon, you would use the following command:

```
smb start
```

You can also use the restart parameter. Note that sometimes daemons need to include their path name. Therefore, the smb command might be:

```
/etc/rc.d/init.d/smb start
```

Before moving on to the next section, we need to discuss the xinet daemon (xinetd). This daemon maintains multiple network services including telnet, wu-ftp, SWAT, Linuxconf-web, and rsh commands to maintain system security. All these services' configuration files are stored in the /etc/xinetd.d directory. To stop, start, or restart xinetd, which in turn will stop, start, or restart the services that it manages, you would use the following commands:

```
service xinetd stop
service xinetd start
service xinetd restart
```

Several utilities allow you to manage your daemons. They include ntsysv, chkconfig, tksysv, and Linuxconf.

12.5 USER ACCOUNTS

A **user account** enables a user to log onto a computer with an identity that can be authenticated and authorized for access to computer resources. Users can be either actual people (accounts tied to a particular physical user) or logical users (accounts that exist for applications so that they can do particular things). In Linux, every account has an owner attached to it. At the very least, an account needs one root user; however, most Linux distributions ship with several special users set up.

Any file created is assigned a user and group when it is made, as well as being assigned separate read, write, and execute permissions for the file's owner, the group assigned to the file, and any other users on that host. The user and group of a particular file, as well as the permissions on that file, can be changed by root or, to a lesser extent, by the creator of the file.

Each account is associated with the following:

- **User name**—The user name is the name by which the account is known to people. Each user who logs onto the computer must have his or her own unique user ID (UID).
- **Password**—Linux accounts are first protected by a password. A person attempting to log in must provide both a user name and a password. While the user name is generally public knowledge, passwords should be kept secret.
- **Home directory**—Every account has a home directory associated with it so that a user can store data files. Typically, each user has his or her own home directory, although it is possible to configure a system so that two or more users share a home directory.
- **Default shell**—When using a Linux computer at the command prompt, Linux presents users with a program known as a shell. The default shell (typically bash) specifies which shell the user will use by default.
- **Program-specific files**—Some programs generate files that are associated with a particular user. Many programs create configuration files to keep user settings and options and to remember where the user left off within certain programs.

The user ID identifies the user behind the scenes. The user creation utilities automatically choose the next available number when adding a new user. The range of UIDs available depends on which distribution is in use. Typically, a certain range of lower numbers is reserved for administrative accounts, and numbers beyond that are available for individual users. The user range is easily determined by looking in /etc/passwd to see which numbers were assigned to the user accounts already created.

12.5.1   User Names

Most versions of Linux support user names consisting of any combination of uppercase and lowercase letters, numbers, and most punctuation symbols including periods and spaces. User names are case-sensitive although it is recommended to keep them all low-ercase for simplicity. You should note that some punctuation symbols and spaces may cause problems for certain Linux utilities. User names must begin with a letter, and they can be up to 32 characters in length. Some utilities may truncate the user names to 8 char-acters. Therefore, some administrators try to limit user names to 8 characters.

For a network, especially for a network with lots of users, it is recommended to have a consistent naming convention so that it will be easier to find names that follow a con-sistent pattern. For example, you can use the first initial, middle initial, and last name or the first name and last initial. Of course, if you have a hundred employees or more, you might need to use something like first name, middle initial, and last name to differentiate all the users. You should create a document stating the naming scheme of user accounts and provide this document to all administrators for whom you might create accounts. To identify the type of employee, add additional letters at the beginning or at the end of the name. For example, you can add a T- in front of the name or an X at the end of the name to indicate that the person is under contract or a temporary employee.

12.5.2   Passwords

The purpose of a password is to ensure that those who attempt to access computer re-sources are who they say they are. Therefore, for a secure system, it is important to require passwords for everyone.

In UNIX, the password is encrypted by default using the AT&T-developed (and Na-tional Security Agency–approved) algorithm called the Data Encryption Standard (DES). The **Data Encryption Standard (DES)** was developed in 1975 and was standardized by ANSI in 1981 as ANSI X.3.92. It is a popular symmetric-key encryption method that uses block cipher. The key used in DES is based on a 56-bit binary number, which allows for 72,057,594,037,927,936 encryption keys. Of these 72 quadrillion encryption keys, a key is chosen at random.

When you first enter the password in Linux, the password is encrypted in a way that it cannot be decrypted. When a person logs in, the password is entered, and it is encrypted in the same way. It is then compared with the stored encrypted password to see if the two match. If they match, the user is allowed access to the system. So in reality, the operat-ing system does not know the password, but only the encrypted form of the password.

Unfortunately, the U.S. government has banned the export of DES outside the United States. FreeBSD had to find a way to both comply with U.S. law and retain compatibility with all the other UNIX variants that still used DES. The solution was to divide up the en-cryption libraries so that U.S. users could install the DES libraries and use DES but inter-national users still had an encryption method that could be exported abroad. This is how

FreeBSD came to use MD5 as its default encryption method. MD5 is similar to DES and is believed to be a bit more secure, so installing DES is offered primarily for compatibility reasons. While DES passwords are limited to 8 characters, MD5 allows passwords to be as long as 256 characters.

It is pretty easy to identify which encryption method FreeBSD is set up to use. Examining the encrypted passwords in the /etc/passwd file is one way. Passwords encrypted with the MD5 hash are longer than those encrypted with the DES hash and also begin with the characters 1. DES password strings do not have any particular identifying characteristics. However, they are shorter than MD5 passwords, and they are coded in a 64-character alphabet that does not include the $ character, so a relatively short string that does not begin with a dollar sign is very likely a DES password.

When a person is having a problem logging in, the first thing you should suggest to the user is to check the capitalization of the user name and the password and to check the Caps Lock key on the keyboard. You would be amazed how often this is the problem.

12.5.3   The /etc/passwd File

When user accounts are created, they are placed in the /etc/passwd file. Although administrators often use tools to avoid editing this file directly, it is important to be familiar with the file as part of system maintenance.

The syntax of an entry in the /etc/passwd file is:

```
username:password:UID:GUID:name:home:shellUser
```

If the password entry is an x, it means that shadow passwords are in use and that the user's password is stored in the /etc/shadow file. If the account does not have a password at all, it will be indicated with an asterisk (*). This is typically used for daemons and system programs such as shutdown. If the password is a series of characters, then the password is encrypted into the /etc/passwd file.

NOTE: Do not edit this password manually. Use the passwd command instead.

The name entry is the real name assigned to the user. Since the user name is not always intuitive to others, the name entry is used often by programs to display the user's name.

The shell entry is the path to the shell used by default for this account. The default for Linux is /bin/bash unless the administrator uses a custom shell instead of the general user.

12.5.4   Shadow Passwords

Traditionally, UNIX and Linux systems have stored passwords and other account information in an encrypted form in the /etc/passwd file. Various tools need access to this file for nonpassword information, so it needs to be readable to all users. The solution is to create shadow passwords. This is accomplished by taking the encrypted password entries from the password file and placing them in a separate file called shadow. This way, the regular password file would continue to be readable by all users on the system, and the actual encrypted password entries would be readable only by the root user (the login prompt is run with root permission).

NOTE: During the installation of Red Hat Linux, shadow password protection for your system is enabled by default, as are MD5 passwords.

Shadow passwords offer a few distinct advantages over the previous standard of storing passwords on UNIX and Linux systems, including:

- Improved system security by moving the encrypted passwords (normally found in /etc/passwd) to /etc/shadow, which is readable only by root
- Information concerning password aging (how long it has been since a password was last changed)
- Control over how long a password can remain unchanged before the user is required to change it
- The ability to use the /etc/login.defs file to enforce a security policy, especially concerning password aging

If you did not enable shadow passwords and your version of Linux supports shadow passwords, you can convert your regular passwords to shadow passwords by using the pwconv program.

If shadow passwords are enabled on the system, the /etc/shadow contains the password information. The syntax for the /etc/shadow file is:

```
username:password:lchange:mchange:change:wchange:disable:
expire:system
```

The lchange is the number of days since January 1, 1970, since the password was last changed. The mchange is the number of days before the password can be changed again. This value is set to zero by default. Change is the number of days until the password must be changed. Wchange is the number of days before the password is set to expire that the system will begin warning the user to change the password.

Once the password reaches the time when it must be changed, the disable value comes into play—it is the count of days before the account is automatically disabled if the password is not changed. Expire is the count of days since January 1, 1970, which equates to the date the account is to expire. System is reserved for the system.

12.5.5 Home Directories

The **home directory** is a folder used to hold or store a user's personal documents. In addition, it offers a place for configuration files that are unique to the user, which allows each user to work in a customized environment. Therefore, a user can log on and use applications with the same settings as before without having them changed by another user even if the two users are logged in at the same time.

Typically, most servers have the home directories at /home, and the person's home directory is named as his or her login name. Thus, if your login name were jsmith, your home directory would be /home/jsmith. The exception to this is the system accounts, such as a root user's account. The home directory for root is traditionally / with most variants of UNIX. Many Linux installations use /root.

The decision to place home directories under /home is strictly arbitrary—but it does make organizational sense. The system really doesn't care where you place home directories so long as the location for each user is specified in the password file (/etc/passwd). Some people may break up the /home directory by department, thereby creating /home/engineering, /home/accounting, /home/admin, etc., and then have people located under each department.

NOTE: Of course, since many users can deplete a lot of available disk space, the /home directory frequently resides on a separate partition or hard drive.

12.5.6   The Root Account

The superuser, which basically is the root account, is the person who has access to everything on the system. Therefore, on the Linux machine, the root account is all powerful. But with that power comes great responsibility. Many novice system administrators constantly work on the system while logged in as root. This is unwise and dangerous.

Since the root account has ultimate power, if a person logged in as the root account types in the wrong command, the person can wipe out large portions of the file system or accidentally break down the security of a bunch of files or break a number of programs. As a result of mistakes or intentional abuse combined with poor backups and documentation, you could lose days of productivity while repairing the damage. Most systems contain restrictions on root logins, so they can only be done from the console. This helps prevent outsiders from gaining access to a system over the network by using a stolen password.

One way to give you temporary superuser privileges or to take on another user's identity is to use the su command. When the su command is executed at the command prompt, you will be prompted for the root password. If you type that password correctly, subsequent commands will be executed as root. To turn back into a normal user, you just have to type exit.

To take on another user, you just have to type su followed by the user's name. If you are already logged in as the root, you can take on another user's identity without the user's password. This will allow you to switch to the user to figure out problems with his or her account, including problems with accessing computer and network resources (files, printers, and so on).

A safer command is the sudo command, which is similar to the su command except it allows you to execute a single command as root. The /etc/sudoers file contains a list of users who may use sudo, and with what commands.

If you forget the root password, you can follow these steps to set up a new root password:

1. Power up your PC as usual. At the LILO boot prompt, type the name of the Linux boot partition and then type single:

```
linux single
```

This causes Linux to start up as usual, but run in a single-user mode that does not require you to log in. After Linux starts, you will see the bash# command-line prompt.

2. Use the passwd command to change the root password and enter the new password.

NOTE: The passwd command will ask you to verify the password.

When complete, the passwd command changes the password and displays the message:

```
passwd: all authentication tokens updated successfully
```

3. Reboot the PC by pressing Ctrl+Alt+Del. After Linux starts, it displays the familiar login screen. Log in as normal.

As you can see, it is not difficult to break into a Linux computer. Therefore, you will have to rely on BIOS passwords and physical security to help maintain your system security.

12.5.7   Creating, Modifying, and Deleting Users

Adding users can be accomplished through the useradd utility.

NOTE: This program is called adduser in some distributions.

In its simplest form, you may type just useradd *username,* where *username* is the user name you want to create. Its basic syntax is:

```
useradd [-c comment] [-d home-dir] [-e expire-date] [-f
inactive-days] [-g initial-group] [-G group[,...]]
   [-m [-k skeleton-dir] | -M] [-p encrypted-password]
[-s shell]
   [-u UID [-o]][-r] [-n] username
```

NOTE: Some of these parameters modify settings that are valid only when the system uses shadow passwords. (See table 12.8.)

To administer user accounts, you will need to be logged in as root. The example that follows shows how to add a new user account. It is for a user named "jsmith" and home directory of "/home2/jsmith/".

```
useradd jsmith -d /home2/jsmith
```

If you want to create a new user called jsmith, make the new user a member of the audit group and the manager group, and make the manager group the primary group, you would use the following command:

```
useradd -g manager -G manager,audit jsmith
```

NOTE: Once a user account has been created, a password for the account needs to be set. This is done with the passwd command (shown later in this chapter).

The usermod program is used to modify existing users. It is very similar to the useradd program and has all the parameters used with the useradd command. In addition, it has the following parameters:

- Usermod allows the addition of a –m parameter when used with –d. –d alone changes the user's home directory, but it does not move any files. Adding –m causes usermod to move the user's files to the new location.
- Usermod supports a –l parameter, which changes the user's login name to the specified value. For instance, usermod jsmith johns changes jsmith to johns.

NOTE: If you change a user's name, you will need to change the owner of all files that the user previously owned including those that are in his or her home directory. This is done with the chown command.

Table 12.8 useradd Options

Option	Description
-c comment	The comment field for the user. Some administrators store public information like a user's office or telephone number in this field. Others store just the user's real name, or no information at all.
-d home-dir	The account's home directory. It defaults to /home/username on most systems.
-e expire-date	The date on which the account will be disabled, expressed in the form YYYY-MM-DD. Many systems will accept alternative forms, such as MM-DD-YYYY, or a single value representing the number of days since January 1, 1970. The default is for an account that does not expire. This option is most useful in environments in which users' accounts are inherently time-limited, such as accounts for students taking particular classes or temporary employees.
-f inactive-days	The number of days after a password expires after which the account becomes completely disabled. A value of -1 disables this feature and is the default.
-g initial-group	The name or GID of the user's default group. The default for this value varies from one distribution to another.
-G group[, . . .]	The names or GIDs of one or more groups to which the user belongs. These groups need not be the default group, and more than one may be specified by separating them with commas.
-m [-k skeleton-dir]	The system automatically creates the user's home directory if –m is specified. Normally, default configuration files are copied from /etc/skel (such as .Xdefaults, .bash_logout, .bash_profile, .bashrc files) to every new user's home directory when new accounts are created, but you may specify another template directory with the –k option. Many distributions use –m as the default when running useradd.
-M	This option forces the system to *not* automatically create a home directory, even if /etc/login.defs specifies that this action is the default. The /etc/login.defs file is explained later in this chapter.
-p encrypted-password	This parameter passes the pre-encrypted password for the user to the system. The encrypted-password value will be added, unchanged, to the /etc/passwd or /etc/shadow file.
-s shell	The name of the user's default login shell. On most systems, this defaults to /bin/bash.
-u UID [-o]	Creates an account with the specified user ID (UID) value. This value must be a positive integer, and it is normally above 500 for user accounts. System accounts typically have numbers below 100. The –o option allows the number to be reused so that two user names are associated with a single UID.
-r	This parameter allows the creation of a system account (an account with a value lower than UID_MIN, as specified in the /etc/login.defs, which is normally 100, 500, or 1000). *NOTE:* useradd also doesn't create a home directory for system accounts.
-n	In some distributions, including Red Hat, the system creates a group with the same name as the specified user name. This parameter disables this.

Table 12.9   The Chage Command Option

Options	Description
-l	This option changes chage to display account expiration and password aging information for a particular user.
-m *mindays*	This is the minimum number of days between password changes. 0 indicates a user can change a password multiple times in a day, 1 means a user can change a password once a day, 2 means a user may change a password once every 2 days, and so on.
-M *maxdays*	The maximum number of days during which a password is valid.
-d *lastday*	This is the last day a password was changed. This value is normally maintained automatically by Linux, but you can use this parameter to artificially alter the password change count. *lastday* is expressed in the format YYYY/MM/DD, or as the number of days since January 1, 1970.
-I *inactive*	This is the number of days between password expiration and account disablement. An expired account may not be used or may force the user to change the password immediately upon logging in, depending upon the distribution. A disabled account is completely disabled.
-E *expiredate*	You can set an absolute expiration date with this option. For example, you might use -E 2002/06/01 to have the account expire on June 1, 2002. The date may also be expressed as the number of days since January 1, 1970. The value of −1 represents no expiration date.
-W *warndays*	This is the number of days before account expiration that the system will warn the user of the impending expiration. It's generally a good idea to use this feature to alert users of their situation, particularly if you make heavy use of password change expirations.

■ You may lock (disable) or unlock a user's password with the -L and -U options, respectively. When an account is locked, it inserts an encrypted exclamation point (!), making the user unable to log in.

NOTE: When using the -G option to add a user to new groups, if you do not list any groups that the user was previously a member of, the user will be removed.

The chage command allows you to modify account settings relating to account expiration. You can configure Linux to disable the account if the password hasn't been changed in a specified period of time—this forces users to periodically change their password. You can also have the account expire on a certain date.

The chage utility has the following syntax:

```
chage [-l] [-m mindays] [-M maxdays] [-d lastday]
[-I inactive] [-E expiredate] [-W warndays] username
```

See table 12.9.

To delete a user from the /etc/passwd, you use the userdel *username* command. To delete a user from the /etc/shadow, you use userdel -r *username*. In addition, the –r makes the userdel command remove all the files from the user's home directory and all the files in the home directory. To delete the jsmith account, use the following command:

```
userdel -r jsmith
```

Of course, users may create files outside of their home directories. To find these files, you would use the find command to search for files that belong to a particular UID. For example, if jsmith has a UID of 623, you could locate all of jsmith's files by using the following command:

```
find / -uid 623
```

You can then decide to delete the files or change the ownership of the files.

Before you consider deleting an account, you should first consider disabling it. This can be used when an employee hasn't started yet or someone has left the company and you would like to assign the account to his or her replacement. To disable an account, you can either:

- Open the /etc/passwd or /etc/shadow file and change the password entry to an asterisk (*). To reactivate the account, create a new password for them as root with passwd command.
- Disable the account using Linuxconf.

If you hire someone to replace that person, you can then change the user name and change the owner with the chown of all the files including those that are in his or her home directory.

The easiest command to change passwords is the passwd command. The passwd command has the following syntax:

```
passwd [-k] [-l] [-u [-f]] [-d] [-S] [username]
```

See table 12.10.

Table 12.10   Passwd Command Options

Options	Descriptions
-k	This parameter indicates that the system should update an expired account.
-l	This parameter locks an account by prefixing the encrypted password with an exclamation mark (!). The files are still available.
-u [-f]	The –u parameter unlocks an account by removing a leading exclamation mark. By default, passwd will refuse to create an account without a password. The -f option will force to override this protection.
-d	This parameter removes the password from an account.
-S	This option displays information on the password for an account, indicating whether or not it's set and what type of encryption it uses.

Ordinary users may use passwd to change the password, but most parameters may only be used by root. As a security measure, except for when the root account is executing the passwd command, passwd asks for a user's old password before changing the password.

12.6   GROUPS

A group is a collection of user accounts used to organize them. A group is not an account. By using groups, you can simplify administration by assigning permissions to the group, and everyone listed in the group will be assigned these permissions. Users can be members of multiple groups. Like users, groups also have unique identifiers called group IDs (GIDs). Like the UID, the GID is determined differently among the Linux distributions.

Group membership is controlled through the /etc/group file, which contains a list of groups and the members that belong to the group. In addition to membership defined in the /etc/group file, each user has a default or primary group. The user's primary group is set in the user's configuration in the /etc/passwd file. When users log onto the computer, their group membership is set to their primary groups. When users create files or launch programs, those files and running programs are associated with the person's primary group. A user can still access files that belong to other groups, so long as the user belongs to that group and the group access permissions allow the access.

The group commands are similar to the user commands; however, instead of working on individual users, they work on groups listed in the /etc/group file. Note that changing group information does not cause user information to be automatically changed. For example, if you remove a group whose GID is 100 and a user's default group is specified as 100, the user's default group would not be updated to reflect the fact that the group no longer exists.

The groupadd command adds groups to the /etc/group file. The command options for the groupadd program are as follows:

```
groupadd [-g gid] [-r] [-f] group
```

See table 12.11.

If you want to add a new group called managers with the GID of 750, you would type in the following command:

```
groupadd -g 750 manager
```

To add users to the group, you would use useradd when you create new users or the usermod for existing users.

The groupmod command allows you to modify the parameters of an existing group. The options for this command are:

```
groupmod -g gid -n group-name group
```

where the -g option allows you to change the GID of the group and the -n option allows you to specify a new group of a group. Additionally, of course, you need to specify the name of the existing group as the last parameter.

For example, if you have the project1 group and you want to change its name to the widget group, you would issue the following command:

Table 12.11   Groupadd Command Options

Option	Description
-g *gid*	Specifies the GID for the new group as gid. By default, this value is automatically chosen by finding the first available value.
-r	Tells groupadd that the group being added is a system group and should have the first available system group GID.
-f	When adding a new group, Linux will exit without an error if the specified group to add already exists. By using the option, the program will not change the group settings before exiting. This is useful in scripting cases where you want the script to continue if the group already exists.
group	This option is required. It specifies the name of the group you want to add.

```
groupmod -n widget project1
```

To delete a group, you would use the groupdel command. The syntax for this command is:

```
groupdel group
```

For example, if you want to remove the widget group, you would type in the following command:

```
groupdel widget
```

Remember that when a user creates a file, the file will be owned by the user and by the user's primary group. If the user wants to assign group ownership of the file to another group that the user is a member of, the user can switch groups by using the newgrp command before he or she creates the file:

```
newgrp group
```

There is an extra level of groups that is often ignored. Within this level exists group administrators, passwords, and more. This comes in handy with workgroups and other project-oriented groupings where the system administrator really does not need to be involved beyond the group creation state. The project manager can be made responsible for the group itself and who belongs to it. Once the group is created, you can assign someone group administrator status with the command format:

```
gpasswd -A groupadmin group
```

The system administrator can also add members to the group with:

```
gpasswd -M user group
```

After this, the group administrator may choose from a set of commands. To add a group member, use the following command:

```
gpasswd -a user group
```

To remove someone, use:

```
gpasswd -d user group
```

To add a password to the group to prevent nonmembers from joining it with the new-grp command, you would use the gpasswd group command, which then prompts you for the new password.

12.7   CENTRALIZED AUTHENTICATION USING NIS

If users on your network are working on more than one machine, you will need to create a login/password pair for them on each machine. A more elegant solution is to maintain a centralized database known as Network Information Service (NIS) that client machines refer to for authentication. If you are familiar with Microsoft networks, this is similar to Microsoft's domain controller.

NIS makes it possible to share the data of critical files across the local area network. Typically, files, such as /etc/passwd and /etc/group, which ideally would remain uniform across all hosts, are shared via NIS. In this way, every network machine that has a corresponding NIS client can read the data contained in these shared files and use the network versions of these files as extensions to the local versions.

For example, when you need to log onto a machine on your network, you need a login/password pair that is valid on that machine. This can become a problem over a larger network where you may have people using more than one machine. An example of this would be a computer lab where people are going to be working on different machines most of the time. You will then be forced to create logins for each user on every machine that he or she is likely to use. NIS steps in here to provide centralized authentication. All the logins are created on a single machine, which client machines access to authenticate users.

Once you have centralized your authentication, you should also make the home directory of the users available on the machine that they log on by using the Network File System (NFS). NFS allows you to export a directory for mounting on other machines so that it will appear as a local directory on the client machine. This is completely transparent to the user.

NIS and its associated tools are available across nearly all the distributions. For an NIS client machine, all you need is ypbind and yp-tools. The ypserv package is only required if you're setting up the machine as a server. You will also need the portmap daemon, which is used to manage RPC requests. This is used by NIS as well as NFS and is present in most Linux distributions. Look for a package named portmap on your distribution CD. Fortunately, these are typically installed during the installation of Linux.

12.8   FILE SECURITY

Besides login security, when you think of security, you think of file security where you determine who can access what directories and files and how they can access those directories and files. Therefore, it is obvious that if you are going to be setting up a network server or if you want to secure a Linux workstation, you will need to understand file security and how to configure file security.

There are three components to Linux's file permission handling:

- **User name (or UID)**—A user name (or UID) is associated with each file on the computer. This is frequently referred to as the file owner.
- **Group (or GID)**—Every file is associated with a particular GID, which links the file to a group. This is something referred to as the group owner. Normally, the group of a file is one of the groups to which the file's owner belongs, but root may change the file's group to one unassociated with the file's owner.
- **File access permission**—The file access permission is a code that indicates who may access the file, relative to the file's owner, the file's group, and all others.

You can see all three elements by using the ls -l command on a file. (See figure 12.10.) The output of this command has several different components, each with a specific meaning.

- **Permission string**—The first component (for example, `-rwxr-xr-x` in figure 12.10) is the permission string. Along with the user and group names, it's what determines whether you may access a file. As displayed by ls -l, the permission string is a series of codes. Sometimes the first character of this string is omitted, particularly when discussing ordinary files, but it's always present in a ls -l listing.
- **Number of hard links**—Linux supports hard links in its file systems. A hard link allows one file to be referred to by two or more different file names. Internally, Linux uses a data structure known as an inode to keep track of the file, and multiple file names point to the same inode. The number 1 in figure 12.10 output means that just one file name points to this file; it has no hard links. Larger numbers indicate that hard links exist. For instance, 3 means that the file may be referred to by three different file names.
- **Owner**—The next field, root, is the owner of the file. In the case of a long user name, the user name may be truncated.
- **Group**—The manager is the group to which the example file belongs. Many system files belong to the root owner and root group.
- **File size**—The next field, 33026 in figure 12.10, is the size of the file in bytes.

Figure 12.10   Using the ls -1 Command to View the File Permissions

```
[root@host1-1 data]# ls -l
total 148
drwxr-xr-x    2 root       root          4096 Apr 15 17:38 archive
-rwxr-xr-x    1 user1      manager      33026 Apr 15 17:36 file1.txt
-rwxrwxrwx    1 user1      manager      35812 Apr 15 17:36 file2.txt
-rwxr-----    1 user1      user1        48738 Apr 15 17:36 file3.txt
-rwxrwx---    1 root       root         23196 Apr 15 17:36 file4.txt
```

- **Creation time**—The next field contains the file creation time and date. If the file is older than a year, you'll see the year rather than the creation time, although the time is still stored with the file.
- **File name**—The final field is the name of the file. When using the ls command, if the complete path to the file is used, the complete path appears in the output.

12.8.1   File Access Codes

The file access control string is 10 characters in length. The first character is the file type code, which determines how Linux will interpret the file—such as ordinary data, a directory, or a special file type. See table 12.12.

The remaining nine characters represent the permissions (for example, rwxr-xr-x), which are broken into three groups of three characters. The first group controls the file owner's access to the file, the second controls the group's access to the file, and the third controls all other users' access to the file. This is often referred to as world permissions.

In each of the three groupings, the permission string determines the presence or absence of each of three types of access: read, write, and execute. The absence of the permission is denoted by a hyphen (-) in the permission string. The presence of the permission is indicated by a letter (r for read, w for write, and x for execute).

- **Read permission (r)**—For a file, the read permission enables you to read the file. For a directory, the read permission allows the ls command to list the names of the

Table 12.12   File Type Codes

Code	Meaning
-	Normal data file—It may be text, an executable program, graphics, compressed data, or just about any other type of data.
d	Directory—Disk directories are files just like any other, but they are marked as a directory, and they contain file names and pointers to disk inodes.
l	Symbolic link—The file contains the name of another file or directory. When Linux accesses the symbolic link, it tries to read the linked-to file.
p	Named pipe—A pipe allows two running Linux programs to communicate with each other. One opens the pipe for reading, and the other opens it for writing, allowing data to be transferred between the programs.
s	Socket—A socket is similar to a named pipe, but it permits network and bidirectional links.
b	Block device—A hardware device to and from which data is transferred in blocks of more than 1 byte. Disk devices (hard disks, floppies, CD-ROMs, and so on) are common block devices.
c	Character device—A hardware device to and from which data is transferred in units of 1 byte. Examples include parallel and serial port devices.

files in the directory. You must also have execute permission for the directory name to use the –l option of the ls command or to change to that directory. It has a value of 4.

- **Write permission (w)**—For a file, the write permission means you can modify the file. For a directory, you can create or delete files inside that directory. It has a value of 2.
- **Execute permission (x)**—For a file, the execute permission means you can type the name of the file and execute it. For a directory, the execute permission means you can change to that directory (with the cd command). It has a value of 1.

You cannot view or copy the file unless you also have read permission. This means that files containing executable Linux commands, called shell scripts, must be both executable and readable by the person executing them. Programs written in a compiled language such as C, however, can have only executable permissions, to protect them from being copied where they shouldn't be copied.

If you have the execute command and you do not have the read permission for the directory, you can change into the directory but ls –l will not work. You can list directories and files in that directory, but you cannot see additional information about the file or directory by just doing an ls –l command. This is a highly desirable characteristic for directories, so you'll almost never find a directory on which the execute bit is not set in conjunction with the read bit.

Directories can be confusing with respect to write permission. Recall that directories are files that are interpreted in a special way. As such, if a user can write to a directory, that user can create, delete, or rename files in the directory, even if the user isn't the owner of those files and does not have permission to write to those files.

Thus, the example of a permission string of rwxr-xr-x means that the file's owner, the file's group, and all other users can read and execute the file. Only the file's owner has write permission to the file. You can easily exclude those who don't belong to the file's group, or even all but the file's owner, by changing the permission string.

Individual permissions, such as execute access for the file's owner, are often referred to as permission bits. This is because Linux encodes this information in binary form. Because it is binary, the permission information can be expressed as a single 9-bit number. This number is usually expressed in octal (base 8) form because a base 8 number is 3 bits in length. This means that the base 8 representation of a permission string is three digits long, one digit each for the owner, group, and world permissions, respectively. The read, write, and execute permissions each correspond to one of these bits. The result is that you can determine owner, group, and world permissions by adding base 8 numbers, 1 for execute permission, 2 for write permission, and 4 for read permission. See table 12.13 for more examples.

Permissions								
Owner			Group			World		
r	w	x	r	w	x	r	w	x
4	2	1	4	2	1	4	2	1

Table 12.13   Example Permissions and Their Likely Uses

Permission String	Octal Code	Meaning
rwxrwxrwx	777	Read, write, and execute permissions for all users.
rwxr-xr-x	755	Read and execute permissions for all users. The file's owner also has write permission.
rwxr-x---	750	Read and execute permissions for the owner and group. The file's owner also has write permission. Nongroup members have no access to the file.
rwx------	700	Read, write, and execute permissions for the file's owner only; all others have no access to the file.
rw-rw-rw-	666	Read and write permissions for all users. No execute permission for anybody.
rw-rw-r--	664	Read and write permissions for the owner and group, and read-only permission for world. No execute permission for anybody.
rw-rw----	660	Read and write permissions to the owner and group. No permissions to world.
rw-r--r--	644	Read and write permissions to the owner. Read-only permission to all others.
rw-r----	640	Read and write permissions to the owner, and read-only permission to the group. No permissions to others.
rw-------	600	Read and write permissions to the owner. No permissions to anybody else.
r--------	400	Read-only permission to the owner. No permissions to anybody else.

Example:

If you give the owner of the file all rights (read, write, and execute), you give the group read and execute, and you give everyone else no rights, the base 8 rights would be expressed as 750.

Permissions								
Owner			**Group**			**World**		
r	w	x	r	–	x	r	–	–
4	2	1	4	–	1	–	–	–
7			5			0		

Many of the permission rules do not apply to root. The superuser can read or write any file on the computer, even files that grant access to nobody (that is, those that have 000 permissions). The superuser still needs an execute bit to be set to run a program file, but the superuser has the power to change the permissions on any file, so this limitation isn't very substantial.

A few special permissions options are also supported, and they may be indicated by changes to the permission string:

- **Set user ID (SUID)**—The SUID option is used in conjunction with executable files, and it tells Linux to run the program with the permissions of the user who runs the program. For instance, if a file is owned by root and has its SUID bit set, the program runs with root privileges and can therefore read any file on the computer. Some servers and other system programs run in this way, which is often called SUID root.

NOTE: The SUID option represents a security risk if it is not carefully controlled. SUID programs are indicated by an s in the owner's execute bit position of the permission string, as in `rwsr-xr-x`.

- **Set group ID (SGID)**—The SGID option is similar to the SUID option, but it sets the group of the running program to the group of the file. It's indicated by an s in the group execute bit position of the permission string, as in `rwxr-sr-x`.

NOTE: The SUID option represents a security risk if it is not carefully controlled.

- **Sticky bit**—In modern Linux implementations, it is used to protect files from being deleted by those who don't own the files. When this bit is present on a directory, the directory's files can only be deleted by their owners, the directory's owner, or root. The sticky bit is indicated by a t in the world execute bit position, as in `rwxr-xr-t`.

A file's owner and root are the only users who may adjust a file's permissions. Even if other users have write access to a directory in which a file resides and write access to the file itself, they may not change the file's permissions (but they may modify or even delete the file). To understand why this is so, you need to know that the file permissions are stored as part of the file's inode, which isn't part of the directory entry. Read/write access to the directory entry, or even the file itself, doesn't give a user the right to change the inode structures (except indirectly—for instance, if a write changes the file's size or a file directory eliminates the need for the inode).

12.8.2   Setting Default Permissions

When a user creates a file, that file has default ownership and permissions. The default owner is, understandably, the user who creates the file. The default group is the primary group of the user that created the file. The default permissions, however, are configurable. These are defined by the user mask (umask), which is an octal file creation mask. The umask contains the bits that are off by default when a new file is created. This umask takes as input an octal value that represents the bits to be removed from 777 permissions for directories or from 666 permissions for files.

NOTE: You cannot set the execute bit of a file with a user mask. The user mask is set by the umask command.

Example:

If you have a umask of 022 and you create a new file, the 022 will be subtracted from 666.

- Owner permissions:
  - If you look at the first 6 of the 666, you can see that it represents 42−, which represents rw permissions.
  - If you look at the 0 of the 022, you notice that 0 represents −, which represents no permissions.
  - If you subtract bit by bit, you get $4 - 0 = 4, 2 - 0 = 2$, and $0 - 0 = 0$, which gives you a 6 for the resultant group permission.
- Group permissions:
  - If you look at the second 6 of the 666, you can see that it represents 42−, which represents rw permissions.
  - If you look at the first 2 of the 022, you notice that 2 represents − 2−, which represents the write permissions.
  - If you subtract bit by bit, you get $4 - 0 = 4, 2 - 2 = 0$, and $0 - 0 = 0$, which gives you a 4 for the resultant owner permission.
- World permissions:
  - If you look at the last 6 of the 666, you can see that it represents 42−, which represents rw permissions.
  - If you look at the second 2 of the 022, you notice that 2 represents − 2−, which represents the write permissions.
  - If you subtract bit by bit, you get $4 - 0 = 4, 2 - 2 = 0$, and $0 - 0 = 0$, which gives you a 4 for the resultant owner permission.

All Permissions								
Owner			**Group**			**World**		
r	w	x	r	w	x	r	w	x
4	2	–	4	2	–	4	2	–
	6			6			6	

Umask								
Owner			**Group**			**World**		
r	w	x	r	w	x	r	w	x
–	–	–	–	2	–	–	2	–
	0			2			2	

Resulting Permissions								
Owner			**Group**			**World**		
r	w	x	r	w	x	r	w	x
4	2	–	4	–	–	4	–	–
	6			4			4	

Therefore, with a umask of 022, for any file created, the owner of the file will automatically have read and write permissions. The group that the file is assigned to will have the read permission, and everyone else will have the read permission.

If you had a umask of 002, the owner and the group will have read and write permissions, and everyone will only have read permission.

Example:

Let's say you want the owner to have the read and write permissions, the group only to have read permission, and everyone to have no permissions. You would need to keep the following in mind:

■ Giving the owner the read and write permissions means that you don't want to shut off any of the permissions for the owner. Therefore, they will be represented by – = 0.

- Giving the group the read permission for the file means that you want to shut off the write permission for the group. Therefore, -w- will be represented by $-2-=2$.
- The no permissions for the file means that you want to shut off the read and write permission, which will be represented by $42-=6$.

Therefore, the umask should be 026. It would be set with the following command:

```
umask 026
```

To find what the current umask is, type umask alone, without any parameters. Typing umask –S produces the umask expressed symbolically, rather than in octal form. You may also specify a umask in this way when you want to change it, but in this case, you specify the bits that you do want set. For instance umask u=rwx, g=rx, o=rx is equivalent to umask 022.

Ordinary users can enter the umask command to change the permissions on new files they create. The superuser can also modify the default setting for all users by modifying a system configuration file. Typically, /etc/profile contains one or more umask commands. Setting the umask in /etc/profile might or might not actually have an effect, because it can be overridden at other points, such as the user's own configuration files. Nonetheless, setting the umask in /etc/profile or other system files can be a useful procedure if you want to change the default system policy. Most Linux distributions use a default umask of 002 or 022.

12.9   GROUP STRATEGIES

The traditional way of creating user accounts is that you create one or more groups in /etc/group and then assign one of these groups to be the primary group for each new user account that is added. For example, you may create one group for the accounting department and another group for the information systems department in your organization. When you create accounts for new hires in the accounting department, each user account receives a unique user account, but the same group as the other accountants. Note that user accounts can belong to more than one group; the primary group for an account is simply the default group for that account.

The purpose of having groups is to allow users who are members of a particular group to share files. Typically, this is done by creating a directory and changing the group ownership of the directory to that of the group that is going to share files in the directory. For example, suppose you have a group set up for the users in the accounting department called accgrp, and you would like to create a shared file called accshared under /home. You would do the following:

```
mkdir /home/accshared
chown nobody.accgrp /home/accshared
chmod 775 /home/accshared
```

Any user who is a member of the accgrp group can now create files in the /home/accshared directory.

12.9.1   User Private Group

Several problems can arise when sharing files. For example, suppose that a user, who happens to be a member of accgrp but whose primary group is something other than accgrp, creates a file in the /home/accshared directory. The user who created the file will be the owner of the file, and the group ownership of the file will be the user's primary group. Unless the user who created the file remembers to use the chgrp command on the file to set the ownership of the file to the accgrp group, other users in that group may not have the necessary permissions to access the file.

To solve this problem, Red Hat uses the user private group (UPG) scheme as the default behavior for creating accounts under Red Hat Linux. Every user has a primary group; the user is the only member of that group. The solution to this particular problem is to use something called the set group id bit or setgid bit. The setgid bit is applied to a directory with the chmod command. When the setgid bit is set for a directory, any files that are created in that directory automatically have their group ownership set to be that of the group owner for the directory. The command to set the setgid bit for the /home/accshared directory is:

```
chmod g+s /home/accshared
```

or, alternately,

```
chmod 2775 /home/accshared
```

Setting the setgid bit solves the problem of making sure all files that are created in a shared directory belong to the correct group. The other problem that can arise has to do with permission settings. Typically, most users run with a umask setting of 022, which specifies that any files that they create are not modifiable by any account that is a member of the user's group or by anyone else on the system. If the user is creating a file in a shared directory and he or she wishes to allow others update access to the file, he or she will have to remember to issue a chmod command to make the file group writable. The way around this problem is to set the default umask (usually in /etc/profile) to be 002. With this umask, any files that are created will be modifiable by the user and any member of the group assigned the file.

At this point, you might think that with a umask of 002, anyone who is a member of that user's primary group will automatically have write access to any file that the user creates in his or her home directory. But you must remember that the advantage behind the user private group scheme is since every user account is the only member in his or her private group, having the umask set to 002 has no effect on file security.

12.9.2   Project Groups

One approach to group configuration is to create separate groups for specific purposes or projects. As most companies have multiple projects, most employees work on just one product, and for security reasons, you don't want users working on one product to have access to information relating to other products. In such an environment, a Linux system may be well served by having main user groups, one for each product. Most users will be members of precisely one group. If you configure the system with a umask that denies world access, those who don't belong to a specific product's group won't be able to read files relating to that product. You can set read or read/write group permissions to allow group members to easily exchange files. Of course, individual users may use chmod to customize permissions on particular files and directories. If a user needs access to files associated with multiple products, you can assign that user to as many groups as are needed to accommodate the need. For instance, a supervisor might have access to all three groups.

12.9.3   Multiple-Group Membership

On today's networks, you will have many people who are members of multiple groups. This means that users will be able to do the following things:

- Read files belonging to any of the user's groups, provided that the file has group read permission.
- Write files belonging to any of the user's groups, provided that the file has group write permission.
- Run programs belonging to any of the user's group, provided that the file has group execute permission.
- Change the group ownership of any of the user's own files to any of the groups to which the user belongs.
- Use newgrp to make the user's primary group any of the groups to which the user belongs. Files created thereafter will have the selected group as the group owner.

Since you can only be logged in under one group at a time, when the user creates files for another project that is not the user's primary group, the user will become the owner of the file and his or her primary group will become the group owner of the file. Unless you can teach all your users to use the newgrp command to temporarily change their primary group before creating such files or to use the chgrp command to change group ownership of a group (both of these tasks will be difficult for novice users), you are going to have problems.

Therefore, when having users assigned to several groups, it is extremely important to use user private groups (UPGs). By using the UPG scheme, groups are automatically assigned to files created within that directory, which makes managing group projects that share a common directory very simple.

For example, let's say you have a big project called devel, with many people editing the devel files in the devel directory. Make a group called devel, chgrp the devel directory to devel group, and add all the devel users to the devel group. If you then set the setgid bit for the devel directory, all devel users will be able to edit the devel files and create new files in the devel directory. The files they create will always retain their devel group status, so other devel users will always be able to edit them.

If you have multiple projects like devel and users who are working on multiple projects, these users will never have to change their umask or group when they move from project to project. If set correctly, the setgid bit on each project's main directory "selects" the proper group for all files created in that directory.

12.10   CONFIGURING THE NETWORK CONNECTION IN LINUX

To enable network connection, you must first activate the network device by loading the appropriate network driver and then loading and configuring the proper network protocols. The utilities used to do this will vary from Linux distribution to Linux distribution.

12.10.1   Configuring the Network Device

Before you can connect to the network, you must first activate your network device by loading the appropriate network driver (kernel compiled with the driver or loaded as a module). Most distributions use modules as a default since it makes it easier to probe for cards.

If the driver is configured as a module, and you have autoloading modules set up, you will need to tell the kernel the mapping between device names and the module to load in the /etc/conf.modules file. For example, if your eth0 device is a 3Com 3C905 Ethernet card, you would add the following line to your /etc/conf.modules file:

```
alias eth0 3c59x
```

where 3c59x is the name of the device driver. Typically, Ethernet devices register themselves as being eth*X*, where *X* is the device number. The first device is eth0, the second is eth1, and so on. ARCnet cards are specified with an arc*X,* and Token Ring cards are specified with a tr*X*.

You will need to set this up for every network card you have. For example, if you have a network card based on the DEC Tulip chipset and an SMC-1211 based on the RealTek 8139 chipset in the same machine, you would need to make sure your /etc/conf.modules file includes the lines:

```
alias eth0 tulip
alias eth1 rtl8139
```

12.10.2   ifconfig

Knowing how to configure your network services by using the command prompt is important for multiple reasons. First, if you cannot start X Window, you have no choice but to use the command prompt. Another reason is for remote administration where you may not able to run a graphical configuration tool from a remote site. Finally, it is always nice to be able to perform network configuration through scripts, and command-line tools are best suited for scripts.

The ifconfig program is responsible for setting up your network interface cards. While the ifconfig is natively accessed through the command prompt, many tools have been written to provide a menu-driven or graphical interface, which in turn accesses the ifconfig for you. For example, Caldera users can configure their network interface through COS, and Red Hat users can use the netcfg program. The ifconfig command is used to configure the IP address and netmask, and it can also be used to display current settings.

The format of the ifconfig command is as follows:

```
ifconfig device IP_address options
```

where *device* is the name of the Ethernet device (for example, eth0 or eth1), *address* is the IP address you wish to apply to the device, and *options* are one of the following:

up	Enables the device.
down	Disables the device.
arp	Enables this device to answer ARP requests.
-arp	Disables the device from answering ARP requests.
mtu *value*	Sets the maximum transmission unit (MTU) of the device to *value*. Under Ethernet, this defaults to 1500.
netmask *mask*	Sets the subnet mask to the interface to *mask*. If a value is not supplied, ifconfig calculates the netmask based on the class of the IP address. A class A address gets a netmask of 255.0.0.0, class B gets 255.255.0.0, and class C gets 255.255.255.0.
broadcast *bcast*	Sets the broadcast address to this interface to *bcast*. If a value is not supplied, ifconfig calculates the broadcast address based on the class of the IP address in a similar manner to netmask.
pointtopoint *ppp_address*	Sets up a point-to-point (PPP) connection where the remote address is *ppp_address*.

In its simplest usage, all you need to do is provide the name of the interface being configured and the static IP address. The ifconfig program will deduce the rest of the information based on the IP address. Thus you could enter the following command at the command prompt to set the eth0 device to the IP address 192.168.1.65:

```
ifconfig eth0 192.168.1.65
```

Because 192.168.1.65 is a class C address, the calculated subnet mask will be 255.255.255.0, and the broadcast address will be 192.168.1.255.

If the IP address you are setting is a class A or class B address that is subnetted differently, you will need to explicitly set the broadcast and netmask addresses on the command line:

```
ifconfig dev ip netmask mask broadcast bcast
```

where *dev* is the network device you are configuring, *ip* is the IP address you are setting it to, *nmask* is the netmask, and *bcast* is the broadcast address. For example:

```
ifconfig eth0 1.1.1.1 netmask 255.255.255.0 broadcast
1.1.1.255
```

You can list all the active devices by running ifconfig with no parameters. (See figure 12.11.) You can list all devices, regardless of whether they are active, by running ifconfig –a.

Figure 12.11 Using the ifconfig Command

```
[root@localhost /root]# ifconfig
eth0       Link encap:Ethernet  HWaddr 00:A0:C9:E7:3B:65
           inet addr:63.197.142.129  Bcast:63.197.142.255  Mask:255.255.255.0
           UP BROADCAST RUNNING MULTICAST  MTU:1500  Metric:1
           RX packets:696 errors:0 dropped:0 overruns:0 frame:0
           TX packets:826 errors:0 dropped:0 overruns:12 carrier:0
           collisions:0 txqueuelen:100
           Interrupt:10 Base address:0xcf80

lo         Link encap:Local Loopback
           inet addr:127.0.0.1  Mask:255.0.0.0
           UP LOOPBACK RUNNING  MTU:16436  Metric:1
           RX packets:10 errors:0 dropped:0 overruns:0 frame:0
           TX packets:10 errors:0 dropped:0 overruns:0 carrier:0
           collisions:0 txqueuelen:0

[root@localhost /root]#
```

12.10.3 Communicating with Other Networks

If your host is connected to a network with multiple subnets, you need a router to communicate with the other networks. As your computer sends a packet onto the network, your computer does not know the correct path to the destination computer. And because it does not know the correct route, the packet will be sent out using the default route, which starts with the router connected to your network. When the router gets the packet, it will then determine which is the best path to use and will forward the packet to the appropriate router. Then that router will do the same, and the process will repeat itself until the packet gets to the destination network, which will then forward the packet to the host destination computer.

There are actually instances where you will need to change your routes manually. Typically, this is necessary when multiple network cards are installed in the same host where each NIC is connected to a different network. You should know how to add a route so that packets can be sent to the appropriate network based on the destination address.

A typical Linux host knows of three routes:

- The loopback route, which simply points toward the loopback device
- The route to the local area network, so that packets destined to hosts within the same LAN are sent directly to them
- The default route or gateway

To change routes, you are going to use the route command. The typical route command is structured as follows:

```
route cmd type addy netmask mask gw gway dev dn
```

See table 12.14.

To set the default route on a sample host that has a single Ethernet device and a router of 192.168.1.1, you would use this command:

```
route add -net default gw 192.168.1.1 dev eth0
```

Table 12.14   Route Command Options

Parameter	Description
cmd	Either *add* or *del* depending on whether you are adding or deleting a routing. If you are deleting a route, the only other parameter you need is *addy*.
Type	Either –net or –host depending on whether *addy* represents a network address or a router address.
addy	The destination network to which you want to offer a route.
netmask *mask*	Sets the netmask of the *addy* address to *mask*.
gw *gway*	Sets the router address for *addy* to *gway*. Typically used for the default route.
dev *dn*	Sends all packets destined to *addy* through the network device *dn* as set by ifconfig.

Figure 12.12   The Route Command Excused Without Any Parameters

```
Destination    Gateway         Genmask         Flags Metric Ref    Use Iface
192.168.1.0    *               255.255.255.0   U     0      0        0 eth0
127.0.0.0      *               255.0.0.0       U     0      0        0 lo
default        192.168.1.254   0.0.0.0         UG    0      0        0 eth0
```

To delete the route destined to 192.168.1.42, you would use:

```
route del 192.168.1.42
```

To display the route table, simply run route without any parameters (see figure 12.12):

```
route
```

Table 12.15 explains the information the route command displays.

Note that route displays the host names to any IP address it can look up and resolve. If there are network difficulties, and DNS or NIS servers become unavailable, the route command will hang on, trying to resolve host names and waiting to see if the servers come back and resolve them. This wait will go on for several minutes until the request times out. To get around this, use the –n option with route, so that the same information is shown but route will make no attempt to perform host name resolution on the IP addresses.

12.10.4   Troubleshooting TCP/IP Commands

To troubleshoot a TCP/IP network, several utilities are available. They are ifconfig, ping, route, and netstat. To isolate a TCP/IP problem in an organized way, you would do the following to determine exactly where the problem is:

1. Use ifconfig and route commands.
2. Ping 127.0.0.1 (loopback address).
3. Ping the IP address of the computer.

Table 12.15 Information Displayed Using the Route Command

Entry	Description
The flags are:	U The connection is up. H The destination is a host. G The destination is a gateway.
Metric	The cost of a route, usually measured in hops. This is meant for systems that have multiple paths to get to the same destination, but one path is preferred. The Linux kernel doesn't use this information, but certain advanced routing protocols do.
Ref	The number of references to this route. This is not used in the Linux kernel. It is here because the route tool itself is cross-platform. Thus it prints this value since other operating systems do use it.
Use	The number of successful route cache lookups. To see this value, use the –F option when invoking route.

4. Ping the IP address of the default gateway (router).
5. Ping the IP address of the remote host.

To verify the TCP/IP configuration, we have already discussed the ifconfig command. You can list all the active devices by running `ifconfig` with no parameters. You can list all devices, regardless of whether they are active, by running `ifconfig -a`. You can also use the route command to verify your default gateway.

Another very useful command to verify a network connection is the ping command. It sends packets to a host computer and receives a report on their round-trip time. If you ping the loopback address of 127.0.0.1 and the ping fails, you need to verify that the TCP/IP software is installed correctly. If you ping your own IP address and it is unsuccessful, you need to check the IP address. If you can't ping the IP address of the default gateway/router, you need to verify the IP address (make sure the address is in the correct subnet) and subnet mask. If you can't ping a remote computer on a different subnet or network, you need to verify the IP address of the default gateway/router, make sure the remote host/computer is functional, and verify the link between routers. For example:

```
ping 127.0.0.1
ping 137.23.34.112
```

The ping command can also be used to ping a host/computer by a host/DNS name. If your ping by address but not by name tells you that the TCP/IP is running fine but the name resolution is not working properly, you must check the HOSTS file and the DNS server to resolve domain names. For example:

```
ping www.microsoft.com
```

If the time takes up to 200 milliseconds, the time is considered very good. If the time is between 200 and 500 milliseconds, the time is considered marginal. And if the time is

Figure 12.13   Using the Ping Cammand to Verify Network Connection With a Computer Represented by an IP Address and a Domain Name

```
[root@localhost /root]# ping 216.148.218.197
PING 216.148.218.197 (216.148.218.197) from 63.197.142.129 : 56(84) bytes of data.
64 bytes from 216.148.218.197: icmp_seq=0 ttl=245 time=22.025 msec
64 bytes from 216.148.218.197: icmp_seq=1 ttl=245 time=19.972 msec
64 bytes from 216.148.218.197: icmp_seq=2 ttl=245 time=19.985 msec
64 bytes from 216.148.218.197: icmp_seq=3 ttl=245 time=19.971 msec
64 bytes from 216.148.218.197: icmp_seq=4 ttl=245 time=19.981 msec

--- 216.148.218.197 ping statistics ---
5 packets transmitted, 5 packets received, 0% packet loss
round-trip min/avg/max/mdev = 19.971/20.386/22.025/0.838 ms
[root@localhost /root]# ping redhat.com
PING redhat.com (216.148.218.197) from 63.197.142.129 : 56(84) bytes of data.
64 bytes from 216.148.218.197: icmp_seq=0 ttl=245 time=26.340 msec
64 bytes from 216.148.218.197: icmp_seq=1 ttl=245 time=19.982 msec
64 bytes from 216.148.218.197: icmp_seq=2 ttl=245 time=19.972 msec
64 bytes from 216.148.218.197: icmp_seq=3 ttl=245 time=19.981 msec
64 bytes from 216.148.218.197: icmp_seq=4 ttl=245 time=19.977 msec

--- redhat.com ping statistics ---
5 packets transmitted, 5 packets received, 0% packet loss
round-trip min/avg/max/mdev = 19.972/21.250/26.340/2.548 ms
[root@localhost /root]# []
```

over 500 milliseconds, the time is unacceptable. A Request timed out message indicates total failure, see figure 12.13.

Viewing the current configuration (ifconfig), pinging the loopback address (ping 127.0.0.1), and pinging the IP address of your computer will verify that the TCP/IP is properly functioning on your PC. By pinging the IP address of the default gateway or router, as well as other local IP computers, you determine if the computer is communicating on the local network. If it cannot connect to the gateway or any other local computer, either you are not connected properly or the IP is misconfigured (IP address, IP subnet mask, or gateway address). If you cannot connect to the gateway, but you can connect to other local computers, check your IP address, IP subnet mask, and gateway address, and check to see if the gateway is functioning by using the ping command at the gateway to connect to your computer and other local computers on your network as well as pinging the other network connections on the gateway/router or pinging computers on other networks. If you cannot ping another local computer, but you can ping the gateway, most likely the other computer is having problems and you need to restart this procedure at that computer. If you can ping the gateway, but you cannot ping a computer on another gateway, you need to check the routers and pathways between the two computers by using the ping command.

Another useful command is the traceroute command, which enables you to determine the path a packet is taking across your network and into other networks. It sends out a packet of information to each hop (gateway/router) individually, and does it three times. Therefore, the traceroute command can help determine where the break is in a network.

If you see any asterisks instead of a host name, that machine is probably not available. This could be caused by network failure, firewall protection, the fact that the machine is not currently running, or the machine not being configured properly.

As an administrator of a network, it is important for you to view the status of your network. In Linux, the netstat command is used. netstat displays network connections, routing tables, interface statistics, masquerade connections, and netlink messages.

Figure 12.14   Using the netstat Command to Show the Interfaces and Statistics

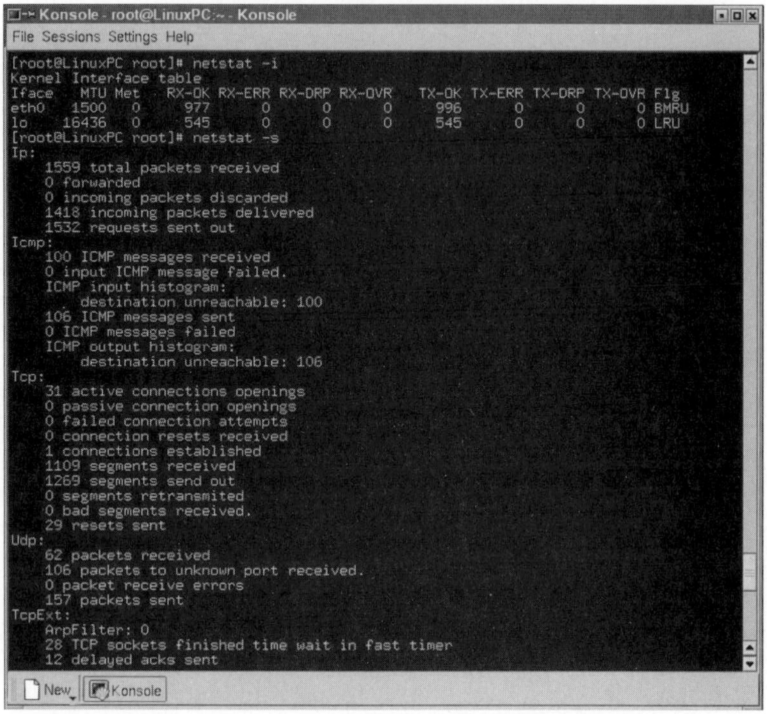

The most useful options for netstat are `netstat -i` to display the current interface status. When you use the option, you should pay attention to sent and received packets, collisions, and errors. The `netstat -s` shows the statistics for connections, packets transmitted and received, errors, and forwarding. (See figure 12.14.)

NOTE: A retransmission percentage of 20–25 percent or more suggests a network bottleneck.

12.11   NETWORK FILE SYSTEM

Network File System (NFS) was developed by Sun Microsystems in the 1980s as a way for UNIX to share files and applications across the network. It allows you to attach a remote drive or directory to a virtual file system and work with it as if it were a local drive.

NFS is a stateless protocol, which means that each request made between the client and server is complete and does not require knowledge of prior transactions. This allows servers to go down and come back up without having to reboot the clients.

When a user accesses an NFS directory on another computer, the NFS access is handled by remote procedure calls (RPCs). RPCs are responsible for handling the requests between the client and server. Whenever a service wants to make itself available on a server, it needs to register itself with the RPC service manager, called portmapper. Portmapper takes care of telling the client where the actual service is located on the server.

Both UNIX and Linux can provide files using NFS, and both can access directories and files using NFS. Windows 2000 and Windows Server 2003 computers can access NFS if you load the UNIX services. For other versions of Windows, you will have to install client software for NFS. Two popular products are provided by Sun and NetManage.

12.11.1   Configuring an NFS Server

The key configuration file for NFS on an NFS server is the /etc/exports file. The exports file specifies which directories are to be shared with which clients (hosts) and each client's access rights.

NOTE: If you do not plan out your export rules well, the NFS can be a massive security hole.

The /etc/exports file has the following format:

```
/directory_to_export    host1(permissions)
host2(permissions) host3(permissions)
   # Comments
   /another_directory_to_export host1(permissions)
host2(permissions)
```

The *directory_to_export* and *another_directory_to_export* are directories that you want to make available to other machines on your network. In these examples, you must supply the absolute path name for this entry. On the same line, you then list which computers can access the specified directory. You can specify the names of hosts in four ways:

- By using their direct host name.
- By using @*group,* where *group* is the specific netgroup. Wild card hosts in the group are ignored.
- By using wild cards in the host name. The asterisk (*) can match an entire network. For example, *.animals.acme.com matches all hosts that end in animals.acme.com.
- By matching IP subnets with address/netmask combinations. For example, to match everything in the 192.168.42.0 network where the netmask is 255.255.255.0, you use 192.168.42.0/24.

After the computers, you can specify the permissions. (See table 12.16.) If the list is longer than the line size permits, you can use the standard backslash (\) continuation character to continue on the next line. Some examples would be:

```
/export/data      daffy(rw,root_squash) bugs(rw, root_
squash)
   /export/reports    *.acme.com(rw,all_squash)
```

The first line creates an NFS statement that allows users from the machine daffy and bugs to mount the /export/data with read and write capability. But by using the root_squash option, root is automatically mapped to be an anonymous user called nobody, which effectively "squashes" the power of the remote root user to the lowest local user,

Table 12.16   Access Options Commonly Used While Setting NFS Exports

Option	Purpose
all_squash	All visitors are set to be anonymous users.
Insecure	No limits on which port NFS mounts can originate from.
Noaccess	The visitor is not allowed to descend into subdirectories.
Ro	The visitor is given read-only access.
root_squash	While accessing the mounted directory, root is automatically mapped as the nobody user, which helps maintain security.
rw	The visitor is given full read/write access (default).
no_root_Squash	Acknowledge and trust the client's root account.
Secure	Requires all NFS mounts to originate from a port below 1024 (default).

preventing remote root users from acting as though they were the root user on the local system. The second line creates an NFS statement that allows anyone from the acme domain to mount the /export/reports directory with read/write access. However, everyone who mounts this item is forced into anonymous user permissions.

It is considered a good convention to place all the directories you want to export in the /export directory. This makes the intent clear and self-documenting. If you need the directory to also exist elsewhere in the directory tree, use symbolic links. For example, if your server is exporting its /usr/local directory, you should place the directory in /export, thereby creating /export/usr/local. Because the server will need access to the /export/usr/local directory, you should create a symbolic link from /usr/local that points to the real location, /export/usr/local.

Unfortunately, NFS is not a very secure method for sharing disks. Although taking some steps to protect yourself from hackers pretending to be a common user provides a moderate level of security, there is not much more you can do. Any time you share a disk with another machine via NFS, you need to give the users of that machine (especially the root user) a certain amount of trust. If you believe that the person you are sharing the disk with is untrustworthy, you need to explore alternatives to NFS for sharing data and disk space. As always, stay up to date on the latest security bulletins coming from the Computer Emergency Response Team (www.cert.org) and keep up with all the patches from your distribution vendor.

After you set up your /etc/exports file, run the exportsfs command with the –r option:

```
exportsfs -r
```

This sends the appropriate signals to the rpc.nfsd and rpc.mountd daemons to reread the /etc/exports file and update their internal tables.

12.11.2 Mounting an NFS Drive or Directory

To mount an NFS drive or directory to your file system, you would use the mount command:

```
mount servername:/exported_dir /dir_to_mount
```

Servername is the name of the server from which you want to mount a file system. The *exported_dir* is the directory listed in its /etc/exports file on the specified server (*servername*), and */dir_to_mount* is the location on your local machine where you want to mount the file system. For example, if the /export/home directory is located on the server1 computer and you want to mount it as the /home directory, you would type in the following command:

```
mount server1:/export/home /home
```

Of course, the directory must exist in your local file system before anything can be mounted there.

You can pass options to the mount command. The most important characteristics are specified in the –o option. These options are listed in table 12.17. An example of these parameters in use would be:

```
mount -o rw,bg,intr, soft,retrans= 6 server1:/export/
home /home
```

To unmount the NFS mount, use the umount command:

```
umount /home
```

12.12 SAMBA

Samba used in Linux uses the Common Internet File System (CIFS), which is basically an updated SMB Protocol to provide file sharing and print sharing among Linux and Microsoft computers. Samba needs two daemons to operate. The smb daemon provides the file and print services to SMB clients by using port 139, and the nmbd daemon provides NetBIOS name serving and browsing support by using port 137. You can also run nmbd interactively to query other name service daemons. smbd and nmbd are located in the /usr/sbin directory.

Before you start configuring Samba, you should make sure that the SMB is automatically loaded during boot. Again, use the Linuxconf, ntsysv, or similar utility to activate the SMB daemon for run levels 3 and 5.

Samba uses the following programs:

- The SMB client program (smbclient) implements a simple FTP-like client on a Linux computer.
- The SMB mounting program (smbmount) enables the mounting of server directories on a Linux machine.

Table 12.17   The –o Option Used with the Mount Command

Characteristics	Description
rw	Read/write.
ro	Read-only.
bg	Background mount. Should the mount initially fail, such as when the server is down or the network connection is having problems, the mount place will place itself in the background and continue trying until it is successful. This is useful for file systems mounted at boot time because it keeps the systems from hanging at the mount if the server is down.
intr	Interruptible mount. If a process is pending I/O on a mounted partition, it will allow the process to be interrupted and the I/O call to be dropped.
soft	By default, NFS operations are hard, meaning that they require the server to acknowledge completion before returning to the calling process. The soft option allows the NFS client to return a failure to the calling process after the retrans number of retries.
retrans	Specifies the maximum number of retried transmissions to a soft-mounted file system.
wsize	Specifies the number of bytes to be written across the network at once. The default is 8192 (for example, wsize= 2048). You shouldn't change this value unless you are sure of what you are doing. Setting this value too low or too high can have a negative impact on your system's performance.
rsize	Specifies the number of bytes to be read across the network at once. Like wsize, the default is 8192 bytes. Setting this value too low or too high can have a negative impact on your system's performance.

■ The testparm utility allows you to test your smb.conf configuration file for proper syntax.

■ The smbstatus utility tells you who is currently using the smbd server.

All of these are located in the /usr/bin directory.

Samba web page

    http://www.samba.org/

    http://us1.samba.org/samba/samba.html

Samba documentation

    http://us1.samba.org/samba/docs/

Samba HOWTO documentation

    http://us1.samba.org/samba/docs/Samba-HOWTO-Collection.html

    http://us1.samba.org/samba/docs/Samba-HOWTO-Collection.pdf

12.12.1   SWAT

The easiest way to configure Samba is to use the Samba Web Administration Tool (SWAT), which uses a browser interface. What makes SWAT a little different from other browser-based administration tools is that SWAT does not rely on a web server like Apache.

If you have Red Hat 7.0 or higher, SWAT is disabled by default. To enable it you need to do the following:

1. Verify that /etc/services contains the following line:
   ```
   swat 901/tcp
   ```

2. Comment out the following line in the /etc/xinetd.d/swat file by adding a pound sign (#) before it:
   ```
   disable = yes
   ```

3. Restart Samba by typing in the following command:
   ```
   /etc/rc.d/init.d/smb restart
   ```

For Red Hat Linux 7.0 or higher, you can start Samba in one of two ways:

1. If you are using the GNOME desktop, you can open the main menu, select Programs, select System, and select Samba Configuration.
2. Open an Internet browser such as Netscape Navigator and type in the `http:// 127.0.0.1:901` URL.

A dialog box will appear asking for a user ID and password. You have to be root in order to configure Samba, so enter root as the user ID and the password for root.

When the main SWAT screen appears, you will first configure the [globals] section by clicking on the Global icon button at the top of the screen. The values shown are being read from the smb.conf file (the SWAT configuration file) that already exists on the system. After you have entered the appropriate values for your system, click the Commit Changes button to save them to the file.

Next, you will create shares by clicking on the Shares icon. This will open the Share Parameters page, as shown in figure 12.15. To create a new share, fill in a name for the share and click the Create Share button. An expanded Share Parameters page will appear. You then enter the appropriate information and click Commit Changes to save them to the smb.conf file.

The Status page can be used to verify if smb and nmb daemons are running. You can then start, stop, or restart these daemons by clicking on the Start/Stop buttons or the Restart button. (See figure 12.16.)

12.12.2   Accessing a Samba Share

Samba shares can be accessed by SMB clients on Windows and Linux platforms. Linux access is gained via the smbmount command. The smbmount command allows you to mount an SMB share. Smbclient provides command-line options to query a server for the

Figure 12.15 Configuring a Share Using a Samba HTML Page

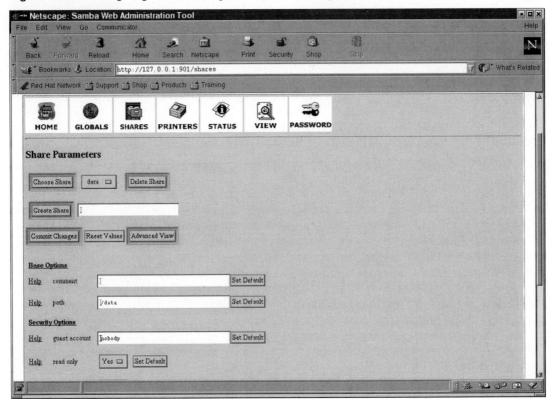

shared directories available or to exchange files. To list all the available shares on the machine with an IP address of 192.168.100.1, use:

```
smbclient -NL 192.168.100.1
```

Any name resolving to the IP address can be substituted for the IP address. The –N parameter tells smbclient not to query for a password if one isn't needed, and the –L parameter requests the list.

Most of the time, you will use the smbmount command, which enables you to mount a Samba share to a local directory. To create a /mnt/test directory on your local workstation, you would run the following command:

```
mount -t smbfs //192.168.100.1/homes /mnt/test -o
username=pregan,dmask=777,fmask=777
```

This command is a smbmount command even though it looks like an ordinary mount command. The –t smbfs tells the mount command to call smbmount to do the work. The preceding command grants all rights to anyone via the directory mask (dmask) and the file mask (fmask) arguments.

Figure 12.16   The Samba HTML Pages Can Also Show the Status of Samba

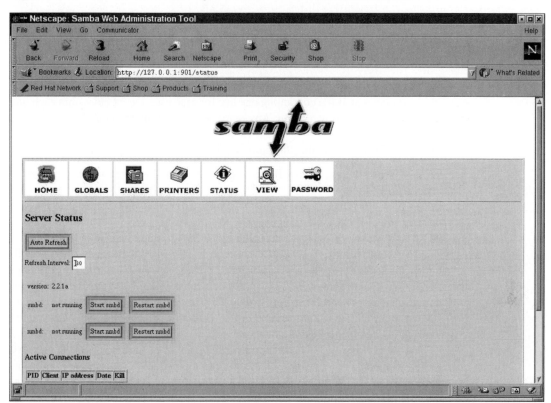

Another way to do the same command is:

```
smbmount //192.168.100.1/homes /mnt/test -o username=
pregan,dmask=777,fmask=777
```

To unmount it, simply run the following command as user root:

```
umount /mnt/test
```

You can also use smbmount to mount shared Windows resources on a Linux computer. The following mounts a Windows share called sharedirectory to the Linux directory /mnt/test directory:

```
mount -t smbfs //computername/sharedirectory /mnt/test
```

This must be done as root. You will be asked for a password. If Windows is in share mode, you would input the password of the share. If you are mounting a user share, use the –U option followed by a valid user name on the Windows computer. Of course, the user must have rights to the share, and the /mnt/test directory must exist.

12.12.3   Using smbpasswd

If the security is set to user, it enforces security by user and password. The smbpasswd is the Samba encrypted password file that contains the user name, Linux user ID, the SMB hashed passwords of the user, account flag information, and the time the password was last changed.

The format of the smbpasswd file used by Samba 2.2 is very similar to the UNIX passwd file. It is an ASCII file containing one line for each user. Each field within each line is separated from the next by a colon. Any entry beginning with the pound symbol (#) is ignored.

The smbpasswd file contains the following information for each user:

- **Name**—This is the user name. It must be a name that already exists in the standard UNIX passwd file.
- **UID**—This is the UNIX UID. It must match the UID field for the same user entry in the standard UNIX passwd file.
- **LANMAN password hash**—This is the LANMAN hash of the user's password, encoded as 32 hex digits. The LANMAN hash is created by DES encrypting a well-known string with the user's password as the DES key. This is the same password used by Windows 95/98 machines. If the user has a null password, this field will contain the characters "NO PASSWORD" as the start of the hex string. If the hex string is equal to 32 "X" characters, then the user's account is marked as disabled and the user will not be able to log onto the Samba server.
- **NT password hash**—This is the Windows NT hash of the user's password, encoded as 32 hex digits. The Windows NT hash is created by taking the user's password as represented in 16-bit, little-endian UNICODE and then applying the MD4 (Internet rfc1321) hashing algorithm to it. The NT password hash is considered more secure than the LANMAN password hash as it preserves the case of the password and uses a much higher-quality hashing algorithm.
- **Account flags**—This section contains flags that describe the attributes of the user's account. In the Samba 2.2 release this field is bracketed by "[" and "]" characters and is always 13 characters in length (including the "[" and "]" characters). The contents of this field may be any of the characters.
    - **U**—This means this is a user account, i.e., an ordinary user. Only user and workstation trust accounts are currently supported in the smbpasswd file.
    - **N**—This means the account has no password (the passwords in the fields LANMAN password hash and NT password hash are ignored). Note that this will only allow users to log on with no password if the null passwords parameter is set in the smb.conf config file.
    - **D**—This means the account is disabled and no SMB/CIFS logins will be allowed for this user.
    - **W**—This means this account is a workstation trust account. This kind of account is used in the Samba PDC code stream to allow Windows NT Workstations and Servers to join a domain hosted by a Samba PDC.
    - Other flags may be added as the code is extended in the future. The rest of this field space is filled in with spaces.

- **Last change time**—This field shows the time the account was last modified. It consists of the characters LCT- (standing for "last change time") followed by a numeric encoding of the UNIX time in seconds since the epoch (1970) that the last change was made.

When you make changes to the smbpasswd file, you will typically use the smbpasswd command:

```
smbpasswd [options] [username] [password]
```

See table 12.18 for a list of smbpasswd options.

Table 12.18   smbpasswd Options

Options	Description
-a	Add user.
-x	Delete user.
-d	Disable user.
-e	Enable user.
-n	Set no password.
-r *remote_machine_name*	Allows users to specify what machine they wish to change their password on. Without this parameter, smbpasswd defaults to the local host. The *remote_machine_name* is the NetBIOS name of the SMB/CIFS server to contact to attempt the password change.
-U username	This option may only be used in conjunction with the *-r* option. When changing a password on a remote machine, it allows the user to specify the user name on that machine whose password will be changed. It is present to allow users who have different user names on different systems to change these passwords.
-R *name_resolve_order*	This option allows the user of smbpasswd to determine what name resolution services to use when looking up the NetBIOS name of the host being connected to. • lmhosts—Look up an IP address in the Samba LMHOSTS file. • host—Do a standard host name to IP address resolution, using the system /etc/hosts, NIS, or DNS lookups. • wins—Query a name with the IP address listed in the WINS server parameter. If no WINS server has been specified, this method will be ignored. • bcast—Do a broadcast on each of the known local interfaces listed in the interfaces parameter. This is the least reliable of the name resolution methods, as it depends on the target host being on a locally connected subnet. The default order is lmhosts, host, wins, and bcast, and without this parameter or any entry in the smb.conf file, the name resolution methods will be attempted in this order.

To add users, you would type the following:

```
smbpasswd -a username
```

You will then be asked to specify the password for the new user.

smbpasswd also allows users to change their encrypted SMB password. Ordinary users can only run the command with no options. The screen will prompt them for their old SMB password and then ask them for their new password twice, to ensure that the new password was typed correctly. No passwords will be echoed on the screen while being typed. If you are the root, you can change other people's passwords by typing:

```
smbpasswd username
```

12.13 vi TEXT EDITOR

If you are not familiar with Linux, then you will not be familiar with the text editor used in Linux. So this section is to introduce you to the basic operations of the vi program.

To start the vi editor, you would type in the following command:

```
vi filename
```

where *filename* is the name of the file that you want to edit. vi automatically sets aside a temporary buffer as a working space for use during this editing session. If the file you specify when calling vi does not already exist in your current working directory, a new file will be created with the name that you have specified. For example, if you wanted to create a new file called testfile, you would type:

```
vi testfile
```

When the vi editor creates a new file, the cursor will be placed on the screen in the home position (upper left-hand corner). The cursor on most terminals is a rectangle that completely covers the character space it is situated upon and reflects your current position within the buffer. Below the cursor along the left side of the screen, a tilde (~) character is positioned in the first character space on each line. The tilde is used to indicate a void area in the file where there are no characters, not even blank spaces. At the bottom of the screen, the editor displays a file information line that presents the name of the file being edited, in this case testfile, followed by "[new file]" to indicate that this is a new file and it has never been opened before. (See figure 12.17.)

The vi editor allows a user to modify and add material to an already existing file. This is done by using the name of a current file when calling vi. vi will then fetch a copy of the contents of the specified file from disk storage, leaving the original undisturbed, and place the copy into the buffer. It is here, in the buffer, where vi permits you to do your editing. The idea of editing in the buffer is very beneficial, because if you really botch up your editing job (such as deleting a major section), the original file on the disk remains unaffected.

Upon opening a file, vi will display the first screenful of text starting at the top of the file, place the cursor in the home position, and print the file information line, which displays the name of the file along with the number of lines and characters it contains, on the

Figure 12.17   Starting vi Editor to Create a New File

```
~

~

~

~

~


"testfile"  [new file]
```

Figure 12.18   Opening a File With vi

```
This  is  some  sample  text  1.

This  is  some  sample  text  2.

This  is  some  sample  text  3.

This  is  some  sample  text  4.

This  is  some  sample  text  5.

This  is  some  sample  text  6.


~

~

~

"sample"  7L,  169C
```

last line of the screen. For example, to edit the file *sample,* which has 7 lines and 169 characters, you would use the following command:

```
vi sample
```

See figure 12.18.

As you can see in figure 12.18, the cursor sits in the home position, overlaying the first character of the file. When the cursor is moved, this character will be exposed, unchanged. In general, most terminal screens will display only 20 to 24 lines of text. If a file has more text than will fit on one screen, the unviewed material remains in the buffer until needed.

vi has two modes: an insert mode and a command mode. Insert mode is used when you want to insert text to the screen. When you are in insert mode, the only two things you can do are to insert text and to rewrite over text using the Backspace key. Command mode is used for issuing special commands to vi. When using command mode, you cannot insert text. However, you can search for a specific string, substitute certain words for other words, quit vi, save the text to a file, and so on.

As mentioned before, when the vi editor is invoked, a temporary working area is created called a buffer. This buffer or work space is only available for your use while you are using the editor. It is into this buffer that vi either places a duplicate copy of the file as it exists on the disk or creates a new file. When you exit the editor, this work area is discarded. To avoid losing all your work, you will have to save your file before exiting vi. Therefore, you should periodically save the file in case something unexpected happens.

To save the file, you need to execute the write command. This is done by typing:

`:w`

As soon as you type the :, vi switches to command mode. The cursor jumps to the bottom of the screen, and it is at the bottom of the screen that the :w command is echoed. Then a copy of all material in the buffer is transported to the disk, overwriting the previous contents of the disk file.

When you have finished your editing, you need to tell vi that you wish to quit the editor and return to your shell. This is done by typing:

`:q`

vi will respond by updating the file information line with the name of your file in quotes, followed by the current number of lines and characters. After exiting the editor, the Linux prompt will again appear on your screen.

vi requires that the buffer be empty of newly edited material when you type :q. Thus if you have made any alterations to the file since the last time you typed :w, vi will not know how you want these changes handled. The editor will print the statement:

`No write since last change (:quit! overrides)`

at the bottom of the screen. If you decide you want to retain these changes, you must type :w before issuing a new :q. Most experienced users get in the habit of combining these two commands into the single command :wq (write and quit).

Sometimes when working on a file, it is desirable to leave the editor without saving the modifications. This is accomplished by typing:

`:q!`

Again, the : switches into command mode. The ! means force. In short, you force vi to quit without saving the changes made to the file.

You might use this command if you started to edit a file and do not like the way the changes are shaping up. When you use this command, vi will immediately discard all alterations made to your file since the last :w. If you have not used a :w since opening the file, all changes since the beginning of the editing session will be abandoned.

NOTE: When you use the :quit! command, the editor believes you know what you are doing and will immediately follow your instructions. You will not have a second chance to retrieve your work once the buffer is closed, the file is discarded, and your changes are lost.

There may be a time when you are editing a file and cannot decide if you want to keep the original disk file and the new edited file that you are currently working on. You can then write the file to a different file name:

```
:w newfilename
```

Command mode is the initial state encountered when vi is started. It should be mentioned that when vi receives instructions on what action is required, the majority of the vi commands are not echoed on the screen. Instead, the program simply executes the response the command requires. In general, the steps given below will be repeated again and again each editing session while in command mode:

1. Place text in the terminal window.
2. Position the cursor.
3. Give the editing command.

After a short time of using the editor, these actions will become second nature to you.

To add or insert text into the text file, you must change to the insert text mode. The easiest way to change to the insert text mode is to press the Insert key or the I key. When you do this, the bottom of the screen will display -- INSERT --. (See figure 12.19.) When -- INSERT -- is displayed, whatever you type will be inserted at the cursor that you are at.

Figure 12.19   Changing to the Insert Text Mode

```
This is some sample text 1.

This is some sample text 2.

This is some sample text 3.

This is some sample text 4.

This is some sample text 5.

This is some sample text 6.

_

~

~

~

-- INSERT --
```

If you press the Insert key again, -- INSERT -- will be replaced with -- REPLACE --. Now whatever you type, will overwrite the text. To go back to -- INSERT --, press the Insert key again. To leave the insert mode, you would press the Esc key.

There are several ways to insert text other than the I command. The a (short for "append") command inserts text beginning after the current cursor position, instead of at the current cursor position. The A command will insert the text at the end of the current line. To begin inserting text at the next line, use the o command (short for "open"). (See table 12.19.)

The most obvious way to move around the file is by using the arrow key or by using the following commands:

h—Move left one character.
j—Move down one line.
k—Move up one line.
l—Move right one character.

After using these four commands for a while, you may decide to become more precise in moving the cursor by adding a number to the command. For example, 3h would move the cursor to the third line from the top of the screen. The command 3l would likewise move the cursor to the third line from the bottom of the screen. For other vi commands, see table 12.19.

Table 12.19   vi Commands

Moving the Cursor		Paging through Text	
Move to the left one character	h	Back one screen	{ctrl-b}
Move down one line	j	Down half a screen	{ctrl-d}
Move up one line	k	Down one screen	{ctrl-f}
Move to the right one character	l	Forward to end of file	G
		Move cursor to specified line	*line no.* G
Moving by Line		Up half a screen	{ctrl-j}
Beginning of current line	0 or ^		
Beginning of first screen line	H	**Ending and Saving Your Editing Sessions**	
Beginning of last screen line	L	Write changes to original file	:w
Beginning of middle screen line	M	Write to specified file	:w *filename*
End of current line	$	Force write to a file	:w! *filename*
Left to beginning of word	b, B	Quit (no changes made)	:q
Right to end of word	e, E	Quit and save changes	ZZ, :wq
Right to beginning of word	w, W	Quit and discard changes	:q!
Beginning of next sentence	)		
Beginning of previous sentence	(		

Continued

Table 12.19 Continued

Searching through Text			Join current line with next line	J
Backward for pattern	*?pattern*		Repeat last modification command	.
Forward for pattern	*/pattern*		Replace current character	r
Repeat previous search	n		Replace text to end of line	R
Reverse direction of previous search	N		Substitute text for character	s
			Undo the previous command	u
Creating Text			Transpose characters	xp

Creating Text	
Append text after cursor	a
Append text after end of line	A
Insert text before cursor	i
Insert text at beginning of line	I
Open new line after current line	o
Open new line before current line	O
Read file in after specified line	:r *filename*

Controlling the Screen Display of Your Session

Repaint the current screen	{ctrl-l}
Display line #, # of lines, etc.	{ctrl-g}

Cut and Copy Commands

Yank (copy) word into buffer	yw
Yank (copy) current line into buffer	yy or Y
Yank (copy) *n* lines	*n*yy or *n*Y
Put buffer text after cursor	p
Put buffer text before cursor	P

Modifying Text	
Change current word	cw, cW
Change current line (cursor to end)	C
Delete character (cursor forward)	x
Delete character (before cursor)	X
Delete word	dw, dW
Delete line	dd
Delete text to end of line	D

Making Corrections during Text Insertions

Overwrite last character	{delete}
Overwrite last word	{ctrl-w}

Some Useful ex Commands for Use in vi

:set all	Display all set options
:set autoindent	Automatically indent following lines to the indentation of previous line
:set ignorecase	Ignore case during pattern matching
:set list	Show special characters in the file
:set number	Display line numbers
:set nonu	Disable display of line numbers
:set shiftwidth=n	Width for shifting operators < >
:set showmode	Display mode when in Insert, Append, or Replace mode
:set wrapmargin=n	Set right margin 80-n for autowrapping lines (inserting new lines). 0 turns it off.

SUMMARY

1. Linux is licensed through GNU, which is short for "GNU's not UNIX."

2. The Linux kernel acts as a mediator for your programs and your hardware.

3. Linux distributors package the Linux kernel with basic support programs, the GNU utilities and compilers, the X Window system, and other useful programs.

4. The kernel numbering system is *major.minor.patch*. If *minor* is an even number, it is a stable version. If *minor* is an odd number, it is a beta.

5. The shell is a program that reads and executes commands from the user. You can think of a shell as a command interpreter.

6. X Window provides a GUI to Linux.

7. Linux supports various file systems for storing data including ext2 file systems (developed specifically for Linux), journaled file systems including ext3 and reiserfs, Microsoft's VFAT file systems (used in Windows 95/98), and the ISO 9660 CD-ROM file systems.

8. The standard layout forms a directory tree, which starts at the root directory (designated as the / directory).

   NOTE: This is different from DOS and the command prompt under Windows, which use the backslash (\) as the root directory and as a separator for directories.

9. Journaling file systems are superior to static file systems when it comes to guaranteeing data integrity and increasing overall file system performance.

10. The virtual file system (VFS) is a kernel software layer that handles all system calls related to the Linux file system, and it must manage all the different file systems that are mounted at any given time.

11. The /proc file system does not store data; rather, its contents are computed on demand according to user file I/O requests.

12. A daemon is a background process or service that monitors and performs many critical system functions and services.

13. Any network services managed by xinetd (including telnet, wu-ftp, SWAT, Linuxconf-web, and rsh).

14. A user account enables a user to log onto a computer with an identity that can be authenticated and authorized for access to computer resources.

15. When user accounts are created, they are placed in the /etc/passwd file.

16. Shadow passwords are created by taking the encrypted password entries from the password file and placing them in a separate file called shadow.

17. The home directory is a folder used to hold or store a user's personal documents.

18. The superuser, which basically is the root account, is the person who has access to everything on the system.

19. Adding users can be accomplished through the useradd utility.

20. The usermod program is used to modify existing users.

21. To delete a user from the /etc/passwd, you use the `userdel username` command.

22. The easiest command to change passwords is the passwd command.

23. A group is a collection of user accounts used to organize collections of accounts.

24. NIS (Network Information System) is a network naming and administration system for smaller networks. It allows the centralizing of user accounts and passwords so that they can be used on numerous machines.

25. To view the user and group owners and view the permissions, you can use the ls –l command.

26. The purpose of having groups is to allow users who are members of a particular group to share files.

27. The ifconfig program is responsible for setting up your network interface cards.

28. Network File System is a way for UNIX to share files and applications across the network. It allows you to attach a remote drive or directory to a virtual file system and work with it as if it were a local drive.

29. Samba used in Linux uses the Common Internet File System (CIFS), which is basically an updated SMB Protocol to provide file sharing and print sharing among Linux and Microsoft computers.

30. vi is a text editor commonly used in Linux.

QUESTIONS

1. The _____ is the central core of the operating system that interacts directly with the hardware.
   a. kernel
   b. worm
   c. program central
   d. shell

2. The component that manages the computer's memory, allocates it to each process, and schedules the work done by the processor is the _____.
   a. kernel
   b. system configuration files
   c. utility programs
   d. shell

3. In Linux, the command interpreter is known as the _____.
   a. kernel
   b. system configuration files
   c. utility programs
   d. shell

4. _____ is a set of standard operating system interfaces based on the UNIX operating system. (Choose the best answer.)
   a. Linux
   b. GNU
   c. Bash
   d. POSIX

5. Linux is a descendant from which operating system? (Choose the best answer.)
   a. MS-DOS
   b. Windows 95
   c. UNIX
   d. Windows 3.11
   e. OS/2

6. Although there are many Linux distributions, the term *Linux* refers to _____. (Choose the best answer.)
   a. Red Hat Linux
   b. Linus Torvalds
   c. The Linux kernel
   d. Free operating system

7. Which one of the following versions indicates a stable version of the Linux kernel? (Choose the best answer.)
   a. 2.3.5
   b. 2.2.5
   c. 2.3a.5
   d. 2.4.5a

8. The graphical interface used with Linux is _____. (Choose the best answer.)
   a. Windows
   b. X Window
   c. Windows for Workgroups
   d. desktop

9. How do you designate the Linux root partition?
   a. /
   b. /root
   c. /bin
   d. /home

10. You are asked to install on a machine that has several versions of UNIX on it. There are many different partitions. You only want to use the same type of partition that Linux already uses. What is the native file system format used by Linux?
    a. ufs
    b. DOS
    c. ext2fs
    d. FAT

11. Suppose you have a single IDE hard drive with three primary partitions. The first two are set aside for MS-DOS, and the third is an extended partition that contains two logical partitions, both used by Linux. Which one of these devices refers to the first Linux partition?
    a. /dev/hda1
    b. /dev/hda3
    c. /dev/hdc1
    d. /dev/hda2
    e. /dev/sda1

12. Which first-level segment of the file system contains a majority of system and server configuration files within its subdirectories?
    a. /var
    b. /bin
    c. /lib
    d. /etc
    e. /sbin

13. What is the default shell in most Linux distributions?
    a. Bash
    b. Tcsh
    c. Bourne shell
    d. Korn
    e. C shell

14. What does the command ifconfig show?
    a. ifconfig is not a valid command.
    b. It shows the current settings for each of your serial interfaces.
    c. It shows the current settings for each of your network interfaces.
    d. It doesn't show anything, but is used as a conditional statement in bash scripting.

15. What is the correct command to the IP address 172.16.1.1 to eth0 with a subnet mask of 255.255.0.0?
    a. ipconfig eth0 172.16.1.1 255.255.0.0
    b. ifconfig eth0 172.16.1.1 255.255.0.0
    c. ipconfig eth0 172.16.1.1 netmask 255.255.0.0
    d. ifconfig eth0 172.16.1.1 netmask 255.255.0.0

16. What is most likely the name of your Ethernet card if you have only one in your Linux computer?
a. NIC1
b. ife0
c. eth1
d. eth0

17. Which of the following types of information is returned by typing ifconfig eth0? (Choose all that apply.)
a. The names of programs that are using eth0
b. The IP address assigned to eth0
c. The hardware address of eth0
d. The host name associated with eth0

18. Which command should you use to verify basic connectivity between computers?
a. ftp
b. ping
c. lynx
d. PING

19. Which Linux command will display all the networks between your computer and another computer?
a. tr
b. tracert
c. route
d. traceroute

20. Which of the following utilities could be used to display the ports for which the server is currently listening?
a. ifconfig
b. ipchains
c. netstat
d. netlist
e. mapnodes

21. Which file systems provide journaling capabilities? (Choose all that apply.)
a. VFAT
b. ext3
c. ext2
d. reiserfs

22. Where does Linux store encrypted user passwords?
a. /etc/passwd
b. /etc/shadow
c. /usr/passwd
d. /root/shadow

23. What is the name of the account created during installation of Linux that allows special administrative privileges?
a. Root
b. Admin
c. Superuser
d. Administrator

24. Which command would create a new user named Pat and make her part of the admin group?
a. useradd pat:admin
b. newuser pat:admin
c. useradd -g admin pat
d. usermod -n pat -g admin

25. Which file permission allows only the user and members of the user's group to change or erase a file, while allowing everyone to read it?
a. -rw-r--rw-
b. -r-xr-xr--
c. -rw-rw-r--
d. -r--rw-rw-

26. What is the octal representation of the permission rw-r--r-?
a. 322
b. 422
c. 741
d. 644

27. What is the umask of permission set rwxr--rw-?
a. 746
b. 031
c. 022
d. None of the above

28. Jason uses a UNIX computer. He wants to modify the UNIX file that stores user password information in an encrypted format. Which file should Jason modify?
a. Hosts
b. Passwd
c. Login
d. Services

29. You are a network engineer for a large company that uses UNIX servers on its WAN. You create a group account for the research department on the Linux computer in Kansas City. Which file on the UNIX computer stores information about the research group?
a. HOSTS
b. LOGIN
c. ALIAS
d. GROUP

30. The service that will make file shares available to Windows computers is _____.
a. Samba
b. Squid
c. Sendmail
d. Lokkit

31. What does NFS provide?
a. Access to a print server
b. Access to remote files
c. Access to processor services
d. Access to local files

32. The command-line interface to Linux is known as a _____.
a. Linux
b. shell
c. X
d. Window
e. NIC
f. DOS

33. You want your Windows 98 clients to connect to a newly installed UNIX server on your network. There are two possible solutions, either of which will work. What are they? (Choose two answers.)
a. Install a third-party NFS client on Windows 98 machines.
b. Install Samba on your UNIX server.
c. Install the XPS Protocol on your clients and server.
d. Install the Microsoft Client for UNIX on your client computers.

HANDS-ON EXERCISES

Exercise 1: Installing Red Hat Linux 7.2 or 7.3

1. With the Red Hat Linux boot disk in drive A and the Red Hat Linux CD #1 in the CD drive, boot the computer.

   NOTE: If your system supports a bootable CD, you can insert the CD into the drive and reboot the computer.

   NOTE: You may need to enter the CMOS setup program to specify to boot from the A drive or the CD-ROM.

2. At the boot: prompt, press the Enter key to install Linux in graphical mode.
3. Select your language and click on the Next button.
4. Select your keyboard and your keyboard layout and click on the Next button.
5. Select the mouse type and click on the Next button. If the mouse has only two buttons, you can enable the Emulate 3 buttons if you desire.
6. Click on the Next button on the Welcome to Red Hat Linux screen.
7. Select the Custom System and click the Next button.
8. Select the Manually partition with fdisk option and click the Next button.
9. Click the hda button to select the first hard drive.
10. At the Command: prompt, type m↵.
11. At the Command: prompt, type p↵. Notice all the partitions on the drive.
12. At the Command: prompt, type d↵. To delete the first partition (hda1), type 1↵. Delete the remaining partitions.
13. To add a new partition, type n↵. Type p↵ to add a primary partition. Type 1↵ for the partition number. When asked for the first cylinder, type 1↵. To specify a 4500-MB partition, type + 4500M↵ when the prompt asks for the last cylinder.
14. To add a second partition to be used as a swap partition, type n↵ to add a new partition. Type p↵ to specify a primary partition. Type 2↵ to specify the second partition. For the first cylinder, press the Enter key to keep the default. When the screen asks for the size of the partition, type in + *XXX*M↵, where *XXX* is the amount of your RAM.
15. At the Command: prompt, type p↵ to view the partitions.
16. At the Command: prompt, type l↵ to view the partition types. Notice that Linux (ext2) is type 83 and the Linux swap type is 82.
17. At the Command: prompt, type t↵ to change the partition type. To choose the second partition, type 2↵. When the screen asks for the hex code, type 82↵. The screen should say Changed system type of partition 2 to 82 (Linux swap).
18. Display the partition table again and notice the two partitions and their types. In addition, notice that no partition is selected as the bootable partition.
19. Display the Help menu.
20. To make the first partition bootable, type a↵ to toggle. When the screen asks for which partition, type 1↵.
21. Display the partition. Notice the asterisks (*) under the Boot column.
22. To save all these changes, you must write the table to disk and exit. Therefore, type in w↵.
23. When you return to the fdisk screen that displays which drive to run fdisk on, click the Next button.
24. If the screen asks if you would like to format this partition as a swap partition, click on the Yes button.

25. For Red Hat Linux install, you still run Disk Druid to mount the drives. Click on the hda1 partition to select it and click the Edit button. Select the / as the mount point, select the ext2 file system, and click the OK button. If a warning appears, click on the Yes button to continue and format these partitions. Click the Next button.

26. To show you how to use Disk Druid, click on the Back button to get back to the Disk Setup screen. Highlight the first partition (hda1) and click the Delete button. When the screen asks if you are sure you want to delete this partition, click the Yes button. Delete the other partitions.

27. Click the New button. Change the partition type to Linux swap. Enter the size of your RAM (MBs) and click on the OK button.

28. Click the New button. Select the / for the mount point, select the ext2 file system, and specify 4500 for the size (MBs). Click on the OK button.

29. Click the New button. Select the /home mount point, select the ext3 file system, and specify 500 MB. Then click on the OK button. Click the Next button.

30. When you get to the Boot Loader Configuration screen, select LILO as the boot loader and click on the Next button.

31. If you have a network card, the Network Configuration box will appear. Uncheck the Configure using DHCP box.

    If you are in a classroom, use the values assigned by your teacher. If not, use the following:

IP address:	192.168.*XXX.1YY,* where *XXX* is your room number and *1YY* is your student number.
Netmask:	255.255.255.0
Host name:	host*XX,* where *XX* is your student number.

    If you have the only computer, use 1.101 for *XXX.YYY.* Enter 192.168.*XXX*.254 for the primary DNS server. If your network is connected to a router, enter the 192.168.*XXX*.254. Click on the Next button.

32. Select no firewall and click on the Next button.

33. Select a default language and select the Next button.

34. Select your time zone and click the Next button.

35. Type in `password` for the root password. Then retype password for the Confirm text box. Be sure to use lowercase.

36. To create a second account, click on the Add button. Type user1 for the user name and type in your full name for the Full Name text box. Then type `password` for your account password and password Confirm text boxes. Click the OK button. Click on the Next button.

37. Keep the MD5 passwords and Enable shadow passwords options enabled and click on the Next button.

38. When the Selecting Package Groups screen appears, scroll down to the bottom of the list and enable the Everything option. Click on the Next button.

39. Select your video card and the amount of video memory. Click the Next button.

40. On the About to Install screen, click the Next button. When the program jumps to the next screen, it may appear to freeze. Give it a couple of minutes and it will start moving.

41. When the screen asks for the second CD, remove the first CD and insert the second CD. Click on the OK button.

    NOTE: If you are using Red Hat 7.3, you will also need to insert a third disk.

42. When the screen gets to the Boot disk Creation screen, remove the Linux Installation boot disk and insert a blank disk. Then click the Next button.

43. Find your monitor and select it. If your monitor is not listed, check the monitor documentation and enter the Horizontal Sync and Vertical Sync values. Click on the Next button.

44. Select an appropriate color depth and screen resolution. Click the Test Setting button. When the screen asks if you can see this message, click the Yes button. If not, when it returns to the configuration screen, verify which type of video card you have and the amount of memory.

45. Leave the GNOME desktop environment selected and select the Text login type. Click the Next button.

46. When you get the Congratulations screen, remove the floppy disk and click on the Exit button. Remove the CD disk and let the computer boot Linux.

Exercise 2: Starting and Shutting Down Linux

1. At the login prompt, type in `root` and press the Enter key.

2. When the screen asks for the password, type in `password` and press the Enter key. Notice that since you logged in as the root, the prompt has a # (pound) sign and that your default directory is the root account /root.

3. At the prompt, type in `logout` and press the Enter key.

4. Login as user1 and use password as the password. Notice the $ is used in the prompt and the default directory is your default name.

5. Log out again.

6. Log in as the root account.

7. To shut down Linux, press the Ctrl+Alt+Del keys.

8. Turn the computer on again and start Linux.

9. Log in as the root account.

10. At the prompt, type `shutdown -h now↵`.

11. Start Linux again and log in as the root account.

12. At the prompt, type `shutdown -r 1↵`.

13. Start Linux again and log in as the root account.

14. At the prompt, type `shutdown -r now↵`.

15. Start Linux again and log in as the root account.

16. At the prompt, type in `startx↵`.

17. Click the Main Menu button on the panel, select the Logout option, and click the Yes button. You may need to press Enter one more time to get back to the prompt.

18. At the prompt, type in `startx↵`.

19. Click the Main Menu button on the panel and select the Logout option. Select the Shut Down option and click the Yes button.

Exercise 3: Navigating the Linux File System

1. Start Linux and log in as the root.

2. Your prompt shows that you are in the /root directory. To go up one directory, type `cd ..↵`. Notice that you are now in the root directory (/) since the /root directory is directly underneath the root directory (/).

3. To go back into the root directory, type `cd root↵`.

4. Currently the home directories for the other users are located under the home directory. To change to the home directory that is located in the root directory, type `cd /home↵`.

5. To list the contents of the home directory, type `ls -a↵`.

6. Change into user1's home directory by using `cd user1`.

7. To move up one directory, use the cd command.

8. Use the pwd command to show your current directory.

9. Change back to the user1 directory.
10. Use the pwd command to show your current directory.
11. To change to the root directory, no matter what directory you are in, type `cd /`↵.
12. To go back to the user1 directory in one command, use the `cd /home/user1` command.
13. Change back to the root directory.
14. To list the contents of the home directory, type `ls -a`↵.
15. Change into the boot directory.
16. Execute the `ls -a` command to view the files in the boot directory. Notice the vmlinuz file, which is the kernel.
17. To change to the bin directory located in the root directory directly from the boot directory, type `cd /bin`↵.
18. Execute the `ls -a` command. You should be able to recognize the external Linux commands such as pwd, mov, sort, kill, cat, rm, rmdir, and others.
19. In one command, change to the mnt directory.
20. Display the contents of the directory. This is where you would typically find the CD drive and the floppy disk drives.

Exercise 4: Working with Files

1. Change into the /tmp directory.
2. Display the contents of the directory.
3. To copy the install.log to the /home/user1 directory, type `cp install.log/home/user1`↵ at the prompt.
4. Change into the user1 directory.
5. To make a copy of the install.log file and call it file1.txt, type `cp install.log file1.txt`↵.
6. Display the contents of the directory.
7. Make three more copies of the install.log and call them file2.txt, file3.txt, and fileA.txt.
8. Display the contents of the directory.
9. To list all files using wild cards, type `ls -a *`↵.
10. To list all the files that begin with file, type `ls -a file*`↵.
11. To list all the files that begin with file that range from 0 through 3, type `ls -a file[1-3]*`.
12. To delete the fileA.txt file, type `rm fileA.txt`↵. When you are asked to remove the file, type y↵.
13. To remove all files that start with file, type `rm -f file*`↵ at the prompt.
14. To look for a file, you will use the find command. Therefore, type `find / -name install.log -print`↵. Notice that it is listed twice, once in the /home/user1 and once in the /tmp directory.
15. You should still be in the user1 directory. To move the install.log file to the home directory, you can type `mv install.log /home` or `mv install.log ..`↵. Therefore, type the `mv install.log /home`↵.
16. Delete the install.log file from the home directory.

Exercise 5: Working with the proc file System

1. Change into the /proc directory.
2. View the contents of the directory.

3. To view your IRQs, type `cat interrupts⏎`. Look at which IRQ the mouse is using. Look to see which IRQs are free.
4. To view your I/O addresses, type `cat ioports |more⏎`.
5. To view your information about your processors, type `cat cpuinfo⏎`.
6. View your information on your memory.
7. View your PCI devices.
8. View your information on your swap file.
9. View the version file to view the version of Linux.
10. To see the drivers that are loaded in memory, view the modules file.

Exercise 6: Using vi Text Editor

1. Change into the user1 directory located in the Home directory.
2. To create a file1.txt using vi, type `vi file1.txt⏎`.
3. To enter text, you will need to put vi in the input mode. Press a to append text after the cursor. -- INSERT -- should appear at the bottom of the screen.
4. Type the following text:

   ```
   This exercise is designed to familiarize the user⏎
   with the basic editing features of vi. We will go⏎
   over commands to move the cursor as well as⏎
   showing the user how to save the files he has created. ⏎
   ```

5. To put the vi back into command mode, press the Escape key.
6. To save the file and exit vi, type `:wq`.
7. Display the contents of the directory.
8. Use the cat command to display the contents of the file1.txt file.
9. To modify the text file, type `vi file1.txt⏎`.
10. Use the arrow keys to move the cursor around the screen.
11. Use the h, k, l, and j keys to move the cursor around the screen.
12. Move the cursor to the space after "he."
13. Press the I key to change to insert mode. Press the space bar and type `or she⏎`.
14. Move the cursor to the h in "has" and press the Enter key.
15. Press the Escape key to go back to command mode.
16. To save your work and remain in vi, type `:w⏎`.
17. Move the cursor to the third line.
18. To delete the entire line, type `dd`.
19. Delete the next line by typing `dd`.
20. To bring the last line back, type `P`.
21. Delete the word "user" in the fourth line.
22. To exit vi without saving changes, type `:q!⏎`.
23. Load the file1.txt file using vi.
24. Move the cursor to the last line and type `yy` to copy the line.
25. Press the Enter key to go to the next line.
26. Press the `p` key several times to paste the text.
27. To undo, type u several times.
28. Quit vi without saving the file.

Exercise 7: Using Network Commands at the Command Prompt

1. At the command prompt, type the following command:

   ```
   ifconfig
   ```

2. Record the IP address, mask, and MAC address (HW address) for the eth0 card.
   IP address: _____
   Mask: _____
   MAC address: _____

3. At the command prompt, type the following command:

   ```
   route
   ```

4. Record the default gateway address:
   Gateway address: _____

5. To down the eth0 network interface, type the command:

   ```
   ifconfig eth0 down
   ```

6. To bring the interface back up, type:

   ```
   ifconfig eth0 up
   ```

7. To specify a different IP address and netmask, type the following command:

   ```
   ifconfig eth0 192.168.XXX.1YY netmask 255.255.255.0
   ```

   where XXX is your room number and YY is your computer number.

8. Type the following command to verify the IP address and subnet mask:

   ```
   ifconfig
   ```

9. To change the default gateway, type in the following command:

   ```
   route add -net default gw 192.168.1.254 dev eth0
   ```

10. Type the following command to verify the default gateway:

    ```
    route
    ```

11. To show the status of the interface, type in the following command:

    ```
    netstat -i
    ```

    Check for any errors or dropped packets.

12. To show more details of the interface and its status, type in the following command:

    ```
    netstat -s |more
    ```

Exercise 8: Troubleshooting Network Problems

1. Run the ifconfig command to verify your IP settings.
2. Run the route command to verify your IP settings.
3. Ping the loopback address of 127.0.0.1.
4. Ping your IP address.
5. Ping the local host computer.
6. Ping the host*XXX-YY* computer, where *XXX* is your room number and *YY* is your computer number.
7. Ping the host*XXX-YY*.acme.cxm computer.

8. Ping your instructor's computer.
9. Ping your partner's computer.
10. If you have a router on your network, ping your gateway or local router connection.
11. Ping the far end of the router. You may have to get the address from your instructor.
12. Ping another computer on another network. You may have to get an address from your instructor.
13. Run the traceroute command to the other computer.
14. If you are connected to the Internet, ping www.redhat.com.
15. Run the traceroute command to www.redhat.com.

Exercise 9: Create and Modify Accounts

1. To create an account called jsmith, type the following command at the command prompt:

   ```
   useradd jsmith
   ```

2. To change the password for jsmith to pass3Word, type in the following command:

   ```
   passwd jsmith
   ```

3. To create an account called psanchez with a home directory called /home/account, type the following command at the command prompt:

   ```
   useradd psanchez -d /home/account
   ```

4. Change the password for psanchez to test5Town.
5. To make it so that jsmith cannot change the password for 7 days, type in the following command:

   ```
   chage -m 7 jsmith
   ```

6. Change to the second virtual console by pressing Ctrl+Alt+F2.
7. Log in as jsmith. Notice the current directory.
8. To change the password to lotus3.Test, type the following command:

   ```
   passwd
   ```

9. In the /home/jsmith directory, create a text1.txt file with your name in it.
10. Log out.
11. Log in as psanchez. Notice the current directory.
12. Create a text file in the /home/account directory called text2.txt file with your name in it.
13. Log out.
14. Change back to the first virtual console.
15. To delete the jsmith account, type the following command:

    ```
    userdel -r jsmith
    ```

16. To view the contents of the /home directory, change to the /home directory and execute the ls –l command. Notice that the jsmith directory is gone.
17. To move the home directory of psanchez, type in the following command:

    ```
    usermod -d /home/psanchez -m psanchez
    ```

18. Do an ls –l listing of the /home and /home/psanchez directory.
19. To lock the psanchez account, type the following command:

    ```
    usermod -L psanchez
    ```

20. Change to the second virtual console.

21. Try to log in as psanchez.

22. Change to the first virtual console.

23. To unlock the psanchez account, type in the following command:

```
usermod -U psanchez
```

24. Change to the second virtual console.

25. Log in as psanchez.

26. Log out.

27. Change back to the first virtual console.

28. Create a new group called test by using the following command:

```
groupadd test
```

29. Add psanchez to the test group by typing in the following command:

```
usermod -G test psanchez
```

30. Delete the test group by typing the following command:

```
groupdel test
```

Exercise 10: Working with File Permissions

1. Create a user2 account.

2. Create a group2 group.

3. Assign user2 to group2.

4. If you don't already have one, create a /data directory.

5. Change into the /data directory.

6. Create three text files called file10.txt, file20.txt, and file30.txt with your name in them.

7. Use the `ls -l` command and record the permissions.

8. Change to the second virtual console.

9. Log in as user2.

10. Change into the data directory.

11. View the contents of the file10.txt by using the cat command.

12. Try to add your age to the file10.txt file.

13. Try to delete the file10.txt file using the rm command.

14. Change back to the first virtual console.

15. Assuming that you are still in the /data directory, to change owner of the files to user2 and group2, type the following command:

```
chown user2.group2 *
```

16. Change to the second virtual console.

17. View the contents of the file10.txt by using the cat command.

18. Try to delete the file.

19. Change back to the first virtual console.

20. Change the permissions by typing the following command:

```
chmod 660 file20.txt
```

21. Use the `ls -l` command to view the permissions.

22. Change back to the second virtual console.

23. Open the file20.txt file and add your age to the file. Save the file.

24. Try to delete the file. Remember that to delete a file, you must have the write permission for the directory that the file is in.
25. Change back to the first virtual console.
26. Change to the root directory. Execute the following command to directory permissions:

    ```
    chmod 660 data
    ```

27. Change back to the second virtual console.
28. Try to perform a listing of the data directory. Remember that for you to see a directory listing, you must have the execute permission for the directory.
29. Change back to the first virtual console.
30. Change the permissions for the data directory by typing in:

    ```
    chmod 770 data
    ```

31. Access the directory listing of the /data directory.
32. Delete the file10.txt.
33. Change back to the first virtual console.
34. At the command prompt, type in the following command to view the current umask:

    ```
    umask
    ```

35. To change the umask, type the following command:

    ```
    umask 007
    ```

36. To create an empty text file called file40.txt, type the following command:

    ```
    touch file40.txt
    ```

37. Use the `ls -l` command to view the permissions.

Exercise 11: Recovering a Lost Root Password

1. Reboot the computer and enter the text mode of LILO.
2. At the command prompt, boot into single mode.
3. At the command prompt, use the passwd command to change the password to testpassword.
4. Reboot the computer and test the new password.
5. Change the password back to password.

Exercise 12: Looking for a Security Problem

1. Log in as root.
2. Execute the following command so that Linux can start recording failed logins into the btmp file:

    ```
    touch /var/log/btmp
    ```

3. Change to the second terminal and try to log in with the user1 account with the wrong password.
4. Log in as user1 with the correct password.
5. Change back to the first terminal (root).
6. View the /var/log/messages file. Look for any errors or services being enabled or disabled or any other warnings or notifications of possible security issues. Don't forget you can also use the less command with these logs.
7. View the /var/log/boot.log. Look at the times the computer was started or stopped. You should also look in the /var/log directory for other boot.log files (such as boot.log.1).

8. View the contents of the /var/log/secure. Look at when users or groups were added or deleted, programs and services started, passwords changed, and other possible security issues. You should also look in the /var/log directory for other secure files (such as secure.1).

9. Execute the `last` command. Look to see when user1 logged on, when user1 logged off, and for how long user1 was logged on.

   NOTE: You can also use last |less.

10. Execute the `lastb` command.

11. Execute the `lastlog` command.

12. To find out if there are any SETUID or SETGID permissions on files, use the following command:

    ```
    find / -perm +4000 -print
    ```

    Look to see if you were not aware of any of these.

13. View the /home/user1/.bash_history file for user1. Look for suspicious commands that user1 should not be doing.

Exercise 13: Setting Up a Samba Share

On Your Computer

1. If you don't already have a /data directory, create it.

2. Create a text file called name.txt in the /data directory with your name in the file and "samba share - /data directory".

3. Make sure that the data directory has the read, write, and execute permissions by using the following command:

   ```
   chmod 777 /data
   ```

4. Make sure that the files in the directory have the read, write, and execute permissions by using the following command:

   ```
   chmod 777 /data/*
   ```

5. Use Linuxconf or ntsysv or a similar utility to activate the SMB daemon for run levels 3 and 5.

6. Open the /etc/xinetd.d/swat file and comment out the following line:

   ```
   disable = yes
   ```

   Save the changes.

7. To restart the xinetd.d so that the changes can be read for the SWAT file, run the following command:

   ```
   service xinetd restart
   ```

8. To start Samba, type in the following command:

   ```
   /etc/rc.d/init.d/smb start
   ```

9. Start X Window. Open an Internet browser such as Netscape Navigator and enter the following URL:

   ```
   http://127.0.0.1:901
   ```

   Log in as root.

10. Click on the Status option at the top of the web page. Notice if the smbd and nmbd daemons are running. If they are not, start them.

11. Click on the View option to view your smb.conf file.
12. Click on the Globals option.
13. Make sure your workgroup is GROUPXXX.
14. For your NetBIOS name, be sure it is host*XXX-YY*, where *XXX* is your room number and *YY* is your computer number. If you are not in a classroom, use host1-1; your partner's/second computer should be host1-2.
15. Currently, the security is set to user. Click on the Security help hyperlink. If you want to mainly set up shares without passwords, what options would you select?
16. Close the Help window. For the Security option, change it to share.
17. Click on the update encrypted help hyperlink. For the best security, should this be set to yes or no?
18. Close the Help window.
19. Set the encrypt password to yes.
20. Click on the Commit Changes button.
21. Click on the Shares option at the top of the screen.
22. In the Create share text box, type data and click on the Create share button.
23. In the Path text box, specify the /data directory.
24. Make sure that the Read-only option is set to yes and that the guest OK option is set to yes.
25. At the top of the screen, click on the Commit Changes button.
26. Open a terminal window and execute the following command:

```
smbclient -NL your_IP_Address
```

On Your Partner's Computer

1. Create a /mnt/sambaremote directory.
2. Execute the following command:

```
chmod 775 /mnt/sambaremote
```

3. Execute the following command for your computer:

```
smbclient -NL your_IP_Address
```

4. Execute the following command:

```
smbclient //your_IP_address/data
```

   When the screen asks for a password, just press Enter.
5. At the SMB prompt, execute the ls command.
6. At the SMB prompt, execute the following command:

```
get name.txt
```

7. At the SMB prompt, execute the exit command.
8. At the command prompt, execute the following command:

```
mount -t smbfs //your_IP_address/data /mnt/sambaremote
```

   When the screen asks, just press Enter.
9. Change to the /mnt/sambaremote directory.
10. Perform the ls command.
11. Try to delete the file.
12. Unmount the /mnt/sambaremote directory.

On Your Computer

1. Open the data share using SWAT.
2. Change the guest OK option to no.
3. Commit the changes.
4. At the command prompt, execute the following command:

```
smbpasswd -a user1
```

Enter `password` for the password.

On Your Partner's Computer

1. Perform the following command:

```
smbmount //your_IP_address/data /mnt/sambaremote -o U=user1
```

Provide the appropriate password.
2. Open the name.txt file in the share, add your age to the file, and try to save the changes.
3. Try to delete the file.

On Your Computer

1. Open the data share using SWAT.
2. Change the Read-Only option to no.
3. Commit the changes.

On Your Partner's Computer

1. Perform the following command:

```
smbmount //your_IP_address/data /mnt/sambaremote -o
username=user1,dmask=777,fmask=777
```

2. Open the name.txt file in the share, add your age to the file, and save the changes.
3. Delete the file.
4. Open the /etc/fstab file and add the following line:

```
//Your_IP_Address/data /mnt/sambaremote smb
username=user1,password=password   1 2
```

Microsoft Windows Computer

1. If you have a Windows computer connected to your network, use Network Neighborhood or the equivalent to browse the network and find your Samba server and its resources.

Exercise 14: Setting Up a DNS Server

1. Use Linuxconf, ntsysv, or some similar utility and activate the named daemon for run levels 3 and 5.
2. Open the /etc/named.conf file and add the following two sections. Remember that *XXX* is your room number and *YY* is your computer number.

```
zone "acme.cxm" {

    type master;

    file "named.acme.cxm";

};


zone "XXX.168.192.in-addr.arpa" {
    type master;

    file "named.192.168.XXX";

};
```

3. Save the named.conf file.
4. In the /var/named directory, create a named.acme.cxm file. In this file, add the following content:

```
@           IN   SOA      acme.cxm. hostmaster.acme.cxm. (

                            200272001   ; Serial

                            28800       ; Refresh

                            14000       ; Retry

                            3600000     ; Expire

                            86400 )     ; Minimum

            IN   NS      hostXXX-YY.acme.cxm.

            IN   MX 10   hostXXX-YY.acme.cxm.

hostXX      IN   A       192.168.XXX.YY

www         IN   CNAME   hostXXX-YY.acme.cxm.
```

5. Save the named.acme.cxm file.
6. In the /var/named directory, create the named.192.168.XXX file. In this file, add the following content:

```
@   IN   SOA   1.168.192.in-addr.arpa. hostmaster.acme.cxm. (

                    200272001       ; Serial

                    28800           ; Refresh

                    14000           ; Retry

                    3600000         ; Expire

                    86400 )         ; Minimum


    IN   NS     hostXXX-YY.acme.cxm.

1   IN   PTR    hostXXX-YY.acme.cxm.
```

7. Save the named.192.168.*XXX* file.
8. Start the named daemon. If it is already started, then restart it.
9. Open Linuxconf. Configure your primary DNS server as the address of your computer.
10. Try to ping host*XXX-YY*.acme.cxm. You should notice that the name resolution worked but the ping command itself failed to contact the computer.
11. Try to ping www.acme.cxm.
12. Use the following command to test the DNS server:

```
nslookup hostXXX-YY.acme.cxm
```

13. Use the following command to test the DNS server:

```
dig www.acme.cxm
```

14. Use the following command to test the DNS server:

```
host hostXXX-YY.acme.cxm
```

15. Try the following command to test the reverse DNS:

```
host 192.168.1.1
```

Exercise 15: Create a Web Page

1. Using a text editor, create a file called webpage.htm with the following content:

```
<HTML>

<HEAD>

<TITLE>THIS IS THE TITLE</TITLE>

<META NAME="KEYWORD" CONTENTS="HOT, IMPORTANT">

<META NAME="DESCRIPTION" CONTENTS="This is a sample website">

</HEAD>


</BODY>This is the body of the web page<BODY>
</HTML>
```

2. Open Netscape Communicator or some other web browser and open the following web page:

```
/data/webpage.htm
```

3. Change the file to the following content and save the changes.

```
<HTML>

<HEAD>

  <TITLE>THIS IS THE TITLE</TITLE>

  <META NAME="KEYWORD" CONTENTS="HOT, IMPORTANT">

  <META NAME="DESCRIPTION" CONTENTS="This is a sample
website">

</HEAD>
```

```
</BODY>
    <CENTER>TITLE</CENTER>
    This is the body of the web page
    <OL>
            <LI>First
            <LI>Second
            <LI>Third
    </OL>

    <FONT COLOR="FF0000">RED
    <FONT COLOR="00FF00">GREEN
    <FONT COLOR="0000FF">BLUE

    <BR><BR>
    <FONT COLOR="000000">END
</BODY>
</HTML>
```

4. Open Netscape Communicator or some other web browser and open the following web page:

   ```
   /data/webpage.htm.
   ```

   If you have your browser still open, you can also click on the Refresh button to reread the web page.

Exercise 16: Install and Configure a Web Server

1. Use Linuxconf or ntsysv or a similar utility to activate the HTTP daemon for run levels 3 and 5.
2. To start HTTP, type the following command at the command prompt:

   ```
   /etc/init.d/httpd start
   ```

3. Type the following command at the command prompt:

   ```
   apacheconf
   ```

4. Click the Virtual Hosts tab and click the Edit button.
   What is the Document root directory?

   _____

5. Click the Cancel button.
6. Click the Performance tuning tab.
   What is the maximum number of connections allowed for the web server?

   _____

7. Start Netscape Communicator or some other web browser and open the following website:

   ```
   http://your_ip_address
   ```

   Notice that it brought up a default web page.

8. Rename the webpage.htm page to index.html.

9. Copy the index.html to the /var/www/html directory, replacing the default web page.

10. If your browser is still running, refresh the browser. If not, start Netscape Communicator or some other web browser and open the following website:

```
http://your_ip_address
```

Exercise 17: Configuring and Starting Sendmail

1. Open the /etc/xinetd.d/ipop3 file and comment out the disable=yes line with a pound sign (#). Save the changes.

2. Restart the xinetd service by using the following command:

```
service xinetd restart
```

3. Make sure that your DNS server has an MX record for your mail server.

4. Test the DNS server for the MX register by using the following command:

```
nslookup -querytype=mx acme.com
```

5. Open the /etc/sendmail.cf file. Change the `Cwlocalhost` line to `Cwacme.cxm` and change the `DMdomainname.cxm` to `DMacme.cxm`.

6. Activate the Sendmail service.

7. Execute the following command:

```
mail -v user1@acme.cxm
```

Enter "Test" for the subject and type in "This is a test for the body." Go to the next line by pressing Enter. Type a period (.) and press the Enter key. When the screen asks for the CC:, press the Enter key.

8. Change to the /var/spool/mail directory and use the cat command on the user1 file.

Exercise 18: Configuring Kmail

1. Log in as user1. Start X Window.

2. Start Kmail. If the screen says it needs to create some directories, click OK.

3. Open the Settings menu and select Configure Kmail.

4. Select the Network option on the left.

5. Under the Sending Mail tab, select the Sendmail option.

6. For the Incoming Mail option, click the Add button. Select POP3 and click the OK button. Type your name for Name. For login, type user1 and type `password` for password. Specify hostXXX-YY.acme.cxm. Click the OK button followed by clicking on the other OK button.

7. At the top of the Kmail Window, click on the Check Mail In button (fourth button).

8. Click on the message you got from root and click on the Reply button.

9. Type Hello and click on the Send button.

10. Log out and exit X Window.

11. Log in as root.

12. Set up root's Kmail.

13. Retrieve root's mail. Notice the additional email that root has been getting.

PART IV

MAINTAINING A NETWORK

CHAPTER 13

Network Security

Topics Covered in this Chapter

Introduction

One of the most important features of any network is network security. The network administrator is ultimately responsible for the network security. Modern network operating systems have many levels of security, and to make a network secure, you must look at all these levels. While the administrator is ultimately responsible for security, it is the job of everyone who uses the network to keep the network secure. And it is a job that must never stop.

Objectives

- Define threats, vulnerabilities and exploits and explain how they relate to each other.
- Establish guidelines for your user accounts, their passwords and how they can access the network.
- Explain how to establish a firewall and/or proxy server to protect your network.
- Define Denial of Service and explain how to protect your system against such attacks.

- Compare and contrast private-key and public-key encryption.
- Compare and contrast digital envelopes, digital signatures and digital certificates.
- Explain how a hash encryption protects data.
- Explain how patches can improve the security of your system.

13.1   SECURITY RISKS

By now, you should understand that security is a high concern and a major responsibility for administrators. Of course, when you examine your environment, you will need to assess the risks you currently face, determine an acceptable level of risk, and maintain risk at or below that level. Risks are reduced by increasing the security of your environment.

As a general rule, the higher the level of security in an organization, the more costly it is to implement. Unfortunately, at a higher level of security, you may reduce the functionality of the network. Sometimes, extra levels of security will result in more complex systems for users. However, if the authentication process is made too complex some customers will not bother to use the system, which could potentially cost more than the attacks the network suffers.

A **threat** is a person, place, or thing that has the potential to access resources and cause harm. The threats can be divided into three categories:

- **Natural and Physical**—includes fire, flooding, storms, earthquakes and power failures
- **Unintentional**—includes uniformed employees and customers
- **Intentional**—includes attackers, terrorists, spies and malicious code

A **vulnerability** is a point where a resource is susceptible to attack. It can be thought of as a weakness. Vulnerabilities are often categorized as:

- **Physical**—such as an unlocked door
- **Natural**—such as a broken fire suppression system or no UPS
- **Hardware and Software**—such as out of date antivirus software
- **Media**—such as electrical interference
- **Communication**—such as unencrypted protocols
- **Human**—such as insecure helpdesk procedures

An **exploit** is a type of attack on a resource that is accessed by a threat that makes use of a vulnerability in your environment. This type of attack is known as an exploit. The exploitation of resources can be performed in many ways. Some of the more common are given in the following table. When a threat uses a vulnerability to attack a resource, some severe consequences can result.

Countermeasures are deployed to counteract threats and vulnerabilities, therefore reducing the risk in your environment. For example, an organization producing fragile electronics may deploy physical security countermeasures such as securing equipment to the building's foundation or adding buffering mechanisms. These countermeasures reduce the likelihood that an earthquake could cause physical damage to their assets. Residual risk is what remains after all countermeasures have been applied to reduce threats and vulnerabilities.

To analyze all the ways a network can be attacked or accessed without proper permission, you use the OSI model to categorize the various methods. Most of the attacks that you hear about in the news occur at the application layer, going after web servers and browsers and the information they have access to, but application-layer attacks on open file shares are also common. (See figure 13.1).

While you can setup an array of defense against specific threats, please keep in mind that a defense could fail the first time that an unanticipated threat shows up. Therefore, you will have to constantly monitor your network.

13.2   ACCOUNT SECURITY

The user name and password constitute the first line of defense since every user needs to have them to get access to the network and its resources. For the password to be effective, the administrator should set guidelines and policies on how the user should use the network. The administrator also needs to provide training for using these guidelines and provide frequent reminders.

13.2.1   User Name Policy

As you know by now, the user name is the name of the account that represents the person or persons who need to access the network, and it is the account to which rights and permissions are given to allow access to those network resources on the network.

To make it easier to administer, the user name should follow a consistent naming convention for all users so that it will be easier to find names. For example, you can use the first initial, middle initial, and last name or the first name and last initial. Of course, if you

Figure 13.1   Types of Attacks as They Relate to the OSI Model

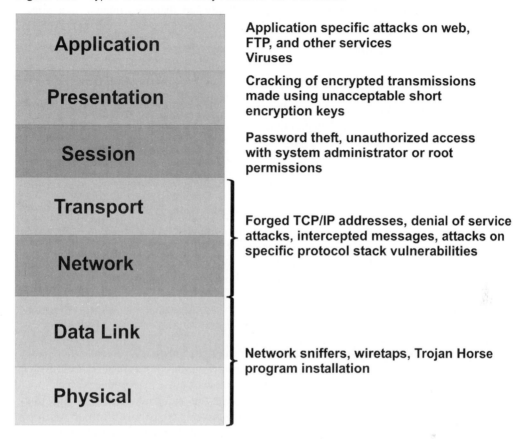

have hundreds of employees, you might need to use a convention like first name, middle initial, and last name to differentiate all the users. You should create a document stating the naming scheme of user accounts and provide this document to all administrators who might create accounts.

To identify the type of employee, add additional letters at the beginning or at the end of the person's name. For example, you can add a T- in front of the name or an X at the end of the name to indicate that the person is under contract or a temporary employee.

While the standard user names make it easier for the administrator, they also make it easier for the hacker. Therefore, to make the network a little bit more secure, you should rename the administrator accounts if possible. For Windows NT, Windows 2000, and Windows Server 2003 servers, the administrator account is administrator. For NetWare, it is admin, and for Linux, it is root.

Lastly, when doing everyday activities, you should not be logged in as the administrator of the server. This will help protect the system from carelessness, and it will restrict the damage that can be caused while logged in as that account. For example, a virus may spread onto the server and then to all the computers on the network you access while logged in as the administrator. If you are not logged in as the administrator, the rights and permissions to many of the files would be read-only and, therefore, immune to a virus.

13.2.2 Password Policies

When it comes to security, the password is more important than the user name. And there are very important things that can be done to make the password more secure. First, establish some common guidelines for the passwords. Second, educate your users to make sure they follow those guidelines and keep giving them reminders on using those passwords. These guidelines should include:

- Always require user names and passwords.
- Don't give your password to anyone.
- Change your password frequently.
- Keep people from seeing you type in your password.
- Do not write your password down near the computer.
- When you leave your computer unattended, log off.
- Use a password-protected screensaver.
- Do not use obvious passwords and do not use easy passwords.
- Use strong passwords.
- If you see a security problem or a potential security problem, report it to the network administrator.

Many of these make common sense. For example, not giving your password to anyone and preventing people from seeing you type in your password are good rules. To keep people from jumping on your computer while you walk away from your desk, you should get in the habit of logging off. And to help protect your computer when you forget to log off, you should use a password-protected screensaver. This is how it works: When you leave, after a short period of time, the screensaver will start. For someone to get past the screensaver, the person would have to type in a password. So this will automatically protect your system when you leave your computer.

One of the easiest ways to hack into a network is by exploiting weak passwords. You would be amazed how many people use easy passwords because they are convenient for the user and easy to remember. Unfortunately, any good hacker will try these first. A weak password would include:

- The word *password*
- The user name
- The person's first name, middle name, or last name
- The name of the person's pet
- The name of a family member
- Birthdates or anniversary dates
- Any text or label on the PC or monitor
- The company's name
- Your occupation
- Your favorite color

■ Typing keys on the keyboard or numeric keypad as they are placed on the keyboard (examples: qwerty, asdfgh, 123, or 147)

■ Any derivatives of the above such as spelling a name backwards

One program that hackers use to break into a server is a cracker program that uses a dictionary attack to figure out the password. A dictionary attack initially referred to finding passwords in a specific list, such as an English dictionary. Today, these lists have been expanded to also include a combination of words, such as *on the fly,* and common words with digits. The program tries by brute force by going through each item in the list until it figures out the password. Depending on the system, the password, and the skills of the attacker, such an attack can be completed in days, hours, or perhaps only a few seconds. With such programs easily available to anyone who wants them, all users should use strong passwords.

Strong passwords are difficult-to-crack passwords that do not have to be difficult to remember and include a combination of numbers, letters (lowercase and uppercase), and special characters. Special characters are those that cannot be considered letters or numbers such as @#$%^&*.

An example of a strong password is:

tqbf4#jotld

Such a password may look hard to remember, but it is not. This password was actually derived from a common typing phrase that includes every letter in the alphabet:

The quick brown fox jumped over the lazy dog.

Take the first letter of each word, put the number 4 and a pound (#) symbol in the middle, and you have a strong password.

To make passwords more secure, many network operating systems have password management features, which were either built into the operating system or installed as an add-on package. For example, some network operating systems have an option to enforce strong passwords.

Another option is to require a minimum number of characters for the password. A strong password should be at least 8 characters if not more. And it shouldn't be any longer than 15 characters because it would then be too difficult for people to remember and to create. Setting a minimum for the number of characters means that a cracker program must try a much larger number of combinations.

To avoid the passwords from being randomly guessed, some of the network operating systems offer an automatic account lockout. For example, you could set up a password such that if the wrong password is tried for an account three times within a 30-minute time period, the system will disable the account for a minimum amount of time or disable it permanently until a network administrator unlocks the account. In either case, such failure should also be logged somewhere. As the administrator, you should then check these logs on a regular basis to see if there are patterns indicating someone is trying to hack into a system without the knowledge of the user.

NOTE: If you are the administrator and you accidentally lock yourself out, you will need another administrator to unlock your account. If you are the only administrator, you may not be able to recover.

Another way to avoid passwords from being randomly guessed is to set up password expirations where an account will automatically expire on a certain date or after a certain amount of days or where the users are forced to change their passwords every so often. This will also help keep the network secure if someone discovers a password since it will be changed from time to time, making the previous password invalid. Most organizations set up passwords to expire every 30 to 45 days. After that, users must reset their passwords immediately or during the allotted grace period. Some systems give the user a few grace logins after the password has expired.

Lastly, some systems will require unique passwords. In this case, a system will remember a certain number of the last passwords used by a user and configure the system so that the user cannot use the same password over and over. When implementing a password history policy, be sure to make the password history large enough to contain at least a year's worth of password changes. For a standard 30-day life-span password, a history of 12 or 13 passwords will suffice.

Before moving on to user management, we should discuss login troubleshooting a little. If a person is having trouble logging in, you should always have the user check to see if the keyboard's caps lock is on. Most of the time, this is the problem, since passwords, and sometimes user names, are case-sensitive. If the user is still having problems, it is most likely easier to change the password for the user and have the user try again. If the problem still persists after changing the password, you should check the password against your encryption scheme. Sometimes, passwords could be sent in clear text when the program is expecting an encrypted password, or the password could be sent encrypted when the program is expecting a clear text password.

13.2.3 User Management

The next step in securing your network would be user management. For example, if you are hiring a temporary employee, most of the network operating systems allow you to create a temporary account and have it automatically disable itself after a certain amount of days or at a certain date. In addition, when someone leaves the company, you either disable or delete a user account.

NOTE: To be sure that the IT department is aware that a person leaves a company, make sure that the company has an exit processor for the employee that includes recovery of all IT equipment and also make sure that the IT department will disable or delete the account. In the same vein, don't forget to set up a checklist for when people leave the company and make sure to gather all keys, pagers, cell phones, company software, laptops, badges, and time cards.

Question:

A safety manager in your company was just fired. What should you do?

Answer:

The first reaction of most novice administrators would be to immediately delete the account. While this will prevent the person from accessing the network and causing all kinds of problems, it is probably not the best solution.

Before you decide, you should consider how much time and work goes into maintaining a user account. First, you have to create the account and fill in addresses, phone numbers, and titles. In addition, the manager has been assigned to groups and been given rights and permissions to files, printers, and other network resources. Therefore, it would be a better decision to disable the account. Then, when a person is hired to replace the safety manager, you can rename the user, reset the password, and reactivate the account. This will save you a lot of time and work. In addition, it will ensure that you didn't miss anything when assigning rights and permissions and that the person will be able to access all the old files.

If you decide to use an anonymous or guest account, you need to treat it with extreme care. Typically, it is not recommended to use such a generic account because you cannot audit it. But if you decide to use such an account, you must make sure that you only give bare minimum access to the network resources.

Next, you should limit the amount of concurrent connections or the number of times that a person can log in at one time. For most users, you should only have them log in once. So if they want to go to another machine to log into, they must first log out of their first machine. This helps avoid the problem of the person leaving a machine unattended.

Lastly, if you have the need, some network operating systems allow you to specify which computer a person can log in from and at what time the person can log in. Therefore, you can specify that someone cannot log in from 7 a.m. to 6 p.m. If a person tries to hack in after hours, that person will be automatically denied even if he or she has a valid user name and password.

13.2.4   Rights and Permissions

When planning the rights and permissions to the network resources, there are two main rules you should follow. First, give only the rights and permissions that are necessary for the individual users to do their job. For example, give them access to the necessary files and give them only the rights they need. For instance, if they need to read a document but don't need to make changes to the document, they only need to have read rights.

Second, make sure that the individual users do have sufficient rights and permissions for what needs to be done. While you want to keep these resources secure, you want to make sure that the users can easily get what they need.

Lastly, you need to understand how rights and permissions are assigned for an operating system to a user and a group. For example, if you assign permission to access a network resource to a group but deny an individual access to the network resource, you need to understand what the resultant rights and permissions will be.

You can secure files that are shared over the network in one of two ways:

- Implement share-level security.
- Implement user-level security.

In a network that uses share-level security, you assign passwords to individual files or other network resources (such as printers) instead of assigning rights to users. You then give these passwords to all users who need access to these resources. All resources are visible from anywhere in the network, and any user who knows the password for a particular network resource can make changes to it. With this type of security, the network support staff will have no way of knowing who is manipulating each resource. Share-level security is best used in smaller networks, where resources are more easily tracked. Windows 9X supports share-level security.

In a network that uses user-level security, rights to network resources (such as files, directories, and printers) are assigned to specific users who gain access to the network through individually assigned user names and passwords. Thus, only users who have a valid user name and password and have been assigned the appropriate rights to network resources can see and access those resources. User-level security provides greater control over who is accessing which resources because users are not supposed to share their user names and passwords with other users. User-level security is, therefore, the preferred method for securing files. Windows NT, Windows 2000, Windows 2003 Server, Windows XP, NetWare, UNIX, and Linux support user-level security.

13.3   PHYSICAL SECURITY

Another thing that most people forget is to secure computer servers and other key components. The server room is the work area of the IT department—this is where the servers and most of the communication devices reside. The room should be secure, with only a handful of people allowed to have access to it. In addition, the server should also be secure. Therefore, you should consider the following:

- The computer should be physically locked when not in use.
- The computer should have a BIOS password set so that you must enter the password.
- You should always log out when the server is not being used.
- You should use passwords on screensavers.

NOTE: Do not select complex screensavers that will take a lot of processing away from the network services.

You should, of course, use a locked door into the server room, and you may consider installing camera equipment to monitor the server room and/or entrance into the server room.

If you have several servers, you should also consider a server rack to hold the servers. In addition, you can also purchase a switch box to connect a single keyboard, mouse, and monitor to several servers. This will allow for less equipment and for a more organized work environment.

13.4   FIREWALLS AND PROXY SERVERS

While you can set up an array of defenses against specific threats, please keep in mind that a defense could fail the first time that an unanticipated threat shows up. Therefore, you will have to constantly monitor your network.

To help monitor the network, two main approaches have proved extremely valuable:

■ **Packet filters**—You can examine traffic at the network layer, looking at the source and destination addresses. The filter can disallow traffic to or from specific addresses or address ranges and can disallow traffic with suspect address patterns.
■ **Firewalls**—You can also examine traffic as high as the application layer, checking ports in message addresses or even checking the internal content of specific application messages. Traffic that fails any of those tests can be rejected.

13.4.1   Packet Filters

As you may recall, TCP/IP addresses are composed of both a machine address and, within the machine, a port number identifying the program to handle the message. The combined address/port information is available in every TCP/IP message (with the exception of broadcasts and some messages exchanged while a TCP/IP address is being assigned via DHCP) and is available for both the sender and the receiver of the message.

Packet filters are based on protecting the network by using an access control list (ACL). This list resides on your router and determines which machine (IP address) can use the router and in what direction. Typically, the router will permit all outgoing traffic. But it will deny any new incoming connections. If a machine has established a connection with a machine on the outside, it will accept those packets. Therefore, it will reject unsolicited connection attempts from the Internet to your computer.

If your packet filter software is capable enough to examine the subnet of the source address based on which physical port delivers the message to the router, you can set up rules to avoid spoofed TCP/IP addresses. The idea behind spoofing is for messages from the Internet to appear to have originated from your LAN; the spoofing filter prevents this by rejecting messages coming on a port with impossible source addresses.

The antispoofing filter is an important part of protecting machines on your network on which you've installed filters to limit particular services to machines on your subnet. For instance, suppose you've installed software on a Linux machine to act as a Windows network file server. You can configure Linux to reject all network traffic originating outside your subnet, preventing computers on the Internet from seeing the file server. If an attacker could pretend to be on your LAN, that safeguard would be bypassed. Defeat that attack with the antispoofing filter.

The antispoofing filter, also known as Egress and Ingress Filtering, will only route outgoing packets if they have a valid internal IP address. By rule, your routers should disregard and drop any outgoing packets that hasn't originated from a valid internal IP address. When you perform this setting change, you will effectively prevent your network from becoming a participant in any spoofing attack. These address filterings are:

Historical Low End Broadcast: 0.0.0.0/8
Limited Broadcast: 255.255.255.255/32
RFC 1918 Private Networks: 10.0.0.0/8
RFC 1918 Private Networks: 172.16.0.0/12
RFC 1918 Private Networks: 192.168.0.0/16

The Loop Back Address: 127.0.0.1
Link Local Networks: 169.254.0.0/16
Class D Addresses: 224.0.0.0/4
Class E Reserved Address: 240.0.0.0/5
Unallocated Address: 248.0.0.0/5

There may be other addresses that need to be blocked by your router. However, the addresses listed above should provide good protection against DoS and/or spoofing attacks.

One disadvantage of a packet filter is that it generally doesn't protect against attacks using UDP. This is because there is no formal connection opened with UDP, as there is with TCP, and therefore the filter cannot reject the opening message.

Besides analyzing the source and destination TCP/IP address, it can also examine the source and destination port numbers and the content of the packet data. This gives it far more power such as it can allow or disallow specific application services such as FTP or web pages and it can allow or disallow access to services based on the content of the information being transferred. You can even combine these functions such as allowing incoming FTP access from the Internet, but only to a specific, designated server.

When you decide which port numbers to block, you need to choose with care. If you decide to block a specific port, any service or application that uses a specific port will not function through the firewall. For example, if you block port 80, you will not be able to contact HTTP web servers that are using the default port 80.

13.4.2   Firewalls

A firewall takes the packet filter one step further. Besides analyzing the source and destination TCP/IP address, it can also examine the source and destination port numbers and the content of the packet data. This gives it far more power—for example, it can allow or disallow specific application services such as FTP or web pages, and it can allow or disallow access to services based on the content of the information being transferred. You can even combine these functions, such as allowing incoming FTP access from the Internet, but only to a specific, designated server.

A packet-filtering firewall can use additional authentication measures, aside from using just the IP addresses. In addition, a packet-filtering firewall may also do what is known as network-address translation or masquerading. This means that the firewall converts the source addresses on outgoing packets so the other host thinks it's connected to the firewall itself, and it converts the destination addresses on incoming packets so they go to the host that requested the connection initially. This has several advantages. For example, a whole network may connect to the Internet using a single IP address. Also, because internal addresses aren't visible to the outside, hosts on the internal network are more secure against attacks from the outside. The downside to this is that any hosts that provide services to the outside must be outside the firewall; otherwise, the firewall must be specifically configured to pass some packets straight to a particular host.

If you set a firewall between the Internet and your private network, it provides no good place to locate publicly accessible servers. If you place the servers out on the Internet in front of the firewall, they will be unprotected. If you have them behind the firewall, you have to create holes in the firewall protection to permit access to the servers. It is those holes that can be exploited on the rest of the network.

13.4.3   Demilitarized Zone

To get around this problem, you create a demilitarized zone (DMZ). Instead of having two ports: the public network on one port of the firewall and the private network on another port, you have three ports. The third port is the DMZ, which is a less secure area than the private network. The DMZ will include the web and FTP servers, and the rest of your computers will be in the private network. (See figure 13.2.)

Each segment—the public, the private, and the DMZ—that is connected to the firewall has a different set of security rules. The private and the public are no different from what we discussed before, On the DMZ LAN, you still want to use antispoofing filters, limit the allowable ports to those used by the servers on the LAN, and disallow access from known attacking sites.

NOTE: Remember that firewalls themselves aren't a guarantee of security. You will have to get in the habit of examining firewall logs for suspicious events, and you have to be vigilant about discovering and applying software security patches.

Figure 13.2   A DMZ

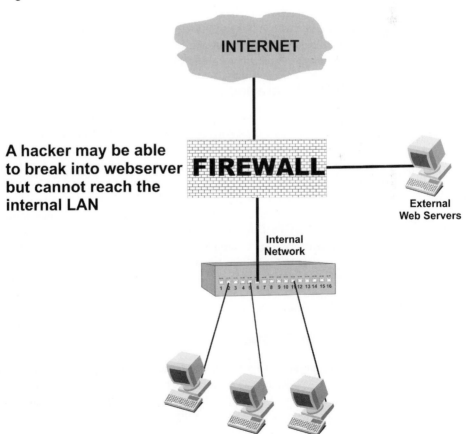

13.4.4   Static and Dynamic Packet Filters

Some firewalls are considered static packet filters, whereas others are considered dynamic packet filters. The static packet filter has a packet-filtering mechanism that allows you to set rules based on protocol and port number to control inbound and outbound access on the external interface.

The dynamic packet filter often referred to as stateful inspection, is a firewall facility that can monitor the state of active connections and use this information to determine which network packets to allow through the firewall. By recording session information such as IP addresses and port numbers into a dynamic state list (also known as a state table), a dynamic packet filter can implement a much tighter security posture than a static packet filter.

For example, assume that you wish to configure your firewall so that all users in your company are allowed out to the Internet, but only replies to users' data requests are let back in. With a static packet filter, you would need to permanently allow in replies from all external addresses, assuming that users were free to visit any site on the Internet. This kind of filter would allow an attacker to sneak information past the filter by making the packet look like a reply (which can be done by indicating "reply" in the packet header). By tracking and matching requests and replies, a dynamic packet filter can screen for replies that don't match a request. When a request is recorded, the dynamic packet filter opens up a small inbound hole so only the expected data reply is let back through. Once the reply is received, the hole is closed. This dramatically increases the security capabilities of the firewall, such as preventing someone from replaying a packet to get access because it is not part of an active connection.

13.4.5   Circuit-level Gateway

While the Application gateway operates on the OSI model's Application layer, the circuit level gateway operates on the network level of the OSI model. A circuit-level gateway translates IP addresses between the Internet and your internal systems. The gateway receives outbound packets and transfers them from the internal network to the external network. Inbound traffic is transferred from the outside network to the internal network.

Circuit-level gateways provide a complete break between your internal network and the Internet. Unlike a packet filter, which simply analyzes and routes traffic, a circuit-level gateway translates packets and transfers them between network interfaces. This helps shield your network from external traffic since the packets appear to have originated from the circuit-level gateway's Internet IP address.

Circuit-level application gateways often provide network address translation (NAT), in which a network host alters the packets of internal network hosts so they can be sent out across the Internet. NAT was explained in chapter 6.

13.4.6   Application Gateway

Application-level gateways take requests for Internet services and forward them to the actual services. Application-level gateways sit between a user on the internal network and a service on the Internet. Instead of talking to each other directly, each system talks to the gateway. Your internal network never directly connects to the Internet.

Application-level gateways help improve security by examining all Application layers, bring context information into the decision process. However, they do this by breaking the client/server model into two connections: one from the client to the firewall and one from the firewall to the server. Unfortunately, since each application gateway requires a different application process, or daemon, you will need to add a different application processor or daemon to support new applications.

13.4.7   Proxy Server

In chapter 6, the proxy server was discussed. If you recall, a proxy server is a server that sits between a client application, such as a web browser, and a real server. It intercepts all requests to the real server to see if it can fulfill the requests itself. If not, it forwards the requests to the real server. If it can fulfill the request itself, the user gets increased performance. In addition, it provides a means for sharing a single Internet connection among a number of workstations. While this has practical limits in performance, it can still be a very effective and inexpensive way to provide Internet services, such as email, throughout an office.

Besides increasing performance, a proxy server sits between a client program (typically a web browser) and some external server (typically another server on the web), which acts as an application-level firewall.

NOTE: Proxy servers are often used with firewalls and sometimes are the same machine or device as a firewall.

The proxy server can monitor and intercept any and all requests that are being sent to the external server or that come in from the Internet connection. Therefore, besides the proxy server improving performance and sharing connections, it can also filter requests.

Filtering requests is the security function and the original reason for having a proxy server. Proxy servers can inspect all traffic (in and out) over an Internet connection and determine if there is anything that should be denied transmission, reception, or access. Since this filtering cuts both ways, a proxy server can be used to keep users out of particular websites (by monitoring for specific URLs) or restrict unauthorized access to the internal network by authenticating users. Before a connection is made, the server can ask the user to log in. To a web user this makes every site look like it requires a login. Because proxy servers are handling all communications, they can log everything the user does. For HTTP (web) proxies this includes logging every URL. For FTP proxies this includes every downloaded file. A proxy can also examine the content of transmissions for "inappropriate" words or scan for viruses, although this may impose serious overhead on performance.

13.4.8   Configuring a Browser through a Proxy Server

To the user, the proxy server is *almost* invisible. All Internet requests and returned responses appear to be made directly with the addressed Internet server. The reason the proxy server is not totally invisible is that the IP address of the proxy server has to be specified as a configuration option to the browser or other protocol program.

Figure 13.3 Configuring the proxy selection and proxy bypass settings in Internet Explorer

For the browser to go through a proxy server, the browser must be configured. These configuration settings include automatic configuration, configuration through scripts, or configuration by manually specifying the settings. Automatic configuration and configuration through scripts both enable you to change settings after you deploy Internet Explorer. By providing a pointer to configuration files on a server, you can change the settings globally without having to change each user's computer. This can help reduce administrative overhead and potentially reduce help desk calls about browser settings.

To configure the proxy selection and proxy bypass settings in Internet Explorer:

1. Open the Tools menu and then click Internet Options.
2. Click the Connections tab and then click LAN Settings.
3. In the Proxy Server area, select the Use a proxy server check box.
4. Click Advanced and then fill in the proxy location and port number for each Internet Protocol that is supported.

Also see figure 13.3.

NOTE: In most cases, only a single proxy server is used for all protocols. In those cases, enter the proxy location and port number for the HTTP setting, and then select the Use the same proxy server for all protocols check box. If you want to manually set the addresses, enable the Use a proxy server and specify the address and port number of the proxy server. If you need to specify different addresses and/or port numbers for the various Internet services, click on the Advanced button.

For Netscape, the same basic options are available if you:

1. Open the Edit menu and select the Preferences option.
2. Find and open the Advanced option.
3. Click on the Proxy option.

Also see figure 13.4.

Figure 13.4   Configuring the proxy settings for Netscape

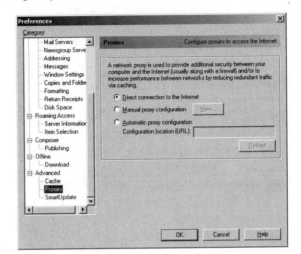

13.4.9   Advanced Firewall Features

Most firewalls today are a hybrid of stateful inspection, circuit-level gateways, and application-level gateways. Only packets dealing with acceptable activities are allowed in and out of your internal network. Some firewalls provide advanced firewalls features that make them more effective in your perimeter.

The firewall is a logical place to install authentication mechanism to help overcome the limitations of TCP/IP. You can also use a reverse lookup on an IP address to try and verify that the user is actually at his or her reported location. This activity helps identify and prevent spoofing attacks.

Firewalls can also provide user authentication. Some application-level gateways contain an internal user account database or integrate with UNIX and Windows domain accounts. These accounts can be used to limit activities and services by user or provide detailed logs of user activity. Because individual users can be easily identified, more granular rule sets can be implemented.

You can also use authentication for remote access. Allowing employees to connect to the internal network from home or while traveling increases productivity, but how can you ensure that the person making the request is who he or she is supposed to be. Most firewalls support third-party authentication methods to provide strong authentication.

Lastly, almost every firewall performs some type of logging. Most packet filters do not enable logging by default because it degrades performance of traffic analysis. So if you want logging, make sure it is enabled and active. Extensive logging can help you track and potentially capture a hacker. Since your firewall is the single point of entry to your network, an attacker will have to pass through it. Your log files should provide information to show you what the hacker was up to on your systems.

13.4.10   Personal Firewalls

Today, many operating systems offer personal firewalls to protect an individual computer when connecting to the Internet. For example, Windows XP has the Internet Connection Firewall (ICF) (see figure 13.5), and Red Hat Linux has GNOME Lokkit. Much like a full-blown firewall, a personal firewall restricts what information is communicated from your home or small office network to the Internet and from the Internet to your computer.

Most of the personal firewalls are considered "stateful" firewalls. A stateful firewall both monitors all aspects of the communications that cross its path and inspects the source and destination address of each message it handles. To prevent unsolicited traffic from the public side of the connection from entering the private side, the private firewall keeps a table of all communications that have originated from the computer. Some of these firewalls can be used in conjunction with Internet sharing, which in turn will keep track of all traffic originating from the computers sharing the Internet link. All inbound traffic from the Internet is compared against the entries in the table. Inbound Internet traffic is only allowed to reach the computers in your network when there is a matching entry in the table that shows that the communication exchange began from within your computer or private network. Communications that originate from a source on the Internet are dropped by the firewall unless an entry using the configuration options of the personal firewall is made to allow passage.

With these personal firewalls, the firewall can be configured to allow unsolicited traffic from the Internet to be forwarded by the computer to the private network. For example, if you are hosting an HTTP web server service and have enabled the HTTP service on your ICF computer, unsolicited HTTP traffic will be forwarded by the computer with the personal firewall to the HTTP web server.

Figure 13.5   Internet Connection Firewall (personal firewall) used in Windows XP

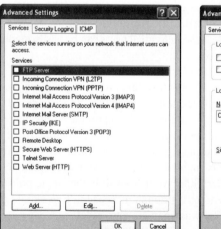

13.5   DENIAL OF SERVICE

Denial of service (DoS) is a type of attack on a network that is designed to bring the network to its knees by flooding it with useless traffic. Many DoS attacks, such as the Ping of Death and Teardrop attacks, exploit limitations in the TCP/IPs. In the worst cases, for example, a website accessed by millions of people can occasionally be forced to temporarily cease operation. A denial-of-service attack can also destroy programming and files in a computer system. Although usually intentional and malicious, a denial-of-service attack can sometimes happen accidentally. (See table 13.1.)

For all known DoS attacks, there are software fixes that system administrators can install to limit the damage caused by the attacks. But like viruses, new DoS attacks are constantly being dreamed up by hackers.

Distributed Denial of Service (DDoS) attacks involve installing programs known as zombies on various computers in advance of the attack. A command is issued to these zombies, which launch the attack on behalf of the attacker, thus hiding their tracks. The zombies themselves are often installed using worms. The real danger from a DDoS attack is that the attacker uses many victim computers as host computers to control other zombies that initiate the attack. When the system that is overwhelmed tries to trace back the attack, it receives a set of spoofed addresses generated by a series of zombies.

The following defensive steps will help you prevent these types of attacks:

- Keep systems updated with the latest security patches.
- Block large ping packets at the router and firewall, stopping them from reaching the perimeter network.
- Apply anti-spoof filters on the router; that is, block any incoming packet that has a source address equal to an address on the internal network.
- Filter the ICMP messages on the firewall and router (although this could affect some management tools).
- Develop a defense plan with your Internet service provider (ISP) that enables a rapid response to an attack that targets the bandwidth between your ISP and your perimeter network.
- Disable the response to directed broadcasts.
- Apply proper router and firewall filtering.
- Use an intruder detection system to check for unusual traffic and generate an alert if it detects any. Configure the system to generate an alert if it detects ICMP_ECHOREPLY without associated ICMP_ECHO packets.
- Each week, more DoS attacks are documented and added to bug tracking databases. You should ensure that you always remain current on these attacks and how you can guard against them.

13.6   ENCRYPTION AND DECRYPTION

In networks, sensitive data exists as it is transmitted across the network, sent as email messages, and stored on files on the disk. **Encryption** is the process of disguising a message or data in what appears to be meaningless data (cipher text) to hide and protect the sensitive

Table 13.1 Forms of Denial-of-Service Attacks

Form	Description
Buffer overflow attacks	The most common kind of DoS attack is simply to send more traffic to a network address than the programmers who planned its data buffers anticipated someone might send. The attacker may be aware that the target system has a weakness that can be exploited, or the attacker may simply try the attack in case it might work. A few of the better-known attacks based on the buffer characteristics of a program or system include: ■ Sending email messages that have attachments with 256-character file names to Netscape and Microsoft mail programs ■ Sending oversized Internet Control Message Protocol (ICMP) packets (this is also known as the Ping of Death) ■ Sending to a user of the Pine email program a message with a "From" address larger than 256 characters
SYN attack	When a session is initiated between the Transport Control Program (TCP) client and server in a network, a very small buffer space exists to handle the usually rapid "handshaking" exchange of messages that sets up the session. The session-establishing packets include a SYN field that identifies the sequence in the message exchange. An attacker can send a number of connection requests very rapidly and then fail to respond to the reply. This leaves the first packet in the buffer so that other, legitimate connection requests can't be accommodated. Although the packet in the buffer is dropped after a certain period of time without a reply, the effect of many of these bogus connection requests is to make it difficult for legitimate requests for a session to get established. In general, this problem depends on the operating system providing correct settings or allowing the network administrator to tune the size of the buffer and the timeout period.
Teardrop attack	This type of denial-of-service attack exploits the way that the Internet Protocol (IP) requires a packet that is too large for the next router to handle to be divided into fragments. The fragment packet identifies an offset to the beginning of the first packet that enables the entire packet to be reassembled by the receiving system. In the Teardrop attack, the attacker's IP puts a confusing offset value in the second or later fragment. If the receiving operating system does not have a plan for this situation, it can cause the system to crash.
Smurf attack	In this attack, the perpetrator sends an IP ping (or "echo my message back to me") request to a receiving site. The ping packet specifies that it be broadcast to a number of hosts within the receiving site's local network. The packet also indicates that the request is from another site, the target site that is to receive the denial of service. (Sending a packet with someone else's return address in it is called spoofing the return address.) The result will be lots of ping replies flooding to the innocent, spoofed host. If the flood is great enough, the spoofed host will no longer be able to receive or distinguish real traffic.

Continued

Table 13.1   Continued

Form	Description
WinNuke	WinNuke is a Windows program that sends special TCP/IP packets with an invalid TCP header. Windows will crash when it receives one of the packets because of the way the Windows TCP/IP stack handles bad data in the TCP header. Instead of returning an error code or rejecting the bad data, it sends the computer to the Blue Screen of Death.
Viruses	Computer viruses, which replicate across a network in various ways, can be viewed as DoS attacks where the victim is not usually specifically targeted but simply a host unlucky enough to get the virus. Depending on the particular virus, the denial of service can range from being hardly noticeable all the way through disastrous.
Physical infrastructure attacks	Here someone may simply snip a fiber-optic cable. This kind of attack is usually mitigated by the fact that traffic can sometimes quickly be rerouted.

data from unauthorized access. **Decryption** is the process of converting data from encrypted format back to its original format. **Cryptography** is the art of protecting information by transforming it (encrypting it) into cipher text.

There are, in general, three types of cryptographic schemes typically used to accomplish these goals:

- **private-key (also known as secret key or symmetric) cryptography**—Uses a single key for both encryption and decryption
- **public-key (or asymmetric) cryptography**—Uses one key for encryption and another key for decryption
- **hash functions**—Uses a mathematical transformation to irreversibly encrypt information

To encrypt and decrypt a file, you must use a key. A **key** is a string of bits used to map text into a code and a code back to text. You can think of the key as a super-decoder ring used to translate text messages to a code and back to text. There are two types of keys: private keys and public keys.

13.6.1   Private-Key Encryption

The most basic form of encryption is the **private-key encryption.** In private-encryption, a single key is used for both encryption and decryption. The sender uses the key (or some set of rules) to encrypt the plain text and sends the cipher text to the receiver. The receiver applies the same key (or ruleset) to decrypt the message and recover the plain text. Because a single key is used for both functions, private-key encryption is also called symmetric encryption.

This requires that each individual must possess a copy of the key. Of course, for this to work as intended, you must have a secure way to transport the key to other people. Since you must keep multiple single keys per person, this can get very cumbersome. Private-key algorithms are generally very fast and easily implemented in hardware. Therefore, they are commonly used for bulk data encryption.

There are two general categories of private-key algorithms: block and stream cipher. A block cipher encrypts one block of data at a time. A stream cipher encrypts each byte of the data stream individually.

13.6.2   Public-Key Encryption

Public-key encryption, also known as the asymmetric algorithm, uses two distinct but mathematically related keys, public and private. The public key is the nonsecret key that is available to anyone you choose, or is made available to everyone by posting it in a public place. It is often made available through a digital certificate. The private key is kept in a secure location used only by you. When data needs to be sent, it is protected with a secret-key encryption that was encrypted with the public key of the recipient of the data. The encrypted secret key is then transmitted to the recipient along with the encrypted data. The recipient will use the private key to decrypt the secret key. The secret key will then be used to decrypt the message itself.

For example, say you want to send data to someone. You would retrieve his or her public key and encrypt the data. You encrypt the data and the secret key, and send both the data and the secret key. Since the recipient's private key is the only thing that can decrypt the secret key, which is the only thing that can decrypt the message, the data can be sent over an insecure communications channel.

While public keys may be stored anywhere for easy retrieval by computers, your private key should be carefully guarded. Since it is not convenient (or possible) for most folks to commit these keys to memory, they are stored on your personal computer's hard disk, encrypted using conventional cryptography, and a password. Thus, anyone in possession of your computer would still not be able to use your private key since it is scrambled with your password.

A disadvantage of public-key encryption is that it's slower than private-key encryption because it requires more overhead.

13.6.3   Digital Envelopes, Signatures, and Certificates

A **digital envelope** is a type of security that encrypts the message using symmetric encryption and encrypts the key to decode the message using public-key encryption. This technique overcomes the problem of public-key encryption being slower than symmetric encryption, because only the key is protected with public-key encryption, providing little overhead.

A **digital signature** is a digital code that can be attached to an electronically transmitted message that uniquely identifies the sender. Like a written signature, the purpose of a digital signature is to guarantee that the individual sending the message really is who he or she claims to be.

A **digital certificate** is an attachment to an electronic message used for security purposes such as for authentication. It is also used to verify that a user sending a message is who he or she claims to be and to provide the receiver with a means to encode a reply. An individual wishing to send an encrypted message applies for a digital certificate from a **Certificate Authority (CA).** The CA issues an encrypted digital certificate containing the applicant's public key and a variety of other identification information. The CA makes its own public key readily available through print publicity or the Internet. The recipient of an encrypted message uses the CA's public key to decode the digital certificate attached to the message, verifies it as issued by the CA, and then obtains the sender's public key and identification information held within the certificate. With this information, the recipient can send an encrypted replay.

13.6.4   Hash Encryption

Hash encryption, also called message digests and one-way encryption, uses algorithms that don't really use a key. Instead, it converts data from a variable length to a fixed length piece of data called a hash value. Theoretically, no two data will produce the same hash value. The hash process is irreversible. You cannot recover the original data by reversing the hash process.

Users must often utilize hashes, when there is information they never want decrypted or read. Hash algorithms are typically used to provide a digital fingerprint of a file or message contents, often used to ensure that the file has not been altered by an intruder or virus. Hash functions are also commonly employed by many operating systems to encrypt passwords.

Among the most common hash functions in use today are a family of Message Digest (MD) algorithms, all of which are byte-oriented schemes that produce a 128-bit hash value from an arbitrary-length message.

13.6.5   RSA and Data Encryption Standards

The **RSA standard** (created by Ron *R*ivst, Adi *S*hamir, and Leonard *A*dleman) defines the mathematical properties used in public-key encryption and digital signatures. The key length for this algorithm can range from 512 to 2048, making it a very secure encryption algorithm. This algorithm uses a number known as the public modulus to come up with the public and private keys. The number is formed by multiplying two prime numbers. The security of this algorithm derives from the fact that while finding large prime numbers is relatively easy, factoring the result of multiplying two large prime numbers is not easy. If the prime numbers used are large enough, the problem approaches being computationally impossible. The RSA algorithm has become the de facto standard for industrial-strength encryption, especially for data sent over the Internet.

The **Data Encryption Standard (DES)** was developed in 1975 and was standardized by ANSI in 1981 as ANSI X.3.92. It is a popular symmetric-key encryption method that uses block cipher. The key used in DES is based on a 56-bit binary number, which allows for 72,057,594,037,927,936 encryption keys. Of these 72 quadrillion encryption keys, a key is chosen at random. If you want to send an encrypted file from one person (source person) to another person (target person), the source person will encrypt the secret key with the target person's public key, which was obtained from his or her certificate. Because the target person's key was used to encrypt the secret key, only the target person using his or her private key will be able to decrypt the DES secret key and decrypt the DES-encrypted data.

NOTE: The U.S. government has banned the export of DES outside of the United States.

Triple DES is a stronger alternative to regular DES, and it is used extensively in conjunction with virtual private network implementations. Triple DES encrypts a block of data using a DES secret key. The encrypted data is encrypted again using a second DES secret key. Finally, the encrypted data is encrypted a third time using yet another secret key. Triple DES is of particular importance as the DES algorithm keeps being broken in shorter and shorter times.

The replacement for DES might be the NSA's recent algorithm called skipjack, officially called the Escrowed Encryption Standard (EES). It is defined in FIPS 185, and uses an 80-bit key rather than the DES 56-bit key. Skipjack was supposed to be integrated into a clipper chip.

A clipper chip is an encryption chip designed under the auspices of the U.S. government. The government's idea was to enforce use of this chip in all devices that might use encryption, including computers, modems, telephones, and televisions. The government would control the encryption algorithm, thereby giving it the ability to decrypt any messages it recovered. The purported goal of this plan was to enable the U.S. government to carry out surveillance on enemies of the state even if they used encryption to protect their messages. However, the clipper chip created a fierce backlash from both public interest organizations and the computer industry in general. The government eventually retracted its original plan but has since promoted two other plans called Clipper 2 and Clipper 3, respectively. The Clipper 3 plan allows the use of any encryption technology but stipulates that government law enforcement agencies be able to recover any keys exported out of the country.

Another type of encryption worth mentioning is the **Pretty Good Privacy (PGP).** The name denotes that nothing is 100 percent secure. PGP is one of the most common ways to protect messages on the Internet because it is effective, easy to use, and free. PGP is based on the public-key method, which, as noted earlier, uses two keys: a public key that you disseminate to anyone from whom you want to receive a message and a private key that you use to decrypt messages you receive.

13.6.6   X.509 Certificates

X.509 certificates are the most widely used digital certificates.

NOTE: X.509 is actually an ITU recommendation, which means that it hasn't been officially defined or approved.

An X.509 certificate contains information that identifies the user, as well as information about the organization that issued the certificate, including the serial number, validity period, issuer name, issuer signature, and subject (or user) name. The subject can be an individual, a business, a school, or some other organization, including a certificate authority. The X.509 describes two levels of authentication: simple authentication based on a password to verify user identity and strong authentication using credentials formed using cryptographic techniques. Of course, it is recommended to use the strong authentication to provide secure services.

13.7   SECURITY PROTOCOLS

Authentication, which is the layer of network security, is the process by which the system validates the user's logon information. Authentication is crucial to secure communication. Users must be able to prove their identity to those with whom they communicate and must be able to verify the identity of others. Typically when a user logs on, the user is then authenticated for all network resources that the user has permission to use. The user only has to log on once although the network resources may be located throughout several computers.

13.7.1   NTLM Authentication

In **NTLM authentication** used in Windows NT domains, the client selects a string of bytes, uses the password to perform a one-way encryption of the string, and sends both the original string and the encrypted one to the server. The server receives the original string and uses the password from the account database to perform the same one-way encryption. If the result matches the encrypted string sent by the client, the server concludes that the client knows the user name/password pair.

13.7.2   Kerberos Authentication

Kerberos employs a client/server architecture and provides user-to-server authentication rather than host-to-host authentication developed at MIT for the TCP/IP network. Its purpose is to allow users and services to authenticate themselves to each other without allowing other users to capture the network packet on the network and resending it so that they can be authenticated over an insecure media such as the Internet. Therefore, Kerberos provides strong authentication for client/server applications by using symmetric secret-key cryptography. After a client and server have used Kerberos to prove their identity, they can also encrypt all of their communications to assure privacy and data integrity as they go about their business.

With Kerberos, security and authentication is based on secret key technology where every host on the network has its own secret key. Of course, it would be clearly unmanageable if every host had to know the keys of all other hosts so a secure, trusted host somewhere on the network, known as a Key Distribution Center (KDC), knows the keys for all of the hosts (or at least some of the hosts within a portion of the network, called a realm). In this way, when a new node is brought, only the KDC and the new node need to be configured with the node's key. Keys can be distributed physically or by some other secure means.

The Kerberos Server/KDC has two main functions, known as the Authentication Server (AS) and Ticket-Granting Server (TGS). When a user requests a service, his or her identity must be established. To do this, a ticket is presented to the server, along with proof that the ticket was originally issued to the user, not a captured network packet containing the password or ticket that is resent.

When a user logs in, the client transmits the user name to the authentication server, along with the identity of the service the user desires to connect to, for example, a file server. The authentication server constructs a ticket, which contains a randomly generated key, encrypted with a file server's secret key, and sends it to the client as part of its credentials, which includes the session key encrypted with the client's key. If the user typed the right password, then the client can decrypt the session key, present the ticket to the file server, and use the shared secret session key to communicate between them. Tickets are time stamped, and typically have an expiration time of only a few hours.

If the client specifies that it wants the server to prove its identity too, the server adds one of the timestamps the client sent in the authenticator, encrypts the result in the session key, and sends the result back to the client. At the end of the exchange, the server is certain that the client is who it says it is. If mutual authentication occurs, the client is also convinced that the server is authentic. Moreover, the client and server shares a key which no one else knows, and can safely assume that a reasonable recent message encrypted in that key originated with the other party.

As you can see through out the entire process of getting authentication the password was never sent on the wire. Therefore, a packet that contains the password cannot be compromised. In addition, many of these tickets and requests are time stamped, to keep track of when the packet was made so that packets are not captured and resent at a later time. Lastly, the encryption algorithm is different for every client.

The current shipping version of this protocol is Kerberos V5 (described in RFC 1510), although Kerberos V4 still exists and is seeing some use. While the details of their operation, functional capabilities, and message formats are different, the conceptual overview above pretty much holds for both. One primary difference is that Kerberos V4 uses only DES to generate keys and encrypt messages, while V5 allows other schemes to be employed (although DES is still the most widely algorithm used).

13.7.3   SSL and S-HTTP

Most companies with connections to the Internet have implemented firewall solutions to protect the corporate network from unauthorized users coming in—and unauthorized Internet sessions going out—via the Internet. But while firewalls guard against intrusion from unauthorized use, they can guarantee neither the security of the link between a workstation and server on opposite sides of a firewall nor the security of the actual message being conveyed.

To create this level of security, two related Internet protocols, Secure Sockets Layer (SSL) and Secure HyperText Transfer Protocol (S-HTTP), ensure that sensitive information passing through an Internet link is safe from prying eyes. Whereas SSL is designed to establish a secure connection between two computers, S-HTTP is designed to send individual messages securely. Since SSL and S-HTTP have very different designs and goals, it is possible to use the two protocols together.

Secure Sockets Layer (SSL) operates at the transport layer, such as at the HTTP, FTP, NNTP, and SMTP level. When both a client (usually in the form of a web browser) and a server support SSL, any data transmitted between the two becomes encrypted. Microsoft Internet Explorer and Netscape Navigator both support SSL at the browser level.

When an SSL client wants to communicate with an SSL-compliant server, the client initiates a request to the server, which in turn sends an X.509 standard certificate back to the client. The certificate includes the server's public key and the server's preferred cryptographic algorithms, or ciphers.

The client then creates a key to be used for that session, encrypts the key with the public key sent by the server, and sends the newly created session key to the server. After it receives this key, the server authenticates itself by sending a message encrypted with the key back to the client, proving that the message is coming from the proper server. After this handshake process, which results in the client and server agreeing on the security level, all data transfer between that client and that server for a particular session is encrypted using the session key. When a secure link has been created, the first part of the URL will change from http:// to https://. SSL supports several cryptographic algorithms to handle the authentication and encryption routines.

Transport Layer Security (TLS) is the successor to SSL. TLS is composed of two layers: the TLS Record Protocol and the TLS Handshake Protocol. The TLS Record Protocol provides connection security with some encryption method such as the Data Encryption Standard (DES). The TLS Record Protocol can also be used without encryption. The TLS Handshake Protocol allows the server and client to authenticate each other and to negotiate an encryption algorithm and cryptographic keys before data is exchanged.

The TLS Protocol is based on Netscape's SSL 3.0 Protocol; however, TLS and SSL are not interoperable. The TLS Protocol does contain a mechanism that allows TLS implementation to back down to SSL 3.0. The most recent browser versions support TLS. The TLS Working Group, established in 1996, continues to work on the TLS Protocol and related applications.

Secure Hypertext Transfer Protocol (S-HTTP) is simply an extension of HTTP. S-HTTP is created by SSL running under HTTP. The protocol was developed as an implementation of the RSA encryption standard. While SSL operates at the transport layer, S-HTTP supports secure end-to-end transactions by adding cryptography to messages at the application layer. Of course, while SSL is application-independent, S-HTTP is tied to HTTP.

An S-HTTP message consists of three parts:

- The HTTP message
- The sender's cryptographic preferences
- The receiver's preferences

The sender integrates both preferences, which results in a list of cryptographic enhancements to be applied to the message.

S-HTTP, like SSL, can be used to provide electronic commerce without customers worrying about who might intercept their credit card number or other personal information. To decrypt an S-HTTP message, the recipient must look at the message headers, which designate which cryptographic methods were used to encrypt the message. Then, to decrypt the message, the recipient uses a combination of his or her previously stated and current cryptographic preferences and the sender's previously stated cryptographic preferences. S-HTTP doesn't require that the client possess a public-key certificate, which means secure transactions can take place at any time without individuals needed to provide a key (as in session encryption with SSL).

13.7.4   Security Protocols for Virtual Tunnels

A **virtual private network (VPN)** uses a public or shared network (such as the Internet or a campus intranet) to create a secure, private network connection between a client and a server. The VPN client cloaks each packet in a wrapper that allows it to sneak (or tunnel) unnoticed through the shared network. When the packet gets to its destination, the VPN server removes the wrapper, deciphers the packet inside, and processes the data.

There are two varieties of VPNs, and they differ primarily in their approach to protecting your data: PPTP and L2TP. The oldest and simplest type of VPN uses the point-to-point tunneling protocol (PPTP).

PPTP's data encryption algorithm, Microsoft point-to-point encryption (MPPE), uses the client's login password to generate the encryption key. This is controversial because hackers are always finding ways to acquire passwords. In addition, early versions of Microsoft PPTP had flaws that could expose tunneled data to inspection by hackers. Microsoft has since patched PPTP for all versions of Windows. To provide authentication, MPPE requires the use of EAP-TLS, MS-CHAP v1, or MS-CHAP v2. MPPE is available on Windows 2000, Windows NT 4.0, and Windows 9X clients.

The more secure alternative to PPTP is L2TP (Layer 2 Tunneling Protocol). L2TP is another Microsoft development merging elements of PPTP with Layer 2 Forwarding, a Cisco Systems Inc. packet encapsulation scheme. L2TP alone is not secure, so it is almost invariably paired with a fast-growing encryption standard called IPSec (Internet Protocol Security), which supports end-to-end encryption between clients and servers.

IPSec offers several advantages. Because it supports end-to-end encryption between clients and servers, it is superior to MPPE, which only supports link encryption. And if implemented properly, IPSec is virtually impenetrable. Ideally, IPSec encryption employs the triple Data Encryption Standard (3DES) based on ANSI X.509 security certificates. Electronic certificates, issued internally or by a public authority such as Verisign Inc., irrefutably identify the client and server. Triple DES encryption (ANSI X9.52) stiffens standard 56-bit encryption keys, which can be broken only with considerable effort, by applying the encryption algorithm three times.

13.7.5   IPSec

As mentioned previously, **IP Security,** more commonly known as **IPSec Protocol,** is a protocol that provide data integrity, authentication and privacy to data being sent from one point to another point. IPSec works at the IP level and works transparently to protect the data from end-to-end. Besides being used to create virtual tunnels over an insecure network such as the Internet, it will be built into IPv6 and it is recommended to be used with wireless technology since the airways transmissions can be easily captured. IPSec is documented in a series of Internet RFCs, the overall IPSec implementation is guided by "Security Architecture for the Internet Protocol," RFC 2401.

IPSec is not a single protocol, but a suite of protocols, which can provide either message authentication and/or encryption. IPSec defines a new set of headers to be added to IP datagrams. These new headers are placed after the IP header and before the Layer 4 Protocol (typically TCP or UDP). The new protocols/headers that provide information for securing the payload of the IP packet are the authentication header (AH) and the encapsulating security payload (ESP).

While AH and ESP provide the means to protect data from tampering, preventing eavesdropping and verifying the origin of the data, it is the Internet Key Exchange (IKE), which defines the method for the secure exchange of the initial encryption keys between the two endpoints. IKE allows nodes to agree on authentication methods, encryption methods, the keys to use and the keys' lifespan.

The information negotiated by IKE is stored in a Security Association (SA). The SA is like a contract laying out the rules of the VPN connection for the duration of the SA. An SA is assigned a 32-bit number that, when used in conjunction with the destination IP address, uniquely identifies the SA. This number is called the Security Parameters Index or SPI.

Perfect Forward Secrecy (PFS) ensures that no part of a previous encryption key plays a part in generating a new encryption key. This basically means that when a new encryption key is generated, the old key does not play any part in the generation of the new key. Under normal circumstances, for performance reasons, an old key will play a part in generating a new key. When using PFS, this requires a little more overhead, as it requires reauthentication every time a new key is generated.

When the **Authentication header (AH)** is added to an IP datagram, the header will ensure the integrity and authenticity of the data, including the fields in the outer IP header. It does not provide confidentiality protection. Thus, while AH provides a method for ensuring the integrity of the packet, it does nothing for keeping its contents secret. In addition, while it provides a mechanism to ensure the integrity of the IP header and the payload of the IP packet that will be transported across an untrusted link, such as the Internet, when used by itself, AH cannot provide a total guarantee of the entire IP header because some of the fields in the IP header are changed by routers as the packet passes through the network. For a truly secure VPN connection, you should use ESP.

AH uses a keyed-hash function rather than digital signatures, because digital signature technology is too slow and would greatly reduce network throughput. If any part of the datagram is changed during transit, this will be detected by the receiver when it performs the same one-way hash function on the datagram and compares the value of the message digest that the sender has supplied. The fact that the one-way hash also involves the use of a secret shared between the two systems means that authenticity can be guaranteed. In addition, AH may also enforce antireplay protection by requiring that a receiving host set the replay bit in the header to indicate that the packet has been seen. Without it, an attacker may be able to resend the same packet many times.

When added to an IP datagram, the **Encapsulating security payload (ESP)** header protects the confidentiality, integrity, and authenticity of the data by performing encryption at the IP packet layer. While the default algorithm for IPSec is 56-bit DES, it does support a variety of symmetric encryption algorithms so that it can provide interoperability among different IPSec products.

ESP operates in two modes: transport and tunnel. When used in transport mode, the ESP header is placed between the IP header and the upper-level (transport layer) header. Any information following the ESP header, including the transport layer headers, is encrypted according to the method described by the SA, and the packet is sent on its way. This method does not use a gateway, so the clear text IP header at the front of the packet contains the actual destination address of the encapsulated datagram. However, ESP can be used, as just mentioned, in conjunction with AH to protect the integrity of the IP header information. You would typically use this for encrypting the contents of the IP packet on network connections of limited bandwidth. At the receiving end of the communication path, this clear text header information is saved, the contents of the encrypted packet are decrypted and reassembled with the correct IP header information, and the packet is sent on its way onto the network.

When operating in tunnel mode, ESP is used between two IPSec gateways, such as a set of routers or firewalls. When used in tunnel mode, the ESP header information is inserted directly before the IP or other protocol datagram that is to be protected. Therefore, the entire IP datagram, including the IP header (which has the true source address and destination address), and its payload (usually the upper-level protocol such as TCP or UPD) is encrypted and encapsulated by the ESP protocol. At the destination gateway, this outer wrapper of information is removed, the contents of the packet are decrypted, and the original IP packet is sent out onto the network to which the gateway is attached. This way, the internals of the network is hidden from would-be hackers.

The datagram being protected is encrypted (according to the methods set up by the SA), and additional headers are added in clear text format so that the new IP datagram can be transported to the appropriate gateway. In other words, the original protocol datagram is encrypted, the ESP header is added, and finally, a new IP datagram is created to transport this conglomeration to its destination gateway point. At the receiving gateway, this outer IP header information is stripped off, and according to the parameters, defined by the SA, the protected payload of original datagram is decrypted.

ESP uses both a header and a trailer to encapsulate datagrams that it protects. The header consists of an SPI, which is used to identify the security association, and a sequence number to identify the packets to, ensure they arrive in the correct order, and to ensure that no duplicate packets are received. The trailer consists of padding from 0 to 255 bytes to make sure that the datagram ends on a 32-bit boundary. This is followed by a field that specifies the length of the padding that was attached so that it can be removed by the receiver. Following this field is a Next Header field, which is used to identify the protocol that is enveloped as the payload.

Additionally, ESP can include an authentication trailer that contains data used to verify the identity of the sender and the integrity of the message. This Integrity Check Value (ICV) is calculated based on the ESP header information, as well as the payload and the ESP trailer. The layout of an ESP datagram is shown in figure 13.6.

AH and ESP can be used independently or together, although for most applications, one of them is sufficient. IPSec also uses other existing encryption standards to make up a protocol suite.

Any routers or switches in the data path between the communicating hosts will simply forward the encrypted and/or authenticated IP packets to their destination. However, if there is a firewall or filtering router, IP forwarding must be enabled for the following IP protocols and UDP port for IPSec to function:

- **IP Protocol ID of 51**—Both inbound and outbound filters should be set to pass AH traffic.
- **IP Protocol ID of 50**—Both inbound and outbound filters should be set to pass ESP traffic.
- **UDP Port 500**—Both inbound and outbound filters should be set to pass ISAKMP traffic.

Figure 13.6   ESP Datagram

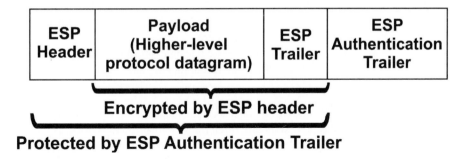

Currently, it is not possible to use IPSec through a NAT or application proxy. Even though the IP header is left intact, the encryption and authentication do not allow for other fields in the packet to be changed. But, there is current work on a new version of IPSec that would work with NAT. When version IPv6 is fully implemented, this will not be an issue.

13.8   DoD SECURITY STANDARDS

The U.S. Department of Defense (DoD) gave responsibility for computer security to the National Security Agency (NSA) in 1981, and the National Computing Security Center (NCSC) was formed. The DoD published a series of books dealing with computer and network security issues known as the Rainbow Series.

The NCSC first released A Trusted Computer System Evaluation Criteria (TCSEC) in 1983 for stand-alone, nonnetworked computers, which is unofficially referred to as the Orange Book. The Orange Book defines the standard parameters of a trusted computer in several classes, indicated by a letter and a number. The higher the letter, the higher the certification. For example, class A is the highest class, and class D is the lowest class.

According to TCSEC, system security is evaluated at one of four broad levels, ranging from class D to class A1, each level building on the previous one, with added security measures at each level and partial level.

- Class D is defined as minimum security; systems evaluated at this level have failed to meet higher-level criteria.
- Class C1 is defined as discretionary security protection; systems evaluated at this level meet security requirements by controlling user access to data.

NOTE: C1 has been discontinued as a certification.

- Class C2, defined as controlled access protection, adds to C1 requirements additional user accountability features, such as login procedures.
- Class B1 is defined as labeled security protection; systems evaluated at this level also have a stated policy model and specifically labeled data.

- Class B2, defined as structured protection, adds to B1 requirements a more explicit and formal security policy.
- Class B3, defined as security domains, adds stringent engineering and monitoring requirements and is highly secure.
- Class A1 is defined as verified design; systems evaluated at this level are functionally equivalent to B3 systems but include more formal analysis of function to assure security.

In 1987, the NCSC released enhanced testing criteria based on the Orange Book standard. The new standard, often referred to as the Red Book, is the Trusted Network Interpretation Environment Guidelines. The Red Book also uses the D through A levels. As with the C2 class in the Trusted Computer implantation, the C2 class is the highest class for generic network operating systems. Higher-level classes require that operating systems be specifically written to incorporate security-level information as the data is input.

The most publicized class is C2, controlled access protection, which must provide a unique user account for each person on the network and provide accountability for the information the user uses. Additionally, the network communications must be secure. Higher-level classes require that operating systems be specifically written to incorporate security-level information as the data is input.

Currently, several network operating systems are under evaluation for C2 Trusted Network certification. The following operating systems are C2-level certified for Trusted Computer (Orange Book):

- Windows NT Workstation and Windows NT Server version 3.5 with Service Pack 3
- Windows NT Workstation and Windows NT Server version 4.0 with Service Pack 6a and C2 Update

If the computer on which Windows NT Server is installed is connected to a network, it loses the C2 Trusted Computer certification. The only currently available network operating system that has achieved C2 Trusted Network certification is:

- Novell IntranetWare [NetWare 4.11 Server] with IntranetWare Support Pack 3A and Directory Services Update DS.NLM v5.90, DSREPAIR.NLM v4.48 and ROLLCALL.NLM v4.10

Information Technology Security Evaluation Criteria (ITSEC) is a European criterion similar to the TCSEC, but with some important differences. ITSEC emphasizes the integrity and availability of products and systems and introduces the distinctions of effectiveness and correctness. The TCSEC is primarily concerned with security policy, accountability, and assurance. Various European certification bodies grant ratings based upon the ITSEC. Examples of ITSEC ratings are level E2, which is a measure of effectiveness, and class F-C2, which is a measure of functionality. A combined E2/F-C2 evaluation is similar in scope to the class C2 TCSEC evaluation.

To verify security certification or check out officially released documents, go to the following website:

http://www.radium.ncsc.mil/tpep/epl/index.html

13.9   AUDITING AND INTRUDER DETECTION

Trying to detect intruders can be a very daunting task, and it often requires a lot of hard work and a thorough working knowledge of the network operating system and the network as a whole. You should always see what logs (especially the security logs such as those found in Event Viewer) are available with your network operating system (and firewalls and proxy servers too). For example, you can check the log that lists all the failed login attempts. If you see that for several days at late hours, the same account has attempted to log in but has failed, it could be that someone is trying to figure out the password and trying to log in but is failing. Sometimes, security logging is enabled by default; other times, it has to be enabled. Windows 2000 and Windows Server 2003 have an auditing service by using group policies that allows you to audit a large array of items, including who accesses which files. These auditing features can also be arranged to check everyone including the administrators in an attempt to keep the administrators honest as well.

Besides looking at your audit logs, you can also look at your firewall logs. A typical firewall will generate large amounts of log information. The firewall logs can tell you port scans and unauthorized connection attempts, identify activity from compromised systems and much more. Of course, the real trick in using the logs is knowing what to look for.

Traffic moving through a firewall is part of a connection. A connection has 2 basic components; a pair of IP addresses and a pair of port numbers. The IP addresses identify each computer involved in the communication. The port numbers identify what services or applications are being utilized. More specifically, it is typically the destination port number that will indicate what applications/services are being used. Of course, knowing what port numbers are associated with what services helps identify malicious activity occurring on the firewall.

- **IP addresses that are rejected**—Although a site will be probed from many places and many times, knowing that a probe is occurring and what is being probed for proves useful information when trying to secure a network.
- **Unsuccessful logins**—Knowing when someone is trying to gain access to critical systems proves useful to help secure a network.
- **Outbound activity from internal servers**—If there is traffic originating from an internal server, having a good understanding of the normal activity on that server will help an administrator determine if the server has been compromised.
- **Source routed packets**—Source routed packets may indicate that someone is trying to gain access to the internal network. Since many networks have an address range that is unreachable from the Internet (such as 10.x.x.x), source routed packets can be used to gain access to a machine with a private address since there is usually a machine exposed to the Internet that has access to the private address range.

In addition to the security information documented above, you can get a lot out of the firewall logs if you have looked at your logs before so that you can be familiar with normal everyday activities. This way, if you know what is normal, it will be easier to identify malicious activity when it occurs.

A honeypot is a computer set up as a sacrificial lamb on the network. The system is not locked down and has open ports and services enabled. This is to entice a would-be attacker to this computer instead of attacking authentic computers on the network. The honeypot contains no real company information, and thus will not be at risk if and when it is attacked. The administrator can monitor the honeypot so they can see how an attack is occurring without putting your other systems in harms way and it may give the administrator an opportunity to track down the attacker. The longer the hacker stays at the honeypot, the more will be disclosed about his or her techniques.

On your network, you should use virus protection software. Of course, for this to be effective, you will need to keep the software up-to-date. In addition, it would be advantageous to have solid knowledge of what Trojans are in circulation, what ports they are using, how they operate and what their general purpose in life is. A nice list of Trojans and their associated ports can be found at http://www.simovits.com/nyheter9902.html.

13.10   PENETRATION TESTING

Penetration testing is the process of probing and identifying security vulnerabilities in a network and the extent to which they might be exploited by outside parties. It is a necessary tool for determining the current security posture of your network. Such a test should determine both the existence and extent of any risk.

The normal pattern for a malicious user or a person to gain information on a target host or network starts with basic reconnaissance. This could be as simple as visiting an organization's website or sites or using public tools to learn more information about the target's domain registrations. After the attacker has gained enough information to their satisfaction, the next logical step is to scan for open ports and services on the target host(s) or network. The scanning process may yield very important information such as ports open through the router and firewall, available services and applications on hosts or network appliances, and possibly the version of the operation system or application. After an attacker has mapped out available hosts, ports, applications, and services, the next step is to test for vulnerabilities that may exist on the target host or network.

When a vulnerability is found and the hacker has gained access to a host, he or she will attempt to keep access and cover their tracks. Covering of tracks most always involves the tampering of logs or logging servers. The defense in-depth strategy is one of a layered approach and assumes the perimeter network can be compromised. With this in mind, it is critical to protect logs and logging servers. In the case of an actual intrusion, many times all an organization is left with is their logs. Protect them accordingly because this may be your only evidence of the incident.

13.10.1   Reconnaissance

The reconnaissance phase can be done many different ways depending on the goal of the attacker. Some of the common available tools are:

- **Nslookup**—Available on Unix and Windows Platforms
- **Whois**—Available via any Internet browser client

- **ARIN**—Available via any Internet browser client
- **Dig**—Available on most Unix platforms and some websites via a form
- **Web Based Tools**—Hundreds if not thousands of sites offer various recon tools
- **Target Website**—The client's website often reveals too much information
- **Social Engineering**—People are an organization's greatest asset, as well as their greatest risk

13.10.2   Scanning

After the penetration engineer or attacker gathers the preliminary information via the reconnaissance phase, they will try and identify systems that are alive. The live systems will probe for available services. The process of scanning can involve many tools and varying techniques depending on what the goal of the attacker is and the configuration of the target host or network. Remember, each port has an associated service that may be exploitable or contain vulnerabilities.

The fundamental goal of scanning is to identify potential targets for security holes and vulnerabilities of the target host or network. Nmap is probably the best known and most flexible scanning tool available today. It is one of the most advanced port scanners available today. Nmap provides options for fragmentation, spoofing, use of decoy IP addresses, stealth scans, and many other features. Nmap could be downloaded from the http://www.insecure.org/nmap/website. Another good software package is GFI LANguard Network Security Scanner (evaluation software available from the http://www.gfisoftware.com/website).

Below is a list of some common tools to perform scanning:

- **Nmap**—Powerful tool available for UNIX that finds ports and services
- **GFI LANguard Network Security Scanner**—Powerful tool available for Windows that finds ports and services
- **Telnet**—Can report information about an application or service; i.e., version, platform
- **Ping**—Available on most every platform and operating system to test for IP connectivity
- **Traceroute**—Maps out the hops of the network to the target device or system
- **Hping2**—Powerful Unix based tool used to gain important information about a network
- **Netcat**—Some have quoted this application as the "Swiss Army knife" of network utilities
- **Queso**—Can be used for operating system fingerprinting

13.10.3   Vulnerability Testing

Vulnerability testing is the act of determining which security holes and vulnerabilities may be applicable to the target network or host. The penetration tester or attacker will attempt to identify machines within the target network of all open ports and the operating systems as well as running applications including the operating system, patch level, and service pack applied.

The vulnerability testing phase is started after some interesting hosts are identified via the nmap scans or another scanning tool and is preceded by the reconnaissance phase. Nmap will identify if a host is alive or not and what ports and services are available even if ICMP is completely disabled on the target network to a high degree of accuracy.

One of the best vulnerability scanners available today just happens to be free. Nessus is available at the following URL: http://www.nessus.org. The Nessus tool is well supported by the security community and is comparable to commercial products such as ISS Internet Security Scanner and CyberCop by CA.

Other free vulnerability scanners include; SARA available at http://www-arc.com/sara/, a special version of SARA is available to specifically test for the SANS/FBI Top 20 most critical Internet security vulnerabilities located at http://www.sans.org/top20.htm. SARA and SAINT are both predecessors of SATAN an early security administrator's tool for analyzing networks.

Once an attacker has gained a list of potential vulnerabilities for specific hosts on the target network they will take this list of vulnerabilities and search for specific exploit to utilize on their victim. Several vulnerability databases are available to anyone on the Internet.

Vulnerability Databases

ISS X-Force—http://www.iss.net/security_center/
Security Focus Database—http://online.securityfocus.com/archive/1
InfoSysSec Database—http://www.infosyssec.com/
Exploit World—http://www.insecure.com/sploits.html

For newer Microsoft operating systems, Microsoft now offers the Microsoft Baseline Security Analyzer (MBSA). It provides a streamlined method of identifying common security misconfigurations. MBSA includes a graphical and command line interface that can perform local or remote scans of Windows systems. MBSA will scan for missing hotfixes and vulnerabilities in the following products: Windows NT 4.0, Windows 2000, Windows Server 2003, Windows XP, Internet Information Server (IIS) 4.0 and 5.0, SQL Server 7.0 and 2000, Internet Explorer (IE) 5.01 and later, and Office 2000 and 2002. MBSA creates and stores individual XML security reports for each computer scanned and will display the reports in the graphical user interface in HTML.

Baseline Security Analyzer Home Page and Download Instructions
http://www.microsoft.com/technet/security/tools/Tools/mbsahome.asp

Baseline Security Analyzer White Paper
http://www.microsoft.com/technet/ security/tools/tools/mbsawp.asp

13.11   PATCHES

The last part of security is to make sure that you have the most updated patches for your network operating system and that you check with the distributor and vendor of the network operating system (as well as the operating systems on the network) for security holes and the patches to fix the security holes.

The Computer Emergency Response Team (CERT) Coordination Center was started in December 1988 by the Defense Advanced Research Projects Agency, which was part of the U.S. Department of Defense, after the Morris worm disabled about 10 percent of all computers connected to the Internet. CERT studies Internet security vulnerabilities, provides services to websites that have been attacked, and publishes security alerts. The CERT Coordination Center undertakes research activities that include the area of WAN computing and involve developing improved Internet security. The organization also provides training to incident-response professionals.

CERT website:
> http://www.cert.org/

Microsoft Baseline Security Analyzer
> http://www.microsoft.com/technet/security/tools/tools/mbsahome.asp

Symantec Security Advisory List
> http://securityresponse.symantec.com/avcenter/security/Advisories.html

SUMMARY

1. The user name and password constitute the first line of defense since every user needs to have them to get access to the network and its resources.

2. With cracker programs easily available to anyone who wants them, all users should use strong passwords, which are difficult-to-crack passwords.

3. To make passwords more secure, many network operating systems have password management features, which were either built into the operating system or installed as an add-on package.

4. Another option is to require a minimum number of characters for the password.

5. If you decide to use an anonymous or guest account, you need to treat it with extreme care.

6. Give only the rights and permissions that are necessary for the individual users to do their job. In addition, make sure that the individual users do have sufficient rights and permissions for what needs to be done.

7. The server room should be secure, with only a handful of people allowed to have access to it.

8. The server should also be secure.

9. Packet filters are based on protecting the network by using an access control list (ACL). This list resides on your router and determines which machine (IP address) can use the router and in what direction.

10. A firewall analyzes the source and destination TCP/IP address, and it can also examine the source and destination port numbers and the content of the packet data.

11. To get around the problem of having holes in a firewall, you can create a demilitarized zone (DMZ). Instead of having two ports: the public network on one port of the firewall and the private network on another port, you have three ports. The third port is the DMZ, which is a less secure area than the private network.

12. The proxy server can monitor and intercept any and all requests that are being sent to the external server or that come in from the Internet connection.

13. Today, many operating systems offer personal firewalls to protect an individual computer when connecting to the Internet.

14. Denial of service (DoS) is a type of attack on a network that is designed to bring the network to its knees by flooding it with useless traffic.

15. Encryption is the process of disguising a message or data in what appears to be meaningless data (cipher text) to hide and protect the sensitive data from unauthorized access.

16. Decryption is the process of converting data from encrypted format back to its original format.

17. Cryptography is the art of protecting information by transforming it (encrypting it) into cipher text.

18. To encrypt and decrypt a file, you must use a key. A key is a string of bits used to map text into a code and a code back to text.

19. The most basic form of encryption is private-key encryption, also known as the symmetric algorithm. This form of encryption requires that each person possess a copy of the key.

20. Public-key encryption, also known as the asymmetric algorithm, uses two distinct but mathematically related keys, public and private. The public key is the nonsecret key that is available to anyone you choose, or is made available to everyone by posting it in a public place. It is often made available through a digital certificate. The private key is kept in a secure location used only by you.

21. A digital envelope is a type of security that encrypts the message using symmetric encryption and encrypts the key to decode the message using public-key encryption.

22. A digital signature is a digital code that can be attached to an electronically transmitted message that uniquely identifies the sender. Like a written signature, the purpose of a digital signature is to guarantee that the individual sending the message really is who he or she claims to be.

23. A digital certificate is an attachment to an electronic message used for security purposes such as for authentication. It is also used to verify that a user sending a message is who he or she claims to be and to provide the receiver with a means to encode a reply.

24. A Certificate Authority (CA) issues an encrypted digital certificate containing the applicant's public key and a variety of other identification information.

25. X.509 certificates are the most widely used digital certificates.

26. Authentication, which is the layer of network security, is the process by which the system validates the user's logon information. Authentication is crucial to secure communication.

27. In NTLM authentication, the client selects a string of bytes, uses the password to perform a one-way encryption of the string, and sends both the original string and the encrypted one to the server.

28. Kerberos is an authentication service. Its purpose is to allow users and services to authenticate themselves to each other without allowing other users to capture the network packet on the network and resending it so that they can be authenticated over an unsecured medium such as the Internet.

29. Trying to detect intruders can be a very daunting task, and it often requires a lot of hard work and a thorough working knowledge of the network operating system and the network as a whole.

30. The last part of security is to make sure that you have the most updated patches for your network operating system and that you check with the distributor and vendor of the network operating system (as well as the operating systems on the network) for security holes and the patches to fix the security holes.

QUESTIONS

1. Which device can prevent Internet hackers from accessing a LAN?
   a. A firewall
   b. A gateway
   c. A multistation access unit (MAU)
   d. A router

2. Pat administers a LAN for the Acme Corporation. Pat has recently connected the LAN to the Internet, but Skip, the CEO, is concerned that hackers will infiltrate the LAN and steal important trade secrets. Skip has also noticed that Internet access is extremely slow, and she wants Pat to improve it. What can Pat use to meet Skip's requirements?
   a. A firewall
   b. A router
   c. An IP proxy
   d. A transceiver

3. For what purpose does a network operating system use access control lists?
   a. To determine which users can use resources on the network
   b. To locate hosts on the network

c. To establish communications between network interface cards (NICs)

d. To determine which TCP/IP service receives data packets

4. Which of the following passwords provides you with the strongest network security?

a. user1   b. User1

c. USER1   d. uSEr1#

5. Pat administers a network, and he wants to implement a strong password policy. How should Pat accomplish this task? (Choose all correct answers.)

a. He should configure passwords to expire every 30 days.

b. He should provide users with password hints.

c. He should configure the network operating system to lock users out of their accounts after three unsuccessful logon attempts.

d. He should require users to create passwords that contain only alphabetical characters.

e. He should require a minimum password length of 8 characters.

6. You administer a LAN for your company. A hacker has tried to gain access to the LAN, and you have been directed to increase network security by implementing a password policy. What should you do to increase network security? (Choose two answers.)

a. Require users to create passwords of 25 or more characters.

b. Require users to use only alphabetical characters in their passwords.

c. Require users to change their passwords each month.

d. Store a password history for each user.

7. You administer a network for a small company. A network user, Pat, cannot access the network through his user account. You successfully access the network with Pat's user name and password. Which of the following could explain why Pat cannot access the network through his user account?

a. Pat's user account is locked.

b. Pat's password has expired.

c. Pat is trying to use a password in the password history.

d. Pat has activated caps lock on his computer.

8. You want to create a strong password to ensure that hackers cannot gain access to the network through your user account. In this scenario, which password is the strongest?

a. password

b. At#CJo$TM

c. 241LaneAve

d. SSN999-99-9999

9. Spoofing is a method of _____.

a. encrypting a data packet

b. tracking transactions to verify authenticity

c. filtering bad packets

d. fooling a firewall into thinking the packet is friendly

e. translating one IP address to another

10. Web servers use what to protect transaction information?

a. QVP   b. DSL

c. LST   d. SSL

11. Within your company's private network, you suspect that a user from marketing is trying to hack into the server in the accounting department. You decide to install a firewall between marketing and accounting to stop the unauthorized use of the network. How good a solution is this?

a. Perfect solution

b. Good solution, but there are better options

c. Somewhat helpful, in that network packets can be better monitored

d. Worthless

12. What is DES?

a. Data Encoded Subnetwork

b. Double-Edged Signaling

c. Dynamic Encoded Synchronization

d. Data Encryption Standard

13. Which of the following is a public-key encryption utility?

a. PGP   b. QSA

c. RSA   d. GPR

14. You manage a small LAN that has a limited-bandwidth connection to the Internet. Your users have been complaining of slow access lately. Which of the following could you place on your network to help improve access?

a. Firewall

b. Multiport switch

c. Internet router

d. Proxy server

15. Which of the following securely transmits data between corporate sites through the Internet?
    a. CVP
    b. QVP
    c. NVP
    d. VPN
16. What is SSL?
    a. Single Socket Layering
    b. Simple Socket Layer
    c. Simple Socket Layering
    d. Secure Sockets Layer
17. What is IPSec?
    a. Internet Protocol sections
    b. An IP secondary protocol service
    c. IP secondaries
    d. An IP encryption protocol
18. What is L2TP?
    a. Layer 2 Tunneling Protocol
    b. Level 2 Transition Protocol
    c. Level 2 Transport Protocol
    d. Layer 2 Transport Protocol
19. You want your users' passwords on the network to be secure. Therefore, you are going to implement a password policy. Which of the following would probably be less of a good idea than the others?
    a. Require the creation of a new password every 90 days.
    b. Set accounts to automatically disable themselves upon three failed login attempts that occur within 30 minutes.
    c. Require passwords to use at least one number in addition to using letters.
    d. Require passwords to be at least 14 characters long.
20. A hardware device that prevents unauthorized connections to your internal network is called what?
    a. Sniffer
    b. Packet wall
    c. Fluke
    d. Firewall
21. You are setting up a network and need to decide on the type of security you are going to implement to protect the network's resources from use by unauthorized persons. Which do you select?
    a. User level
    b. MAC level
    c. Route level
    d. Share level
22. A good rule of thumb is that users should be required to change their passwords every _____ days.
    a. 90
    b. 60
    c. 120
    d. 30
23. C2 Red Book certification is given by the DoD when an operating system is trusted to be connected to a network. Which of the following has been given this certification?
    a. Windows 98
    b. Windows 2000
    c. UNIX
    d. Netware 4.11
24. The Department of Defense has published a series of books dealing with network security issues. Which of the following series is the one it publishes?
    a. Bluebook Series
    b. Redbook Series
    c. Blackbook Series
    d. Rainbow Series
25. You've just fired an employee. At the exit interview, you collect the following items from him or her:
    ■ Office keys
    ■ Pager
    ■ Company software
    ■ Laptop
    What else do you need to get? (Choose two answers.)
    a. Badge
    b. Cleaning deposit
    c. Time card
    d. Security deposit
26. What is the best place to put your email router, public web server, and FTP server in your network?
    a. PRL
    b. CMA
    c. TXV
    d. DMZ
27. Your users on your network are constantly reusing old passwords whenever the system asks them to change their password. You want to put a stop to this. How do you do it?
    a. Set the password history size on all accounts.
    b. Set up user account expiration dates.
    c. Set a password length on all accounts.
    d. Set a password expiration date on all accounts.
28. Your boss is considering replacing one of her company's routers with one that can act as an IP proxy as well as do packet filtering. She believes it will help improve network performance. Is she correct?
    a. Yes
    b. No
29. The Department of Defense has certified which of the following workstation operating systems as being secure? (Choose all that apply.)
    a. Windows NT 4.0
    b. Windows XP
    c. Windows NT 3.5
    d. Windows 2000

30. Packet filtering allows a router or firewall to check and see if packets being received match predetermined criteria, and drops those packets that don't match or fit the current communication session that is established. A list has to be kept on the router which gives it this ability. What is this list known as? (Pick two answers.)

  a. Session table
  b. State table
  c. Dynamic packet list
  d. Dynamic state list

31. What is the practice of sending packets that have fake source addresses called?

  a. IP translation
  b. Packet hopping
  c. Protocol tunneling
  d. IP spoofing

32. The DoD assigned the role of handling computer security certifications to which federal government agency or department?

  a. HHS
  b. NASA
  c. Itself
  d. NSA

33. Which of the following is the strongest bit encryption that the United States currently allows to be exported?

  a. 128 bits
  b. 80 bits
  c. 256 bits
  d. 40 bits

34. The practice of sending abnormally large ICMP packets to a system with the intent to bring it down is called _____.

  a. the Ping of Death
  b. WinNuke
  c. IP spoofing
  d. SYN flood

35. Which of the following are good ways to strengthen users' passwords? (Choose two answers.)

  a. Require them to use their social security number.
  b. Require them to use the names of relatives and pets.

  c. Require them to use passwords that are at least 8 characters long.
  d. Require them to use special characters that are neither letters nor numbers.

36. When an employee is fired or quits, what is the best thing to do to his or her network login account?

  a. Delete it.
  b. Terminate it.
  c. Disable it.
  d. Crush it.

37. Windows NT Server is C2 Orange Book–certified by the Department of Defense. What would cause a computer running this operating system to lose its C2 certification?

  a. Having more than one account with administrative privileges on it
  b. Having an administrative account with a password of less than 8 characters
  c. Being disconnected from the network
  d. Being connected to a network

38. What is skipjack?

  a. A method of counting live network ports
  b. An encryption algorithm
  c. A method of omitting certain unneeded network ports
  d. An extended RJ-45 plug

39. One of your users complains about not being able to log in with the same ID and password that worked fine yesterday. What would you check first?

  a. The Cap Lock key
  b. The workstation's patch cable
  c. The workstation's NIC
  d. The user database to see if the account is locked or disabled

40. All of your users are running some version of Windows NT or 2000 on their workstations. For security reasons, what should you require your users to do if they need to leave their workstations, even if only for a few minutes? (Choose two answers.)

  a. Turn off their computer.
  b. Lock their workstation.
  c. Turn off their monitor.
  d. Log out.

CHAPTER 14

Disaster Prevention and Recovery

Topics Covered in this Chapter

Introduction

When establishing your network, your network must be reliable and must perform adequately for your needs. Of course, even in the best of cases, disaster will happen from time to time. Thus as network administrator, you will have to plan for those disasters so that you can minimize the number of disasters and minimize the effect of those disasters.

Objectives

- Identify the purpose and characteristics of network-attached storage.
- Identify the purpose and characteristics of fault tolerance.
- Identify the purpose and characteristics of disaster recovery.
- Define the features, capabilities, and implementation of RAID 0, RAID 1, and RAID 5 and explain how each relates to fault tolerance or high availability.

- Describe the benefits of hardware RAID over software RAID.
- Describe clustering, scalability, and high availability.
- Define backup, restore, and disaster recovery concepts.
- Perform a full backup.
- Verify a backup.
- Describe adaptive fault tolerance, adapter load balancing, and adapter teaming.

14.1 FAULT TOLERANCE AND DISASTER RECOVERY NEEDS

The cost of server downtime is staggering. For a network of a medium-size business with 300 employees and 100 million dollars of business, the cost of going down 24 times a year, 3 hours at a time (less than 1 percent downtime), can amount to more than 3 million dollars per year in lost revenues, user salaries, and server outage costs.

In a hot site, every computer system and piece of information has a redundant copy. This level of fault tolerance is used when systems must be up 100 percent of the time. Hot sites are strictly fault-tolerant implementations, not disaster recovery implementations. Of course, you can imagine that these types of system are costly. To make a system fault-tolerant, you must overcome the failure points as shown in table 14.1.

In a warm site (also called a nearline site), the network service and data are available most of the time (more than 85 percent of the time). The data and services are less critical than those in a hot site. With hot-site technologies, all fault tolerance procedures are automatic and are controlled by the NOS. Warm-site technology requires a little more administrative intervention, but warm sites aren't as expensive as hot sites.

Table 14.1 Common Points of Failures in a Server

Failure Point	Failure Solution
Network hub and network card	Redundant network cards and hubs
Power problems	Uninterruptible power supply (UPS) Redundant power supplies Putting cluster nodes on separate electric circuits
Disk	Hardware RAID
Other server hardware, such as CPU or memory	Failover clustering
Server software, such as the operating system or specific applications	Failover clustering
Wide area network (WAN) links, such as routers and dedicated lines	Redundant links over the WAN, to provide secondary access to remote connections
Dial-up connection	Multiple modems

The most cost-effective and most commonly used warm-site technology is a duplicate server. A duplicate server is one that is currently not being used and is available to replace any server that fails. When the server fails, the administrator installs the new server and restores the data; the network services are available to users with a minimum of downtime. Using a duplicate server is a disaster recovery method because the entire server is replaced, but in a shorter time than if all the components had to be ordered and configured at the time of the system failure.

NOTE: Corporate networks don't often use duplicate servers. That's because some major disadvantages are associated with using them, including keeping a current backup of data on the duplicate server and the fact that you can lose data since the last backup.

A cold site does not guarantee server uptime and has little or no fault tolerance. Instead, it relies on efficient disaster recovery methods to ensure data integrity, especially the backup of the data.

14.2 RAID

RAID is short for "redundant array of inexpensive disks." It is a category of disk drive that employs two or more drives in combination for fault tolerance and performance. RAID disk drives are used frequently on servers but aren't generally necessary for personal computers. Ideally, you use a RAID system to make sure no data is lost and to recover data from failed disk drives without shutting your system down.

Table 14.2 Types of RAID

Type	Description
RAID 0–Disk striping	Data striping is the spreading out of blocks of each file across multiple disks. It offers no fault tolerance, but it increases performance. Level 0 is the fastest and most efficient form of RAID.
RAID 1–Disk mirroring/duplexing	Disk mirroring duplicates a partition onto two hard drives. When information is written, it is written to both hard drives simultaneously. It increases performance and provides fault tolerance. Disk duplexing is a form of disk mirroring. Disk mirroring uses two hard drives connected to the same card; disk duplexing uses two controller cards, two cables, and two hard drives.
RAID 2–Disk striping with ECC	Level 2 uses data striping plus ECC to detect errors. It is rarely used today since ECC is embedded in almost all modern disk drives.
RAID 3–ECC stored as parity	Level 3 dedicates one disk to error correction data. It provides good performance and some level of fault tolerance.
RAID 4–Disk striping with large blocks	Level 4 offers no advantages over RAID 5 and does not support multiple simultaneously written operations.
RAID 5–Disk striping with parity	RAID 5 uses disk striping and includes byte correction on one of the disks. If one disk goes bad, the system will continue to function. After the faulty disk is replaced, the information on the replaced disk can be rebuilt. This system requires at least three drives. It offers excellent performance and good fault tolerance.

RAID was originally defined as a memory architecture that uses a subsystem of two or more hard disk drives treated as a single, larger logical drive. The purpose of this proposed architecture was to take advantage of the data redundancy inherent in the multiple-drive design as well as to capitalize on the lower costs of smaller drives. When RAID was first proposed, it was considerably cheaper to buy five 200-MB hard drives than one 1-GB drive. Of course, today this is not true any more, so the current focus for RAID is data integrity and reliability instead of cost saving.

Originally RAID comprised six levels (RAID 0 through RAID 5). (See table 14.2.) A few more levels have been added to combine the features of other levels. Although RAID 6 follows the general numbering process, most new levels break with the number sequence, usually for marketing reasons. Not all RAID levels are commercially available, and some are supported by only a few products. Only RAID 0, 1, and 5 are supported by Windows Servers, Linux, and NetWare without additional hardware and software from RAID vendors. Those three levels, though, can be supported by a variety of hard disk and controller combinations.

RAID 0 is the base of RAID technology. RAID 0 stripes data across all drives. With striping, all available hard drives are combined into a single, large virtual file system, with the file system's blocks arrayed so they are spread evenly across all the drives. For example, if you have three 500-MB hard drives, RAID 0 provides for a 1.5-GB virtual hard drive (sometimes referred to as a volume). When you store files, they are written across all three drives. When a large file, such as a 100-MB multimedia presentation, is saved to the virtual drive, a part of it may be written to the first drive, the next chunk to the second, more to the third, and perhaps more wrapping back to the first drive to start the sequence again. The exact manner in which the chunks of data move from physical drive to physical drive depends on the way the virtual drive has been set up, which includes considering drive capacity and the way in which blocks are allocated on each drive. No parity control is used with RAID 0; therefore, it really is a true form of RAID. RAID 0 does have several advantages, though. Most important is that striping provides some increase in performance through load balancing.

RAID 1 is known as disk mirroring or disk shadowing. With RAID 1, each hard drive on the system has a duplicate drive that contains an exact copy of the first drive's contents. Since every bit written to the file system is duplicated, data redundancy exists with RAID 1. If one drive in the RAID 1 array fails or develops a problem of any kind (such as a bad sector), the mirror drive can take over and maintain all normal file system operations while the faulty drive is diagnosed and fixed. RAID 1 also includes disk duplexing, which is the same as disk mirroring except the two drives are on different controller cards so that the drive, controller card, and cable are redundant.

Many RAID 1 disk controllers have software routines that will automatically take a faulty drive offline, run diagnostics on it, and if possible, reformat the drive and copy all data back from the mirror image—all while the file system proceeds as if nothing has happened. Users are usually unaware of faults with RAID 1 controllers. Alert messages can be triggered when a fault occurs.

One big disadvantage of RAID 1 is its use of disks. If you have two 2-GB drives, you can have a total file system of only 2 GB (the other 2 GB are mirrored). You're only getting half the disk space you're paying for, but you do have fully redundant drives. In case of catastrophic failure of a drive, controller, or motherboard, you can remove a mirror drive and boot on another controller or server.

RAID 1 offers an increase in read performance in most implementations, as the controller card allows both drives (primary and mirror) to be read at the same time, resulting in a faster read operation. Write operations are not faster, though, because data must be written to two drives. In many RAID 1 systems that do not use separate drive controllers for the primary and mirror drives, writing can even slow down because the system must perform two complete write operations in sequence.

Implementations of RAID 1 usually require two drives of similar size. If you use a 1.5-GB drive and a 2-GB drive, for example, the extra 0.5 GB on the second drive is wasted. Some controllers let you combine drives of different sizes, with the extra space used for nonmirrored partitions.

RAID 5 is very similar to RAID 0, but one of the hard drives is used for parity (error correction) to provide fault tolerance. To increase performance, spread the error correction drive across all hard drives in the array to avoid having the one drive do all the work in calculating the parity bits. RAID 5 is supported by NT 4.0 and most RAID vendors because it is a good compromise between data integrity, speed, and cost. RAID 5 has better performance than RAID 1 (mirroring). RAID 5 usually requires at least three drives, with more drives preferable.

NOTE: The overhead RAID 5 imposes on RAM can be significant, too, so Microsoft recommends at least an additional 16 MB of RAM when RAID 5 is used.

As with RAID 1, though, drives of disparate capacities may result in a lot of unused disk space because most RAID 5 systems use the smallest drive capacity in the array for all RAID 5 drives. Extra disk space can be used for unstriped partitions, but these are not protected by the RAID system.

Since SCSI drives have command queuing and typically have a higher throughput, SCSI drives are the best choice for RAID systems. According to RAID vendors, SCSI subsystems represent more than 95 percent of the RAID market. Some RAID systems support hot-swappable drives, where a drive can be removed without powering down the system. SCSI controller cards that support RAID through hardware will allow for better performance since they will do much of the calculations that would have been done through the processor as instructed by the software. Lastly, if performance is critical on some systems, I would recommend a system with a RAID controller card that has a relatively large amount of RAM to be used as cache for the controller card.

14.3   REDUNDANT NETWORK CARDS AND HUBS

To make sure that the link that connects to your network has fault tolerance, you have several options. They include:

- Adaptive load balancing (ALB)
- Adapter fault tolerance (AFT)
- 802.3ad link aggregation

Adaptive load balancing technology can increase server bandwidth up to 800 Mbps over Fast Ethernet or 8 Gbps over Gigabit Ethernet, by automatically balancing traffic across as many as eight network adapters. Essentially, each additional adapter adds another 100 Mbits/s or 1000 Mbits/s link to the network. Once ALB is configured, all outgoing server traffic will be balanced across the adapter team. Incoming traffic is carried by a single adapter. In most environments, this is a highly effective solution, since server traffic is primarily outbound. Since the distribution of traffic among the adapters is automatic, there is no need to segment or reconfigure the network. The existing IP address of the server is shared by all the adapters in the server, and traffic is always balanced between them. All of the adapters in a team must be connected to a switch.

NOTE: All adapters in the team should be the same speed.

They can be connected to a single switch or hub, or to two or more switches or hubs, as long as they are on the same network segment. In addition, the teamed adapters provide automatic emergency backup links to the network. If one server link goes down, due to a broken cable, a bad switch port, or a failed adapter, the other adapter(s) automatically accepts the additional load. There is no interrupt in server operation. Some software and drivers also support a network alert to inform IT staff of the problem.

With two or more adapters installed, adapter fault tolerance can be configured to establish an automatic backup link between the server and the network. Should the primary link fail, the secondary link kicks in within milliseconds, in a manner that is transparent to application and users. The redundant link that AFT establishes between the server and the network includes a redundant adapter, a cable, and hub or switch port connection. If there is any problem along the primary link, the secondary link immediately takes over. AFT can also initiate a network alert. When the primary link is fixed, it will automatically revert to the higher-performance link.

Unlike most redundant link technologies, AFT supports mixed-speed teaming using any combination of adapters. With this capability, a relatively inexpensive 100-Mbits/s backup link can be used to safeguard a high-speed Gigabit Ethernet connection. The inexpensive backup may not be able to support the full traffic load as effectively, but it can allow business-critical applications to stay online until the higher-speed link is fixed.

Link aggregation, also referred to as trunking, is a technique that allows parallel physical links between switches or between a switch and a server to be used simultaneously, multiplying the bandwidth between the devices and providing fault tolerance. To meet customer requirements for this type of functionality, several companies support proprietary link aggregation schemes. In 2000, IEEE released 802.3ad as an industry standard for link aggregation, which allows for balancing traffic (up to 8 networks cards and 16 Gbps full duplex of bandwidth) among multiple switches from a single server. Different from ALB and AFT, link aggregation requires a switch that supports it.

PCI HotPlug and Active PC developed by Compaq and IBM, respectively, allow the adapter to be replaced without interrupting network service. If an adapter fails, AFT automatically moves server traffic onto the redundant link and generates a network alert. Both of these enable you to replace the failed adapter without bringing down the server.

14.4   OVERCOMING POWER PROBLEMS

When you turn on your PC, you expect the power to be there. Unfortunately, the power you get from the power company is not always 120 volts AC. The voltage level may drop or increase. While the power supply can handle many of these power fluctuations, other power fluctuations may shut down or damage your computer, corrupt your data, and/or lose any unsaved work.

Studies done by IBM show a typical computer is subject to more than 120 power problems per month. The most common of these are voltage sags. Obvious power problems such as blackouts and lightning make up only 12 percent of the power problems. American Power Conversion, a leading company that manufactures equipment to overcome power problems, states that data loss caused by power problems occurs 45.3 percent of the time, making it the largest cause. Bad power can cause frozen computers and keyboards, errors in data transmissions, corrupt or lost data, frequently aborted modem transfers, and total failure of a computer or computer component.

Figure 14.1   Power Irregularities

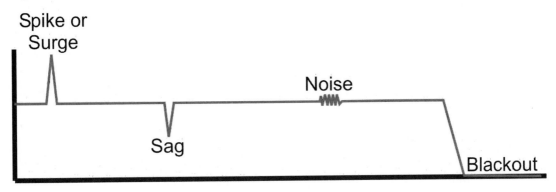

14.4.1   Power Irregularities

Power line irregularities can be classified into two categories: overvoltages and under-voltages. The most dangerous is the overvoltage, which is rated at more than 10 percent additional voltage than what the power supply is rated. The worst of these is a spike that lasts only a nanosecond but that measures as high as 25,000 volts (normally caused by lightning). A **spike** is sometimes known as a transient. A longer-duration overvoltage is called a **surge,** which can stretch into milliseconds. Spikes and surges can visibly damage the electronic components, or they can cause microdamage, which cannot be seen. Besides being caused by lightning, overvoltage can also occur when overburdened power grids switch from one source to another or when a high-powered electric motor tries to grab power. (See figure 14.1.)

Undervoltages (including total power failure) make up 87 percent of all power problems. When an undervoltage occurs, the computer gets less voltage than needed to run properly. Most PCs are designed to withstand prolonged voltage dips of about 20 percent without shutting down. Power outages and short drops in power typically do not physically damage the computer. Unfortunately, the computer does lose data and data can become corrupt.

Undervoltage can be broken into three categories: sags, brownouts, and blackouts. **Sags,** which usually are not a problem, are very short drops lasting only a few milliseconds. **Brownouts,** on the other hand, last longer than sags and can force the computer to shut down, can introduce memory errors, and can cause unsaved work to be lost.

NOTE: Brownouts or power failures of 200 milliseconds are sufficient to cause power problems with the PC.

Brownouts can be caused by damaged power lines and by equipment that draws massive amounts of power (air conditioners, copy machines, laser printers, and coffee makers). **Blackouts** are total power failures.

14.4.2   Noise

In addition to the overvoltages and the undervoltages on the power lines, the computer may experience electrical noise or radio frequency interference caused by telephones, motors, fluorescent lights, and radio transmitters. Noise can introduce errors into executable programs and files.

NOTE: To limit the chance of AC line noise, you should install the computer on its own power circuit.

14.4.3   Power Protection Devices

Most of these voltage fluctuations can be prevented from doing any damage. To protect the computer from overvoltages and undervoltages, several devices can be used. They are surge protectors, line conditioners, standby power supplies, and uninterruptible power supplies.

The most common of these is the **surge protector.** A surge protector is designed to prevent most short-duration, high-intensity spikes and surges from reaching your PC by absorbing excess voltages.

The most common surge protector uses the metal oxide varistor (MOV). A MOV looks like a brightly colored plastic-coated disk capacitor. The MOV works by siphoning electricity to ground when the voltage exceeds 200 volts. Consequently, the voltage spike is "clipped." The excess electricity is then converted into heat. Other devices used to suppress overvoltages are gas discharge tubes, pellet arrestors, and coaxial arrestors. The better surge protectors use a combination of these.

When purchasing a surge protector, you should consider the following:

- **Energy absorption**—Surge protectors are rated by the amount of energy that can be absorbed (measured in joules): 200 joules is basic, 400 is good, and 600+ is excellent.
- **IEEE 587A voltage let-through**—Underwriters Laboratories has established the UL 1449 standard for surge suppressors. It rates suppressors by the amount of voltage that is allowed to pass through to the protected equipment. There are three levels of rated protection: 330 V, 400 V, and 500 V. The lower the number, the better the protection.
- **UL listing**—Underwriters Laboratories Inc. is an independent testing laboratory that certifies electrical equipment specifications. A UL listing indicates that the surge protector meets national electrical code and safety standards.
- **Protection indicator**—An LED indicates if the MOVs are working or not.
- **Circuit breaker or fuse**—Most suppressors will have either a fuse or a resettable circuit breaker, which will blow or trip if there is a short circuit or severe surge that causes excessive current to flow. Breakers are better because fuses, once blown, have to be replaced.
- **Protection guarantee**—This is usually an equipment protection guarantee. It says that if your equipment is damaged when plugged into the suppressor, the manufacturer will pay to have the equipment repaired or replaced.

Some surge protectors will have site wiring fault lights that illuminate only when there is a wiring fault in the circuit to which the surge protector is connected. Otherwise, this light should be off at all times. In addition, new protectors protect much more than

Figure 14.2   UPS

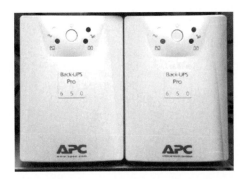

power cables. These may include an RJ-45, RJ-11, or other ports to protect network cards, modems, and the entire system from extremely high surges that occur when, for example, a telephone pole is hit by lightning.

There are several drawbacks to the surge protector. First, it will only protect against overvoltages, not undervoltages. In addition, the life expectancy of a MOV is limited. With every spike, the MOV gets weaker and weaker until it can't protect the PC any more. When buying a surge protector, make sure the surge protector has some kind of indicator (an LED light or a beep) to inform you when the surge protector can no longer protect. In addition, you need to be extremely careful when buying a surge protector. Make sure it's not a device that only gives you extra power connections like an extension cord.

Another type of surge protector is the phone line surge protector. If you have a modem, fax/modem, or fax connection connected in your PC, you should consider getting one. It will prevent surges and spikes, which travel through your telephone lines.

The next level of protection is the **line conditioner.** It uses the inductance of transformers to filter out noise and capacitors (and other circuits) to "fill in" during brownouts. In addition, most line conditioners include surge protection.

The last two forms of protection are based on battery backup systems. They are the **standby power supply (SPS)** and the **uninterruptible power supply (UPS).** (See figure 14.2.) The standby power supply consists of a battery hooked up parallel to the PC. When the SPS detects a power fluctuation, the system will switch over to the battery. Of course, the SPS requires a small but measurable amount of time to switch over (usually one-half of one cycle of the AC or less than 10 milliseconds). Most SPSs will include built-in surge protection devices.

Although similar to the standby power supply, the uninterruptible power supply differs because the battery is connected in series with the PC. The AC power is connected directly to the battery. Since the battery always provides clean DC power, the PC is protected against overvoltages and undervoltages.

For the SPS and UPS, when DC power is sent from the battery, the DC power has to be converted back to AC power before reaching the PC's power supply. Most SPSs and UPSs will generate a sine wave. Be aware, though, that some poorer-quality SPSs and UPSs generate a square wave instead of a sine wave and should be avoided.

NOTE: If you do not use a UPS or SPS for long periods of time, you should not discharge the battery. If you do, the battery may lose some of its capacity to store power or may be unable to accept a charge at all. Of course, always check the manufacturer's documentation.

A UPS is ideal to use in local area networks (LANs) and wide area networks (WANs), which consist of many computers connected together usually with cable. The computers on the network are divided into servers and workstations. Since the main function of the server is to provide file services (multiple users accessing data files, large databases, and application programs), the servers access a disk constantly. If a power disturbance occurs when a file is open or a file is reading or writing, the file can easily become lost or corrupted. If the File Allocation Table is corrupted, it could lead to losing the entire disk. It is probably not cost-effective to have UPSs for every PC, but it is important that each server has a UPS to help protect against power-related problems.

Uninterruptible power supplies are usually not designed to keep a PC running for hours without power. Instead, they are usually used to give the user or users enough time to save all files and to properly shut down the PC. In addition, you should not connect laser printers to a UPS since the laser printers have large current demands and can generate line noise.

Most SPSs and UPSs used for servers will have a system management port. This is usually a standard serial port or USB port that allows the SPS or UPS to connect to the host computer it is protecting. The host computer runs management software that gathers statistics about the power the SPS or UPS is using and providing. When a power failure occurs, this port is used to signal from the SPS or UPS, informing the management software on the host computer that the power to the SPS or UPS has failed. The management software can then initiate a graceful shutdown of the computer, including sending out messages to its users to save all their work and log off the system.

14.4.4   Redundant Power Supplies

One advanced feature available on mission-critical computers such as servers is a redundant power supply. The system will have two or more power supplies within the system, each of which is capable of powering the entire system by itself. If for some reason there is a failure in one of the units, the other one will seamlessly take over to prevent the loss of power to the PC. You can usually even replace the damaged unit without taking the machine down. This is called hot swapping, and it is an essential productivity backup for use in servers and other machines used by a number of people. Unfortunately, these types of systems are not cheap.

14.5   CLUSTERING

Clustering is connecting two or more computers, known as nodes, together in such a way that they behave like a single computer. It is used for parallel processing, for load balancing, and for fault tolerance. The computers that form the cluster are physically connected by cable and are logically connected by cluster software. As far as the user is concerned, the cluster appears as a single system to end users.

Network load balancing clusters distribute client connections over multiple servers. Internet clients access the cluster using a single IP address (or a set of addresses for a multihomed host). The clients are unable to distinguish the cluster from a single server. Server programs do not identify that they are running in a cluster. However, a network load balancing cluster differs significantly from a single host running a single server program, because it provides uninterrupted service even if a cluster host fails. The cluster also can respond more quickly to client requests than a single host (for load-balanced ports).

In a failover configuration, two or more computers serve as functional backups for each other. If one should fail, the other automatically takes over the processing normally performed by the failed system, thus eliminating downtime. If the server is to share common data, the cluster servers are connected to at least one shared SCSI bus with a storage device connected to both servers, and they are also connected to at least one storage device that is not shared. Of course, failover clusters are highly desirable for supporting mission-critical applications.

A **storage area network (SAN)** is a high-speed subnetwork of shared storage devices. A SAN's architecture works in a way that makes all storage devices available to all servers on a LAN. If an individual application in a server cluster fails (but the node does not), the cluster service will typically try to restart the application on the same node. If that fails, it moves the application's resources and restarts them on another node of the server cluster. This process is called failover.

Network-attached storage (NAS) is hard disk storage that is set up with its own network address rather than being attached to the department computer that is serving applications to a network's workstation users. By removing storage access and its management from the department server, both application programming and files can be served faster because they are not competing for the same processor resources. The network-attached storage device is attached to a local area network (typically, an Ethernet network) and assigned an IP address. File requests are mapped by the main server to the NAS file server. Network-attached storage includes multidisk RAID systems as well as software for configuring and mapping file locations to the network-attached device. NAS can be part of a SAN. While low cost may be a reason to choose a NAS, NAS devices are not upgradable.

14.6   PLANNING DISASTER RECOVERY

When a computer has a problem, the user's first response is "Oh no, not now!" I am sure that you will agree that there is no good time for a computer to break down. When a network fails, the failure can affect many people and can literally cost a company thousands of dollars of business or productivity for every hour the network is down.

One of your primary jobs as administrator is to deal with those disasters and to plan ahead to minimize the frequency of the failures and the degree of the failure. This means you should:

1. Establish a backup plan and perform the backing up of data.
2. Document the network so that you and your team can find information quickly about the network.

3. Maintain a log listing all problems and their solutions to be used to determine trends, to plan for network and personnel resources, and to make it easy to look up solutions for when the same problem occurs again.

14.7   TAPE DRIVES

Tape drives read and write to a long magnetic tape. They are relatively inexpensive and offer large storage capacities, making tape backup drives ideal for backing up hard drives on a regular basis. To back up a hard drive, you just insert a tape into the drive, start a backup software package, and select the drive/files you want to back up and it will be done. If a drive or file is lost, the backup software can be used to restore the data from the tape to the hard drive. If the right tape drive and backup software are chosen, the drive could automatically back up the hard drive at night, when it is being used least. The only thing you would have to remember is to replace the tape each day.

NOTE: Sometimes tapes fail, and there have been times when people think they selected a drive or file to be backed up only to find that they have a blank tape when disaster occurs. Therefore, it is important to occasionally test the tapes by choosing an unimportant file and restoring it to the hard drive to make sure the tape is viable.

The tape player drags a magnetic tape across a head, and the player reads the information from the magnetic tape. If the tape head is dirty, the computer may report a tape error. Cleaning the tape drive will normally resolve this problem. If cleaning the tape drive is not effective, then you should try using a new tape. You should replace the tape drive if these troubleshooting attempts fail. If the system reports that no tape drive is present, be sure that the drive is on and plugged in.

14.7.1   The Tape Media

Before the IBM PC, magnetic tapes were used on older mainframes as a primary storage device and a backup storage device. Eventually the magnetic tape evolved into the floppy disk followed by the hard drive. When the IBM PC was introduced, it included a drive port for a cassette tape storage device.

Floppy disks, which are probably the closest to a tape, consist of a Mylar platter coated with a magnetic substance to hold magnetic fields. In addition, floppy disks are random-access devices. This means that no matter where the data is located on the disk, the read/write heads can move directly to the proper sector and start to read or write.

Instead of using Mylar platters, tapes use a long polyester substrate that is coated with a layer of magnetic material. Unlike a floppy disk, a tape stores and retrieves data sequentially. Therefore, when a file needs to be retrieved, the search has to start at the beginning of the tape and read each area of the tape before it gets to the correct file. Since it takes time to find the appropriate file, tapes are completely inappropriate as a PC's primary storage device.

14.7.2   Recording Methods

Tapes are divided into parallel tracks laid across the tape. The number of tracks varies with the drive and the standard it follows. The data is recorded either parallel, serpentine, or helical scan.

Parallel recording spreads the data throughout the different tracks. As an example, if a tape is divided into 9 tracks, a byte of information with parity could be spread out through all 9 tracks (1 bit per track). Newer tape systems may lay 18 or 36 tracks across the tape, allowing 2 or 4 bytes of information. While the tape offers high transfer rates, data retrieval time is slow because the tape drive might have to fast-forward across the entire tape before retrieving the data. In addition, the read/write assembly is quite complicated since it must consist of several poles and gaps, one for each track. Unfortunately, the complexity also drives the cost of the drive up.

Most PC tape systems use serpentine recording. While the tape is still divided into tracks, it will write the data onto one track, reach the end of the track, move to the next track, and write to the second track. It will keep repeating this process until it runs out of tracks. A serpentine tape can access data quickly by moving its head between the different tracks. Since the read/write assembly requires only one pole and gap, the drives are cheaper.

The newest method of recording is helical scan. Much like a VCR read/write head, tape backup drives with helical scan use read/write heads mounted at an angle on a cylindrical drum. The tape is partially wrapped around the drum. As the tape slides across the drum, the read/write heads rotate. As each head approaches the tape, the heads take swipes at the tape, reading or writing the data. The tape is moved only slightly between swipes, allowing data to be packed very tightly. In addition, since each head is skewed slightly from the others, the heads respond well to signals written in the same orientation, but not well to the other signals. Therefore, blank spaces are not needed. Lastly, if two more heads are added to the drum, data can be read immediately after it is written. Therefore, if any errors are detected, the data can be rewritten immediately on the next piece of tape.

14.7.3   Tape Standards

Tapes come in different sizes and shapes and offer different speeds and capacities. As a result, several standards have been developed, including the quarter-inch cartridge (QIC) and the digital audiotape (DAT).

In 1972, the 3M Company introduced the first quarter-inch tape cartridge designed for data storage. The cartridge measured 6 in. × 4 in. × 5/8 in. Although, the cartridge became the standard, each tape drive manufacturer used different encoding methods, varied the number of tracks, and varied the data density on the tape, causing all kinds of compatibility problems. (See figure 14.3.)

As a result, in 1982, a group of manufacturers formed the QIC Committee to standardize tape drive construction and application. The full-size quarter-inch cartridge standardized by the QIC Committee is also referred to as the DC 6000 cartridge. The DC stands for "data cartridge."

Figure 14.3   A DAT and DC QIC Tape

The first tape that the QIC Committee approved was the QIC-24, which used serpentine recording. It had 9 tracks and a density of 8000 bits/in., giving a total storage capacity of 60 MB. It achieved 90 in./s and offered 720 Kbits/s with the QIC-02 interface.

Throughout the years, the data density was increased and more tracks were added. The last QIC-1000-DC packed 30 tracks across the tape at 36,000 bits/in., allowing up to 1.2 GB per tape cartridge. The speed was also increased 2.8 Mbits/s. Yet the QIC-1000-DC drives could read previous QIC tapes.

By 1989, the QIC Committee revamped the original QIC standard by using 1,7 RRL encoding and higher coercivity. This allowed higher bit density, a high number of tracks, and faster transfer rates. By 1995, the QIC introduced the QIC-5210-DC, which had 144 tracks and 76,200 bits/in., allowing for 25 GB. (See table 14.3.)

Since the full-size QIC is too large to fit into a drive bay, the QIC Committee created the minicartridge, which was 3.25 in. × 2.5 in. × 1.59 in. The minicartridges are also referred to as the DC 2000 cartridges.

The QIC-40-MC was the first standard adopted. It fit into the 5.25 drive bay, and it connected to the computer by using the floppy drive controller. Since floppy disk drives use MFM encoding, so did the QIC-40-MC. Different from previous tapes, QIC specified the format of the data on the tape, which included how sectors were assigned to files and FAT to list bad sectors. This requires the tapes to be formatted, which takes time to complete. Consequently, you could buy tapes formatted or unformatted. Another advantage was that the tapes could be accessed randomly. Although the tape had to be moved to the proper sector, it did not have to read each file sequentially although it did have to move the tape. (See table 14.4.)

In 1995, a number of tape and drive manufacturers, including Conner, Iomega, HP, 3M, and Sony, introduced Travan technology. Instead of using the standard QIC size, the cartridge measured 0.5 in. × 3.6 in. × 2.8 in. The front was smaller, which was inserted into the drive, while the back was larger to contain the tape spools. The different sizes allowed longer tapes and more data capacities. In addition, the Travan drive accepts standard DC 2000 cartridges and QIC-Wide cartridges. (See table 14.5.)

The newest type of tape is the digital audiotape, which uses the same technology as VCR tapes (helical scan). The 8-mm DAT tapes allow capacities up to 35 GB or larger.

Table 14.3   QIC Data Cartridges

QIC Standard Number	Capacity without Compression	Tracks	Interface	Original Adoption Date
QIC-24-DC	45 MB or 60 MB*	9	SCSI or QIC-02	4-83
QIC-120-DC	125 MB	15	SCSI or QIC-02	10-85
QIC-150-DC	150 MB or 250 MB*	18	SCSI or QIC-02	2-87
QIC-525-DC	320 MB or 525 MB*	26	SCSI or SCSI-2	5-89
QIC-1350-DC	1.35 GB	30	SCSI-2	5-89
QIC-1000-DC	1.2 GB	30	SCSI or SCSI-2	10-90
QIC-6000C	6 GB	96	SCSI-2	2-91
QIC-2100-DC	2.1 GB	30	SCSI-2	6-91
QIC-5010-DC	13 GB	144	SCSI-2	2-92
QIC-2GB-DC	2.0 GB	42	SCSI-2	6-92
QIC-5GB-DC	5 GB	44	SCSI-2	12-92
QIC-4GB-DC	4 GB	45	SCSI-2	3-93
QIC-5210-DC	25 GB	144	SCSI-2	8-95

*Depending on the length of the tape.

The DAT standard has primarily been developed and marketed by Hewlett-Packard. HP chairs the DDS (Digital Data Storage) Manufacturers Group, which led the development of the DDS standards.

Data is not recorded on the tape in the MFM or RLL format, but rather bits of data received by the tape drive are assigned numerical values, or digits. Then these digits are translated into a stream of electronic pulses that are placed on the tape. Later, when information is being restored to a computer system from the tape, the DAT tape drive translates these digits back into binary bits that can be stored on the computer.

Digital data storage (DDS) tapes are currently the newest standard of the digital audiotape. DDS-3 can hold 24 GB (or equivalent), or more than 40 CD-ROMs, and supports a data transfer rate (DTR) of 2 Mbits/s, some drives can even support up to 40 GB or larger. A DDS is slightly larger than a credit card. In a DDS drive, the tape barely creeps along, requiring about 3 seconds to move an inch. The head drum spins rapidly at 2000 revolutions/min, putting down 1869 tracks across a linear inch of tape which allow 61 Kbits/in. The main advantage of the DAT is its access speed and capacity. The standard DDS protocols are shown in table 14.6, all of which are backwards-compatible.

Table 14.4 QIC Minicartridges

	Capacity without Compression	Tracks	Interface	Original Adoption Date
QIC-40-MC	40 MB or 60 MB[*]	20	Floppy or optional card	6-86
QIC-80-MC	80 MB or 120 MB[*]	28	Floppy or optional card	2-88
QIC-128-MC	86 MB or 128 MB	32	SCSI or QIC	5-89
QIC-3030-MC	555 MB	40	SCSI-2 or QIC	4-91
QIC-3020-MC	500 MB	40	Floppy or IDE	6-91
QIC-3070-MC	4 GB	144	SCSI-2 or QIC	2-92
QIC-3010-MC	255 MB	40	Floppy or IDE	6-93
QIC-3040-MC	840 MB	42/52	SCSI-2 or QIC	12-93
QIC-3080-MC	1.6 GB	60	SCSI-2 or QIC	1-94
QIC-3110-MC	2 GB	48	SCSI-2 or QIC	1-94
QIC-3230-MC	15.5 GB	180	SCSI-2 or QIC	6-95
QIC-3095-MC	4 GB	72	SCSI-2 or QIC	12-95

[*]Depending on the length of the tape.

Table 14.5 Travan Technology

	TR-1	TR-2	TR-3	TR-4	TR-5
Capacity: Native Compressed	400 MB 800 MB	800 MB 1.6 GB	1.6 GB 3.2 GB	4 GB 8 GB	10 GB 20 GB
DTR: Minimum Maximum	62.5 KBps 125 KBps	62.5 KBps 125 KBps	125 KBps 250 KBps	60 MB/min 70 MB/min	60 MB/min 110 MB/min
Tracks	36	50	50	72	108
Data Density	14,700 ftpi	22,125 ftpi	44,250 ftpi	50,800 ftpi	50,800 ftpi
Compatibility	QIC 80 (read/write) QIC 40 (read-only)	QIC 3010 (read/write) QIC 80 (read-only)	QIC 3010/ QIC 3020 (read/write) QIC 80 (read-only)	QIC 3080/ QIC 3095 (read/write) QIC 3020 (read-only)	QIC 3220 (read/write) TR-4 QIC 3095 (read-only)

Table 14.6 DDS Drives

Standard	Capacity	Maximum DTR
DDS	2 GB	55 KBps
DDS-1	2–4 GB	0.55–1.1 MBps
DDS-2	4–8 GB	0.55–1.1 MBps
DDS-3	12–24 GB	1.1–2.2 MBps
DDS-4	20–40 GB	2.4–4.8 MBps

Table 14.7 DLTs

Standard	Capacity (n/c)	Interface	Maximum DTR
DLT2000	15–30 GB	SCSI	2.5 MBps
DLT4000	20–40 GB	SCSI	3 MBps
DLT7000	35–70 GB	SCSI	20 MBps

One of the newest tapes is the DLT (digital linear tape). Designed for high-capacity, high-speed, and highly reliable backup, DLTs are ½ in. wide, have capacities of 35–70 GB (or more) compressed, and have a data transfer rate of 5 to 10 Mbits/s or more. Unfortunately, the drives are quite expensive and are used primarily for network server backup. (See table 14.7.)

One of the most significant of the new formats is the next generation of digital linear tapes, otherwise known as Super DLTs. Drives based on Super DLT technology will far exceed the 35-GB native capacity of the DLTtape IV format—with which it aims to be backwards-compatible.

Using a combination of optical and magnetic recording techniques known as laser-guided magnetic recording (LGMR), Super DLT uses lasers to more precisely align the recording heads. At the core of LGMR is an optically assisted servo system referred to as a pivoting optical servo (POS). This combines high-density magnetic read/write data recording with laser servo guiding. Designed for high-duty cycle applications, the POS has a much lower sensitivity to external influences, which allows a much greater track density than is possible with other tape systems. The POS system decreases manufacturing costs and increases user convenience by eliminating the need for preformatting the tape. Furthermore, 10 to 20 percent more capacity is gained by deploying the optical servo on the unused backside of the medium, making the entire recording surface available for actual data.

As indicated by table 14.8, the ultimate goal is to cram up to 1.2 TB of uncompressed data onto a single cartridge with transfer rates rising to an eventual 100 MBps

Table 14.8 SDLTs

	SDLT 220	SDLT 320	SDLT 640	SDLT 1280	SDLT 2400
Native capacity	110 GB	160 GB	320 GB	640 GB	1.2 TB
Compressed capacity (2:1 compression)	220 GB	320 GB	640 GB	1.28 TB	2.4 TB
Native DTR	11 MBps	16 MBps	32 MBps	50+ MBps	100+ MBps
Compressed DTR	22 MBps	32 MBps	64 MBps	100+ MBps	200+ MBps
Media	SDLT I	SDLT I	SDLT II	SDLT III	SDLT IV
Interfaces	Ultra2 SCSI LVD HVD	Ultra2 SCSI Ultra160 SCSI	Ultra320 SCSI Fiber channel	TBD	TBD
Date	Q1 2001	Q1 2002	Q3 2003	Q1 2005	

Table 14.9 8-mm Tapes

Standard	Capacity (n/c)	Interface	Maximum DTR
Standard 8-mm	3.5–7 GB	SCSI	32 MB/min
Standard 8-mm	5–10 GB	SCSI	60 MB/min
Standard 8-mm	7–14 GB	SCSI	60 MB/min
Standard 8-mm	7–14 GB	SCSI	120 MB/min
Mammoth	20–40 GB	SCSI	360 MB/min

uncompressed. Initial products, however, offer a more modest 110 GB, with sustained data transfer rates of 11 MBps in native mode.

Originally, 8-mm tape technology was designed for the video industry to transfer high-quality color images to tape for storage and retrieval. Now 8-mm technology has been adopted by the computer industry as a reliable way to store large amounts of computer data. Similar to DAT but with generally greater capacities, 8 mm also employs helical scan technology. A drawback to the helical scan system is the complicated tape path. Because the tape must be pulled from a cartridge and wrapped tightly around the spinning read/write cylinder, a great deal of stress is placed on the tape. (See table 14.9.)

There are two major protocols, utilizing different compression algorithms and drive technologies, but the basic function is the same. Exabyte Corporation sponsors standard

Table 14.10 AIT Formats

	AIT-1	AIT-2	AIT-3	S-AIT
Native capacity	35 GB	50 GB	100 GB	500 GB
Compressed capacity	90 GB	130 GB	26 GB	1.3 TB
Native DTR	4 MBps	6 MBps	12 MBps	30 MBps
Compressed DTR	10 MBps	15.6 MBps	31.2 MBps	78 MBps
Form factor	3.5 in.	3.5 in.	3.5 in.	5.25 in.
Media type	8-mm AME	8-mm AME	8-mm AME	1/2-in. AME
MTBF (hours)	300,000	300,000	400,000	500,000

8-mm and Mammoth, while Seagate and Sony represent a new 8-mm technology known as advanced intelligent tape (AIT).

Exabyte has been a leader in the tape storage industry for more than a decade, pioneering the use of 8-mm tape for backup, incorporating Sony's camcorder-based mechanisms into more than 1.5 million tape drives. While camcorder-based mechanisms are adequate for low-duty cycle applications, they are less appropriate for today's demanding server-based applications. Introduced in 1996, Mammoth is a more advanced and reliable technology and represents Exabyte's response to the requirements of this midrange server market.

Mammoth features an Exabyte-designed and -manufactured deck that has 40 percent fewer parts than previous 8-mm drives and that was specifically designed to improve reliability by reducing tape wear and tension variation. A solid aluminum deck casting provides the extra accuracy and rigidity needed to maintain tight tolerances. The casting shields the internal elements from dust and contamination and directs heat away from the tape path. A three-point shock-mount system isolates the casting from the sheet-metal housing, providing protection from external forces. With seven custom application-specific integrated circuits (ASICs), Mammoth calibrates itself regularly and, searches for and reports any errors.

The AIT recording technology today offers uncompressed capacities including 25 GB, 35 GB, 50 GB, and 100 GB, while showing scalability to a sixth-generation product with an uncompressed capacity of up to 800 GB (up to 2 TB compressed), all using the compact 8-mm cartridge form factor. The higher compression specifications for capacity and performance for the AIT family are achieved through the incorporation of adaptive lossless data compression (ALDC) technology, which delivers an average 2.6-to-1 compression ratio—significantly more than some competing products. (See table 14.10.)

Advanced intelligent tape (AIT) was the first multisourced tape standard targeted at the midrange server market, which is typically characterized by systems that support 2 to 129 users in a commercial environment. Several breakthroughs make this possible, including a stronger, thinner medium that is more stable and has better coatings than previously available, new head technologies, higher levels of integration, and a unique memory-in-cassette (MIC) feature. The result is multigigabyte, high-performance tape drive systems with very low frequencies of error that are perfect for tape libraries and robotic applications associated with midrange systems backup.

The most unique feature of the AIT format remains Sony's innovative MIC drive interface system. In the AIT-2, this consists of a 64-Kbit memory chip built into the data cartridge. The data contained on the chip includes the tape's system log, search map, and other user-definable information, allowing data to be accessed immediately no matter what section of the tape is being accessed. The ability of the MIC to support multiple partitions and multiple load points drastically reduces the average time to access data to fewer than 20 seconds, compared with an average of over 100 seconds for conventional, competing technologies.

A key factor in the capabilities of AIT technology is the use of the extremely durable and field-proven advanced metal evaporated (AME) tape technology, and AIT-2 uses Sony's latest formulation of AME media providing the patented diamond-like carbon (DLC) coating with a higher-output metalization layer. The consequent benefits are an extremely long head life—one able to withstand thousands of media uses—in a system capable of providing high-density recording in a compact form factor.

In 2001 Sony announced a Write Once Read Many (WORM) variation of its AIT-2 technology format, thereby bringing advanced data security to the 3.5-in form factor. The additional security against inadvertent or malicious deletion or alteration of data makes the format particularly suited for the archival of financial, securities, government, medical, and insurance data. Table 14.10 compares the family of AIT formats, including the S-AIT format which is expected to reach the market by the end of 2002.

14.8   BACKUP

Data is the raw facts, numbers, letters, or symbols that the computer processes into meaningful information. Examples of data include a letter to a company or a client, a report for your boss, a budget proposal of a large project, or an address book of your friends and business associates. Whatever the data is, it can be saved (or written to disk) so that it can be retrieved at any time, it can be printed on paper, or it can be sent to someone else over the telephone lines.

Data stored on a computer or stored on the network is vital to the users and probably the company. The data represents hours of work, and the data is sometimes irreplaceable. Therefore, I consider data as the most important part of the computer. Data loss can be caused by many things including hardware failure, viruses, user error, and malicious users. When disaster occurs, the best method to recover data is backup, backup, backup. When disaster has occurred and the system does not have a backup of its important files, it is often too late to recover the files.

A **backup** of a system consists of an extra copy of data and/or programs. As a technician, consultant, or support person, you need to emphasize at every moment the need to back up on servers and client systems. In addition, it is recommended that the clients save their data files to a server so that you have a single, central location to keep the backup. This may go as far as selecting and installing the equipment, doing the backup, or training other people to do the backup. When doing all of this, be sure to select the equipment and method that will assure that the backup will be completed on a regular basis. Remember that if you have the best equipment and software but no one completes the backup, the equipment and software are wasted.

THE BEST METHOD FOR DATA PROTECTION IS BACKUP, BACKUP, BACKUP.

When creating a plan for a backup, these three steps should be followed:

1. Develop a backup plan.
2. Stick to the backup plan.
3. Test the backup.

When developing a backup plan, you must consider the following:

1. What equipment will be used?
2. How much data needs to be backed up?
3. How long will it take to do the backup?
4. How often must the data be backed up?
5. When will the backup take place?
6. Who will do that backup?

Whatever equipment, person, or method is chosen, you must make sure that the backup will be done. If you choose the best equipment, the best software, and the brightest person, but the backup is not done for whatever reason, you wasted your resources—and you put your data at risk.

Backups can be done with floppy disks, extra hard drives (including network drives), compact disk drives, tape drives, and other forms of removable media. Probably the best method of backing up files would be tape drives. A tape drive can store 20 or more gigabytes.

Question:

How often should the backup be done?

Answer:

How often the backup is done depends on the importance of the data. If you have many customers loaded into a database that is constantly changing or if your files represent the livelihood of your business, you should back the files up everyday. If only a few letters get sent throughout the week and contain nothing vitally important, you can do a backup once a week.

All types of backups can be broken into the categories shown in table 14.11.

Some vendors also recognize another type of backup called copy. The copy backup is a normal backup, but it does not shut off the archive attribute. This is typically used to back up the system before you make a major change to the system. The archive attribute is not shut off so that your normal backup procedures are not affected.

Table 14.11 Types of Backup

Normal/full	The full backup will back up all files selected and shut off the archive file attribute, indicating the file has been backed up.
Incremental	An incremental backup will back up the files selected if the archive file attribute is on (files since the last full or incremental backup). After the file has been backed up, it will shut off the file attribute, indicating that the file has been backed up. *NOTE:* You should not mix incremental and differential backups.
Differential	A differential backup will back up the files selected if the archive file attribute is on (files since the last full backup). Different from the incremental backup, it does not shut off the archive attribute. *NOTE:* You should not mix incremental and differential backups.

Example 1:

You decide to back up the entire hard drive once a week on Friday. You decide to use the full backup method. Therefore, you perform a full backup every Friday. If the hard drive goes bad, you use the last backup to restore the hard drive.

Example 2:

You decide to back up the entire hard drive once a week on Friday. You decide to use the incremental method. Therefore, you perform a full backup on week 1. This will shut off all the archive attributes, indicating that all the files have been backed up. On week 2, week 3, and week 4, you perform incremental backups using a different tape or disk. Since the incremental backup turns the archive attribute off, it backs up only new files and changed files. Therefore, all four backups make up the entire backup. It is much quicker to back up a drive using an incremental backup than a full backup. Of course, if the hard drive fails, you must restore all four backups (backups 1, 2, 3, and 4) to restore the entire hard drive.

Example 3:

You decide to back up the entire hard drive once a week on Friday. You decide to use the differential method. Therefore, you perform a full backup on week 1. This will shut off all the archive attributes, indicating that all the files have been backed up. On week 2, week 3, and week 4, you perform differential backups using a different tape or disk. Since the differential backup does not turn the archive attribute off, it backs up the new files and the changed files since the last full backup. Therefore, the full backup and the last differential backup make up the entire backup. It is much quicker to back up a drive using a differential backup than a full backup but slower than using an incremental backup. If the hard drive fails, you must restore backup 1 and the last differential to restore the entire hard drive.

After the backups are complete, you should check to see if they actually worked. This can be done by picking a nonessential file and restoring it to the hard drive. This way you can discover if the backups are empty or a backup/restore device is faulty.

You should keep more than one backup. Tapes and disks do fail. One technique is to rotate through three sets of backups. If you perform a full backup once a week, you would then use three sets of backup tapes or disks. During week 1, you would use tape/disk 1. During week 2, you would use tape/disk 2, and during week 3, you would use tape/disk 3. On week 4, you start over and use tape/disk 1. If you have to restore a hard drive and the tape or disk fails, you can always go to the tape or disk from the week before. In addition, I would perform monthly backups and store them elsewhere. You would be surprised how many times a person loses a file but may not know it for several weeks. If the data is important enough, you may consider keeping a backup set in a fireproof safe offsite. Lastly, when a system is initially installed and when you make any major changes to the system's configuration, it is always recommended to make two backups before proceeding. This way, if anything goes wrong, you have the ability to restore everything to the way it was before the changes. The reason for the two backups is tapes have been known to go bad on occasion.

Some places use the grandfather, father, son (GFS) backup rotation, which requires 21 tapes based on a 5-day rotation. Each month, you create a grandfather backup, which is stored permanently offsite, never to be reused. Each week, you create a full weekly backup (father), and each day, you create a differential or incremental backup (son).

After completing a backup, you should properly label the tape or disk before removing it and then store it in a secure, safe place. In addition, you should keep a log of what backups have been done. Especially if you need to rebuild the server, the log will prove to be especially helpful since it keeps track of what was backed up and when it was backed up. It will also let you know if someone is forgetting to do the backup.

14.9   VIRUSES

A **virus** is a program designed to replicate and spread, generally without the knowledge or permission of the user. Computer viruses spread by attaching themselves to other programs or to the boot sector of a disk. When an infected file is executed or accessed or the computer is started with an infected disk, the virus spreads into the computer. Some viruses are cute, some are annoying, and others are disastrous. Some of the disastrous symptoms of a virus include the following:

- Computer fails to boot.
- Disks have been formatted.
- The partitions are deleted, or the partition table is corrupt.
- Computer cannot read a disk.
- Data or entire files are corrupt or are disappearing.
- Programs don't run anymore.
- Files become larger.
- System is slower than normal.
- System has less available memory than it should.
- Information being sent to and from a device is intercepted.

Table 14.12   Virus Facts

Viruses can't infect a write-protected disk.	Viruses can infect read-only, hidden, and system files.
Viruses don't typically infect a document (except macro viruses).	Viruses typically infect boot sectors and executable files.
They do not infect compressed files.	A file within a compressed file could have been infected before being compressed.
Viruses don't infect computer hardware such as monitors or chips.	Viruses can change your CMOS values, causing your computer not to boot.
You cannot get a virus just by being on the Internet or a bulletin board.	You can download an infected file.

Question:

How does a virus spread?

Answer:

Since viruses are small programs that are made to replicate themselves, viruses spread very easily. For example, you are handed an infected disk or you download a file from the Internet or a bulletin board. When the disk or file is accessed, the virus replicates itself to RAM. When you access any files on your hard drive, the virus again replicates itself to your hard drive. If you shut off your computer, the virus in the RAM will disappear. Unfortunately, since your hard drive is infected, the RAM becomes infected every time you boot from the hard drive. When you insert and access another disk, the disk also becomes infected. You then hand the disk or send an infected file to someone else, and the cycle repeats itself.

Symantec, the developer of Norton Antivirus, says that most current infections are caused by viruses that are at least 3 years old. Stiller Research, the developer of Integrity Master, states that viruses are widespread but only a relatively small number (about 100) account for 90 percent of all infections. Table 14.12 lists a number of facts about viruses.

14.9.1   Types of Viruses

Computer viruses can be categorized into five types:

1. Boot sector
2. File
3. Multipartite
4. Macro
5. Trojan horse

Every logical drive (hard drive partition and floppy disk) has a boot sector with both bootable and nonbootable components. The boot sector contains specific information relating to the formatting of the disk. It also contains a small program, called the boot program, which loads the operating system files. On hard drives, the first physical sector (side 0, track 0, sector 1) contains the master boot record (MBR) and partition table. The master boot program uses the partition table to find the starting location of the bootable partition (active partition). It then tells the computer to go to the boot sector of the partition and load the boot program. A boot sector virus is transmitted by rebooting the machine from an infected diskette. When the boot sector program on the diskette is read and executed, the virus goes into memory and infects the hard drive, specifically the boot sector or the master boot program.

File infector viruses attach themselves to or replace executable files (usually files with the COM or EXE file name extension), but they can also infect SYS, DRV, BIN, OVL, and OVY files. Uninfected programs become infected when they are opened, including when they are opened from the DOS DIR command. Or the virus simply infects all the files in the directory from which it was run.

A multipartite virus has the characteristics of both boot sector viruses and file viruses. It may start as a boot sector virus and spread to executable files, or it may start from an infected file and spread to the boot sector.

A macro, or formula language, used in word processing, spreadsheets, and other application programs, is a set of instructions that a program executes on command. Macros group several keystrokes into one command or perform complex menu selections. They therefore simplify redundant or complex tasks. The macro viruses are the newest strain and are currently the most common type of virus. Unlike previous viruses, macro viruses are stored in a data document and spread when the infected document is accessed or transferred. Currently, the most vulnerable applications are Microsoft Word and Microsoft Excel. Some macro viruses modify the contents of the document and can even cause documents to be sent out via email.

The fifth type of virus is the Trojan horse virus, which, by definition, is not really a virus since it does not replicate itself. Nonetheless, the Trojan horse virus is a program that appears to be legitimate software, such as a game or useful utility. Unfortunately, when you run the Trojan horse and the trigger event occurs, the program will do its damage, such as formatting your hard drive.

Some viruses can be characterized as polymorphic viruses or stealth viruses. A polymorphic virus mutates, or changes its code, so that it cannot be as easily detected. Stealth viruses try to hide themselves by monitoring and intercepting a system's call. For example, when the system seeks to open an infected file, the stealth virus uninfects the file and allows the operating system to open it. When the operating system closes the file, the virus reinfects the file.

For more information on viruses—how they work, how they affect your computer, and what particular viruses are like—check out the URLs listed in table 14.13.

14.9.2   Worms

A **worm** is a program or algorithm that replicates itself over a computer network and usually performs malicious actions, such as using up the computer's resources and possibly shutting the system down. Typically, a worm enters the computer because of vulnerabilities in the computer's operating system.

Table 14.13   General Information about Viruses

Virus Information Library	http://vil.mcafee.com/default.asp
Virus Encyclopedia	http://www.symantec.com/avcenter/vinfodb.html
Virus hoaxes	http://vil.mcafee.com/hoax.asp http://www.symantec.com/avcenter/hoax.html
Virus Reference Area	http://www.symantec.com/avcenter/refa.html

As noted in chapter 13, the Computer Emergency Response Team (CERT) Coordination Center (CC) was started in December 1988 by the Defense Advanced Research Projects Agency, which was part of the U.S. Department of Defense, after the Morris worm disabled about 10 percent of all computers connected to the Internet. CERT studies Internet security vulnerabilities, provides services to websites that have been attacked, and publishes security alerts. CERT/CC's activities include doing research in the area of WAN computing and developing improved Internet security. The organization also provides training to incident-response professionals.

CERT website:
http://www.cert.org/

Microsoft Baseline Security Analyzer
http://www.microsoft.com/technet/security/tools/tools/mbsahome.asp

Symantec Security Advisory List
http://securityresponse.symantec.com/avcenter/security/Advisories.html

14.9.3   Virus Hoaxes

A virus hoax is a letter or email message warning you about a virus that does not exist. For example, the letter or warning may tell you that certain email messages may harm your computer if you open them. In addition, the letter or message usually tells you to forward the letter or email message to your friends, which creates more network traffic.

14.9.4   Antivirus Software

Antivirus software will detect and remove viruses and help protect the computer against viruses. Whichever software package is chosen, it should include a scanner-disinfector and an interceptor-resident monitor. The scanner-disinfector software will look for known virus patterns in the RAM, the boot sector, and the disk files. If a virus is detected, the software will typically attempt to remove the virus. The interceptor-resident monitor is a piece of software that is loaded and remains in the RAM. Every time a disk is accessed or a file is read, the interceptor-resident monitor software will check the disk or file for the same virus patterns that the scanner-disinfector software does. In addition, some interceptor-resident monitor software will detect viruses within the files as you download them from the Internet or bulletin board.

Unfortunately, scanner-disinfector software has three disadvantages. First, it can detect only viruses that it knows about. Therefore, the antivirus software package must be continually updated. The easiest way to do this is through the Internet. Second, it cannot always remove the virus. Therefore, the file may need to be deleted, or a low-level format may need to be performed on the hard drive. Lastly, if it succeeds in removing the virus, the file or boot sector may still have been damaged. Thus, the infected file still needs to be deleted or replaced, the boot sector may need to be re-created, partitions may need to be re-created, or a low-level format needs to be performed on the disk.

If you think a virus is present even though the interceptor software is installed, boot from a clean write-protected disk. This will ensure that the RAM does not contain a virus. Without changing to or accessing the hard drive, run an updated virus scanner-disinfector from the floppy disk. If a virus is detected and removed, it is then best to reboot the computer when the scanner-disinfector has finished checking the hard drive.

NOTE: Also boot from a bootable floppy and check the hard drive before installing any antivirus software.

14.9.5   Protecting against Viruses

To avoid viruses, you should follow these steps:

1. You should not use pirated software since there is more of a chance for it having a virus.
2. You should treat files downloaded from the Internet and bulletin boards with suspicion.
3. You should not boot from or access a floppy disk of unknown origin.
4. You should educate your fellow users.
5. You should use an updated antivirus software package that constantly detects viruses.
6. You should back up your files on a regular basis.
7. You should not use the administrative accounts for general use.
8. You should not give more rights than what is needed.
9. You should keep your operating system up to date with patches.

14.9.6   Removing a Virus

If you suspect that you have a virus, you need to immediately check your hard drive and disk with a current antivirus software package. If you think that your hard drive is infected, you need to have a noninfected, write-protected bootable floppy disk that contains the antivirus software. You need to boot the computer with the noninfected disk without accessing the hard drive. Next, run the software to check the hard drive. If you think that you have a virus on the floppy disk, boot your computer using the hard drive. If you have been using the possible infected disk on your computer, you need to first check the hard drive for viruses. Lastly, execute the antivirus program to check the floppy drive.

14.10 SERVER PERFORMANCE

Performance is the overall effectiveness of how data moves through the system. To be able to improve performance, you must determine the part of the system that is slowing down the throughput. It could be:

- The speed of the processor
- The speed of the memory
- The speed of the disk system
- The speed of your network adapter system

This limiting factor is referred to as the bottleneck of the system.

Sometimes relieving one bottleneck will cause another bottleneck. For example, when there is a relatively small number of clients, a 256-MB RAM configuration holds its own. But as the number of clients increases, the available cache shrinks, and the file server starts to rely heavily on the disk subsystem. Adding RAM will increase the available file cache and let the server get the most out of its single microprocessor. If the microprocessor is saturated, you won't see a significant improvement in performance if you add RAM. If the hard drive is causing the bottleneck and you add a second microprocessor, you only make matters worse by inundating the already overworked disk with more requests. No matter how many microprocessors and how much RAM you add, you will not be able to increase the server performance until you replace the disk subsystem with a faster one.

NOTE: When a server becomes overworked, another option would be to install another server and move some of the services from one server to the other server.

Therefore, the question is how many microprocessors and how much RAM is enough. Unfortunately, the answer depends on too many factors to give you a simple answer. Some of these factors include the number of clients, the type of server, the type of services offered by the server, the type of files being accessed, and the way the files are being accessed.

Therefore, when first installing a server, you would perform a best guess on hardware requirements based on your prior experience. After the server is installed, you would then keep monitoring the performance of the server to determine if any part of the sever needs to be upgraded. To monitor the server, most servers have intensive utilities that can help you determine the server's performance and bottlenecks. For Windows 2000, it is the Performance Monitor and Network Monitor, and for Novell NetWare, it is the Monitor NLM and Novell's Lanalyzer.

Lastly, when choosing the equipment for a server, you should not choose the newest or the fastest equipment. While you still want to choose items that offer good performance, it is very important that you choose items that are reliable—that have a good track record. Remember that one of the primary goals of a network is to minimize downtime. So reliability is more important than a little bit more performance.

14.10.1 Processor Performance

If the processor has a CPU utilization of approximately 80 percent, that means that 80 percent of the time, the processor is actually doing work. If the CPU utilization is constantly at 80 percent or higher, that indicates that the processor is working too hard and you need to either upgrade your system to a faster processor, increase the number of processors, or decrease the load of the server.

Symmetric multiprocessing (SMP) involves a computer that uses two or more microprocessors that share the same memory. If software is written to use the multiple microprocessors, several programs can be executed at the same time, or multithreaded applications can be executed faster. A multithreaded application is an application that is broken into several smaller parts and executed simultaneously. For example, a word processor uses a thread to respond to keys typed on the keyboard by the user to place characters in a document. Other threads are used to check spelling, paginate the document as you type, and spool a document to the printer in the background. The ability for an OS to use additional microprocessors is known as multiprocessing scalability.

To use multiple processors, the processors need to use the same speed, cache, and steppings (internal version of the processor). In addition, to take advantage of SMP, you also must have an operating that will support SMP. Most modern NOSs support SMP. For Microsoft Windows to support SMP, you will need to use a HAL (hardware abstraction layer), and for Linux, you will have to reconfigure and recompile the kernel.

14.10.2   Memory Performance

To increase performance of the memory system, you must first make sure you have plenty of RAM. If you do not have enough memory, your system will use your virtual memory more often, which will slow performance significantly. Next, as faster processors and new motherboards are released, new forms of memory are also being released. For example, the current standard for memory is SDRAM, which replaced both EDO and FPDRAM. Over the last few years, SDRAM was released in three main speeds: PC-66, PC-100, and PC-133. Recently, SDRAM is starting to be replaced with DDR-SDRAM and RDRAM.

Cache memory is a special ultrafast memory that acts as a buffer between the microprocessor and the slower RAM. The data that is stored in the cache is based on an old computer science principle stating that if the processor recently referred to a location in memory, it is likely that it will refer to it again in the near future. Using a cache to hold recently used memory values saves the processor from going to memory each time to reload them. Today, Level 1 (L1) cache and Level 2 (L2) cache are built into the processor.

Significant improvement can be seen with a larger cache on workstations and servers that are working with large data sets such as large databases and heavy graphics applications. For example, moving from a 512-KB L2 cache to a 1-MB L2 cache increases the performance from 10 to 25 percent, depending on the application and load.

Intel offers processors aimed at the server market. For example, the company offers the Intel Pentium II Xeon processor, Intel Pentium III Xeon processor, and Intel Xeon processor. Different from other processors, some models have a larger cache. In addition, these processors also have other features that help monitor system performance.

14.10.3   Disk System Performance

For a server, you should always use SCSI over IDE or EIDE. SCSI hard drives have a higher bandwidth and reduced latency. SCSI hard drives also use command queuing.

Command queuing allows a device to accept multiple commands and execute them in an order that is more efficient rather than in the order received. You can think of command queuing as being similar to having a busload of people. You can travel to each person's stop in the order that the person entered the bus, or you can try to figure out the best way to drop these people off. With the latter choice, if two people live in a nearby area, you would drop them off one after the other. For SCSI hard drives, this increases the performance of computers running multitasking operating systems and makes it ideal for servers.

NOTE: As discussed earlier in this chapter, RAID is also one way to increase disk performance.

14.10.4   Network Performance

With the system monitoring software, you can analyze the packets that are going in and out of your PC. Network monitoring software, such as Network Monitor (Windows NT and Windows 2000) and Network Lanalyzer (Novell NetWare), enables you to detect and troubleshoot problems on LANs. With network monitoring software, you can identify network traffic patterns and network problems such as computers that make many more requests than other computers, and you can also identify unauthorized users on your network. In addition, you can capture frames (packets) directly from the network, or you can display, filter, save, and print the captured frames so that you can analyze them later. For example, if you suspect that a card is jabbering (a card is throwing out garbage packets on the network using up valuable bandwidth), you can use the network monitoring software to capture these packets and then analyze them to determine the MAC address of the offending link.

SUMMARY

1. In a hot site, every computer system and piece of information has a redundant copy. This level of fault tolerance is used when systems must be up 100 percent of the time.

2. In a warm site (also called a nearline site), the network service and data are available most of the time (more than 85 percent of the time).

3. A cold site does not guarantee server uptime and has little or no fault tolerance.

4. RAID is short for "redundant array of inexpensive disks." It is a category of disk drive that employs two or more drives in combination for fault tolerance and performance.

5. RAID 0 stripes data across all drives. With striping, all available hard drives are combined into a single, large virtual file system, with the file system's blocks arrayed so they are spread evenly across all the drives.

6. RAID 1 is known as disk mirroring or disk shadowing. With RAID 1, each hard drive on the system has a duplicate drive that contains an exact copy of the first drive's contents.

7. RAID 1 also includes disk duplexing, which is the same as disk mirroring except the two drives are on different controller cards so that the drive, controller card, and cable are redundant.

8. To make sure that the link that connects to your network has fault tolerance, you have several options: adaptive load balancing (ALB), adapter fault tolerance (AFT), and 802.3ad link aggregation.

9. A surge protector is designed to prevent most short-duration, high-intensity spikes and surges from reaching your PC by absorbing excess voltages.

10. A line conditioner uses the inductance of transformers to filter out noise and capacitors (and other circuits) to "fill in" during brownouts.

11. The standby power supply consists of a battery hooked up parallel to the PC. When the SPS detects a power fluctuation, the system will switch over to the battery.

12. The uninterruptible power supply has the battery connected in series with the PC and the AC power. Since the battery always provides clean DC power, the PC is protected against overvoltages and undervoltages.

13. One advanced feature available on mission-critical computers such as servers is a redundant power supply.

14. Clustering is connecting two or more computers, known as nodes, together in such a way that they behave like a single computer.

15. Network load balancing clusters distribute client connections over multiple servers.

16. In a failover cluster configuration, two or more computers serve as functional backups for each other.

17. A storage area network (SAN) is a high-speed subnetwork of shared storage devices. A SAN's architecture works in a way that makes all storage devices available to all servers on a LAN.

18. Tape drives read and write to a long magnetic tape. They are relatively inexpensive and offer large storage capacities, making tape backup drives ideal for backing up hard drives on a regular basis.

19. Data is the most important part of the computer.

20. A backup of a system consists of an extra copy of data and/or programs.

21. THE BEST METHOD FOR DATA PROTECTION IS BACKUP, BACKUP, BACKUP.

22. A virus is a program designed to replicate and spread, generally without the knowledge or permission of the user.

23. Some of the disastrous symptoms of a virus include a computer that fails to boot, a computer that cannot access a disk, data that are corrupt, and a system that is slower than normal.

24. Antivirus software will detect and remove viruses and help protect the computer against viruses.

25. Performance is the overall effectiveness of how data moves through the system.

26. To be able to improve performance, you must determine the part of the system that is slowing down the throughput.

QUESTIONS

1. What is the most important part of the computer?
   a. The microprocessor
   b. The RAM
   c. The hard drive
   d. The data

2. The best method for protecting the data is _____.
   a. RAID
   b. a surge protector and UPS
   c. backup, backup, backup
   d. antivirus software

3. Which is ideal for backing up an entire hard drive?
   a. Zip drive   b. RAID
   c. Second hard drive   d. Tape drive

4. After backing up a drive, you should occasionally _____.
   a. restore a nonessential file to the hard drive
   b. reformat the hard drive
   c. shut down the system
   d. reformat the tape

5. The archive attribute is the attribute that indicates whether a file is backed up. Which of the following backups does not shut off the archive attribute?
   a. Full
   b. Differential
   c. Incremental
   d. None of the above

6. Which backup method requires you to provide the full backup and provide a tape for each day you want to go back and restore?
   a. Incremental   b. Full
   c. Differential   d. Daily

7. Your Windows 2000 server is suffering from poor performance due to excessive paging. What is the best way to alleviate the excessive paging on your server?
   a. Add RAM to your server.
   b. Implement a disk management strategy on your server and create multiple paging files.

   c. Upgrade the disk drive containing the paging file to one with a faster data access speed.

   d. Upgrade the CPU on your server.

8. Which of the following devices provides the highest level of protection for your computer equipment?

   a. A surge protector

   b. A UPS

   c. An SPS

   d. A line conditioner

9. You are the network engineer for a large manufacturing company. Recently, you installed a critical Microsoft SQL Server 7.0 computer on the company's LAN in Dallas, Texas. Since you installed the Microsoft SQL Server 7.0 computer, it has stopped responding twice and has spontaneously rebooted several times. You test the Microsoft SQL Server 7.0 computer and determine that it is not causing the problem. You test the power lines and find that fluctuations in the electric power grid are reaching the Microsoft SQL Server 7.0 computer. Which device should you install between the power line and the Microsoft SQL Server 7.0 computer on the LAN in Dallas to ensure that the computer receives a clean, uninterrupted power signal?

   a. A surge protector

   b. A UPS

   c. A line conditioner

   d. An SPS

10. Which device only provides protection from a power spike or surge?

   a. A UPS

   b. A surge protector

   c. An SPS

   d. A line conditioner

11. You administer a network that is connected to the Internet, and you want to provide your network with the strongest possible virus protection. Where should you install the virus protection software? (Choose two answers.)

   a. On one server

   b. On all servers

   c. On one workstation

   d. On all workstations

12. You want to implement the least expensive form of server fault tolerance. Which method should you choose?

   a. The failover clustering method

   b. The cold-site method

   c. The true clustering method

   d. The warm-site method

13. You want to implement the most cost-effective fault tolerance method. Which of the following methods should you use to accomplish this task?

   a. The duplicate server method

   b. The true clustering method

   c. The failover clustering method

   d. The RAID method

14. You want to provide fault tolerance for the servers on your network. Which method should you use?

   a. The failover clustering method

   b. The RAID method

   c. The rollover clustering method

   d. The two-phase commit method

15. Which RAID level provides disk duplexing?

   a. 0   b. 3

   c. 1   d. 5

16. You want to establish disk striping without parity on a server computer. Which RAID level should you use?

   a. 0   b. 3

   c. 1   d. 5

17. Which of the following is characteristic of disk duplexing?

   a. It uses two hard disks.

   b. It stores parity information on the hard disks in the set.

   c. It stores an exact copy of the primary hard disk in the set.

   d. It uses one hard disk controller.

18. You are completing a full backup of the information stored on a network server. What happens to the archive bit that is associated with each file after the backup is complete?

   a. The archive bit is cleared.

   b. The archive bit is deleted.

   c. The archive bit is incremented by 1.

   d. The archive bit is incremented by 2.

19. Which type of software should you update most frequently with software patches?

   a. Presentation software

   b. Virus protection software

   c. Spreadsheet software

   d. Word processing software

20. You use a tape drive to back up important data on a server on the LAN. You start the backup operation, and the computer displays a tape error message. What is the first action you should perform to resolve this problem?

a. Use a new tape.

b. Clean the tape drive.

c. Ensure that the tape drive is turned on.

d. Replace the tape drive.

21. What should you do if you are attempting to back up a Windows NT Server 4.0 computer to an external DAT tape drive, and the computer reports that no tape drive is present?

a. Use a new tape.

b. Replace the tape drive.

c. Clean the tape drive.

d. Ensure that the tape drive is turned on.

22. What are four common tape formats currently in use?

a. Travan, linear, TR1, and audio

b. QIC, DSS, Travan, and Scotch

c. QIC, DAT, DLT, and Travan

d. DAT, DDS, TR3, and TR4

23. Digital audiotape uses a _____ recording method, similar to VCR recording:

a. servo scan   b. reel-to-reel

c. helical scan   d. laser

24. You want to provide a critical file server with the ability to shut down with no power interruption when AC to the building is suddenly cut off. What device do you need to connect to it?

a. CPS   b. SPS

c. VLM   d. UPS

25. A disk array on your network server has data distributed across several drives, providing fast read/write access. You are told by another network admin that if one drive fails, you'll need to restore from backup tape, because the entire volume spanning the disk array would be lost. This array is running what level RAID?

a. 4   b. 5

c. 2   d. 0

26. Two hard drives are installed in a server, each on its own controller. You've decided to configure disk duplexing, where a disk on one controller automatically copies data to another disk on its own controller. What level RAID do you need to configure?

a. 4   b. 1

c. 5   d. 0

27. What is the difference between a UPS and an SPS?

a. An SPS has an external battery.

b. One uses UTP cable and other uses STP cable.

c. A UPS does not have a switching time delay.

d. There is no significant difference between the two.

28. Your boss wants you to back up all data, applications, and operating systems on a server. Which backup methods could you use? (Choose two answers.)

a. Full backup

b. Incremental backup

c. Differential backup

d. Copy backup

29. Because of an upcoming driver and/or firmware update to your server's disk array controller, you want to do a full backup of all data first, in case the controller malfunctions after the new drivers are installed. However, you don't want to throw off your current backup schedule. What backup method should you use?

a. Differential   b. Incremental

c. Copy   d. Full

30. What does RAID level 5 use to implement fault tolerance?

a. Sequence bit   b. Security bit

c. Parallel bit   d. Parity bit

31. When a network problem occurs, which of the following is NOT part of your procedures?

a. Documention

b. Solving

c. Performing backups

d. Troubleshooting

32. You have five hard drives in a RAID 5 array. Approximately how much disk space is available for data?

a. 75   b. 80%

c. 100   d. 50%

33. You have five hard drives in a RAID 5 array. Each drive has a 50-GB capacity. How many gigabytes are available for data storage?

a. 225 GB   b. 200 GB

c. 250 GB   d. 175 GB

34. You have two 20-GB hard drives in your server, which is running RAID 1. How much disk space is available for data storage?

a. 60 GB   b. 40 GB

c. 80 GB   d. 20 GB

35. To ensure security of your company's data, what is the best place to store your backup tapes?

a. Locked in a file cabinet offsite.

b. Locked in a fireproof safe onsite.

c. Locked in the server room onsite.

d. Locked in a fireproof safe offsite.

36. What term means a condition in which voltage quickly increases past normal and then drops back down just as quickly within 1 second (or even within a few milliseconds)?

a. Spike   b. Brownout

c. Surge   d. Blackout

37. What occurs when voltage quickly rises above normal and stays that way for several seconds?

a. Spike   b. Brownout

c. Surge   d. Blackout

38. You need to add extra data storage capability to your network, and you are trying to decide whether to install a Windows 2000 server or an NAS device. What advantage does the NAS device have that might cause you to decide on it?

a. Has higher bandwidth

b. Has far greater storage capability

c. Requires less expensive cabling

d. Costs less

Introduction to the Macintosh OS

With very few exceptions, the Mac OS will not run on any hardware platform except the Macintosh. Through the years, there have been many different Macintosh models, with varying degrees of speed and power. All models since 1994 are based on the PowerPC microprocessor. PowerPC processors use RISC-based computer architecture developed jointly by IBM, Apple Computer, and Motorola Corporation. The name is derived from IBM's name for the architecture, *P*erformance *O*ptimization *w*ith *E*nhanced *R*ISC.

Much like any other piece of software, the Mac OS has gone through several versions throughout the years. The first one—System 1—was released in 1984 with the original Macintosh. It was one of the earliest GUIs, even before Microsoft Windows 3.0. By comparison with modern versions of the Mac OS, its interface, called Finder, was pretty bare. Since the early Macintosh computers did not have color, System 1 did not have color. Color was added in 1986 with the next version, called System 6.

System 7 added multitasking with a program called Multifinder. In addition, it gave users support for TrueType fonts and the ability to share out a disk onto the network so that other Mac users could access it. It also gave users the ability to use virtual memory.

Mac OS 8 was released in 1997. As a result of a partnership with Microsoft, Internet Explorer was installed as the default browser. Apple also increased its cross-platform connectivity with the introduction of an updated version of its PC Exchange product, which now had support for Windows 9X long file names. Finally, the OS contained its own Java Virtual machine for running Java applications.

With Mac OS 9, Apple added multiuser features including the capability of specifying different settings and environments for multiple users of the same Macintosh. Along with that, Apple introduced the Keychain, which stored the various online passwords for a user so that only one password was required when a user went online. Finally, Apple included a network browser so that a user could browse the network easily for a network server.

The most recent version of the Mac OS X (OS 10) was introduced in 2001. Based on a UNIX kernel, it is the first major rewrite of the Mac OS in years that is a very stable platform and in general more powerful. With this rewrite, the Mac OS X interface utilizes more color and graphics than the previous version and allows you to save files directly to PDF format (the format used by Adobe Acrobat).

The Mac OS can perform many functions on a network. In addition to being a client, a Macintosh computer can act as an integrated file, print, mail, and web server using AppleShare. As well, newer versions of AppleShare (today known as AppleShare IP) allow the sharing of directories, files, and printers using the CIFS Protocol so that Microsoft Windows and Linux computers running Samba can access those shared resources. The advantage of having a Macintosh as a server is that it is extremely easy to administer. It is so easy, in fact, that many first-time users have no problems networking Macs and making them into file (or other) servers. From the Apple menu, you just select the control panels and then TCP/IP. If the machine does not have a TCP/IP control panel, then the machine is not using Open Transport (OT) networking; in that event, check for the existence of a MacTCP control panel instead.

With the advent of System 7.5.3, Apple released Open Transport (mentioned above). OT changed the way the Mac OS looks at networking. In addition to increasing the performance of LocalTalk networks, Apple also simplified much software required for setting it up on any network. The easy way to tell whether you are using OT or classic networking is to take a look at your control panels. If you have an AppleTalk control panel, you are using Open Transport. If you have a network control panel, you are using classic networking.

Open Transport/TCP works over Ethernet, Token Ring, AppleTalk (as MacIP), and serial lines such as MacPPP and InterSLIP. Open Transport/TCP is configured using the TCP/IP control panel. Configuration may be done manually or through a BOOTP, DHCP, RARP, or MacIP server.

By default, the TCP/IP control panel comes up in basic mode. Advanced or administration mode may be entered via the Edit menu. These modes allow expert users additional choices as well as the ability to augment information returned from a configuration server or to fill in gaps in the returned information.

The TCP/IP control panel may be used at any time to reconfigure the system. However, TCP will not notice the new configuration until it has unloaded from the system. By default, this takes about 2 minutes after the last application using TCP or UDP has gone away.

To manually set up Open Transport/TCP, follow these steps:

1. Select the interface to use, or pick AppleTalk (MacIP) to run over AppleTalk on the interface selected in the AppleTalk control panel. (See figure A.1.)
2. If an Ethernet interface is selected, a check box will appear offering the use of 802.3. By default, Open Transport/TCP uses Ethernet_II rather than 802.3.
3. Select Manually as the configuration method.
4. Fill in the IP address in dot notation (for example, 128.1.1.1).
5. Fill in the default domain extension to be used on name searches.
6. In Advanced User mode only, you may enter an admin domain. This is used to allow implicit searches.
7. Fill in the subnet mask in dot notation.
8. Fill in the IP address of the default IP router.
9. Fill in the IP address(es) of one or more Domain Name Servers.
10. In Advanced User mode only, you may enter additional search domains.
11. If a Hosts file is required, select it using the Hosts file button. For details about the Hosts file, see the description that follows.

Figure A.1   TCP/IP Configuration for the Apple Macintosh Computer

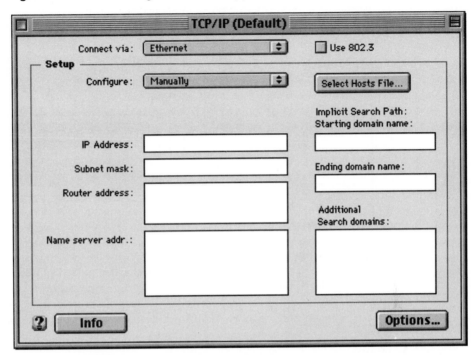

To use a DHCP server to set up Open Transport/TCP, follow these steps:

1. Select the interface to use in the AppleTalk control panel.
2. If an Ethernet interface is selected, a check box will appear offering the use of 802.3. By default, Open Transport/TCP uses Ethernet_II rather than 802.3.
3. Select Using DHCP as the configuration method.
4. In Advanced User mode only, you may enter an admin domain. This is used to allow implicit searches.
5. In Advanced User mode, you may enter a subnet mask, but it is not required. If a value is entered, it will be used if no subnet mask is returned from the DHCP server. Otherwise, any value entered is ignored.
6. In Advanced User mode, the manually entered IP addresses of routers are attached to the end of the (possibly empty) list of IP routers returned by the DHCP server.
7. In Advanced User mode, the manually entered IP addresses of Domain Name Servers are attached to the end of the (possibly empty) list of name servers returned by the DHCP server.
8. In Advanced User mode only, you may enter additional search domains.
9. If a Hosts file is required, select it using the Hosts file button.

Once you have all the hardware connected, the next step is to try to see the other computers or printers on the network. Every Mac's window to the outside world is the

Figure A.2   Using the Chooser to Access Network Resources

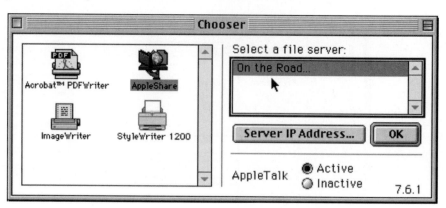

Chooser. (See figure A.2.) The Chooser is where you choose the items that your Mac is connected to, such as printers and other computers. Once you have chosen the type of networking in the AppleTalk or Network control panel, you can open the Chooser; you should be able to see the name of each computer that has been set up to share files on the network by simply selecting the AppleShare icon on the left side. If you are trying to access printers, you will want to choose the printer type that you have—for many Apple laser printers you will be selecting LaserWriter 8.

Apple Macintosh computers do not have a ping command built into the operating system. Therefore, if you need to troubleshoot using the ping command, you will have to download a freeware or shareware software package.

Answers to Odd-Numbered Questions

CHAPTER 1

1. D	11. A	21. E
3. D	13. A	23. A
5. B	15. C	25. True
7. B	17. D	27. B, C, and D
9. B and C	19. D	29. D

CHAPTER 2

1. A	33. A	65. D
3. D	35. D	67. D
5. B	37. A	69. D
7. B	39. D	71. B
9. D	41. D	73. A
11. D	43. A	75. D
13. B	45. A	77. D
15. C	47. B	79. D
17. C	49. A	81. B
19. C	51. A	83. B
21. C	53. D	85. A
23. D	55. A	87. D
25. A	57. D	89. D
27. C	59. A	91. D
29. B and D	61. D	93. E
31. A, B, and D	63. D	95. A

CHAPTER 3

1. D	33. C	65. A
3. A	35. A	67. A and C
5. D	37. B	69. D
7. D	39. C	71. A and B
9. A	41. D	73. D
11. D	43. C	75. D
13. D	45. D	77. A
15. B	47. D	79. D
17. C	49. A and C	81. B
19. D	51. B	83. A, B, and C
21. D	53. A	85. B and D
23. D	55. A	87. A and C
25. A	57. C	89. D
27. B	59. B	91. D
29. C	61. A	93. A
31. D	63. A	

CHAPTER 4

1. A	23. D	45. D
3. D	25. B and D	47. C
5. A	27. D	49. B
7. A	29. D	51. A and C
9. B	31. A	53. B
11. B and D	33. A	55. A
13. D	35. C	57. A, B, and C
15. A	37. D	59. B
17. D	39. D	61. A
19. B and D	41. B	63. Nothing wrong.
21. D	43. B	

CHAPTER 5

1. B and C	13. A	25. B and D
3. A	15. A	27. D
5. B and C	17. A	29. D
7. D	19. C	31. B
9. C	21. C	33. B and D
11. C	23. D	35. A

37. D
39. A
41. A
43. D
45. D
47. D
49. C

51. D
53. D
55. B
57. A
59. B and D
61. D
63. B

65. D
67. A, B, C, and D
69. A
71. A
73. D

CHAPTER 6

1. A
3. A, D, and E
5. B
7. A, C, and D
9. C
11. D
13. A
15. A and D
17. E
19. B
21. C
23. A
25. A
27. A
29. B
31. D
33. D
35. A
37. C
39. D
41. B
43. A

45. B
47. C
49. C
51. A
53. D
55. D
57. C
59. B
61. C
63. D
65. A
67. D
69. C
71. D
73. B
75. A
77. D
79. A
81. a. The mask for the entire network should be 255.255.255.0.

b. On the top router, the first router port should have an address of 141.171.34.254.

c. On the ring network, you have one multi-home computer that is connected to the ring network and the star network. The interface connected to the ring network should have a gateway of 141.171.35.17.

d. The computer on the far right on the star network should have a gateway of 141.171.40.15 or 141.171.40.188.

CHAPTER 7

1. A and B
3. A
5. A and B
7. A
9. C
11. D
13. B

15. A and C
17. C
19. A
21. B
23. The internal IPX address of the server at the bottom is the same as the external

IPX address on the top network.

25. B
27. D
29. D

CHAPTER 8

1. B	21. B	41. D
3. B	23. D	43. C
5. C	25. A	45. C and D
7. D	27. C	47. D
9. A	29. D	49. D
11. A	31. A	51. A
13. B	33. D	53. A
15. C	35. C and D	55. A, B, and C
17. E	37. A and C	
19. A, B, C, D, and E	39. D	

CHAPTER 9

1. D	3. A

CHAPTER 10

1. B	19. A	37. C and D
3. A	21. C	39. C and D
5. A	23. A and C	41. A
7. A	25. A, B, and C	43. D
9. A	27. A	45. D
11. C	29. D	47. B
13. A	31. C	49. D
15. D	33. D	
17. C	35. D	

CHAPTER 11

1. B	11. B and D	21. D
3. A, B, and C	13. A and C	23. B
5. C	15. A, C, and D	25. A
7. C	17. B	
9. A	19. C	

CHAPTER 12

1. A
3. D
5. C
7. B
9. A
11. B

13. A
15. D
17. B and C
19. D
21. B and D
23. A

25. C
27. D
29. D
31. B
33. A and B

CHAPTER 13

1. A
3. A
5. A, C, and E
7. D
9. D
11. D
13. A

15. D
17. D
19. D
21. A
23. D
25. A and C
27. A

29. A and C
31. D
33. D
35. C and D
37. D
39. A

CHAPTER 14

1. D
3. D
5. B
7. A
9. C
11. B and D
13. C

15. C
17. A and C
19. B
21. D
23. C
25. C
27. C

29. C
31. C
33. B
35. D
37. C

Glossary

.NET—A Microsoft operating system platform that incorporates distributed applications that bring users into the next generation of the Internet by conquering the deficiencies of the first generation and giving users a more enriched experience in using the web for both personal and business applications.

100BaseFX—A form of Fast Ethernet that operates over multimode fiber-optic cabling. It provides 100 Mbps of throughput.

100BaseT4—A form of Fast Ethernet that runs over existing Category 3 UTP by using all four pairs. Three pairs are to provide simultaneous transmission, and the fourth pair is used for collision detection. It provides 100 Mbps of throughput.

100BaseTX—A form of Fast Ethernet that uses two pairs of the standard Category 5 UTP to provide 100 Mbps of throughput.

100VG-ANYLAN—A 100-Mbps half-duplex transmission that allows 100 Mbps on a four-pair Category 3 cabling system and allows for voice over IP (VoIP).

10Base2—A simplified version of the 10Base5 Ethernet network. The name describes a 10-Mbps baseband network with a maximum cable segment length of approximately 200 meters (actually 185 meters). Instead of having external transceivers, the transceivers are on the network card, which attaches to the network using a BNC T-connector. The cable used is 50-ohm RG-58 A/U coaxial-type cable. Different from the 10Base5 network, the 10Base2 does not use a drop cable. Also known as thinnet.

10Base5—A form of Ethernet that has a 10-Mbps baseband network with cable segments up to 500 meters long. It uses a 50-ohm RG-8 and RG-11 as a backbone cable. 10Base5 uses physical and logical bus topology. Also known as thicknet.

10BaseT—A form of Ethernet that uses UTP to form a logical bus topology while actually being a physical star topology. Therefore, network devices are connected to a hub (or switch).

386 microprocessor protection model—A model used on the Intel processor. It uses four rings to run programs. Each Ring has a different security level. The inner ring (Ring 0) has the highest security, and Ring 3 has the lowest security.

5-4-3 rule—An important rule specifying that an Ethernet network must not exceed five segments connected by four repeaters. Of these segments, only three of them can be populated by computers. This means that the distance of a computer cannot exceed five segments and four repeaters when communicating with another computer on the network.

802.11 standard—A wireless standard.

802.1X—A standard for passing EAP over a wired or wireless LAN.

A

access control list (ACL)—A set of data that informs a computer's operating system which permissions, or access rights, each user or group has to a specific system object, such as a directory or file. Each object has a unique security attribute that identifies which users have access to it, and the ACL is a list of each object and user access privileges such as read, write, or execute.

access method—The set of rules defining how a computer puts data onto the network cable and takes data from the cable.

access point—Device that acts as routers in wireless LAN (WLAN) and is used to connect to a wired network infrastructure.

acknowledgment (ACK)—A special message that is sent back when a data packet makes it to its destination.

Active Directory—(1) A directory service that combines domains, X.500 naming services, DNSs, X.509 digital certificates, and Kerberos authentication. It stores all information about the network resources and services such as user data, printers, servers, databases, groups, computers, and security policies. In addition, it identifies all resources on a network and makes them accessible to users and applications. (2) A directory service used on Microsoft Windows 2000 or Windows .NET server domains, which uses the "tree" concept for managing resources on a network.

Active Directory integrated zone—A DNS zone defined using the Active Directory, not the zone files.

ad hoc network—A wireless LAN (WLAN) that has wireless stations that communicate directly with each other without using an access point or any connection to a wired network.

Address Resolution Protocol (ARP)—A TCP/IP that is used to obtain hardware addresses (MAC addresses) of hosts located on the same physical network.

addressing—A method used to identify senders and receivers.

administrative share—A shared folder typically used for administrative purposes. To make a shared folder or drive into an administrative share, the share name must have a $ at the end of it. Since the share folder or drive cannot be seen during browsing, you would have to use a UNC name, which includes the share name (including the $).

ADSL Lite—A DSL specification. It is a low-cost, easy-to-install version of ADSL specifically designed for the consumer marketplace. ADSL Lite is a lower-speed version of ADSL (up to 1.544 Mbps downstream, up to 512 Kbps upstream) that will eliminate the need for the telephone company to install and maintain a premises-based POTS splitter. Also known as g.lite.

amplitude—Represents the peak voltage of the sine wave.

analog signal—The opposite of a digital signal. Instead of having a finite number of states, it has an infinite number of values that are constantly changing. Analog signals are typically sinusoidal waveforms, which are characterized by their amplitude and frequency.

Apple Macintosh—A computer that uses one of the easiest-to-use graphical user interfaces. Different from the previous operating systems, with very few exceptions, the Mac OS will not run on any hardware platform except the Macintosh computer.

AppleTalk—A protocol and networking software found on Apple Macintosh computers.

application layer—The highest layer of the OSI reference model, which initiates communication requests. It is responsible for interaction between the layers in the operating system and provides an interface to the system. It provides the user interface to a range of networkwide distributed services including file transfer, printer access, and electronic mail.

application programming interfaces (APIs)—A set of routines, protocols, and tools for building software applications.

application server—A server that is similar to a file and print server except the application server also does some of the processing.

arbitration—See *access method*.

ASCII (American Standard Code for Information Interchange)—A character set that consists of alphanumeric code.

asymmetrical DSL (ADSL)—A form of DSL that transmits an asymmetric data stream, with much more going downstream to the subscriber and much less coming back.

asynchronous—Devices that use intermittent signals. They can occur at any time and at irregular intervals. They do not use a clock or timing signal.

asynchronous transfer mode (ATM)—Both a LAN and a WAN technology, which is generally implemented as a backbone technology. It is a cell-switching and multiplexing technology that combines the benefits of circuit switching and packet switching.

attenuation—When the strength of a signal falls off with distance over a transmission medium. This loss of signal strength is caused by several factors—for example, when the signal is converted to heat due to the resistance of the cable or when the energy is reflected as the signal encounters impedance changes throughout the cable.

AUI (adapter unit interface) connector—A female 15-pin D connector used to connect to an Ethernet 10Base5 network. Also known as a DIX (Digital-Intel-Xerox) connector.

B

backbone cable—(1) Can be used as a main cable segment such as that found in a bus topology network. (2) Can be the main network connection through a building, campus, WAN, or the Internet.

backbone cabling—Provides interconnections between telecommunications closets, equipment rooms, and entrance facilities and includes the backbone cables, intermediate and main cross-connects, terminations, and patch cords for backbone-to-backbone cross-connections.

backbone wiring—The system of cables often designed to handle a higher bandwidth. It is used to interconnect wiring closets, server rooms, and entrance facilities (telephone systems and WAN links from the outside world).

backend—In client/server applications, services provided by the server.

backup—An extra copy of data and/or programs. As a technician, consultant, or support person, you need to emphasize at every moment the need to have backup on servers and client systems.

backup domain controller—A server that contains duplicate copies of the domain database on Windows NT networks.

band—A contiguous group of frequencies that are used for a single purpose.

bandwidth—The amount of data that can be carried on a given transmission medium.

baseband system—A system that uses the transmission medium's entire capacity for a single channel.

basic rate interface (BRI)—A form of ISDN that defines a digital communications line consisting of three independent channels: two bearer (or B) channels, each carrying 64 kilobytes per second, and one data (or D) channel at 16 kilobits per second. For this reason, the ISDN basic rate interface is often referred to as 2B+D.

basket systems—Baskets or trays used on the back of office furniture. They are used to hold and hide network cables without using cable ties.

baud rate—The modulation rate or the number of times per second that a line changes state.

bearer channel (B channels)—A channel used in ISDN lines to transfer data at a bandwidth of 64 Kbps (kilobits per second) for each channel.

bend radius—The radius of the maximum arc into which you can loop a cable before you will impair data transmission. Generally, a cable bend radius is less than four times the diameter of the cable.

binary digit (bit)—A piece of information that can represent the status of an on/off switch. When several bits are combined, they can signify a letter, a digit, a punctuation mark, a special graphical character, or a computer instruction.

binary number system—The simplest number system. It is based on only two digits, a zero (0) and a one (1).

bind—A method that links the protocol stack to the network card.

BIND—The most widely used name server (DNS server).

bindery—A flat database that keeps track of users and groups on Novell NetWare server 3.2 or earlier. It is administered with a menu-based DOS utility known as SYSCON.

blackouts—Total power failures.

blade server—A server that uses server blades, which are slid into the server much like an expansion slot is plugged into a PC.

Bluetooth—A form of short-range radio technology aimed at simplifying communications among devices and the Internet. It also aims to simplify data synchronization between Net devices and other computers.

BNC barrel—A connector that allows connecting two coax cables together.

BNC connector—Short for British Naval Connector, Bayonet Nut Connector, or Bayonet Neill Concelman. It is a type of connector used with coaxial cables. The basic BNC connector is a male type mounted at each end of a cable. This connector has a center pin connected to the center cable conductor and a metal tube connected to the outer cable shield. A rotating ring outside the tube locks the cable to any female connector.

BNC T-connectors—Used with the 10Base2 system. They are female devices for connecting two cables to a network interface card (NIC).

Bootstrap Protocol (BOOTP)—A TCP/IP that enables a booting host to configure itself dynamically.

bridge—A device that works at the data link OSI layer. It connects two LANs and makes them appear as one or is used to connect two segments of the same LAN. The two LANs being connected

can be alike or dissimilar, such as an Ethernet LAN connected to a Token Ring LAN.

broadband system—A system that uses the transmission medium's capacity to provide multiple channels by using frequency-division multiplexing (FDM).

broadcasting—A method that sends unaddressed packets to everyone on the network.

brouter—Short for bridge router. It is a device that functions as both a router and a bridge. A brouter understands how to route specific types of packets (routable protocols), such as TCP/IP packets. For other specified packets (nonroutable protocols), it acts as a bridge that simply forwards the packets to the other networks.

brownouts—Drops in power that can force the computer to shut down, can introduce memory errors, and can cause unsaved work to be lost.

bus topology—A topology that looks like a line, where data is sent along a single cable. The two ends of the cable do not meet, nor do the two ends form a ring or loop.

byte—Equal to 8 bits of data. One byte of information can represent one character.

C

cable modem—A device located in subscribers' homes to create a virtual local area network (LAN) connection over a cable system.

cable system—A system that delivers broadcast television signals efficiently to subscribers' homes using a sealed coaxial cable line.

cable tester—A device designed to test a cable.

cable ties—Small devices used to bundle cables traveling together and to pull the cables off the floor so that they will not get trampled or run over by office furniture.

cable—A physical transmission medium that has a central conductor of wire or fiber surrounded by a plastic jacket, used to carry electrical or light signals between computers and networks.

cabling system—The veins of the network that connect all the computers together and allow them to communicate with each other.

carrier sense multiple access (CSMA)—The most common form of contention (access method) used on networks.

carrier sense multiple access with collision avoidance (CSMA/CA)—The access mechanism used in Apple's LocalTalk network. Collision avoidance uses time slices to make network access smarter and avoid collisions.

carrier sense multiple access with collision detection (CSMA/CD)—The access method utilized in Ethernet and IEEE 802.3. The collision detection approach listens to the network traffic as the card is transmitting. By analyzing network traffic, it detects collisions and initiates retransmissions.

CDFS—Short for CD-ROM file system. The read-only file system used to access resources on a CD-ROM disk. Windows supports CDFS so as to allow CD-ROM file sharing. Because a CD-ROM is read-only, you cannot assign specific permissions to files through CDFS.

Cell Relay—A form of packet-switching network that uses relatively small, fixed-size packets called cells.

cell—A relatively small, fixed-size packet.

cellular topology—A topology used in wireless technology where an area is divided into cells. A broadcast device is located at the center and broadcasts in all directions to form an invisible circle (cell). All network devices located within the cell communicate with the network through the central station or hub, which is interconnected with the rest of the network infrastructure. If the cells are overlapped, devices may roam from cell to cell while maintaining a connection to the network as the devices.

centralized computing—The processing done for many people by a central computer.

Certificate Authority (CA)—A trusted third-party organization or company that issues digital certificates used to create digital signatures and public-key–private-key pairs. The role of the CA in this process is to guarantee that the individual granted the unique certificate is, in fact, who he

or she claims to be. Usually, this means that the CA has an arrangement with a financial institution, such as a credit card company, which provides it with information to confirm an individual's claimed identity. CAs are a critical component in data security and electronic commerce because they guarantee that the two parties exchanging information are really who they claim to be.

Challenge Handshake Authentication Protocol (CHAP)—The most common dial-up authentication protocol. It uses an industry Message Digest 5 (MD5) hashing scheme to encrypt authentication.

channel service unit (CSU)—A device that connects a terminal to a digital line that provides the LAN/WAN connection.

checksum—A form of error control.

circuit switching—A technique that connects the sender and the receiver by a single path for the duration of a conversation. Once a connection is established, a dedicated path exists between both ends. It is always consuming network capacity even when there is no active transmission taking place (such as when a caller is put on hold). Once the connection has been made, the destination device acknowledges that it is ready to carry on a transfer. When the conversation is complete, the connection is terminated.

Classless Interdomain Routing (CIDR)—A new IP addressing scheme that replaces the older system based on classes A, B, and C. With CIDR, a single IP address can be used to designate many unique IP addresses. A CIDR IP address looks like a normal IP address except that it ends with a slash followed by a number, called the IP prefix.

clear text—In cryptography, refers to any message that is not encrypted.

client—A computer that requests services.

Client for Microsoft Networks—Provides the redirector (VREDIR.VXD) to support all Microsoft networking products that use the Server Message Block (SMB) Protocol.

Client for NetWare Networks—A program that runs with the IPX Protocol. It is used to process login scripts, support all NetWare 3.XX command-line utilities and some 4.X/5.X command-line utilities, connect and browse to NetWare servers and access printers on the NetWare server.

client/server network—A network made up of servers and clients. It is typically used on a medium-size or large network.

client software—Software that allows a workstation attached to the network to communicate.

cloud—Represents a logical network with multiple pathways as a black box. The subscribers that connect to the cloud don't worry about the details inside the cloud. Instead, the only thing the subscribers need to know is that they connect at one edge of the cloud and the data is received at the other edge of the cloud.

clustering—Connecting two or more computers, known as nodes, together in such a way that they behave like a single computer. It is used for parallel processing, for load balancing, and for fault tolerance. The computers that form the cluster are physically connected by cable and are logically connected by cluster software. As far as the user is concerned, the cluster appears as a single system to end users.

coaxial cable—A cable that has a center wire surrounded by insulation and then a grounded shield of braided wire (mesh shielding). The copper core carries the electromagnetic signal, and the braided metal shield acts as both a shield against noise and a ground for the signal. The shield minimizes electrical and radio frequency interference and provides a connection to ground. Sometimes referred to as coax.

collision—The situation that occurs when two or more devices attempt to send a signal along the same channel at the same time. The result of a collision is generally a garbled message. All computer networks require some sort of mechanism to either prevent collisions altogether or to recover from collisions when they do occur.

COM+—An extension of the component object model. COM+ is both an object-oriented programming architecture and a set of operating system services. It adds to COM a new set of system services for application components while they are running, such as notifying them of significant events or ensuring they are authorized to run. COM+ is intended to provide a model that makes it relatively easy to create business applications that work well with the Microsoft Transaction Server (MTS) in a Windows NT system.

Common Internet File System (CIFS)—A public or open variation of the Server Message Block Protocol used by Microsoft Windows. It also is currently used by Linux with Samba, Novell Netware 6.0, and AppleTalkIP.

communication server—A remote access server.

component object model (COM)—An object-based programming model designed to promote interoperability by allowing two or more applications or components to easily cooperate with one another, even if they were written by different vendors at different times in different programming languages, or even if they are running on different computers running different operating systems. COM is the foundation technology upon which broader technologies can be built, such as object linking and embedding (OLE) technology and ActiveX.

Computer Emergency Response Team Coordination Center (CERT/CC)—Started in December 1988 by the Defense Advanced Research Projects Agency (which was part of the U.S. Department of Defense) after the Morris worm disabled about 10 percent of all computers connected to the Internet. CERT/CC is located at the Software Engineering Institute, a federally funded research center operated by Carnegie Mellon University. CERT/CC studies Internet security vulnerabilities, provides services to websites that have been attacked, and publishes security alerts. CERT/CC's research activities include the area of WAN computing and focus on developing improved Internet security. The organization also provides training to incident-response professionals.

connectionless—A type of protocol that does not require an exchange of messages with the destination host before data transfer begins, nor does it make a dedicated connection, or virtual circuit, with a destination host. Instead, a connectionless protocol relies upon upper-level, not lower-level, protocols for safe delivery and error handling.

connection-oriented network—A network for which you must establish a connection using an exchange of messages or must have a preestablished pathway between a source point and a destination point before you can transmit packets.

container—In directory services, an object that can hold other objects.

contention—Two or more devices contending (competing) for network access. Any device can transmit whenever it needs to send information. To avoid data collisions (two devices sending data at the same time), specific contention protocols were developed requiring the device to listen to the cable before transmitting data.

continuity check—A test done with a device such as an ohmmeter. Since a wire or fuse is a conductor, essentially you should measure no resistance (0 ohms) to show that there is no break in the wire or fuse.

cooperative multitasking—Each program can control the CPU for as long as it needs it. If a program is not using the CPU, however, it can allow another program to use it temporarily.

crossed pairs—Two wires connected improperly, causing the two wires to be crossed.

crossover cable—A cable that can be used to connect from one network card to another network card or a hub to a hub; reverses the transmit and receive wires.

crosstalk—When signals induct (law of induction) or transfer from one wire to the other.

Cryptography—The art of protecting information by transforming it (encrypting it) into cipher text.

current state—A digital signal periodically measured for the specific state.

cyclic redundancy check (CRC)—A form of error control.

D

daemon—A background process or service that monitors and performs many critical system functions and services.

data—The raw facts, numbers, letters, or symbols that the computer processes into meaningful information.

data channel (D channel)—Used in ISDN lines for transmitting control information.

data circuit-terminating equipment (DCE)—Special communication devices that provide the interface between the DTE and the network. Examples include modems and adapters. The purpose of the DCE is to provide clocking and switching services in a network, and it actually transmits data through the WAN. Therefore, the DCE controls data flowing to or from a computer.

Data Encryption Standard (DES)—A popular symmetric-key encryption method developed in 1975 and standardized by ANSI in 1981 as ANSI X.3.92. DES uses a 56-bit key and is illegal to export out of the United States or Canada if you don't meet the BXA requirements.

Data Link Control (DLC)—A special, non-routable protocol that enables computers running Windows 2000 to communicate with computers running a DLC Protocol stack such as IBM mainframes, IBM AS/400 computers, and Hewlett-Packard network printers that are connected directly to the network using older HP JetDirect network cards.

data link layer—The OSI layer that is responsible for providing error-free data transmission and establishing local connections between two computers or hosts. It divides data it receives from the network layer into distinct frames that can then be transmitted by the physical layer, and it packages raw bits from the physical layer into blocks of data called frames.

data service unit (DSU)—A device that performs all error correction, handshaking, and protective and diagnostic functions for a telecommunications line.

data terminal equipment (DTE)—Devices used on end systems that communicate across the WAN. They control data flowing to or from a computer. They are usually terminals, PCs, and network hosts, which are located on the premises of individual subscribers.

datagram—A packet on an IP network.

de facto standard—"From the fact" standard. A standard that has been accepted by the industry just because it was the most common.

de jure standard—"By law" standard. A standard that has been dictated by an appointed committee.

decimal number system—The most commonly used numbering system. Each position contains 10 different possible digits. Since there are 10 different possible digits, the decimal number system has numbers with base 10. The digits are 0, 1, 2, 3, 4, 5, 6, 7, 8, and 9.

decryption—The process of converting data from encrypted format back to its original format.

demand priority—An access method in which a device makes a request to the hub and the hub grants permission. High-priority packets are serviced before any normal-priority packets.

demarcation point (demarc)—The point where the local loop ends at the customer's premises.

demilitarized zone (DMZ)—An area used by a company that wants to host its own Internet services without sacrificing unauthorized access to its private network. The DMZ sits between the Internet and an internal network's line of defense, usually some combination of firewalls and bastion hosts. Typically, the DMZ contains devices accessible to Internet traffic, such as web (HTTP) servers, FTP servers, SMTP (email) servers, and DNS servers.

demultiplexing—The technique in which the destination computer receives the data stream and separates and rejoins the application's segments. See *multiplexing*.

denial of service (DoS)—A type of attack on a network that is designed to bring the network to its knees by flooding it with useless traffic.

designator—Keeps track of which drive designations are assigned to network resources.

device driver—Program that controls a device. It acts like a translator between the device and programs that use the device. Each device has its own set of specialized commands that only its driver knows. While most programs access devices by using generic commands, the driver accepts the generic commands from the program and translates them into specialized commands for the device.

device manager—The software that allows you to manage your devices and device drivers in Microsoft Windows.

DHCP (Dynamic Host Configuration Protocol)—An extension of BOOTP. It is used to automatically configure a host during bootup on a TCP/IP network and to change settings while the host is attached. There are many parameters that you can automatically set with the DHCP server including IP addresses, subnet masks, DNS servers, WINS servers, and domain names.

DHCP relay agent—A computer that relays DHCP and BOOTP messages between clients and servers on different subnets. This way, you can have a single DHCP server handle several subnets without the DHCP server being connected directly to those subnets.

dial-up networking—A method used when a remote access client makes a nonpermanent, dial-up connection to a physical port on a remote access server by using the service of a telecommunications provider such as an analog phone, ISDN, or X.25.

diffused IR—Infrared that spreads the light over an area to create a cell; limited to individual rooms.

digital certificate—An attachment to an electronic message used for security purposes such as for authentication and for verification to ensure that a user sending a message is who he or she claims to be and to provide the receiver with a means to encode a reply.

digital envelope—A type of security that encrypts the message using symmetric encryption and encrypts a key to decode the message using public-key encryption.

digital multimeter (DMM)—A device that combines several measuring devices including a voltmeter and an ohmmeter.

digital signals—A system based on a binary signal system produced by pulses of light or electric voltages. The site of the pulse is either on/high or off/low to represent 1s and 0s. Digital signals are the language of computers.

digital signature—A digital code that can be attached to an electronically transmitted message that uniquely identifies the sender. Like a written signature, the purpose of a digital signature is to guarantee that the individual sending the message really is who he or she claims to be.

digital subscriber line (DSL)—A special communication line that uses sophisticated modulation technology to maximize the amount of data that can be sent over plain twisted-pair copper wiring, which is already carrying phone service to subscribers' homes.

directed IR—Infrared that uses line-of-sight or point-to-point technology.

directory number (DN)—The 10-digit phone number the telephone company assigns to any analog line.

direct-sequence spread spectrum (DSSS)—A technique that generates a redundant bit pattern for each bit to be transmitted. This bit pattern is called a chip (or chipping code). The intended receiver knows which specific frequencies are valid and deciphers the signal by collecting valid signals and ignoring the spurious signals. The valid signals are then used to reassemble the data.

directory service—A network service that identifies all resources on a network and makes those resources accessible to users and applications.

Resources can include email addresses, computers, and peripheral devices (such as printers).

directory services servers—A server used to locate information about the network such as domains (logical divisions of the network) and other servers.

diskless workstation—A computer that does not have its own disk drive. Instead the computer stores files on a network file server.

distance-vector–based routing—A routing protocol that periodically advertises or broadcasts the routes in their routing tables, but it only sends the information to its neighboring routers.

distributed component object model (DCOM)—A set of Microsoft concepts and program interfaces in which client program objects can request services from server program objects from other computers in a network. Since DCOM uses TCP/IP and HTTP, it can work with a TCP/IP network or the Internet.

distributed computing—Processing done by the individual PCs rather than a central computer.

DIX connector—Short for Digital-Intel-Xerox. A female 15-pin D connector used to connect to an Ethernet 10Base5 network. Also known as an AUI connector.

DNS name space—Hierarchial structure of the DNS database arranged as an inverted logical tree structure.

DNS zone—A portion of the DNS name space whose database records exist and are managed in a particular DNS database file.

domain—A logical unit of computers and network resources that define a security boundary. It is typically found on medium- or large-size networks or networks that require a secure environment. Different from a workgroup, a domain uses one database to share its common security and user account information for all computers within the domain. Therefore, it allows centralized network administration of all users, groups, and resources on the network.

domain controller—A Windows server that contains the domain database.

domain local group—A group used in Windows NT family domains that is usually used to assign rights and permissions to network resources that are in the domain of the local group. Different from a domain local group, it can list user accounts, universal groups, and global groups from any domain and local groups from the same domain.

Domain Name System (DNS)—A service that provides mappings between names and IP addresses, along with distributing network information (i.e., mail servers). Also known as Domain Name Server.

domain user account—In Windows NT family domains, an account that can log onto a domain to gain access to the network resources.

DS0 channel—A channel that can carry 64 Kbps of data, sufficient to carry voice communication.

dumb terminal—A display monitor and keyboard that has no processing capabilities. A dumb terminal is simply an output device that accepts data from another computer such as a main frame.

duplex cables—A cable that has two optical fibers inside a single jacket. The most popular use for a duplex fiber-optic cable is as a fiber-optic LAN backbone cable. Duplex cables are perfect because all LAN connections need a transmission fiber and a reception fiber.

Dynamic Host Configuration Protocol (DHCP)—A TCP/IP server that is used to automatically assign TCP/IP addresses and other related information to clients.

E

E-1 line—A digital line with 32 64-Kbps channels for a bandwidth of 2.048 Mbps.

E-3 line—A digital line with 512 64-Kbps channels for a bandwidth of 34.368 Mbps.

EAP-transport-level security (EAP-TLS)—An EAP type that is used in certificate-based security environments. The EAP-TLS exchange of messages provides mutual authentication, negotiation of the encryption method, and secured private-key exchange between the remote access client and the authenticating server.

E-carrier system—An entire digital system that consists of permanent dedicated point-to-point connections. It is typically used in Europe.

electromagnetic interference (EMI)—Signals caused by large electromagnets used in industrial machinery, motors, fluorescent lighting, and power lines that cause interference to other signals.

electronic mail (email)—A sophisticated tool that allows you to send text messages and file attachments (documents, pictures, sound, and movies) to anyone with an email address.

electrostatic discharge (ESD)—Electricity generated by friction that occurs, for example when your arm slides on a tabletop or when you walk across a carpet. Electronic devices including network cards and computer components can be damaged by electrostatic discharge.

encapsulation—The concept of placing data behind headers (and before trailers) for each layer.

encoding—The process of changing a signal to represent data.

encryption—The process of disguising a message or data in what appears to be meaningless data (cipher text) to hide and protect the sensitive data from unauthorized access.

enterprise—Any large organization that utilizes computers, usually consisting of multiple LANs.

enterprise WAN—A WAN that is owned by one company or organization.

entity—Identifies the hardware and software that fulfills a role or service of a server.

entrance facility—The place where the outside telecommunications service enters the building and interconnects with the building's telecommunications systems. In a campus or multibuilding environment, it may also contain the building's backbone cross-connections.

equipment room—The area in a building where telecommunications equipment is located and the cabling system terminates.

error control—Refers to the notification of lost or damaged data frames.

Ethernet—The most widely used LAN technology today. It uses a logical bus topology.

event viewer—In the Windows NT family, a very useful utility used to view and manage logs of system, program, and security events on your computer. An event viewer gathers information about hardware and software problems and monitors Windows security events.

exchange—The service that provides email on the Microsoft Windows NT, Windows 2000, or Windows 2003 Server.

Extensible Authentication Protocol (EAP)—An authentication protocol that allows new authentication schemes to be plugged in as needed. Therefore, EAP allows third-party vendors to develop custom authentication schemes such as retina scans, voice recognition, fingerprint identification, smart cards, Kerberos, and digital certificates.

Extensible Markup Language (XML)—A specification developed by the W3C. XML is a pared-down version of SGML, designed especially for web documents. It allows designers to create their own customized tags, enabling the definition, transmission, validation, and interpretation of data between applications and between organizations.

external IPX address—An 8-digit (4-byte) hexadecimal number used to identify the network on an IPX network.

F

Fast Ethernet—An extension of the 10BaseT Ethernet standard that transports data at 100 Mbps and yet still keeps using the CSMA/CD Protocol used by 10-Mbps Ethernet.

fat client—A client that performs the bulk of the data processing operations. The data itself is stored on the server. Although a fat client also refers to software, it can also apply to a network computer that has relatively strong processing abilities.

FAT32—A file system that is an enhancement of the FAT/VFAT file system, which uses 32-bit FAT entries. It supports hard drives up to 2 TB.

fault tolerance—The ability of a system to respond gracefully to an unexpected hardware or software failure. There are many levels of fault tolerance, the lowest being the ability to continue operation in the event of a power failure. Many fault-tolerant computer systems mirror all operations—that is, every operation is performed on two or more duplicate systems, so if one fails, the other can take over.

fax server—A server that manages fax messages sent into and out of the network through a fax modem.

Fiber Distributed Data Interface (FDDI)—A MAN Protocol that provides data transport at 100 Mbps and can support up to 500 stations on a single network.

fiber optic—Cable that consists of a bundle of glass or plastic threads, each of which is capable of carrying data signals in the form of modulated pulses of light.

File Allocation Table (FAT)—A simple and reliable file system that uses minimal memory. It supports file names of 11 characters, which include the 8 characters for the file name and 3 characters for the file extension.

file attribute—A characteristic about each file. Attributes can be either on or off. The most common attributes include read-only, hidden, system, and archive.

file server—A server that manages user access to files stored on a server. When a file is accessed on a file server, the file is downloaded to the client's RAM. For example, if you are working on a report using a word processor, the word processor files will be executed from your client computer and the report will be stored on the server. As the report is accessed from the server, it would be downloaded or copied to the RAM of the client computer.

NOTE: All of the processing done on the report is done by the client's microprocessors.

file sharing—A network server that allows you to access files on another computer without using a floppy disk or other forms of removable media. To ensure that the files are secure, most networks can limit both the access to a directory or file and the kind of access (permissions or rights) a person or a group of people have.

File Transfer Protocol (FTP)—A TCP/IP that allows a user to transfer files between local and remote host computers.

firewall—A system designed to prevent unauthorized access to or from a private network. Firewalls can be implemented in both hardware and software, or a combination of both. Firewalls are frequently used to prevent unauthorized Internet users from accessing private networks connected to the Internet, especially intranets. All messages entering or leaving the intranet pass through the firewall, which examines each message and blocks those that do not meet the specified security criteria.

flow control—The process of controlling the rate at which a computer sends data.

forest—In Active Directory, one or more trees connected by two-way, transitive trust relationships, which allow users to access resources in the other domain/tree.

forwarders—DNS servers configured to send all recursive queries to a selected list of servers. Servers used in the list of forwarders provide recursive lookup for any queries that a DNS server receives that it cannot answer based on its own zone records. During the forwarding process, a DNS server configured to use forwarders essentially behaves as if it were a DNS client to its forwarders. Typically forwarders are used on remote DNS servers that use a slow link to access the Internet.

fox and hound—See *tone generator and probe.*

frame—A structured package for moving data that includes not only the raw data, or "payload," but also the sender's and receiver's network addresses and error-checking and control information.

Frame Relay—A packet-switching protocol designed to use high-speed digital backbone links to support modern protocols that provide for error handling and flow control for connecting devices on a wide area network.

Frame Relay Access Device (FRAD)—Multiplexes and formats traffic for entering a Frame Relay network. Sometimes referred to as a Frame Relay Assembler/Dissembler.

frequency—The number of times a single wave will repeat over any period. It is measured in hertz (Hz), or cycles per second.

frequency hopping—A technique that quickly switches between predetermined frequencies, many times each second. Both the transmitter and receiver must follow the same pattern and maintain complex timing intervals to be able to receive and interpret the data being sent.

frequency-division multiplexing (FDM)—A method that uses its transmission medium's capacity to provide multiple channels. Each channel uses a carrier signal, which runs at a different frequency than the other carrier signals used by the other channels.

front end—The interface used in client/server applications that is provided to a user or another program. It is the part that the user or program will see and interact with.

full-duplex dialog—A form of dialog that allows every device to both transmit and receive simultaneously.

Fully Qualified Domain Name (FQDN)—Is used to identify computers on a TCP/IP network. Examples include MICROSOFT.COM and EDUCATION.NOVELL.COM. Sometimes referred to as just a domain name.

G

gateway—Hardware or software that links two different types of networks by repackaging and converting data from one network to another network or from one network operating system to another.

General Public License (GPL)—A form of copyright to protect software from being taken over and being kept from free public use.

giant—See *long frame.*

Gigabit Ethernet—A form of Ethernet that has a bandwidth of a gigabit throughput.

global catalog—In Active Directory, a database that holds a replica of every object. However, instead of storing the entire object, it stores those attributes most frequently used in search operations (such as a user's first and last names).

global group—A group used in Windows NT family domains that usually is used to group people within its own domain. Therefore, it can list user accounts and global groups from the same domain. The global group can be assigned access to resources in any domain.

global WAN—A WAN that is not owned by any one company and could cross national boundaries. The best-known example of a global WAN is the Internet, which connects millions of computers.

GNU—Short for GNU's not UNIX (and pronounced with a hard "g"). It refers to a UNIX-compatible software system developed by the Free Software Foundation (FSF). As GNU was started, hundreds of programmers created new, open-source versions of all major UNIX utility programs, with Linux providing the kernel. Many of the GNU utilities were so powerful that they have become the virtual standard on all UNIX systems. For example, gcc became the dominant C compiler, and GNU emacs became the dominant programmer's text editor.

gopher—A system that predates the World Wide Web for organizing and displaying files on Internet servers. A gopher server presents its contents as a hierarchically structured list of files.

group—A collection of user accounts. Groups are not containers. They list members, but they do not contain the members.

H

half-duplex dialog—A form of dialog that allows each device to both transmit and receive, but not at the same time. Therefore, only one device can transmit at a time.

handshaking—The process that communication devices use to negotiate a common data rate and other transmission parameters.

hardware abstraction layer (HAL)—In the Windows NT family, a library of hardware manipulating routines that hides the hardware interface details. It contains the hardware-specific code that handles I/O interfaces, interrupt controllers, and multiprocessor operations so that it can act as the translator between specific hardware architectures and the rest of the Windows software.

hashing scheme—A method that scrambles information in such a way that it's unique and can't be changed back to the original format.

hexadecimal number system—A number system based on 16 digits (0, 1, 2, 3, 4, 5, 6, 7, 8, 9, A, B, C, D, E, and F). One hexadecimal digit is equivalent to a four-digit binary number (4 bits, or a nibble), and two hexadecimal digits are used to represent a byte (8 bits).

High-Level Data Link Control (HDLC) Protocol—The protocol used with PPPs that encapsulates its data during transmission.

home directory—A folder used to hold or store a user's personal documents.

HomeRF SWAP—A form of Bluetooth that is designed specifically for wireless networks in homes. SWAP is short for Shared Wireless Access Protocol.

hop—The trip a data packet takes from one router to another router or from a router to another intermediate point to another in the network.

hop count—The number of routers that a data packet takes to its destination.

horizontal cabling—The cabling system that extends from the work area receptacle to the horizontal cross-connect in the telecommunications closet. It includes the receptacle and an optional transition connector (such as an undercarpet cable connecting to a round cable).

horizontal wiring system—The system of cables that extend from wall outlets throughout the building to the wiring closet or server room.

host—Computer or other device that connects to the network and is the source or final destination of data.

host name—Name assigned to a specific computer within a domain or subdomain by an administrator to identify the TCP/IP host.

HOSTS file—A text file that lists the IP address followed by the host name used to resolve host names to IP addresses.

hub—A device that works at the physical OSI layer. It is a multiported connection point used to connect network devices via a cable segment. Also known as a concentrator.

hybrid fiber coaxial (HFC) network—A telecommunication technology in which optical fiber cable and coaxial cable are used in different portions of a network to carry broadband content (such as video, data, and voice).

hybrid topology—A topology scheme that combines two of the traditional topologies, usually to create a larger topology. In addition, the hybrid topology allows you to use the strengths of the various topologies to maximize the effectiveness of the network.

hyperlink—An element in an electronic document that links to another place in the same document or to an entirely different document. Typically, you click on the hyperlink to follow the link. Hyperlinks are the most essential ingredient of all hypertext systems, including the World Wide Web.

HyperText Markup Language (HTML)—The authoring language used to create documents on the World Wide Web.

Hypertext Transfer Protocol (HTTP)—A TCP/IP that is the basis for exchange over the World Wide Web.

I

IEEE 802.3—The 802 project standard that dictates the specification for carrier-sense multiple access with collision detection (CSMA/CD) LAN used in Ethernet.

IMAP4—A standards-based message access protocol. It follows the online model of message access, although it does support offline and disconnected modes. IMAP4 offers an array of up-to-date features, including support for the creation and management of remote folders and folder hierarchies, message status flags, new-mail notification, retrieval of individual MIME body parts, and server-based searches to minimize the amount of data that must be transferred over the connection.

Independent Basic Service Set (IBBS)—See ad hoc network.

Independent Computing Architecture (ICA)—A protocol that allows multiple computers to take control of a virtual computer and use it as if it were their desktop (thin client).

indirect IR—See *diffused IR*.

infrared (IR) system—Wireless technology based on infrared light—light that is just below the visible light in the electromagnetic spectrum.

infrastructure mode—A wireless LAN (WLAN) that consists of at least one access point (AP) connected to the wired network infrastructure and a set of wireless end stations.

Integrated Services Digital Network (ISDN)—A planned replacement for POTS so that it can provide voice and data communications worldwide using circuit switching while using the same wiring that is currently being used in homes and businesses. Because ISDN is a digital signal from end to end, it is faster and much more dependable with no line noise. ISDN has the ability to deliver multiple simultaneous connections, in any combination of data, voice, video, or fax, over a single line and allows for multiple devices to be attached to the line.

interference—When undesirable electromagnetic waves affect the desired signal.

internal IPX network number—An 8-digit (4-byte) hexadecimal number. It is used to identify a server on an IPX network.

Internet—A global network connecting millions of computers.

Internet connection sharing (ICS)—A program/server that allow a single dial-up connection to be shared across the network. Typically found on newer Microsoft OSs.

Internet Control Message Protocol (ICMP)—A TCP/IP that sends messages and reports errors regarding the delivery of a packet.

Internet Group Management Protocol (IGMP)—A TCP/IP that is used by IP hosts to report host group membership to local multicast routers.

Internet Information Server (IIS)—The service that provides web services (HTTP and FTP) on a Microsoft Windows NT, Windows 2000, or Windows .NET server.

Internet Protocol (IP)—Connectionless protocol primarily responsible for addressing and routing packets between hosts.

Internet Protocol Security (IPSec)—A set of protocols developed by the IETF to support the secure exchange of packets at the IP layer. IPSec has been deployed widely to implement virtual private networks (VPNs). IPSec supports two encryption modes: transport and tunnel. Transport mode encrypts only the data portion (payload) of each packet; it leaves the header untouched. The more secure tunnel mode encrypts both the header and the payload. On the receiving side, an IPSec-compliant device decrypts each packet. For IPSec to work, the sending and receiving devices must share a public key. This is accomplished through a protocol known as Internet Security Association and Key Management Protocol/Oakley (ISAKMP/Oakley), which allows the receiver to obtain a public key and authenticate the sender using digital certificates.

internetwork—A network that is internal to a company and is private. It is often a network consisting of several LANs linked together. The smaller LANs are known as subnetworks or subnets.

interprocess communication (IPC)—A mechanism that allows bidirectional communication between clients and servers using distributed applications. IPC is a mechanism used by programs and multiprocesses. IPCs allow concurrently running tasks to communicate between themselves on a local computer or between the local computer and a remote computer.

intranet—A network based on TCP/IP, the same protocol that the Internet uses. Unlike the Internet, an intranet belongs to a single organization, accessible only by the organization's members. An intranet's websites look and act just like any other websites, but they are isolated by a firewall to stop illegal access.

NOTE: An Intranet could have access to the Internet, but does not require it.

inverse query—A DNS query that provides the IP address and requests a FQDN.

IP address—A logical address used to uniquely identify a connection on a TCP/IP address (logical address).

IPng—See *IPv6.*

IPv6—A new version of the Internet Protocol (IP) currently being reviewed in IETF standards committees. It is designed to allow the Internet to grow steadily, in terms of both the number of hosts connected and the total amount of data traffic transmitted. Also known as IPng.

IPX—Short for Internetwork Packet Exchange. A networking protocol used by the Novell NetWare operating systems. Like UDP/IP, IPX is a datagram protocol used for connectionless communications. Higher-level protocols, such as SPX and NCP, are used for additional error recovery services.

ISP—Short for Internet service provider. A company that provides access to the Internet.

iterative query—A DNS query that sends the best answer it currently has back as a response. The best answer will be the address being sought or an address of a server that would have a better idea of its address.

J

jabber—An error in which a faulty device (usually a NIC) continuously transmits corrupted or meaningless data onto a network. This may stop the entire network from transmitting data because other devices will perceive the network as busy.

jitter—A term used to describe instability in a signal wave. It is caused by signal interference.

journaling file system—A file system that can keep track of either the changes to a file's "metadata" (information such as ownership, creation dates, and so on) or to the data blocks associated with a file, or to both so that the file system becomes more resistant against corruption or damage.

K

Kerberos—An authentication system designed to enable two parties to exchange private information across an otherwise open network. It works by assigning a unique key, called a ticket, to each user that logs onto the network. The ticket is then embedded in messages to identify the sender of the message.

Kerberos Security—A security protocol that is used for distributed security within a domain tree/forest.

kernel—The central module of an operating system. It is the part of the operating system that loads first, and it remains in RAM. Because it stays in memory, it is important for the kernel to be as small as possible while still providing all the essential services required by other parts of the operating system and applications. Typically, the kernel is responsible for memory management, process and task management, and disk management.

kernel mode—A Windows NT family mode that runs in Ring 0 of the Intel 386 microprocessor protection model. While user-mode components are protected by the OS, the kernel-mode components are protected by the processor. It has direct access to all hardware and all memory including the address space of all user-mode processes. Kernel mode includes the Windows NT family executive, hardware abstraction layer (HAL), and the microkernel.

key—A string of bits used to map text into a code and a code back to text. You can think of the key as a super-decoder ring. There are two types of keys: public keys and private keys.

L

last mile—The telephone line that runs from your home or office to the telephone company's central office (CO) or neighborhood switching station (often a small building with no windows). Also known as local loop and subscriber loop.

layer 3 switch—A device that combines a router and a switch. It has been optimized for high-performance LAN support and is not meant to service wide area connections.

Layer Two Tunneling Protocol (L2TP)—An extension to PPP that enables ISPs to operate virtual private networks (VPNs). L2TP merges the best features of two other tunneling protocols: PPTP from Microsoft and L2F from Cisco Systems. Like PPTP, L2TP requires that the ISP's routers support the protocol.

Lightweight Directory Access Protocol (LDAP)—A set of protocols for accessing information directories. LDAP is based on the standards contained within the X.500 standard, but it is significantly simpler.

line conditioner—A device that uses the inductance of transformers to filter out noise and capacitors (and other circuits) to "fill in" brownouts. In addition, most line conditioners include surge protection.

Line Printer Daemon (LPD)—A TCP/IP that provides printing.

link-state algorithm—A routing protocol that sends updates directly (or by using multicast traffic) to all routers within the network. Each router, however, sends only the portion of the routing table that describes the state of its own links. In essence, link-state algorithms send small updates everywhere. Also known as shortest-path-first algorithm.

Linux—(Pronounced LIH-nuhks with a short "i.") A UNIX-like operating system that was designed to provide personal computer users a free or very low-cost operating system comparable to traditional and usually more expensive UNIX systems.

Linux Documentation Project (LDP)—A loose team of writers, proofreaders, and editors who are working on a set of definitive Linux manuals.

LMHOSTS file—A text file similar to a HOSTS file used to resolve computer names to IP addresses.

load sharing—See *round-robin.*

local area network (LAN)—A network that has computers that are connected within a geographically close area, such as a room, a building, or a group of adjacent buildings.

local loop—The telephone line that runs from your home or office to the telephone company's central office (CO) or neighborhood switching station (often a small building with no windows). Also known as subscriber loop and last mile.

local procedure call (LPC)—A mechanism used to transfer information between applications on the same computer.

local user account—In Windows NT family domains, an account that allows users to log on at and gain resources on only the computer where you create the local user account. The local user account is stored in the local security database. When using a local user account, you will not be able to access any of the network resources. In addition, for security reasons, you cannot log on as a local user account on a domain/ domain controller.

LocalTalk—Apple's own network interface and cable system. Since AppleTalk can only operate at 230 Kbps, it is not commonly used.

logical link control (LLC)—The sublayer of the data link layer (OSI model) that manages the data link between two computers within the same subnet.

logical topology—Part of the data link layer. Describes how the data flows through the physical topology or the actual pathway of the data.

long frame—A frame greater than 1518 bytes. Long frames are typically formed by faulty or out-of-specification LAN drivers. Also known as a giant.

M

Mac OS—An operating system that runs the Macintosh computer.

mail server—A server that manages electronic messages (email) between users.

mailslot—Connectionless messaging, which means that messaging is not guaranteed. Connectionless messaging is useful for identifying other computers or services on a network, such as the browser service offered in Windows 2000.

mainframes—Large centralized computers used to store and organize data.

Management Information Base (MIB)—See *MIB.*

media access control (MAC) address—A hardware address (physical address) identifying a node on the network. It is a unique hardware address (unique on the LAN/subnet) burned onto a ROM chip assigned by the hardware vendors or selected with jumpers or DIP switches.

media access control (MAC) sublayer—The lower sublayer of the data link layer (OSI model) that communicates directly with the network adapter card. It defines the network logical topology, which is the actual pathway (ring or bus) of the data signals being sent. In addition, it allows multiple devices to use the same medium, and it determines how the network card gets access to or control of the network medium so that two devices don't trample over each other.

member server—A server that is not a domain controller and does not have a copy of the domain database.

mesh topology—A topology where every computer is linked to every other computer.

message transfer agent (MTA)—The program responsible for receiving incoming emails and delivering the messages to individual users. The MTA transfers messages between computers. Hidden from the average user, it is responsible for routing messages to their proper destinations. MTAs receive messages from both MUAs and other MTAs, although single-user machines more often retrieve mail messages using POP. The MTA is commonly referred to as the mail server program. UNIX Sendmail and Microsoft Exchange Server are two examples of MTAs.

Messaging Application Programming Interface (MAPI)—A system built into Microsoft Windows that enables different email applications to work together to distribute mail. As long as both applications are MAPI-enabled, they can share mail messages with each other.

metric—A standard of measurement, such as a hop count, that is used by routing algorithms to determine the optimal path to a destination.

metropolitan area network (MAN)—A network designed for a town or city, usually using high-speed connections such as fiber optics.

MIB—Short for management information base. A database of objects that can be monitored by a network management system. Both SNMP and RMON use standardized MIB formats that allow any SNMP and RMON tools to monitor any device defined by a MIB.

micron (μm)—A unit of length equal to one millionth (10^{-6}) meter.

Microsoft Challenge Handshake Authentication Protocol (MS-CHAP)—An authentication protocol that is Microsoft's proprietary version of CHAP. Unlike PAP and SPAP, it lets you encrypt data that is sent using PPP or PPTP connections using Microsoft Point-to-Point Encryption (MPPE).

microwave—A form of electromagnetic energy that operates at a higher frequency (low-GHZ frequency range) than radio wave communications. Since it provides higher bandwidths than those available using radio waves, it is currently one of the most popular long distance transmission technologies.

modem (modulator-demodulator)—A device that enables a computer to transmit data over telephone lines. Since the computer information is stored and processed digitally, and the telephone lines transmit data using analog waves, the modem converts digital signals to analog signals (modulates) and analog signals to digital signals (demodulates).

modulation—The process of changing a signal to represent data.

MT-RJ connector—A fiber-optic connector similar to an RJ-45 connector. It offers a new, small form factor, two-fiber connector that is lower in cost and smaller than the duplex SC interface.

multifiber cable—A cable that has anywhere from three to several hundred optical fibers in it, typically in a multiple of two.

multimode fiber (MMF)—Fiber-optic cable that is capable of transmitting multiple modes (independent light paths) at various wavelengths or phases.

multiplexer (mux)—A device that sends and receives several data signals at different frequencies.

multiplexing—The technique that allows data from different applications to share a single data stream.

multipoint connection—A connection that links three or more devices together through a single communication medium.

Multipurpose Internet Mail Extension (MIME)—A specification for formatting non-ASCII messages so that they can be sent over the Internet. Many email clients now support MIME, which enables them to send and receive graphics, audio, and video files via the Internet mail system. In addition, MIME supports messages in character sets other than ASCII.

multistation access unit—(**MAU** or **MSAU**—A physical-layer device unique to Token Ring networks that acts like a hub. While a hub defines a logical bus, the MAU defines a logical ring.

N

name server—A program that translates names from one form into another. Name servers are often associated with Domain Name Services that translate host and domain names to IP addresses.

name space—The set of names in a naming system.

named pipe—Connection-oriented messaging carried out by using pipes that set up a virtual circuit between two points to maintain reliable and sequential data transfer. A pipe connects two processes so that the output of one can be used as input to the other.

narrowband radio system—A radio system that transmits and receives user information on a specific radio frequency. Narrowband radio keeps the radio signal frequency as narrow as possible—just wide enough to pass the information.

NDIS—Short for network device interface specification, developed by Microsoft. It allows multiple protocols to use a single network adapter card at the same time.

nearest active upstream neighbor (NAUN)—A node in a Token Ring network, where each node of a Token Ring network acts as if it is a repeater that receives token and data frames from its neighbor.

NetBEUI—Short for NetBIOS Enhanced User Interface. It provides the transport and network layers for the NetBIOS Protocol, usually found on Microsoft networks. NetBEUI is a protocol used to transport data packets between two nodes. While NetBEUI is smaller than TCP/IP or IPX and is extremely quick, it will only send packets within the same network. Unfortunately, NetBEUI is not a routable protocol, which means that it cannot send a packet to a computer on another network.

NetWare—A popular local area network (LAN) operating system developed by the Novell Corporation. NetWare is a software product that runs on a variety of different types of LANs, from Ethernets to IBM Token Ring networks. It provides users and programmers with a consistent interface that is independent of the actual hardware used to transmit messages.

NetWare Directory Services (NDS)—A global, distributed, replicated database that keeps track of users and resources and provides controlled access to network resources. It is typically found on Novell NetWare networks.

NetWare loadable module (NLM)—Software that enhances or provides additional functions in a NetWare 3.x or higher server. Examples include support for database engines, workstations, network protocols, fax, and print servers.

network—Two or more computers connected together to share resources such as files or a printer. For a network to function, it requires a network service to share or access a common medium or pathway to connect the computers. To bring it all together, protocols give the entire system common communication rules.

network address—An address that uniquely identifies a network.

Network Address Translation (NAT)—A method of connecting multiple computers to the Internet (or any other IP network) using one IP address. With a NAT gateway running on this single computer, it is possible to share that single address between multiple local computers and connect them all at the same time. The outside world is unaware of this division and thinks that only one computer is connected.

network analyzer—See *protocol analyzer.*

network-attached storage (NAS)—Hard disk storage that is set up with its own network address rather than being attached to the department computer that is serving applications to a network's workstation users.

Network Basic Input/Output System (Net-BIOS)—A common program that runs on most Microsoft networks. It is used by applications to communicate with NetBIOS-compliant transports such as NetBEUI, IPX, or TCP/IP. It is responsible for establishing logical names (computer names) on the network, establishing a logical connection between the two computers, and supporting reliable data transfer between computers that have established a session.

network card—See *network interface card.*

Network File System (NFS)—A TCP/IP that provides transparent remote access to shared files across networks.

network interface card (NIC)—A device used to connect computers to the network by using a special expansion card (or components built into the motherboard). The network card will then communicate by sending signals through a cable (twisted-pair, coaxial, or fiber optics) or by using wireless technology (infrared or radio waves). The role of the network card is to prepare and send data to another computer, receive data from another computer, and control the flow of data between the computer and the cabling system.

network layer—The OSI model layer that is concerned with the addressing and routing necessary to move data (known as packets or datagrams) from one network (or subnet) to another. This includes establishing, maintaining, and terminating connections between networks, making routing decisions, and relaying data from one network to another.

network operating system (NOS)—An operating system that includes special functions for connecting computers and devices into a local area network (LAN), to manage the resources and services of the network and to provide network security for multiple users.

network prefix—The network number that is derived from the network bits of the IP address.

nodes—Devices connected to the computer including networked computers, routers, and network printers.

noise—Interference or static that destroys the integrity of signals. Noise can come from a variety of sources, including radio waves, nearby electrical wires, lightning and other power fluctuations, and bad connections.

Novell Directory Services (NDS)—A directory service based primarily on the X.500 Internet directory standard used on Novell NetWare servers 4.0 and higher. All the network resources are represented as objects and placed into a hierarchical structure, called an NDS tree.

Novell Netware—A network operating system that helped bring networking to PCs. Unfortunately, since then, it has mostly been replaced by Microsoft servers.

NTFS—A file system for the Windows NT family OSs designed for both the server and workstation. It provides a combination of performance, reliability, security, and compatibility.

NTLM authentication—An authentication process in which the client selects a string of bytes, uses the password to perform a one-way encryption of the string, and sends both the original string and the encrypted one to the server. The server receives the original string and uses the password from the account database to perform the same one-way encryption. If the result matches the encrypted string sent by the client, the server concludes that the client knows the user name/password pair.

NWLink—Microsoft's NDIS-compliant, 32-bit implementation of the Internetwork Packet Exchange (IPX), Sequenced Packet Exchange (SPX), and NetBIOS Protocols used in Novell networks. NWLink is a standard network protocol that supports routing and can support NetWare client/server applications, where NetWare-aware Sockets-based applications communicate with IPX/SPX Sockets-based applications.

O

object—A distinct, named set of attributes or characteristics that represent a network resource, including computers, people, groups, and printers.

Octet—A grouping of eight bits.

ODI—Short for open data link interface. Developed by Apple and Novell, it allows multiple protocols to use a single network adapter card.

ohmmeter—A device that can check wires and connectors and measure the resistance of an electronic device.

open—Refers to a conductor with a break in it or wires that are unconnected, which prevents electricity from flowing.

open architecture—A specification of the system or standard that is public.

Open Shortest Path First (OSPF)—A Link State Route Discovery Protocol where each router periodically advertises itself to other routers.

Open Systems Interconnection (OSI) reference model—The world's most prominent networking architecture model.

optical time domain reflectometer (OTDR)—The fiber-optic equivalent of the TDR that is used to test copper cables. The OTDR transmits a calibrated signal pulse over the cable to be tested and monitors the signal that returns to the unit. Instead of measuring signal reflections caused by electrical impedance as a TDR does, however, the OTDR measures the signal returned by backscatter, a phenomenon that affects all fiber-optic cables.

organizational unit (OU)—In directory services, a container used to hold and organize objects including users, groups, computers, and other organizational units.

OS/2—One of the first 32-bit GUI operating systems. Originally developed by IBM and Microsoft before Microsoft's Windows NT, it is now owned and controlled by IBM. It uses the same domain concept used by Windows NT.

P

packet—A piece of a message transmitted over a packet-switching network. One of the key features of a packet is that it contains the destination address in addition to the data.

packet switching—A technique in which messages are broken into small parts called packets. Each packet is tagged with source, destination, and intermediary node addresses as appropriate. Packets can have a defined maximum length and can be stored in RAM instead of on hard disk. Packets can take a variety of possible paths through the network in an attempt to keep the network connections filled at all times.

Password Authentication Protocol (PAP)—An authentication protocol in which passwords are sent across the link as unencrypted plain text.

patch—A temporary fix to a program bug. A patch is an actual piece of object code that is inserted into (patched into) an executable program.

patch panel—A panel with numerous RJ-45 ports. The wall jacks are connected from the back of the patch panel to the individual RJ-45 ports. You can then use patch cables to connect the port in the front of the patch panel to a computer or a hub. As a result, you can connect multiple computers with a hub located in the wiring closet or server room.

peer-to-peer network—A network that has no dedicated servers. Instead, all computers are equal. Therefore, they provide services and request services. Since a person's resources are kept on his or her own machine, a user manages his or her own shared resources. Sometimes referred to as a workgroup.

permanent virtual circuit (PVC)—A permanently established virtual circuit that consists of one mode, data transfer. PVCs are used in situations in which data transfer between devices is constant.

permission—Defines the type of access granted to an object or object attribute. The permissions available for an object depend on the type of object.

personal computer (PC)—A computer meant to be used by one person. The first personal computer produced by IBM was called the PC, and increasingly the term *PC* came to mean IBM or IBM-compatible personal computers, to the exclusion of other types of personal computers, such as Macintoshes. In recent years, the term *PC* applies to any personal computer based on an Intel microprocessor or on an Intel-compatible microprocessor.

phase—Measured in degrees to specify how close two sinusoidal waves are to each other.

physical layer—The OSI model that is responsible for the actual transmission of the bits sent across a physical medium. It allows signals, such as electrical signals, optical signals, or radio signals, to be exchanged among communicating machines. Therefore, it defines the electrical, physical, and procedural characteristics required to establish, maintain, and deactivate physical links.

physical topology—Part of the physical layer. It describes how the network actually appears.

ping—A utility to determine whether a specific IP address is accessible. It works by sending a packet to the specified address and waiting for a reply. Ping is used primarily to troubleshoot Internet connections.

plain old telephone service (POTS)—The PSTN/standard telephone service that most homes use.

plenum—The space above the ceiling and below the floor used to circulate air throughout the workplace.

plenum cable—A special cable that gives off little or no toxic fumes when burned.

point-to-point topology—A topology that connects two nodes directly together.

Point-to-Point Protocol (PPP)—The predominant protocol for modem-based access to the Internet that provides full-duplex, bidirectional operations between hosts and can encapsulate multiple-network-layer LAN protocols to connect to private networks.

Point-to-Point Tunneling Protocol (PPTP)—A new technology for creating virtual private networks (VPNs), developed jointly by Microsoft Corporation, U.S. Robotics, and several remote access vendor companies, known collectively as the PPTP Forum. A VPN is a private network of computers that uses the public Internet to connect some nodes. Because the Internet is essentially an open network, PPTP is used to ensure that messages transmitted from one VPN node to another are secure. With PPTP, users can dial in to their corporate network via the Internet. PPTP uses the Internet as the connection between remote users and a local network, as well as between local networks.

policies—In Windows NT family networks, a tool used by administrators to define and control

how programs, network resources, and the operating system behave for users and computers in a domain or Active Directory structure.

polling—An access method that has a single device (sometimes referred to as a channel-access administrator) designated as the primary device, which polls or asks each of the secondary devices known as slaves if they have information to be transmitted.

POP3—The newest version of the Post Office Protocol, which can be used with or without SMTP.

port—An address on a host where an application makes itself available to incoming data.

portal—A logical integration between wired LANs and 802.11.

POSIX—A set of IEEE and ISO standards that define an interface between programs and UNIX. By following the POSIX standard, developers have some assurance that their software can be easily ported or translated to a POSIX-compliant operating system.

Post Office Protocol (POP)—A TCP/IP that defines a simple interface between a user's mail client software and email server. It is used to download mail from the server to the client and allows the user to manage his or her mailbox.

PPPoE (PPP over Ethernet)—A protocol that was designed to bring the security and metering benefits of PPP to Ethernet connections such as those used in DSL.

preemptive multitasking—The operating system parcels out CPU time slices to each program.

presentation layer—The OSI model layer that ensures that information sent by an application-layer protocol of one system will be readable by the application-layer protocol on a remote system. It also provides encryption/decryption and compression/decompression of data and network redirectors.

Pretty Good Privacy (PGP)—A technique for encrypting messages that is based on the public-key method. To encrypt a message using PGP, you need the PGP encryption package, which is available free from a number of sources.

primary domain controller (PDC)—The server that contains the master copy of the domain database on Windows NT networks.

primary name server—A DNS name server that stores and maintains the zone file locally. Changes to a zone, such as adding domains or hosts, are done by changing files at the primary name server.

primary rate interface (PRI)—A form of ISDN that includes 23 B channels (30 in Europe) and one 64-KB D channel. PRI service is generally transmitted through a T-1 line (or an E-1 line in Europe).

print server—A server that manages user access to printer resources connected to the network, allowing one printer to be used by many people.

print sharing—A network service that allows several people to send documents to a centrally located printer in the office.

private network—A network that is not connected to the Internet.

private-key encryption—The most basic form of encryption that requires that each person possess a copy of the key. Of course, for this to work as intended, you must have a secure way to transport the key to other people. You must keep multiple single keys per person, which can get very cumbersome. Private-key algorithms are generally very fast and easily implemented in hardware. Therefore, they are commonly used for bulk data encryption.

process—An executing program.

Project 802—A set of standards that have several areas of responsibility including the network card, the wide area network components, and the media components.

propagation delay—The amount of time that passes between the point when a signal is transmitted and the point when it is received at the opposite end of the copper or optical cable.

proprietary system—A system or architecture that is privately owned and controlled by a company and has not divulged a system or architecture specifications that would allow other companies to duplicate the product.

protocol analyzer—A software or a hardware/software device that allows you to capture or receive all the packets on your media, store them in a trace buffer, and then show a breakdown of each of the packets by protocol in the order they appeared. Therefore, it can help you analyze all levels of the OSI model to determine the cause of a problem. Also known as a network analyzer.

protocol suite—A set of protocols that work together.

protocols—The rules or standards that allow computers to connect to one another and enable computers and peripheral devices to exchange information with as little error as possible.

proxy—Any device that acts on behalf of another.

proxy server—A server that performs a function on behalf of other computers. It is typically used to provide local intranet clients with access to the Internet while keeping the local intranet free from intruders.

public-key encryption—The nonsecret key that is available to anyone you choose or is made available to everyone by posting it in a public place. It is often made available through a digital certificate. The private key is kept in a secure location used only by you. When data needs to be sent, it is protected with a secret key encryption that was encrypted with the public key of the recipient of the data. The encrypted secret key is then transmitted to the recipient along with the encrypted data. The recipient will use the private key to decrypt the secret key. The secret key will then be used to decrypt the message itself.

public switched telephone network (PSTN)—The international telephone system based on copper wires (UTP cabling) carrying analog voice data.

pull partner—A WINS server that requests new database entries from its partner. The pull occurs at configured time intervals or in response to an update notification from a push partner.

punch-down block—A device used to connect several cable runs to each other without going through a hub.

punch-down tool—A tool used to connect wires to a punch-down block.

push partner—A WINS server that sends update notification messages. The update notification occurs after a configurable number of changes to the WINS database.

Q

quality of service (QoS)—Guaranteed bandwidth that has connection-oriented networks that provide sufficient bandwidth for audio and video without the jitters or pauses and with the transfer of important data within a timely manner.

query—A request for information from a database.

R

rackmount cabinet—A cabinet designed to hold several servers.

radio frequency (RF)—Signals that reside between 10 KHz and 1 GHz of the electromagnetic spectrum. It can be used to transmit data through the air.

radio frequency interference (RFI)—Signals from transmission sources, such as a radio station, which may cause interference with other signals.

recursive query—A query that asks the DNS server to respond with the requested data or with an error stating that the requested data doesn't exist or that the domain name specified doesn't exist.

redirector—Intercepts file input/output requests and directs them to a drive or resource on another computer.

Redundant Array of Independent (or Inexpensive) Disk (RAID)—A category of disk drives that employ two or more drives in combination for fault tolerance and performance. RAID disk

drives are used frequently on servers but aren't generally necessary for personal computers.

reflective IR—See *diffused IR*.

remote access server—(1) A server that hosts modems for inbound requests to connect to the network. (2) The computer and associated software that is set up to handle users seeking access to the network remotely.

remote access service (RAS)—A service that allows users to connect remotely using various protocols and connection types.

Remote Authentication Dial-In User Service (RADIUS)—An industry standard client/server protocol and software that enables remote access servers to communicate with a central server to authenticate dial-in users and authorize their access to the requested system or service for authenticating remote users.

remote procedure call (RPC)—A mechanism used to transfer information between applications that are on separate computers.

repeater—A device that works at the physical OSI layer. It is used to regenerate or replicate a signal or to move packets from one physical medium to another.

Requests for Comments (RFCs)—The series of documents that specify the TCP/IP standards.

requester—A program or part of a program that requests services.

resolver—A client that uses a name server.

Reverse Address Resolution Protocol (RARP)—A TCP/IP that permits a physical address, such as an Ethernet address, to be translated into an IP address.

reverse query—A DNS query used by a resolver that knows an IP address and wants to know the host name.

right—(1) Authorizes a user to access a file or directory on a network server. (2) Authorizes a user to perform certain actions on a computer, such as logging onto a system interactively, logging on locally to the computer, backing up files and directories, performing a system shutdown, or adding or removing a device driver.

ring topology—A topology that has all devices connected to one another in a closed loop. Each device is connected directly to two other devices.

riser cable—Cable intended for use in vertical shafts that run between floors.

risers—Vertical connections between floors.

RJ-45—A connector that supports 10BaseT/ 100 BaseTX (UTP) cabling.

root domain—The top of the DNS tree. It is sometimes shown as a period (.) or as empty quotation marks (""), indicating a null value.

round-robin—Rotates the order of resource records data returned in a query answer in which multiple resource records exist of the same resource record type for a queried DNS domain name. Since the client is required to try the first IP address listed, a DNS server configured to perform round-robin rotates the order of the Host (A) resource records when answering client requests. Also known as load sharing.

router—A device that works at the network OSI layer. It connects two or more LANs. In addition, it can break a large network into smaller, more manageable subnets. As multiple LANs are connected together, multiple routes are created to get from one LAN to another. Routers then share status and routing information with other routers so that they can provide better traffic management and bypass slow connections.

Router Information Protocol (RIP)—A distance-vector route discovery protocol where the entire routing table is periodically sent to the other routers.

RSA—A public-key encryption technology based on the fact that there is no efficient way to factor very large numbers. Deducing an RSA key, therefore, requires an extraordinary amount of computer processing power and time. The RSA algorithm has become the de facto standard for industrial-strength encryption, especially for data sent over the Internet. It is built into many software products, including Netscape Navigator and Microsoft Internet Explorer.

runt—See *short frame.*

S

sags—They usually are not a problem. Very short drops in power lasting only a few milliseconds.

Samba—Software used on a Linux computer that provides file and print sharing like a Windows NT server using the CIFS Protocol.

satellite systems—A microwave system that provides far bigger areas of coverage than can be achieved using other technologies. The microwave dishes are aligned to geostationary satellites that can relay signals either between sites directly or via another satellite. The huge distances covered by the signal result in propagation delays of up to 5 seconds.

schema—(Pronounced SKEE-ma.) The structure of a database system, described in a formal language supported by the database management system (DBMS). In a relational database, the schema defines the tables, the fields in each table, and the relationships between fields and tables. Schema are generally stored in a data dictionary. Although a schema is defined in text database language, the term is often used to refer to a graphical depiction of the database structure.

screened twisted-pair (ScTP) cable—A is hybrid of STP and UTP cable, which contains four pairs of 24 AWG, 100-ohm wire surrounded by a foil shield or wrapper and a drain wire for bonding purposes. Sometimes referred to as foil twisted pair (FTP) cable because the foil surrounds all four conductors.

search engine—A program that searches documents for specified keywords and returns a list of the documents where the keywords were found.

second extended file system (ext2)—The native Linux file system. It is very similar to other modern UNIX file systems but most closely resembles the Berkeley Fast Filesystem used by BSD systems. The maximum size of an ext2 file system is 4 TB, while the maximum file size is currently limited to 2 GB by the Linux kernel.

second-level domain names—Part of the DNS system, they are variable-length names registered to an individual or organization for use on the Internet.

secondary name server—A DNS name server that gets the data from its zone from another name server, either a primary name server or another secondary name server. The process of obtaining this zone information across the network is referred to as a zone transfer.

Secure Hypertext Transport Protocol (S-HTTP)—An extension to HTTP to support sending data securely over the World Wide Web.

Secure Sockets Layer (SSL)—A protocol developed by Netscape for transmitting private documents via the Internet. SSL works by using a public key to encrypt data that's transferred over the SSL connection. Both Netscape Navigator and Internet Explorer support SSL, and many websites use the protocol to obtain confidential user information, such as credit card numbers. By convention, URLs that require an SSL connection start with https: instead of http:.

segment—(1) A single cable such as a backbone cable. (2) A cable that connects a hub and a computer.

Sendmail—The most widely used mail transfer agent (MTA).

Serial Line Interface Protocol (SLIP)—A simple protocol in which you send packets down a serial link delimited with special END characters.

server—A service provider that provides access to network resources.

server blade—A single circuit board populated with components such as processors, memory, and network connections that are usually found on multiple boards. Server blades are designed to slide into existing servers (known as a blade servers) much like an expansion slot is plugged into a PC. With server blades, many computers can be placed in a small area.

Server Message Block (SMB)—The SMB Protocol jointly developed by Microsoft, Intel, and IBM. It defines a series of commands used to pass information between networked computers. Clients connected to a network using NetBIOS over TCP/IP, NetBEUI, or IPX/SPX can send SMB commands.
Note: Microsoft refers to NetBIOS over TCP/IP as NBT.

server rack—A rack designed to hold several servers.

service—A program, routine, or process that performs a specific system function to support other programs.

service access point (SAP)—Used by the LLC to identify which protocol it is.

Service Advertising Packet (SAP)—An IPX Protocol used to advertise the services of all known servers on the network, including file servers, print servers, and so on. Servers periodically broadcast their service information while listening for SAPs on the network and storing the service information. Clients then access a service information table when they need to access a network service.

Service Profile Identifier (SPID)—A directory number and additional identifier used to identify the ISDN device to the telephone network.

session—A reliable dialog between two computers.

session layer—The OSI model layer that allows remote users to establish, manage, and terminate a connection (session).

share-level security model—A model that a network administrator uses to assign passwords to network resources.

sharing—The process of making a drive, directory, or printer available to users on the network.

shell—A term for the interactive user interface with an operating system. The shell is the layer of programming that understands and executes the commands a user enters. In some systems, the shell is called a command interpreter. A shell usually implies an interface with a command syntax (think of the DOS operating system and its C:> prompts and user commands such as dir and edit).

shielded twisted pair (STP)—Similar to unshielded twisted pair except that it's usually surrounded by a braided shield that serves to reduce both EMI sensitivity and radio emissions.

Shiva Password Authentication Protocol (SPAP)—An authentication protocol that sends a password across a link in reversibly encrypted form. It is typically used when connecting to a Shiva LanRover or when a Shiva client connects to a Windows 2000–based remote access server.

short—Occurs when a circuit has a zero or abnormally low resistance path between two points, resulting in excessive current. In networking cables, a short is an unintentional connection made between two conductors (such as wires) or pins/contacts.

short frame—A frame that is less than 64 bytes with a valid CRC. Short frames are typically formed by faulty or out-of-specification LAN drivers. Also known as a runt.

S-HTTP—See *Secure Hypertext Transport Protocol.*

signaling—The method for using electrical, light energy, or radio waves to communicate.

Simple Mail Transfer Protocol (SMTP)—A TCP/IP for the exchange of electronic mail over the Internet. It is used to send mail between email servers on the Internet or to allow an email client to send mail to a server.

Simple Network Management Protocol (SNMP)—A TCP/IP that defines procedures and management information databases for managing TCP/IP-based network devices.

simplex dialog—A form of dialog that allows communications on the transmission channel to occur in only one direction. Essentially, one device is allowed to transmit and all the other devices receive.

simplex fiber-optic cable—A type of cable that has only one optical fiber inside the cable jacket. Since simplex cables only have one fiber inside them, there is usually a larger buffer and a thicker jacket to make the cable easier to handle.

single-mode fiber (SMF)—Fiber-optic cable that can transmit light in only one mode, but the narrower diameter (compared to the diameter of multimode cabling) yields less dispersion, resulting in longer transmission distances.

small office/home office (SOHO)—A small network used primarily in a home office that might be part of a larger corporation but yet remain apart from it. SOHO networks are usually peer-to-peer networks.

SNMP agent—A client or device that returns the appropriate information to an SNMP manager.

SNMP community—A collection of hosts grouped together for administrative purposes. Deciding what computers should belong to the same community is generally, but not always, determined by the physical proximity of the computers. Communities are identified by the names you assign to them.

SNMP manager—The console through which the network administrator performs network management functions.

socket—A logical address assigned to a specific process running on a host computer. It forms a virtual connection between the host and client, and it identifies a specific upper-layer software process or protocol.

solid cable—Cable used for the cabling that exists throughout the building. This should include cables that lead from the wall jacks to the server room or wiring closet.

SONET—Short for Synchronous Optical Network. It is the North American equivalent of SDH that specifies synchronous data transmission over fiber-optic cables.

source-route bridging (SRB)—A bridging method used on Token Ring networks that is responsible for determining the path to the destination node.

spanning tree algorithm (STA)—A method used with Ethernet bridges that designates a loop-free subset of the network's topology by placing those bridge ports that, if active, would create loops into a standby (blocking condition) mode.

spike—An overvoltage that occurs that may damage the computer. Sometimes known as a transient.

split pair—Incorrect pinouts that cause data-carrying wires to be twisted together resulting in additional crosstalk. Split pairs can be the result of mistakes during the installation. The solution is to reattach the connectors at both ends using either the T568-A or T568-B pinouts.

spread-spectrum signals—Signals that are distributed over a wide range of frequencies and then collected onto their original frequency at the receiver. Different from narrowband signals, spread-spectrum signals uses wider bands, which transmit at a much lower spectral power density (measured in watts per hertz).

SPX—Short for Sequenced Packet Exchange. A transport-layer protocol (layer 4 of the OSI model) used in Novell NetWare networks. The SPX layer sits on top of the IPX layer (layer 3) and provides connection-oriented services between two nodes on the network. SPX is used primarily by client/server applications.

SQL server—A server that is a database management system (DBMS) that can respond to queries from client machines formatted in the SQL language, a database language.

standard—Either a dictated specification that a PC, hardware, or software must follow or a PC, hardware, or software that has become popular.

standby power supply—A device that consists of the battery hooked up parallel to the PC. When the SPS detects a power fluctuation, the system will switch over to the battery to power the PC. Of course, the SPS requires a small but measurable amount of time to switch over (usually one-half of one cycle of the AC, or less than 10 milliseconds). Most SPSs will include built-in surge protection devices.

star topology—The most popular topology in use. Each network device connects to a central

point such as a hub, which acts as a multipoint connector.

static routing—A method that uses table mappings established by the network administrator prior to the beginning of routing. These mappings do not change unless the network administrator alters them.

statistical time-division multiplexing (STDM)—A modified method of time-division multiplexing that analyzes the amount of data that each device needs to transmit and determines on the fly how much time each device should be allocated for data transmission on the cable or line. As a result, the STDM uses the bandwidth more efficiently.

storage area network (SAN)—A high-speed subnetwork of shared storage devices. A SAN's architecture works in a way that makes all storage devices available to all servers on a LAN. If an individual application in a server cluster fails (but the node does not), the cluster service will typically try to restart the application on the same node. If that fails, it moves the application's resources and restarts them on another node of the server cluster. This process is called failover.

straight tip (ST) connector—Probably the most widely used fiber-optic connector. It uses a BNC attachment mechanism similar to the thinnet connector mechanism.

straight-through cable—A cable that can be used to connect a network card to a hub. It has the same sequence of colored wires at both ends of the cable.

stranded cable—Cable that is typically used as a patch cable between patch panels and hubs and between the computers and wall jacks. Since the stranded wire isn't as firm as solid wire, it is a little easier to work with.

subdomain names—Part of the DNS system. They are additional names that an organization can create that are derived from the registered second-level domain name. The subdomain allows an organization to divide a domain into departments or geographical locations, making the partitions of the domain name space more manageable. A subdomain must have a contiguous domain name space. This means that the domain name of a zone (child domain) is the name of that zone added to the name of the domain or parent domain.

subnet, or subnetwork—A simple network or smaller network used to form a larger network.

subnet mask—Numbers used to define which bits represent the network address (including the subnet number) and which bits represent the host address.

subscriber (SC) connector—Typically, a latched connector. This makes it impossible for the connector to be pulled out without releasing the connector's latch (usually by pressing some kind of button or release). Sometimes known as the square connector.

subscriber loop—The telephone line that runs from your home or office to the telephone company's central office (CO) or neighborhood switching station (often a small building with no windows). Also known as local loop and last mile.

supernetting—The process of combining multiple IP address ranges into a single IP network such as combining several class C networks.

surge—An overvoltage that can stretch into milliseconds and that may damage the computer.

surge protector—A device designed to prevent most short-duration, high-intensity spikes and surges from reaching your PC by absorbing excess voltages.

Switched Virtual Circuit (SVC)—See *temporary virtual circuit.*

switch—See *switching hub.*

Switched Multimegabit Data Service (SMDS)—A high-speed, cell-relay, wide area network (WAN) service designed for LAN interconnection through the public telephone network.

switching hub—A fast multiported bridge that builds a table of the MAC addresses of all the connected stations. It then reads the destination address of each packet and forwards the packet to the correct port. Sometimes referred to as switch or a layer 2 switch.

symmetric multiprocessing (SMP)—Processing for a computer that uses two or more microprocessors that share the same memory. If software is written to use the multiple microprocessors, several programs can be executed at the same time, or multithreaded applications can be executed faster.

synchronous—Refers to devices that use a timing or clock signal to coordinate communications between them.

Synchronous Digital Hierarchy (SDH)—An international standard that specifies synchronous data transmission over fiber-optic cables.

SYS volume—The primary volume on a NetWare server.

T

T-1 line—A digital line that has 24 64-Kbps channels for a bandwidth of 1.544 Mbps.

T-3 line—A digital line that has 672 64-Kbps channels for a bandwidth of 44.736 Mbps.

tape drive—A device that reads from and writes to a long magnetic tape. Tape drives are relatively inexpensive and offer large storage capacities, making tape backup drives ideal for backing up hard drives on a regular basis.

T-carrier system—The first successful system that converted the analog voice signal to a digital bit stream. While the T-carrier system was originally designed to carry voice calls between telephone company central offices, today it is used to transfer voice, data, and video signals between different sites and to connect to the Internet.

telecommunication network (telnet)—A virtual terminal protocol (terminal emulation) allowing a user to log onto another TCP/IP host to access network resources (RFC 854).

telecommunications closet—The floor serving facilities for horizontal cable distribution. It can be used for intermediate and main cross-connects.

telephony server—A server that functions as an intelligent answering machine for the network. It can also perform call center and call-routing functions.

temporary virtual circuit—A virtual circuit that is dynamically established on demand and terminated when transmission is complete. Communication over an temporary virtual circuit consists of three phases: circuit establishment, data transfer, and circuit termination.

terminal adapter—A device that connects a computer to an external digital communications line, such as an ISDN line. A terminal adapter is a bit like a modem but since the terminal adapter receives a digital signal instead of an analog signal, it just needs to pass the digital signal forward.

terminal emulation—See *telecommunication network*.

terrestrial systems—A microwave system that uses relay towers to provide an unobstructed path over an extended distance. These line-of-sight systems use unidirectional parabolic dishes that must be aligned carefully.

thicknet—See *10Base5*.

thin client—A computer that is between a dumb terminal and a PC. A thin client is designed to be especially small so that the bulk of the data processing occurs on the server.

thinnet—See *10Base2*.

thread—In programming, a part of a program that can execute independently of other parts. Operating systems that support multithreading enable programmers to design programs whose threaded parts can execute concurrently.

time-division multiplexing (TDM)—A method that divides a single channel into short time slots, allowing multiple devices to be assigned time slots.

time domain reflectometer (TDR)—The primary tool used to determine the length of a copper cable and to locate the impedance variations that are caused by opens, shorts, damaged cables, and interference with other systems. The TDR works much like radar, by transmitting a signal on a cable with the opposite end left open and measuring the amount of time it takes for the signal's reflec-

tion to return to the transmitter. When you have this elapsed-time measure, called the nominal velocity of propagation (NVP), and you know the speed at which electrons move through the cable, you can determine the length of the cable.

token—A packet that is passed around the network in an orderly fashion from one device to the next to inform devices they can transmit data.

token passing—An access method that specifies that a network device only communicates over the network when it has the token (a special data packet that is generated by the first computer that comes online in a Token Ring network). The token is passed from one station to another around a ring. When a station gets a free token and transmits a packet, it travels in one direction around the ring, passing all the other stations along the way.

Token Ring—A network technology that uses a ring logical topology. For computers to access the network, they use a token.

tone generator and probe—A device that consists of a unit (tone generator) that you connect to a cable with a standard jack or an individual wire with alligator clips, which transmits a signal over the cable or wire. The other unit is a penlike probe that emits an audible tone when touched to the other end of the cable or wire or even its insulating sheath. Sometimes called a "fox and hound" wire tracer.

top-level domains—Immediately below the root domain found on the top of the NDS tree. They indicate a country, region, or type of organization. Three-letter codes indicate the type of organization. For example, COM indicates "commercial" (business) and EDU stands for "educational institution."

topology—Describes the appearance or layout of the network. Depending on how you look at the network, there is the physical topology and the logical topology.

traceroute—A utility that traces a packet from your computer to an Internet host, showing how many hops the packet requires to reach the host and how long each hop takes. If you're visiting a website and pages are appearing slowly, you can use traceroute to figure out where the longest delays are occurring.

transceivers—Devices that both transmit and receive analog or digital signals.

transition state—A digital signal that represents data by measuring the signal transitions from high to low or low to high. A transition indicates a binary 1, whereas the absence of a transition represents a binary 0.

Transmission Control Protocol (TCP)—A protocol that provides connection-oriented, reliable communications for applications that typically transfer large amounts of data at one time or that require an acknowledgment for data received.

transport layer—An OSI model layer that can be described as the middle layer that connects the lower and upper layers together. In addition, it is responsible for reliable transparent transfer of data (known as segments) between two end points. Since it provides end-to-end recovery of lost and corrupted packets and flow control, it deals with end-to-end error handling, division of messages into smaller packets, numbers of the messages, and repackaging of messages.

Transport Layer Security (TLS)—The successor to the Secure Sockets Layer (SSL), TLS is composed of two layers: the TLS Record Protocol and the TLS Handshake Protocol. The TLS Record Protocol provides connection security with some encryption method such as the Data Encryption Standard (DES). The TLS Record Protocol can also be used without encryption. The TLS Handshake Protocol allows the server and client to authenticate each other and to negotiate an encryption algorithm and cryptographic keys before data is exchanged.

trap—An unsolicited message sent by an SNMP agent to an SNMP management system when the agent detects that a certain type of event has occurred locally on the managed host. The SNMP

manager console that receives a trap message is known as a trap destination. For example, a trap message might be sent when a system restarts or when a router links goes down.

Trivial File Transfer Protocol (TFTP)—A simple form of the File Transfer Protocol (FTP). TFTP uses the User Datagram Protocol (UDP) and provides no security features. It is often used by servers to boot diskless workstations, X-terminals, and routers.

trunk—The single cable usually designed to carry the bulk of the network traffic to other sites or to connect multiple networks or buildings at a site.

trust relationship—A relationship between domains that makes it possible for users in one domain to access resources in another domain. The domain that grants access to its resources is known as the trusting domain. The domain that accesses the resources is known as the trusted domain.

tunnel—The logical connection through which the packets travel in a virtual private network (VPN).

tunneling—The method for transferring data packets over the Internet or other public network, providing the security and features formerly available only on private networks. A tunneling protocol encapsulates the data packet in a header that provides routing information to enable the encapsulated payload to securely traverse the network.

twisted pair—Two insulated copper wires twisted around each other. While each pair acts as if it were a single communication link, twisted pair are usually bundled together into a cable and wrapped in a protective sheath.

U

Uniform Naming Convention (UNC)—See *Universal Naming Convention*.

Uniform Resource Locator (URL)—The global address of documents and other resources on the World Wide Web.

uninterruptible power supply—A device that has a battery connected in series with the PC. The AC power is connected directly to the battery. Since the battery always provides clean power, the PC is protected against overvoltages and undervoltages.

universal asynchronous receiver/transmitter (UART)—A single IC chip that is the translator between the serial device and the system bus and is the component that processes, transmits, and receives data.

Universal Naming Convention (UNC)—A PC format for specifying the location of resources on a local area network (LAN). UNC uses the \\server-name\shared-resource-pathname format. Also known as Uniform Naming Convention.

universal security group—A Windows NT family group that is only available in native mode. It can contain users, universal groups, and global groups from any domain, and it can be assigned rights and permissions to any network resource in any domain in the domain tree or forest.

UNIX—A multiuser, multitasking operating system. It is the grandfather of network operating systems, developed at Bell Labs in the early 1970s.

unshielded twisted pair (UTP)—The same type of cable that is used with telephones. It's the most common cable used in networks. UTP cable consists of two pairs or four pairs of twisted wires.

user account—An account that represents the user and enables a user to log onto a network.

User Datagram Protocol (UDP)—A TCP/IP that provides connectionless communications and does not guarantee that packets will be delivered. Applications that use UDP typically transfer small amounts of data at one time. Reliable delivery is the responsibility of the application.

user-level security model—A model in which a user has a user account that includes a user name and password. The user account is provided with an access control list (ACL) each time the user logs onto the network, which specifies which resources the user can access. The user-level security model allows a network administrator to manage network security from a central network location. An administrator would normally assign users to group accounts and then grant the group accounts permissions to resources on the network.

user mode—A Windows NT family mode that runs in Ring 3 of the Intel 386 microprocessor protection model, which is protected by the operating system. It is a less privileged processor mode that has no direct access to hardware and can only access its own address space. Since programs running in Ring 3 have very little privilege to programs running in Ring 0 (kernel mode), the programs in user mode should not be able to cause problems with components in kernel mode.

user profile—In Windows, a collection of folders and data that stores the user's current desktop environment and application settings.

V

value-added network (VAN)—A network with special services such as electronic data interchange (EDI) or financial services such as credit card authorization or ATM transactions.

vampire tap—A mechanical device that uses conducting teeth to penetrate the insulation and attach directly to the wire conductor.

vertical cabling system—Cable system that connects between floors.

VFAT—An enhanced version of the FAT structure, which allows Windows to support long file names (LFNs) up to 255 characters. Someone who refers to FAT probably means VFAT.

vi—A text editor used on Linux machines.

virtual circuit—A logical circuit created to ensure reliable communications between two network devices. To provide this, it provides a bidirectional communications path from one device to another and is uniquely identified by some type of identifier. A number of virtual circuits can be multiplexed into a single physical circuit for transmission across the network.

virtual file system (VFS)—A kernel software layer used on Linux machines that handles all system calls related to the Linux file system.

virtual local area networks (VLANs)—A collection of nodes grouped together in a single broadcast domain that is based on something other than physical location. A VLAN is a switched network that is logically segmented on an organizational basis, by functions, project teams, or applications rather than on a physical or geographical basis.

virtual private network (VPN)—A protocol that uses a public or shared network (such as the Internet or a campus intranet) to create a secure, private network connection between a client and a server. The VPN client cloaks each packet in a wrapper that allows it to sneak (or tunnel) unnoticed through the shared network. When the packet gets to its destination, the VPN server removes the wrapper, deciphers the packet inside, and processes the data.

virus—A program designed to replicate and spread, generally without the knowledge or permission of the user.

virus hoax—A letter or email message warning you about a virus that does not exist.

voice over IP (VoIP)—Voice communication sent over an IP network.

voltmeter—A device that measures voltage output or voltage signal.

volume—A fixed amount of storage on a disk or tape. The term *volume* is often used as a synonym for the storage medium itself, but it is possible for a single disk to contain more than one volume or for a volume to span more than one disk.

W

web browser—The client program/software that you run on your local machine to gain access to a web server. It receives the HTML commands, interprets the HTML, and displays the results.

web server—A computer equipped with the server software that uses Internet protocols such as HyperText Transfer Protocol (HTTP) and File Transfer Protocol (FTP) to respond to web client requests on a TCP/IP network via web browsers.

wide area network (WAN)—A network that uses long-range telecommunication links to connect the network computers over long distances and often consists of two or more smaller LANs. Typically, the LANs are connected through public networks, such as the public telephone system.

Wi-Fi—Short for wireless fidelity; it's another name for IEEE 802.11b. Wi-Fi is used in place of 802.11b in the same way that Ethernet is used in place of IEEE 802.3.

Windows 2000—A newer operating system based on the Windows NT architecture.

Windows Driver Model (WDM)—A driver technology developed by Microsoft to create drivers that are source-code-compatible for Windows 98, 2000, Me, and XP. WDM works by channeling some of the work of the device driver into portions of the code that are integrated into the operating system. These portions of code handle all the low-level buffer management, including DMA and plug and play device enumeration. The WDM device driver becomes more streamlined with less code and works at greater efficiency.

Windows Internet Naming Service (WINS)—A system that contains a database of IP addresses and NetBIOS (computer names) that update dynamically. It is used to resolve IP addresses from the computer names.

Windows NT—An operating system that can act as a high-performance network operating system, which is robust in features and services, security, performance, and upgradability. While NT stands for "new technology", it is several years old.

Windows NT Directory Service (NTDS)—The system of domains and trusts for a Windows NT Server network. To manage the users and groups of a domain and to set up trusts with other domains (trusts allow users to access resources in other domains), you use the User Manager for Domains program.

WINS proxy agent—A WINS-enabled computer configured to act on behalf of other host computers that cannot directly use WINS. WINS proxies help resolve NetBIOS name queries for computers located on a subnet where there is no WINS server. The proxies do this by hearing broadcasts on the subnet of the proxy agent and forwarding those responses directly to a WINS server. This keeps the broadcast local and yet gets responses from a WINS server without using the P node.

WINS replication partner—A computer that duplicates WINS databases to other WINS servers.

Winsock—A programming interface and supporting program that handles input/output requests for Internet applications in a Windows operating system. It's called Winsock because it's an adaptation for Windows of the Berkeley UNIX sockets interface.

wire map tester—A device that uses wire map testing, which transmits signals through each wire in a copper twisted-pair cable to determine if it is connected to the correct pin at the other end.

Wired Equivalent Privacy (WEP)—A form of encryption used by wireless communication.

wireless LAN (WLAN)—A local area network without wires that transfers data through the air using radio frequencies.

work area—The area that includes the station equipment, patch cable, and adapters (such as a media filter).

workgroup—A peer-to-peer network.

workstation—(1) In networking, the term *workstation* refers to any computer connected to a local area network. It could be a workstation or a personal computer. (2) The type of computer used for engineering applications (CAD/CAM), desktop publishing, software development, and other types of applications that require a moderate amount of computing power and relatively high-quality graphics capabilities.

World Wide Web—A system of Internet servers that support specially formatted documents.

worm—A program or algorithm that replicates itself over a computer network and usually performs malicious actions, such as using up the computer's resources and possibly shutting the system down. Typically, a worm enters the computer because of vulnerabilities available in the computer's operating system.

X

X Window—An open standard GUI interface.

X.400—A set of standards relating to the exchange of electronic messages (messages can be email, fax, voice mail, telex, etc.). It was designed to let you exchange email and files with the confidence that no one besides the sender and the recipient will ever see the message, that delivery is assured, and that proof of delivery is available if desired.

X.500—A directory service where objects are organized similar to the files and folders on a hard drive.

X.509 certificates—The most widely used digital certificates.

XML—See *Extensible Markup Language (XML)*.

Z

zone transfer—The process of obtaining DNS zone information across the network.

Index